THE SUN AND THE HELIOSPHERE
IN THREE DIMENSIONS

# ASTROPHYSICS AND
# SPACE SCIENCE LIBRARY

A SERIES OF BOOKS ON THE RECENT DEVELOPMENTS
OF SPACE SCIENCE AND OF GENERAL GEOPHYSICS AND ASTROPHYSICS
PUBLISHED IN CONNECTION WITH THE JOURNAL
SPACE SCIENCE REVIEWS

VOLUME 123

PROCEEDINGS

# THE SUN
# AND THE HELIOSPHERE
# IN THREE DIMENSIONS

PROCEEDINGS OF THE XIXth ESLAB SYMPOSIUM,
HELD IN LES DIABLERETS,
SWITZERLAND, 4–6 JUNE 1985

Edited by

## R. G. MARSDEN

*Solar and Heliospheric Science Division,*
*Space Science Department of ESA,*
*Noordwijk, The Netherlands*

## D. REIDEL PUBLISHING COMPANY

A MEMBER OF THE KLUWER  ACADEMIC PUBLISHERS GROUP

DORDRECHT / BOSTON / LANCASTER / TOKYO

**Library of Congress Cataloging in Publication Data**

ESLAB Symposium (19th : 1985 : Les Diablerets, Switzerland)
   The sun and the heliosphere in three dimensions.

   (Astrophysics and space science library; v. 123)
   Includes index.
   1.   Sun—Congresses.   2.   Heliosphere—Congresses.   3.   Cosmic
dust—Congresses.   I.   Marsden, R. G. (Richard George), 1951-
II.   Title.   III.   Series.
QB520.E74      1985         523.7         86-474
ISBN 90-277-2198-X

Published by D. Reidel Publishing Company,
P.O. Box 17, 3300 AA Dordrecht, Holland.

Sold and distributed in the U.S.A. and Canada
by Kluwer Academic Publishers,
190 Old Derby Street, Hingham, MA 02043, U.S.A.

In all other countries, sold and distributed
by Kluwer Academic Publishers Group,
P.O. Box 322, 3300 AH Dordrecht, Holland.

Printed in The Netherlands

# TABLE OF CONTENTS

Foreword

Opening Address                                                              1

SECTION I: THE CORONA

Coronal Magnetic Fields - a Mini Survey
    R.M. MacQueen                                                            5

Origins of the Solar Wind in the Corona
    G.L. Withbroe                                                           19

The Heliospheric Energy Source
    E.N. Parker                                                             33

Coronal Spectroscopy and Imaging from Spartan during the
Polar Passage of Ulysses
    J.L. Kohl, H. Weiser, G.L. Withbroe, G. Noci and
    R.H. Munro                                                              39

The Structure and Rotation of the Solar Corona : Implications
for the Heliosphere
    D.G. Sime                                                               45

OVI Diagnostics of Solar Wind Generation
    G. Noci, J.L. Kohl and G.L. Withbroe                                    53

Solar Wind Observations Near the Sun
    J.W. Armstrong, W.A. Coles, M. Kojima and B.J. Rickett                  59

Three-Dimensional Reconnection after a Prominence Eruption
    R.A. Kopp and G. Poletto                                                65

SECTION II: SOLAR HARD X-RAYS

Stereoscopic Measurements of Hard Solar X-Rays and Related
Topics
    K. Hurley                                                               73

Temporal Evolution of an Energetic Electron Population in
an Inhomogeneous Medium.  Application to X-Ray Bursts
    N. Vilmer, A.L. MacKinnon, and G. Trottet                               87

Influence of Solar Flares on the X-Ray Corona
  D.M. Rust and D.A. Batchelor                                          93

## SECTION III: CORONAL TRANSIENTS AND MASS EJECTIONS

Coronal Transients at High Heliographic Latitudes
  D.G. Sime                                                            101

The Solar Cycle Dependence of Coronal Mass Ejections
  R.A. Howard, N.R. Sheeley, Jr., D.J. Michels
  and M.J. Koomen                                                      107

Helios Images of Coronal Mass Ejections
  B.V. Jackson                                                         113

Relationship of Coronal Transients to Interplanetary Shocks:
3-D Aspects
  R. Schwenn                                                           119

3-Dimensional Configurations of Interplanetary Disturbances
Associated with Coronal Mass Ejections
  T. Watanabe, T. Kakinuma and M. Kojima                               123

The 3-Dimensional Extent at 1 AU of Interplanetary Disturbances
Associated with Disappearing Solar Filaments
  C.S. Wright                                                          129

3-Dimensional, Time-Dependent, MHD Model of a Solar Flare-
Generated Interplanetary Shock Wave
  M. Dryer, S.T. Wu and S.M. Han                                       135

## SECTION IV: SOLAR WIND

Interplanetary Scintillation Observations of the Solar Wind
at High Latitudes
  W.A. Coles and B.J. Rickett                                          143

Exploration of Heliosphere by Interplanetary Scintillation
  R.V. Bhonsle, S.K. Alurkar, S.S. Degaonkar, H.O. Vats and
  A.K. Sharma                                                          153

Evolution of Turbulence and Waves in the Solar Wind in Radius
and Latitude
  A. Barnes                                                            159

A Computer Simulation Study of the Microscopic Structure of a
Typical Current Sheet in the Solar Wind
  M. Roth                                                              167

Solar Wind Composition and What We Expect to Learn from Out-
of-Ecliptic Measurements
  J. Geiss and P. Bochsler                                             173

Comets and Three-Dimensional Solar Wind Structure
    J.C. Brandt                                                                      187

Structure and Dynamics of Corotating and Transient Streams in
Three Dimensions
    L.F. Burlaga                                                                     191

Propagation of Solar Wind Features: A Model Comparison Using
Voyager Data
    N.I. Kömle, H.I.M. Lichtenegger and H.O. Rucker                                  205

SECTION V: HELIOSPHERIC STRUCTURE

3-D Coronal and Heliospheric Structure from Radio Observations
    J.L. Bougeret, S. Hoang and J.L. Steinberg                                       213

Latitude Distribution of Interplanetary Magnetic Field Lines
Rooted in Active Regions
    G.A. Dulk, J.L. Steinberg, S. Hoang and A. Lecacheux                             229

Heliospheric Structure and Multispacecraft Observations of
Type III Radio Bursts
    J.L. Steinberg, S. Hoang and A. Lecacheux                                        235

The Relationship of the Large-Scale Solar Field to the Interplanetary Magnetic Field : What Will Ulysses Find ?
    J.T. Hoeksema                                                                    241

The Large-Scale Structure of the Heliospheric Magnetic Field
    A. Balogh                                                                        255

The Heliospheric Current Sheet : 3-Dimensional Structure and
Solar Cycle Changes
    E.J. Smith, J.A. Slavin and B. Thomas                                           267

Solar Wind Speed Azimuthal Variation Along the Heliospheric
Current Sheet
    S.T. Suess, P.H. Scherrer and J.T. Hoeksema                                     275

3-Dimensional Structure of the Heliosphere as Inferred from
Observations with a Japanese Halley Spacecraft
    T. Saito, K. Yumoto, K. Hirao, I. Aoyama and E.J. Smith                          281

3-Dimensional Structure of the Heliospheric Current Sheet
    C.D. Fry and S.-I. Akasofu                                                       287

SECTION VI: ENERGETIC PARTICLES

Separation and Analysis of Temporal and Spatial Variations
in the 10 April 1969 Solar Flare Particle Event
    R. Reinhard, E.C. Roelof and R.E. Gold                                           297

Acceleration of Energetic Particles at Solar Wind Shocks
    M.A. Lee                                                          305

Shock Acceleration of Nucleons at $\geq 16°$ Solar Latitude
Associated with Interplanetary Corotating Interaction Regions
    J.A. Simpson, E.J. Smith and B. Tsurutani                        319

Latitude Dependence of Co-Rotating Shock Acceleration in the
Outer Heliosphere
    R.E. Gold, L.J. Lanzerotti, C.G. Maclennan and
    S.M. Krimigis                                                    325

Three-Dimensional Gradients of Solar Particles Inside 5 AU
    E.C. Roelof                                                      331

A Spatially Confined, Long-Lived Stream of Solar Particles
    K.A. Anderson and W.M. Dougherty                                 341

Super-Events in the Inner Solar System and Their Relation to
the Solar Cycle
    R. Müller-Mellin, K. Röhrs and G. Wibberenz                      349

The Maximum Entropy Principle in Cosmic Ray Transport Theory
    P. Hick, G. Stevens and J. van Rooijen                           355

SECTION VII: COSMIC RAYS

Modulation of Galactic Cosmic Rays in the Heliosphere
    R.B. McKibben                                                    361

Effects of 3-Dimensional Heliospheric Structures on Cosmic-
Ray Modulation
    J.R. Jokipii                                                     375

Measurement of Radial and Latitudinal Gradients of Cosmic
Ray Intensity During the Decreasing Phase of Sunspot Cycle 21
    D. Venkatesan, R.B. Decker and S.M. Krimigis                     389

North/South Asymmetry in Solar Activity and its Effects on
the High Energy Cosmic Ray Diurnal Variation
    M.A. Shea, D.F. Smart, D.B. Swinson and J.E. Humble              395

The Anomalous Component, Its Variation With Latitude and
Related Aspects of Modulation
    L.A. Fisk                                                        401

Pick-Up Ions in the Solar Wind as a Source of Suprathermal
Particles
    D. Hovestadt, E. Möbius, B. Klecker, G. Gloeckler,
    F.M. Ipavich and M. Scholer                                      413

## SECTION VIII: INTERSTELLAR GAS AND INTERPLANETARY DUST

Neutral Interstellar Gases in the Heliosphere: New Aspects
of the Problem
   H.J. Fahr                                                            421

Interstellar Gas Parameters and Solar Wind Anisotropies
Deduced from H and HE Observations in the Solar System
   J.L. Bertaux, R. Lallement and E. Chassefière                       435

The 3-Dimensional Structure of the Interplanetary Dust Cloud
   R.H. Giese and G. Kinateder                                          441

The Interaction of Solid Particles with the Interplanetary
Medium
   G.E. Morfill, E. Grün and C. Leinert                                 455

## SECTION IX: ULYSSES

The Ulysses Mission
   R.G. Marsden, K-P. Wenzel and E.J. Smith                            477

## SECTION X: SUMMARY

Summary Remarks
   L.A. Fisk                                                           493

List of Participants                                                   503

Subject Index                                                          511

Participants in the XIXth ESLAB Symposium, held in Les Diablerets, Switzerland, 4-6 June, 1985.

# FOREWORD

The 19th ESLAB Symposium on 'The Sun and the Heliosphere in Three Dimensions' was held in Les Diablerets (Switzerland) on 4-6 June 1985. Organised almost exactly ten years after the Goddard Space Flight Center Symposium dealing with the Sun and the interplanetary medium in three dimensions, the aim of this Symposium was not only to review the progress made in understanding the three-dimensional structure and dynamics of the heliosphere, but also to look ahead to the scientific return to be expected from the Ulysses mission. Scheduled for launch in May 1986, the scientific instrumentation on board Ulysses will shed light on the conditions and processes occurring away from the ecliptic plane, thereby adding literally a new dimension to our understanding of the only stellar plasmasphere to which we have direct access.

The scientific programme of the Symposium was built around a series of invited review papers dealing with aspects of the corona and its influence on the interplanetary medium via transient ejecta, the solar wind, energetic solar particles and galactic cosmic rays, interplanetary dust and neutral gas. These invited talks were supplemented by a number of contributed and poster papers. With the exception of three contributed talks and Wibberenz' review of coronal propagation and acceleration of energetic particles, all papers presented at the Symposium are included in this volume. In addition, a paper summarising the scientific objectives and implementation of the Ulysses mission has been included. The invited paper given by Schwenn on coronal transients and interplanetary shocks is presented here in the form of an extended abstract, since the full review paper is to appear elsewhere.

The organisers would like to thank the members of the Scientific Programme Committee, W. I. Axford, D. Bohlin, L. Fisk, J. Geiss, M. Pick and E. J. Smith, for their efforts in helping to establish the scientific programme. Special thanks are due to the Symposium Secretaries, Wendy Collins-Rolfe and Anne v. d. Eijkel, for their patience and hard work which guaranteed the smooth running of the meeting. Additional thanks are due to Wendy Collins-Rolfe for invaluable assistance in the preparation of the final camera-ready manuscript. The technical assistance of Cecil Tranquille during the sessions is gratefully acknowledged.

The very favourable atmosphere experienced in Les Diablerets was due in large part to the hospitality of, and excellent collaboration with, the local authorities.  We acknowledge the support of Mr. P. Messeiller, Director of the Les Diablerets Office of Tourism, and Mr. P. Fontana for the operation of the techinical installations at the Conference Centre.  Last but not least, we would like to express our appreciation to Mr. K. Wartner, Director of the Eurotel in Les Diablerets, for the excellent and friendly service provided.  Mr. Wartner's enthousiasm and personal engagement in organising the social events, together with the friendliness and attentiveness of his staff, contributed in no small part to the success of the meeting.

The 19th ESLAB Symposium was organised by the Space Science Department of ESA and co-sponsored by NASA and COSPAR.


        Richard G. Marsden                    Klaus-Peter Wenzel

# OPENING ADDRESS

ESA's Space Science Department, which used to be called ESLAB, provides the study and project scientists for ESA's scientific satellite programme. Each year this Department organises a Symposium and our meeting in Les Diablerets is the nineteenth such happening.

Frequently, and particularly in recent years, the ESLAB Symposium has been used to present the results arriving from one of the ESA spacecraft. On some occasions we have organised the symposium around a topic not in the ESA programme so that we could better understand what the community wanted and so that our project scientists could adjust their internal research to better equip themselves to support the ESA programme. Sometimes we arrange the symposium - and this symposium on "The Sun and Heliosphere in Three Dimensions" is such an arrangement - in order to prepare to handle the science return from a spacecraft due for launch in the near future.

Since the beginning of the space age there have been those who realised that since the narrow slice of the heliosphere in which the earth finds itself is probably so unrepresentative of the heliosphere as a whole, a mission out of the ecliptic plane had to be organised. The required technology has been available for some time but the multidisciplinary nature of the investigations to be carried out, while assuring broad support, led to a situation where each specialist committee passed the responsibility to another. At last, after a late start and several serious hiccups during development, we have reached a situation where we expect an ESA spacecraft to be launched by NASA in mid-1986, to begin an adventure in regions so far unexplored. We expect to be over the southern solar pole in 1990 and I look forward eagerly to hearing you discuss the results - perhaps in the 1991 ESLAB Symposium.

It is my privilege and pleasure to welcome you here and to wish you a profitable stay in this wonderful location.

D. Edgar Page
Head, Space Science
Department of ESA

SECTION I:  THE CORONA

# CORONAL MAGNETIC FIELDS--A MINI SURVEY

R.M. MacQueen
High Altitude Observatory
National Center for Atmospheric Research
P.O. Box 3000
Boulder, Colorado  80307

ABSTRACT.  Some recent progress in understanding the nature of the
evolution of the global coronal magnetic field is reviewed.  Particularly, the efforts of Hoeksema (1984) in defining the character of
the evolving modes of the potential coronal field are compared with
the currently-known evidence for evolution of coronal white light
structures.  Recent work in examining the soft x-ray intensity of
coronal holes over a major portion of the solar cycle is noted, as
are two new studies investigating the relation of coronal mass ejections to the ambient global coronal magnetic field.

It has long been supposed that the form of the solar corona
reflects the structure of the global solar magnetic field. The origin of this latter field apparently resides in a solar dynamo-like
process,  and the surface features and coronal extension of this
field reflect its coupling with solar rotation and atmospheric properties. Thus, the changes that are evident in the appearance of the
solar corona reflect variations in the evolution of the large-scale
solar fields or in any modification of the fields by the solar atmosphere as they emerge from the solar interior.
   In this brief discussion, we concentrate on summarizing some of
the general properties of the solar corona and their role as a guide
to understanding the nature of global solar magnetic fields. In particular, we will stress the nature of the evolution of the coronal
fields. We will note briefly: some recent efforts concerning the
solar cyclic variation of computed magnetic fields and white light
structures; observed solar cyclic variations of x-ray properties of
coronal holes and their relationship to underlying photospheric
fields; an interesting possibility concerning the relation of
coronal mass ejection transients with changes in the global solar
field; and, finally, the role of that global field in modifying the
direction of propagation of mass ejections.

*R. G. Marsden (ed.), The Sun and the Heliosphere in Three Dimensions, 5–18.*

## 1.   CALCULATIONS OF THE CORONAL MAGNETIC FIELD

Direct observations of the coronal magnetic field are, at present,
not possible. Although techniques employing interpretation of the
Stokes parameters of certain forbidden emission lines allow the
direction of coronal magnetic field lines to be determined (Charvin,
1965; Querfeld 1982), they are limited to the innermost (within 2
$R_\odot$ heliocentric distance) coronal regions and present certain ambi-
guities in analysis. Measurements of the strength of coronal magnetic fields are limited to special circumstances of coronal
activity and are subject to the formidable uncertainties of interpreting the signals from coronal metric wavelength radio emission
(Dulk and McLean, 1978). Thus, at the present time, we are limited
to specifying the global coronal magnetic field via computations of
the outward extension of surface photospheric fields.

It is a fundamental hypothesis that these computed fields--
which utilize only the longitudinal component  of the  surface
field--realistically map the coronal extension of the surface
fields. Procedures for computing the coronal potential fields from
the surface field were developed a number of years ago (Schatten, et
al., 1969; Altschuler and Newkirk, 1969). The boundary conditions
for the solution of the Laplace equation for the potential magnetic
field are twofold. At the inner surface--the photosphere--the fields
are taken to be specified by the observations of the longitudinal
component with magnetographs. At the outer boundary--which may be
chosen at will, but has typically been taken to be 2.5 $R_\odot$ --the magnetic field is assumed to be purely radial, as an approximation to
eclipse observations which generally show only open white light
structures at such heights.  The amplitude of each mode (the latter
specified by the parameter $\ell$ ) of an orthonormal expansion in terms
of the associated Legendre polynomials can be computed from the
scalar potential. Then, the form of the coronal field can be determined.

No quantitative tests of the ability of the potential fields to
reflect observed specific coronal structures have been developed,
but potential field calculations have been shown to reasonably
reflect the global configuration of the white light corona (see the
next section).  Hoeksema (1984, 1986) has recently investigated how
the computed potential field varies with time by studying the modal
ordering of the global field over a major portion of the solar
cycle.  Figure 1 shows the ordering with time of the amplitude of
the $\ell$ =0, 1 and 2 modes of the magnetic field expansion (the monopole, dipole and quadrupole terms, respectively) over the period
1976-1983. The octopolar and higher order terms are seen for the
same period in Figure 2.  (However, note the change in scale between
the amplitudes presented in the two figures). We note firstly that
near the period of solar minimum activity (circa 1976) the dipolar
term dominates all higher order terms by on the order of 5 to 1; as
we approach higher levels of solar activity, however, this dominance
decreases until, during 1979 and 1980, the quadrupolar term actually

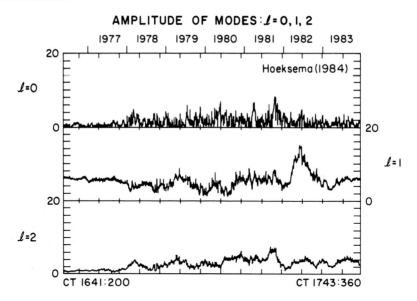

Figure 1.  The amplitudes of the monopolar, dipolar and quadrupolar
components of the coronal potential magnetic field,
from Hoeksema (1984).

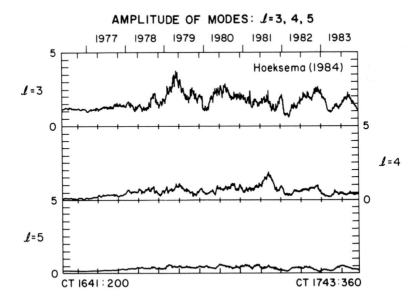

Figure 2.  The amplitudes of higher order components of the coronal
potential magnetic field, from Hoeksema (1984).

exceeds the dipolar term for periods of time. The octopolar term
also is important during periods of high activity, since its magnitude becomes roughly equivalent to both the dipolar and quadrupolar
terms at such times.  The other, higher order terms are negligible
throughout the period investigated. (For a discussion of the monopolar term, which is a measure of the zero offset of the model calculations, see Hoeksema (1984)).

     Before turning to the issue of how this ordering of modes in
the harmonic decomposition reflects observable global coronal conditions, it is interesting to note the variations in the amplitude of
the modes with time. The dipolar amplitude exhibits variations from
year-to-year of on the order of 2-4 microtesla, while for one period
(between 1981 and 1982) the variation is on the order of 10
microtesla. On average, however, it appears that even semiannual
variations of on the order of 50% are common. Roughly the same percent variations in both the quadrupolar and octopolar amplitudes
seem to be the rule rather than the exception.

     Is there a relation between variations of this sort and the
behavior of the observable corona? No studies of this relationship
have been carried out for the period investigated by Hoeksema, but
an earlier, qualitative study by Levine (1977) is relevant.  Levine
examined the modal components of the surface magnetic field during
the period 1973-74, at the declining phase of solar activity. The
period chosen coincided with the period of observations of the
Skylab mission. Levine found that superposed on an overall declining
trend of the relative total energy in all harmonic modes of the photospheric magnetic field there occurred a significant (2 times)
increase and subsequent decline of the total energy over a three
solar rotation period (Figure 3).  This period of great change in
the relative energy of the magnetic field was found to coincide with
the change in the structure of the observed white light corona, as
evidenced by examination of the signal produced by the rotation of
coronal equatorial features past the solar limb (MacQueen and
Poland, 1977). Levine noted that the dominant mode of the magnetic
field shifted from octopolar to dipolar over a solar rotation,
retained its dipolar character for two rotations, and then shifted
to a dominant quadrupolar nature over the next rotation. However,
the actual amplitudes of each of the modes was not specified. In any
event, this example provides one possible link between variations in
the solar surface magnetic fields and the temporal behavior of the
coronal forms. It seems clear that future studies of the correlation
of global coronal forms with the computed modal components of the
calculated coronal fields will prove fruitful in illuminating the
spatial and temporal scales involved.


2.   RELATION TO LONG-LIVED CORONAL FORMS

In the above discussion, we noted that there appears to be a relationship between short-term variations in the computed magnetic
fields and changes in the appearance of coronal structures. We now

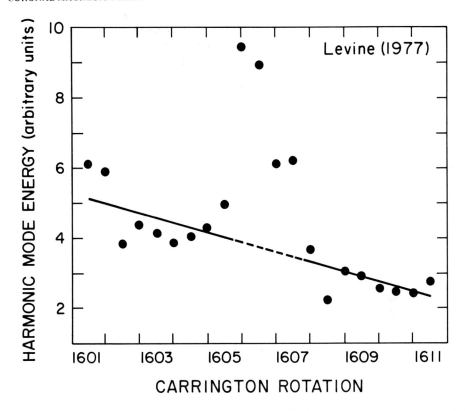

Figure 3. The relative total energy of the photospheric magnetic field during 1973-74, from Levine (1977).

turn to the relation between long-lived coronal forms and the computed global magnetic fields. As the basis for the comparison, we will employ discussions of the behavior of the coronal forms which have been developed over the past few years from examination of eclipse and k-coronameter observations (Hundhausen, 1977; Hundhausen, et al., 1981).

At the outset, one may observe that the temporal evolution of the modes quantitatively examined by Hoeksema (1984) coincides, in the broadest sense, with our expectations concerning the long-term evolution of the corona. For example, near the period of solar minimum activity, the corona is known to assume a relatively simple form wherein it is dominated by equatorial streamers and large polar open field regions--coronal holes. This pattern is reminiscent of that produced by a simple dipolar magnetic field--which, the computations show, dominates the harmonic structure of the corona near

solar minimum. The axis of the dipolar field does not align with the
rotation axis-- thus the term "tilted dipole" (Hundhausen, 1977). As
activity intensifies into the solar maximum period, the form of the
corona becomes less dominated by the dipolar component and extends
over a wider range of solar latitude than at solar minimum.  In this
sense, the solar corona becomes more circularly symmetric. As the
extent of the polar coronal holes recedes, they are less dominant
than at solar minimum.  Thus, in the context of the computed global
magnetic fields, we might expect an increasingly significant contribution of magnetic fields of a wide variety of scales as activity
increases. This appears to be borne out by Hoeksema's computations
of the character of the computed field.

   Although images of the structure of the solar corona obtained
during periods of total solar eclipse have verified this general
behavior of the corona with the solar cycle, it is instructive to
examine the evolution of the solar corona at various times of the
solar cycle in a systematic way. This has been possible with the use
of the observations of the corona which have been obtained from
space-borne coronagraphs and from the ground with K-coronameters
(Hundhausen, 1977; Hundhausen, et al., 1981; Sime, 1985).

   As noted above, the simplest situation exists at periods near
solar minimum activity. Figure 4 illustrates the appearance of the
solar corona derived from K-coronameter latitudinal scans at
heliocentric height 1.5 $R_0$ during 1974. The daily scans are

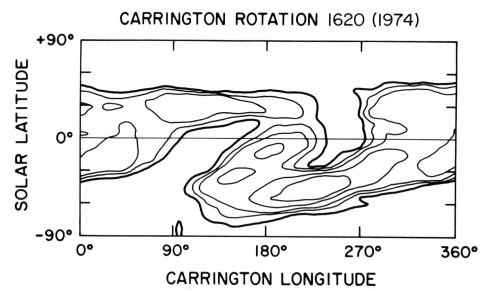

Figure 4.  Synoptic map of the coronal pB product for
Carrington rotation 1620 (1974).

displayed in the form of a "synoptic map" in which the scans are
transformed into a coordinate system of the solar heliographic latitude and the solar longitudes which corresponded to those at a particular solar limb on the day in question. The resultant map is thus
constructed from all available scans for the Carrington rotation (in
this case, rotation number 1620) under study. Note that the contours
of the selected polarization brightness product of the corona at
this time are confined to within latitudes of $\pm$ 45 degrees and exhibit an undulatory, smooth appearance typical of this phase of the
solar cycle. The centroid of maximum brightness describes a serpen-
tine pattern which encircles the solar globe in a manner similar to
the seam of a baseball. This centroid of brightness has been identified as reflective of the position of the "neutral sheet" separating
the polarities of the global potential magnetic field (Pneuman, et
al., 1978; Wilcox and Hundhausen, 1983), and considerable success
has marked the attempts to identify the polarity of the interplanetary magnetic field via the radial outward projection of the neutral
sheet, at least at periods of low solar activity (Burlaga, et al.,
1978; Burlaga, et al., 1981; Bruno, et al., 1982; Behannon, et al.,
1983).

The observed structure of the brightness patterns obtained near
periods of solar minimum activity provided the basis for initial
comparisons of the coronal form with the calculated potential fields
(Howard and Koomen, 1974; Pneuman, et al., 1978). They also provided
the stimulus for the concept that the interplanetary plasma parameters may be best described in terms of a tilted dipolar heliomagnetic coordinate system (Zhao and Hundhausen, 1981).

Contrast the appearance of the corona in 1974 (Figure 4) with
that near solar maximum 1980 (Figure 5). Little similarity remains
in the pattern of contours of the same pB levels as employed in Figure 4. No overall pattern, so evident near solar minimum, can be
identified, and the coronal brightness patterns extend over all
latitudes. The structure of the corona is highly disordered, and
the identification of the position of a neutral sheet presents a
real challenge! The appearance, revealed by the synoptic maps, of
the solar maximum corona is apparently typical of each maximum: at
least, K-coronameter observations of the corona during the 1969 maximum show similar traits.

The long series of K-coronameter observations, encompassing
almost two solar cycles, augmented with the detailed outer coronal
observations from spacecraft, permit an overall scenario of evolution to be portrayed. Near solar maximum, the coronal structures are
of small scale, disordered, and encompass nearly the entire globe.
The appearance and growth of polar coronal holes compresses the
coronal features into lower and lower latitudes, finally resulting
in a coronal pattern of some simplicity, near the solar equator,
separating the solar magnetic hemispheres. At such times, potential
field approximations are particularly applicable, even quantitatively. Thus, in general, we see that the appearance of the corona
traces a history in concert with the computed evolution of the
coronal magnetic field, wherein the dominant dipolar term evident

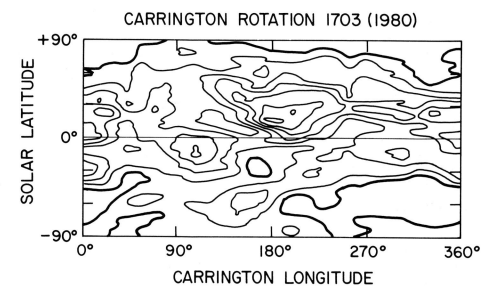

Figure 5.   Synoptic map for Carrington rotation 1703 (1980).

near solar minimum yields to higher order multipolar terms as solar
activity increases.

     Hoeksema (1984) has pointed out some of the similarities and
agreements with the computed behavior of the coronal magnetic fields
and the ordering of their interplanetary extension. Much work
remains, however, on the relation of coronal structures with these
computed fields, and current (early) indications are that similar
phases of different solar cycles may progress with greatly different
rates of evolution. For example, the compression of the tilted
dipole during the current solar cycle 21 apparently is proceeding at
a rate about two times that of the previous cycle. Thus, considerable effort yet remains before we can be confident of the quantitative linkages between computations and observations.


3.   CORONAL HOLES AND EVOLUTION

The above picture of the solar cyclic evolution of coronal holes is
based primarily upon observations of the coronal electron density
alone, such as available from K-coronameter or coronagraph observations. However, a relatively unique data set now has been obtained
by the group at the American Science and Engineering Company that
provides a more exact specification of the solar cyclic properties
of coronal holes. In a preliminary report, Webb and Davis (1985)
have presented a compilation of results from a number of rocket
flights and the Skylab mission.  They have measured the soft x-ray
intensity of coronal hole regions present at several times from 1973

to 1980--encompassing more than one-half a solar cycle.  In addition, they have examined solar surface magnetic field data within
each coronal hole region.  They find that the soft x-ray intensity
in the coronal hole regions varies with time during the solar cycle,
increasing by about a factor of 50 from 1973 to 1980. Over the same
period, selected sample background coronal regions increase in
intensity by on the order of 3-10 times, and the solar surface magnetic flux within the coronal hole regions increase by about 3
times. These are the first results which tie together long-term
trends in coronal hole intensity and surface magnetic field
behavior.  They appear to indicate that coronal holes partake of the
same solar cycle strengthening of the solar magnetic fields typical
of the sun as a whole (Howard, 1971), and these new quantitative
data will allow estimates, for example, of the change of the coronal
base pressure with time and of the interaction of coronal holes with
coronal structures over the solar cycle. The relation between the
evolving conditions on the sun and those observed in interplanetary
space are yet unclear, although Slavin and Smith (1983) have pointed
out variations in the interplanetary flux observed during the past
solar cycle, and initial observations of differences between solar
cycles 20 and 21.

## 4.   GLOBAL CORONAL MAGNETIC FIELDS AND MASS EJECTIONS

The discovery that coronal mass ejections were a frequent and important manifestation of solar activity raised a number of questions
concerning their relation to the ambient global coronal magnetic
field. In particular, the interaction between the ambient field and
the mass ejection has been a subject of some conjecture. Of course,
the fact that global coronal magnetic field patterns do persist from
one rotation to the next indicates that, at least on the global
scale, mass ejections are not so disruptive as to greatly modify
such patterns.
     On the other hand, little is known about the interaction of
mass ejections and the ambient field on the less than global scale.
Since one school of thought concerning the initiation of mass ejections involves an active role for magnetic fields in accelerating
the lower coronal mass from the sun, at least in that view the
ambient coronal fields must undergo some modification, and irrespec-
tive of the nature of the driving force for a mass ejection, one
possibility is that the outward motion of the ejection might be
expected to cause a field modification.  Thus, whether possible
changes in the ambient magnetic fields result from motions of the
fields in driving the ejecta or are an effect of the propagation has
remained an open issue, as, indeed, has been the issue of whether
any such changes occur.
     Two studies have been recently undertaken which bear on these
issues.  The first (Newkirk, 1984) has involved an examination of
the temporal variations of the computed total magnetic field energy
on the basis of potential field models and comparison of these

variations with the frequency of occurrence of coronal mass ejection
transients. The sample period chosen for this study was that encompassing the Skylab mission period (1973-74), for which rather complete observations of mass ejections exist, and for which solar surface magnetic field data are available.  Newkirk has computed the
total energy in the global potential magnetic field, during the
Skylab mission period of 1973-74. The computation has been carried
out for each seven day period during the mission. He finds that the
total magnetic field energy varies by about a factor of two from
rotation to rotation during this time. Interestingly, when the absolute magnitude of the variations in the total energy content is compared with the appropriate running average of the number of transients with time, there seems to be a correlation between periods
when the rate of change of total magnetic field energy is maximized
and the frequency of transient mass ejections (Figure 6). Whether a
result of cause or effect of transients, this represents the first
linkage between a global property of the corona and transient
activity. At present, Newkirk and collaborators are attempting to
refine the analysis of the field energy variations and extend the
study to include proper geometrical specification of the global
potential magnetic field which may be compared with the frequency
and location of coronal mass ejections during the same period.

A second study has involved the examination of the direction of
propagation of coronal mass ejection transients--in particular, a
study of possible differences in the direction of propagation of
mass ejections between near solar minimum (Skylab results) and at
solar maximum (SMM results) (MacQueen, et al., 1985).  Measurements
were carried out on 29 ejections from the former period and 19
events from the latter epoch. In agreement with an earlier study by
Hildner (1977), the new measurements show that, on average, Skylab
epoch mass ejections undergo a 2.2 degree equatorward deflection. On
the other hand, ejections observed during the SMM period show no
significant deviation from the radial direction. Further, no differences between flare-associated or eruptive prominence-associated
events could be detected for either period. These results strongly
suggest that coronal mass ejection events are influenced by the
background coronal magnetic field and flow patterns present during
their occurrence. In particular, it appears that the non-radial
(equatorward) forces affecting the Skylab epoch mass ejections arise
from the presence of the large-scale dipolar magnetic field and flow
configuration present at that time. On the other hand, the SMM epoch
events occur at times when the dipolar field and flow patterns are
no longer dominant.  The results lead one to suggest that the role
of global coronal conditions in influencing the propagation of
coronal mass ejection transients is significant. Despite the fact
that ejections are a significant perturbation to the local coronal
conditions, they appear to be influenced themselves by the prevailing coronal magnetic field and flow conditions.

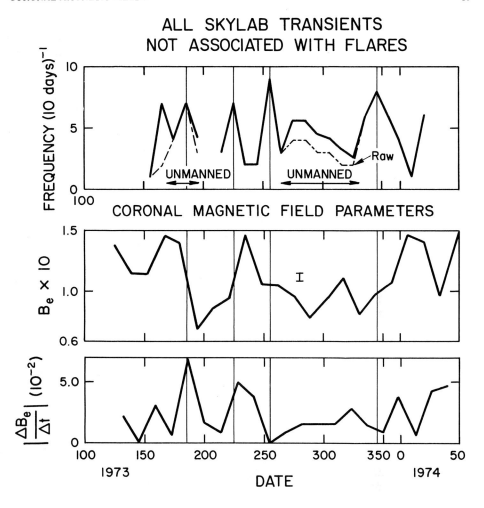

Figure 6. Preliminary results comparing the absolute magnitude
of the computed magnetic field energy (lower panel) with the
frequency of occurrence of non-flare associated coronal transients
during 1973 and 1974. The dashed lines in the upper panel refer
to uncorrected (for duty cycle) frequency of transients.


Taken together (and recognizing that the study of the variation
of the coronal magnetic field energy is yet in a preliminary stage)
these results would appear to suggest two particularly simple (but
hardly exclusive) possibilities. Either (a) coronal magnetic field
changes are causally related to mass ejections, or (b) the two studies are possibly sensing different spatial domains (either radial,
or longitudinal, or both). The latter view would suggest that the
calculated magnetic field energy is responsive to spatial scales not

relevant to the propagation of coronal mass ejections. As in so many other areas, additional efforts and insights are required to distinguish between these possibilities, and/or pose alternatives.

## 5. SUMMARY

In this brief review we have attempted to survey--albeit tersely--the current status of efforts in relating our current understanding of the coronal magnetic field with some of the observed properties of the coronal structure. Until (and if) we are able to directly measure ambient coronal magnetic fields we will be forced to rely on indirect inferences and computed properties of those fields, and we must continually reflect upon the fact that such inferences and calculations are based upon a number of simplifying assumptions which may not correspond well to the real physical situation. However, even given these caveats, it is important to realize that the last decade has seen substantial progress in our attempts to relate computed and observed properties of the solar corona. We now have a much clearer view of how the organization of the global coronal field varies with time, and major progress has been made in relating global solar and interplanetary properties. Substantive questions concerning the nature of the interplanetary sector structure and its solar counterpart have been resolved. As a result, we now have a better appreciation of how the study of the global properties of the solar corona can lead to insight into a number of physical processes in the heliosphere.

As we understand more fully how the global magnetic field evolves, we will come closer toward understanding the nature of the solar dynamo. Then we will have the central link in the connection of the sun and other stars, and we will have achieved a giant step toward understanding general stellar evolution.

## References

Altschuler, M.D. and Newkirk, G.A.: 1969, Solar Phys. 9, 131.

Behannon, K.W., Burlaga, L.F. and Hundhausen, A.J.: 1983, J. Geophys. Res. 88, 7837.

Bruno, R., Burlaga, L.F. and Hundhausen, A.J.: 1982, J. Geophys. Res. 87, 10337.

Burlaga, L.F., Behannon, K.W., Hansen, S.F., Pneuman, G.W. and Feldman, W.C.: 1978, J. Geophys. Res. 83, 4177.

Burlaga, L.F., Hundhausen, A.J. and Zhao, X.-P.: 1981, J. Geophys.
    Res. 86, 8893.

Charvin, P.: 1965, Annales d'Astrophys. 28, 877.

Dulk, G.A. and McLean, D.J.: 1978, Solar Phys. 57, 279.

Hildner, E.: 1977, in Studies of Travelling Interplanetary
    Phenomena, (M. A. Shea, et al., eds.), D. Reidel, Dordrecht.

Hoeksema, J.T.: 1984, Thesis, Department of Applied Physics, Stan-
    ford University, Palo Alto, California.

Hoeksema, J.T.: 1986, 19th Eslab Symposium Proceedings,      .

Howard, R.: 1971, Pub. Astron. Soc. Pac. 83, 550.

Howard, R.A. and Koomen, M.J.: 1974, Solar Phys. 37, 469.

Hundhausen, A.J.: 1977, in Coronal Holes and High Speed Streams,
    J.B. Zirker (ed), Colo. Ass. U. Press, 225.

Hundhausen, A.J., Hansen, R.T. and Hansen, S.F.: 1981, J. Geophys.
    Res. 86, 2079.

Levine, R.H.: 1977, Solar Phys. 54, 327.

MacQueen, R.M. and Poland, A.I.: 1977, Solar Phys. 55, 143.

MacQueen, R.M., Hundhausen, A.J. and Conover, C.W.: 1985, J. Geo-
    phys. Res., in press.

MacQueen, R.M. and Newkirk, G.A.: 1984, EOS 65, 1034.

Pneuman, G.W., Hansen, S.F. and Hansen, R.T.: 1978, Solar Phys. 59,
    313.

Querfeld, C.W.: 1982, Astrophys. J. 255, 764.

Schatten, K.H., Wilcox, J.M. and Ness, N.F.: 1969, Solar Phys. 6,
    442.

Sime, D.G.: 1985, _Future Missions in Solar_, Heliospheric and Space
    Plasma Physics, ESA SP-235,23.

Slavin, J.A. and Smith, E.J.: 1983 _Solar Wind Five_, NASA Conference
    Publication 2280, 323.

Webb, D. and Davis, J.: 1985 _Bull. Am. Astron. Soc._ 17, 636.

Wilcox, J.M. and Hundhausen, A.J.: 1983, _J. Geophys. Res._ 88, 8095.

Zhao, Xue-pu and Hundhausen, A.J.: 1983, _J. Geophys. Res._ 88, 451.

# ORIGINS OF THE SOLAR WIND IN THE CORONA

George L. Withbroe
Harvard-Smithsonian Center for Astrophysics
60 Garden Street
Cambridge, MA 02138
USA

ABSTRACT. Coronal holes are the most well-established coronal source of steady-state solar wind. Coronal mass ejections associated with flares and/or eruptive prominences are another clearly identified source of solar wind, a transient component that accounts for approximately 5% of the total solar mass loss. The role of other coronal structures in the generation of the solar wind is less clear. Streamers and the interfaces between streamers and other coronal regions are potential sources of low speed wind, a hypothesis consistent with existing observations. However, measurements of mass flows in these features are needed to confirm this. The necessary data can be obtained using recently developed remote sensing techniques, particularly if acquired in conjunction with Ulysses. Small-scale dynamical phenomena observed at the base of the corona (spicules, macrospicules and high speed jets) and small-scale structures observed in polar coronal holes (polar plumes) may or may not play a significant role in supplying mass, momentum and energy to the solar wind. New measurements are required to determine the role, if any, of small-scale structures in the generation of the solar wind.

## 1. INTRODUCTION

In this paper we briefly review current knowledge concerning the coronal origins of the solar wind. We consider sources whose role in the generation of the solar wind appears to be well established, such as coronal holes, and sources whose role is speculative at the present time. We also discuss some of the techniques which should lead to improvements in our knowledge concerning this subject.

## 2. CORONAL HOLES

Coronal holes were the first coronal regions identified as sources of steady-state solar wind. An example of a large coronal hole is shown in the x-ray photograph presented in Figure 1. In 1973 Krieger et al.

R. G. Marsden (ed.), The Sun and the Heliosphere in Three Dimensions, 19–32.
© 1986 by D. Reidel Publishing Company.

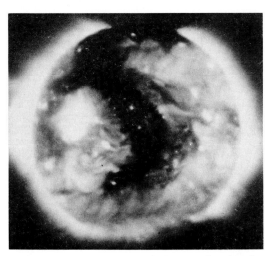

Figure 1.  Skylab soft x-ray photograph showing a large coronal hole
extending from the north pole into equatorial regions. (Courtesy
American Science and Engineering and Harvard College Observatory)


traced a recurrent high speed solar wind stream back to the sun
assuming constant velocity and found that it mapped into the location
of an equatorial coronal hole.  Subsequently the connection between
recurrent high speed streams and corona holes was solidified using in
situ measurements of the solar wind and observations of coronal holes
from OSO-7, Skylab and the ground (cf. reviews by Zirker 1977, 1981 and
references cited therein).  These studies made use of the fact that
many coronal holes, especially during the declining phase of the solar
cycle, are long-lived, lasting many solar rotations, and thus can be
clearly related to recurrent high speed solar wind streams.

Since coronal holes are located at the solar poles throughout most
of the solar cycle (except possibly for times near solar maximum), one
expects the polar regions of the sun to be sources of high speed solar
wind most of the time.  This is confirmed by observations as is
illustrated in Figure 2 (from a review by Rickett and Coles 1983).  The
top portion of the figure gives, as a function of time, the latitudinal
extent of polar coronal holes determined from optical measurements of
the electron-scattered white light corona.  The lower graph gives the
corresponding location of the contour separating high latitude regions
of solar wind with speeds greater than 500 km/s and low latitude
regions with lower speed winds. These wind speeds were determined via
measurements of scintillations of remote astronomical radio sources
(cf. reviews by Rickett and Coles 1983, Coles and Rickett 1985).
Results such as these provide clear evidence that polar coronal holes
are usually sources of high speed solar wind.

Coronal holes are regions with open magnetic field configurations
where the field lines extend from the solar surface out into

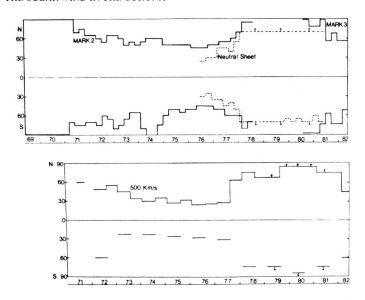

Figure 2.  Upper panel shows 6-mon average boundary in latitude of
polar coronal holes from HAO Mark II and III coronagraphic data.  Also
shown are the extreme latitudes of the magnetic neutral sheet.  Lower
panel shows the 500 km/s contour of 6-mon average solar wind speed
denoting boundaries of fast polar streams.  (Rickett and Coles 1983)

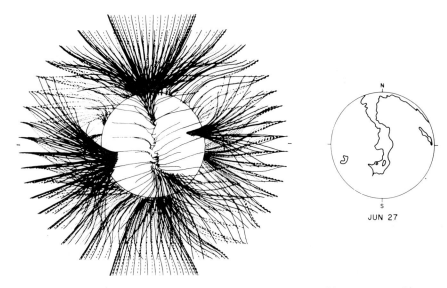

Figure 3.  Comparison of calculated magnetic field configuration for
field lines swept out by solar wind (left side of figure) and a contour
map (right) giving locations of coronal holes determined from x-ray
photographs. (Levine 1977)

interplanetary space.  Since reliable measurements of coronal magnetic
fields in coronal holes do not exist, it has been necessary to infer
the magnetic structure from observations of coronal structures in and
adjacent to coronal holes (e.g. Munro and Jackson 1977) and via
calculations of coronal fields from measurements of magnetic fields at
the photospheric base of the corona (cf. Levine 1977, 1982).  Figure 3
illustrates the large scale configuration of the calculated coronal
magnetic field.  Only magnetic field lines extending into
interplanetary space are plotted.  We see that the footpoints coincide
with the locations of coronal holes determined from x-ray data.  It
must be emphasized that such calculations assume a potential field (no
electric currents in the corona) and thus are only approximations to
the actual fields.  The calculated field configurations also depend
upon the assumed location of the source surface (where the plasma
outflow and field lines become radial).  Hence, they may not indicate
the origins of all the open field lines.  Evidence that this is the
case is provided by the finding that observed areas of coronal holes at
the solar surface typically are larger than those derived by
calculations of the magnetic structure (cf. Levine 1977).

        Current knowledge concerning relationships between coronal holes
and the solar wind can be summarized as follows:  There is strong
evidence that both equatorial and polar coronal holes are sources of
high speed solar wind.  However, it has not been established that
coronal holes are always sources of high speed solar wind or that they
are the only non-transient source of high speed wind.  There is limited
evidence that other regions can be sources of high speed wind (e.g.
Nolte et al. 1977, Burlaga et al. 1978, Sheeley and Harvey 1981) and
that polar coronal holes at solar maximum may not be sources of high
speed wind (see Section 6).  Finally, our knowledge concerning the
details of the coronal magnetic structure inside and around coronal
holes is limited, particularly concerning the locations of the
footpoints of the open field lines.


## 3.  CORONAL MASS EJECTIONS

Coronal mass ejections are the second well-established coronal source
of solar wind.  Figure 4 shows a large mass ejection observed during
Skylab.  These phenomena are typically associated with solar flares
and/or eruptive prominences.  They appear to occur most frequently at
solar maximum at a rate of approximately 2 per day on the average.  At
solar minimum the rate appears to be a factor of about 6 smaller
(Howard et al. 1985, Sheeley et al. 1985).  The speeds typically range
between 200 and 1000 km/s.  Comparisons of coronagraphic observations
by the P-78 spacecraft and in situ measurements of the solar wind by
Helios indicate that coronal mass ejections are often associated with
interplanetary shocks (see review by Schwenn 1986).  Although coronal
mass ejections are clearly an important source of the variable
component of the solar wind, they contribute only about 5% of the total
mass outflow from the sun.  For more information on these dynamic

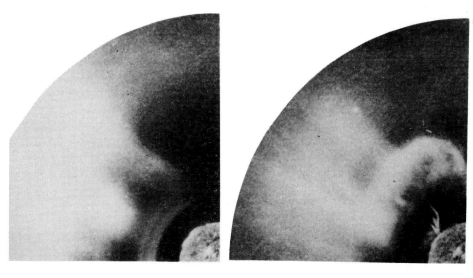

1334 UT                                    1441 UT

Figure 4.  Eruptive prominence and large coronal mass ejection observed
by Skylab telescopes (courtesy Naval Research Laboratory and High
Altitude Observatory).

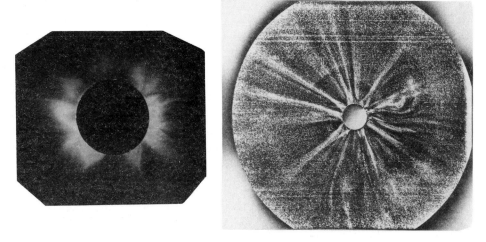

Figure 5.  Photographs of white light corona obtained at 1980 solar
eclipse (courtesy High Altitude Observatory and Los Alamos Scientific
Laboratory).

coronal phenomena see reviews by MacQueen (1980), Dryer (1982), Wagner
(1984), Sheeley et al. (1985) and other papers in this volume.


4.  STREAMERS

Streamers are another source of solar wind.  However, much less is
known about their contribution than is the case for coronal holes and
coronal mass ejections.  Figure 5 shows two views of streamers observed
during the 1980 solar eclipse.  The left-hand photograph, obtained by
the HAO eclipse camera, shows the coronal structure within
approximately 2 solar radii from sun-center, while the right-hand
photograph, obtained by a Los Alamos camera carried on an aircraft
flying along the eclipse path, shows the coronal structure out to about
10 solar radii.  From (1) comparisons between observed coronal
structures and magnetic field configurations calculated from
measurements of photospheric fields and (2) the finding that streamers
overlie neutral lines where the large scale photospheric field reverses
sign, it has been possible to infer the configurations of the coronal
magnetic field in typical streamers.  The inner, bright regions within
about 1.5 solar radii are believed to be regions with closed magnetic
fields, while the surrounding outer envelope and the nearly radial
extensions ("points" of these helmet-like structures) are believed to
be regions with open magnetic fields where the solar wind carries the
field lines into interplanetary space.  The ray-like or fan-like
(depending on the feature and its orientation with respect to the
observer) extensions of streamers most likely are regions of oppositely
directed magnetic fields separated by current sheets.

     Over the years a number of individuals have suggested that
streamers and the interface areas between streamers and other regions
are a likely source of the low speed solar wind (cf. Hundhausen 1972,
Gosling et al. 1981 and references cited therein).  The strongest
evidence for this is that magnetic polarity reversals detected in the
solar wind trace back to the locations of streamers.  In addition,
polarity reversals are often associated with maxima in the proton (and
hence electron) density as would be expected.  Figure 6 (Gosling et al.
1981) shows one possible simple phenomenological model for a streamer
and streamer/coronal hole interface.  Similar models have been proposed
by other investigators.  Such models account for some of the observed
features of the solar wind, including wind speed (high in coronal
holes, low in streamers and the interface region), proton density (low
in coronal holes and high speed wind, high in streamers and low speed
wind) and magnetic field configuration (unipolar in coronal holes,
opposite polarities separated by a neutral sheet in streamers).
Gosling et al. also suggest that the low speed wind originating from
streamers has low helium densities.

     Although streamers are a likely source of the low speed solar
wind, there is much we do not know.  We do not know the relative
contributions to the solar wind mass flux of streamers and the

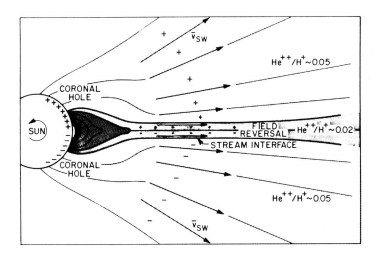

Figure 6.  Idealized schematic view of an equatorial streamer (from
Gosling et al. 1981).  Plasma within the streamer is denser, flows
slower and has a lower helium abundance than plasma outside the
streamer.  The steep radial gradient of coronal density at all
latitudes is masked as if by a radial gradient filter.

interfaces between streamers and other regions.  Nor do we know with
any certainty the surface origins of the open field lines associated
with streamers and these interfaces.  Do these field lines sometimes
terminate in active regions as suggested by some calculations of
coronal magnetic fields (e.g. Levine 1977, 1982), observations of the
location of Type III radio sources (Dulk et al. 1986), and the
appearance of magnetic field configurations inferred from x-ray
photographs of active regions (e.g. Svestka et al. 1977)?  Do they
terminate primarily within the boundaries of coronal holes or in
"quiet" regions of the corona (regions other than coronal holes and
active regions)?  What lies at the footpoint of the interface between
two streamers?  Are there regions where open and closed magnetic fields
are mixed (cf. Kahler et al. 1983)?  In short, there is much to be
learned about the origins of the low speed solar wind and the role that
streamers have in its generation.  As discussed below in Section 6,
answers to some of the above questions can be provided by remote
sensing techniques under development.


5.  FINE SCALE PHENOMENA AND STRUCTURES

Up to this point we have been concerned with large-scale coronal
structures such as coronal holes and streamers.  It is possible that
smaller scale structures may play a significant role in the generation
of the solar wind.  Until the role, if any, of these structures is
understood, our understanding of the physics of solar wind generation

will be incomplete, perhaps incorrect.  Figure 7 contains XUV
spectroheliograms obtained during the Skylab mission.  The bottom
spectroheliogram shows the location of a polar coronal hole (darker
area about the pole).  Note the faint ray-like structures extending
above the surface in this region.  These are polar plumes, features
which appear to have open magnetic field configurations (based on the
forms observed in XUV and white light observations, see Saito 1965,
Bohlin 1977) and can contain 10 to 20% of the coronal mass in a polar
coronal hole (Ahmad and Withbroe 1977).  Do they contribute a
comparable amount of mass to the solar wind?  To be more specific, what
is the relative contribution to the mass outflow of the plumes and
interplume regions?  Some plumes have coronal bright points at their
bases, features which usually mark the location of small-scale bipolar
magnetic regions (cf. Golub et al. 1974, Ahmad and Withbroe 1977, Ahmad
and Webb 1978).  Coronal bright points found in coronal holes appear to
be heated impulsively by a stochastic process, perhaps by rapid
magnetic reconnection (Habbal and Withbroe 1981).  Are the overlying
polar plumes a source of varying solar wind?  Is this wind driven by
thermal or magnetic forces (Mullan and Ahmad 1982)?  Coronal bright
points often flare (exhibit rapid flare-like phenomena where the x-ray
brightness increases by a factor of 10 for several minutes).  Since
ordinary active region flares often are associated with coronal
transients, are bright points flares associated with mini-transients?

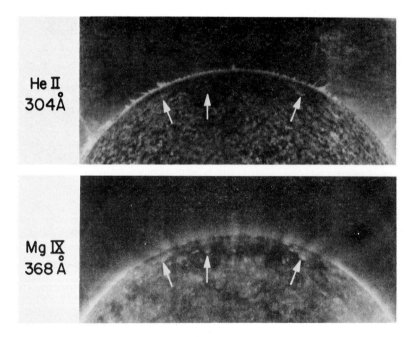

Figure 7.  Spectroheliograms obtained by NRL experiment on Skylab
showing polar plumes (arrows) in a polar coronal hole (Mg IX image) and
jet-like macrospicules at the limb (He II image).

A second class of small-scale features which may or may not play a
role in the generation of the solar wind are the short-lived (minutes)
jets observed at the base of the corona. When observed with high
spatial resolution, the atmosphere at the base of the corona is found
to be dynamic with a large fraction of the surface covered by
chromospheric and transition region material moving up and down in
structures with characteristic sizes of ~1000 km. Three types of
"jets" are observed, the ubiquitous spicules with typical velocities of
25 km/s, macrospicules, large spicules best observed in coronal holes
with velocities up to 150 km/s, and high speed jets with velocities of
the order of 400 km/s (see Beckers 1972, Bohlin 1977, Withbroe et al.
1976, Brueckner and Bartoe 1983). The role, if any, played by these
features in the generation of the solar wind is unknown. Spicules may
be the source of the mass lost in the solar wind outflow; the upward
mass flux provided by spicules is approximately 100 times that lost in
the solar wind outflow. Hence, most of this mass, if heated to coronal
temperatures, must cool and fall back into the chromosphere, perhaps
causing the systematic red shifts observed in spectral lines formed in
the chromospheric-transition region (cf. Athay and Holzer 1982), with
only a small fraction being carried out in the solar wind. However, it
is not known what fraction of the mass observed in spicules get heated
to coronal temperatures ($10^6$ K). There is no direct empirical evidence
that significant amounts of spicular material are heated to coronal
temperatures, and some limited evidence that most of the mass in
spicules remains at lower temperatures (Withbroe 1983).

The high speed jets detected by the HRTS rocket experiment are
perhaps a more interesting phenomena (Brueckner and Bartoe 1983).
These small scale, energetic events have been observed in spectral
lines formed at transition region temperatures (T ~ $10^5$ K) and have
characteristic velocities of ~400 km/s. Given their observed birth
rates, lifetimes, masses and energies, it appears that they could
supply sufficient mass and energy to supply the solar wind and thus be
the ultimate source of the solar wind as suggested by Brueckner and
Bartoe (1983). Alternatively this approximate equality between the
mass and energy requirements for the solar wind and that associated
with high speed jets could be a coincidence with no causal connection
between the two phenomena. The former hypothesis, if true, would
require major revisions in concepts on how the solar wind is generated.
However, until improved observations yield more information about
possible links between the high speed jets and the mass and energy flow
in the overlying corona, the proposed connection between high speed
jets and the solar wind must be considered speculative.

To summarize: the role of small-scale phenomena (spicules, high
speed jets, polar plumes, coronal bright point flares) in the
generation of the solar wind is unknown at the present time. They may
or may not have a significant role in supplying mass, momentum and/or
energy to the solar wind. New observations, particularly by
instruments or groups of instruments, which can simultaneously observe
mass and energy flows in the chromosphere, transition region and corona

with high spatial and temporal resolutions are needed.  This requires a
new generation of high resolution coronal instruments with appropriate
spectroscopic diagnostic capabilities (for measuring coronal mass and
energy flows) to complement instruments monitoring the chromosphere and
transition region (such as the existing HRTS instrument and the very
high resolution Solar Optical Telescope, SOT).


## 6.  NEW REMOTE SENSING TECHNIQUES AND INSTRUMENTS

Recently developed remote sensing techniques can provide new
information concerning the coronal origins of the solar wind.  The
first of these techniques is being developed at the University of
Colorado (Rottman et al. 1981, 1982).  Rottman and collaborators have
developed an EUV instrument which makes use of measurements of Doppler
shifts of spectral lines formed in the chromospheric-coronal transition
region and low corona to search for spectroscopic signatures of sources
of solar wind on the disk.  Figure 8 (Orrall et al. 1983) shows an
example of an application of this technique.  The map on the left
shows the locations of two coronal holes (shaded areas) and the paths
(horizontal lines) along which spectral line profiles of O V and Mg X
lines were measured.  The graph on the right shows the position of the
center of the Mg X (and O V) line as a function of distance along the
disk.  Red shifts are up, blue shifts are down.  We see that the Mg X
line is blue-shifted at the locations of the coronal holes, as would be
expected for a solar region which is a source of strong solar wind
outflow.  Based on data acquired on several rocket flights this
behavior appears to be typical of coronal holes.

    This experiment offers great promise for improving our knowledge
about where on the solar surface solar wind outflows originate.  The
current version of the instrument has flown on short-duration sounding
rocket flights in which there was only enough observing time to make a
few spatial scans across the disk.  The instrument is to be modified so
that it can acquire several days of observations in orbit during
flights of the Space Shuttle.  This will provide sufficient observing
time that maps of the entire solar disk can be acquired.  It will also
permit the spatial and temporal resolution of the instrument to be
improved sufficiently that it can be used to search for waves and
impulsive phenomena believed to deposit heat and momentum in the
corona.

    A second new technique, which is complementary to that discussed
above, is being implemented in a joint program of the Center for
Astrophysics and High Altitude Observatory (Kohl et al. 1980, 1986).
The technique makes use of measurements made by a pair of instruments
(white light coronagraph and UV coronal spectrometer) to measure flow
velocities as a function of height above the solar limb.  The technique
depends upon a phenomenon known as Doppler-dimming.  The intensity of
resonantly scattered emission line radiation depends upon the outflow
velocity of the solar wind.  This radiation is produced by the

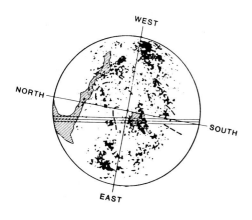
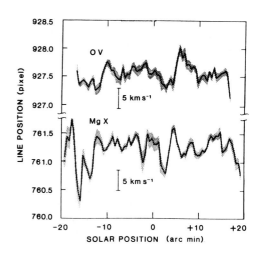

Figure 8.  On the left is a map of the solar disk for 23 November 1981.
The dark areas mark locations of enhanced chromospheric emission, the
cross-hatched regions are coronal holes.  The plots on the right show
the position of line center for O V $\lambda$629 and Mg X $\lambda$625 along the path
of the scans (indicated by the horizontal lines on the disk map).  Red
shifts are up, blue shifts down.  (Orrall et al. 1983).

scattering of radiation from the solar surface by the coronal plasma
(e.g. chromospheric Lyman alpha radiation scattered by hydrogen atoms
in the corona).  Since the resonant scattering cross-section of the
coronal material is highly wavelength dependent (strongly peaked at the
wavelength of the spectral line, e.g. at $\lambda$1216 for hydrogen Lyman
alpha) and the radiation source (e.g. chromospheric Lyman alpha $\lambda$1216)
is narrow, the amount of scattering depends on magnitude of the Doppler
shift and hence flow velocity between the radiation source and
scattering particles.  For hydrogen Lyman alpha this means that
Doppler-dimming becomes important for flow velocities greater than
about 100 km/s, while for heavier ions such as O VI Doppler dimming
becomes important for velocities greater than 30 km/s.  By comparing
the intensities of spectral features which are affected by Doppler
dimming (e.g. resonantly scattered spectral lines) with spectral
features unaffected by Doppler dimming (e.g. electron-scattered white
light radiation), one can determine flow velocities (see Kohl and
Withbroe 1982, Withbroe et al. 1982a, Kohl et al. 1983).

Applications of this Doppler-dimming technique to measurements
acquired on several sounding rocket flights have provided empirical
evidence that the outflow velocities were subsonic for r < 4 solar
radii in a "quiet unstructured" region of the corona and in a polar
coronal hole observed at solar maximum in 1980 and supersonic at r ~ 2
solar radii in a polar coronal hole observed in 1982 (see Withbroe et
al. 1982b, 1985, Kohl et al. 1984).  Based on the empirical limits on
flow velocities, densities, temperatures and velocity amplitude of MHD

waves measured in the 1980 polar region, it appears that this region was more likely a source of low speed wind than high speed wind (Esser et al. 1985, Withbroe et al. 1985). Given the much higher flow velocities that appeared to be present in the 1982 polar region, it was most likely a source of high speed wind. From observations obtained thus far it appears that the physical conditions, and possibly the outflows, in polar coronal holes may differ from region to region. Hence, as indicated in Section 2, one cannot conclude at the present time that all coronal holes are sources of high speed solar wind even though this appears to be the case for most of these features.

The above instrument package (white light coronagraph and UV coronal spectrometer) is presently being converted for flight on the Space Shuttle as part of the SPARTAN program (Kohl et al. 1986). SPARTAN is an instrument carrier which provides accomodations for rocket-class instruments so that they can be operated for a period of several days in orbit during a Shuttle mission. The first solar SPARTAN mission is currently scheduled for the fall of 1986 and will provide measurements of densities (electron, proton, O VI and Fe XII), flow velocities (protons and O VI ions), random line-of-sight velocities of electrons and protons (hence information of electron temperatures, proton temperatures and limits on MHD wave velocity amplitudes) between 1.5 and 4 solar radii and observations of coronal structure between 1.5 and 6 solar radii. It is hoped that at least one flight of the instrument package can be made during a polar passage of Ulysses so that for the first time it will be possible to make a direct connection between physical parameters of the solar wind measured at the coronal origin and far from the sun. This should lead to an improved understanding of the coronal origins of the solar wind and the role played by streamers in solar wind generation. Such measurements can also provide critical constraints on mechanisms for plasma heating, solar wind acceleration, transport and dissipation of momentum and energy and generation of differences in chemical composition.

More advanced instruments for spectroscopically probing the physics of the corona and origins of the solar wind have been proposed for the Solar and Heliospheric Observatory (Huber and Malinovsky-Arduini, 1985) and Pinhole/Occulter Facility (Tandberg-Hanssen et al. 1983).

## 7. SUMMARY

The most well-established coronal sources of steady-state solar wind are coronal holes. There is a substantial body of empirical evidence indicating that recurrent high speed solar wind streams originate in these regions. Coronal mass ejections associated with flares and/or eruptive prominences are another clearly identified source of solar wind, a transient component that accounts for approximately 5% of the total solar mass loss. The role of other coronal structures in the generation of the solar wind is less clear. Streamers and the

interfaces between streamers and other coronal regions are potential
sources of low speed wind, a hypothesis consistent with existing
observations. However, improved measurements are needed, particularly
of mass flows in streamers and their surroundings. Future applications
of recently developed remote sensing techniques could provide valuable
data for this purpose, particularly if made in conjunction with
Ulysses. Small-scale dynamical phenomena observed at the base of the
corona (spicules, macrospicules and high speed jets) and small-scale
structures observed in polar coronal holes (polar plumes) may or may
not play a significant role in supplying mass, momentum and energy to
the solar wind. New measurements are required to determine the role, if
any, of small-scale features in the generation of the solar wind.


## ACKNOWLEDGEMENTS

I wish to express my appreciation to J. L. Kohl and S. R. Habbal for
stimulating discussions on the subject of this review. Support was
provided by NASA under grants NAGW-249 and NAG 5-613.


## REFERENCES

Ahmad, I. A. and Webb, D.: 1978, Solar Phy. 58, 323.
Ahmad, I. A. and Withbroe, G. L.: 1977, Solar Phys. 53, 397.
Athay, R. G. and Holzer, T. E. 1982, Astrophys. J. 255, 743.
Beckers, J. M.: 1972, Ann. Rev. Astron. Astrophys. 10, 73.
Bohlin, J. D.: 1977, in J. B. Zirker (ed.), Coronal Holes and High
    Speed Solar Wind Streams, Colorado Assoc. Univ. Press, Boulder,
    CO, p. 27.
Brueckner, G. E. and Bartoe, D. F.: 1983, Astrophys. J. 272, 329.
Burlaga, L. F., Behannon, K. W., Hansen, S. F., Pneuman, G. W. and
    Feldman, W. C.: 1978, J. Geophys. Res. 83, 4177.
Coles, W. A. and Rickett, B. J.: 1986, this volume.
Dryer, M.: 1982, Space Sci. Rev. 33, 233.
Dulk, G. A., Steinberg, J. L., Hoang, S. and Lecacheux, A.: 1986, this
    volume.
Esser, R., Leer, E., Habbal, S. R. and Withbroe, G. L.: 1985, submitted
    to J. Geophys. Res..
Golub, L., Krieger, A. S., Silk, J. K., Timothy, A. F. and Vaiana, G.
    S.: 1974, Astrophys. J. 189, L93.
Gosling, J. T. et al..: 1981, J. Geophys. Res. 86, 5438.
Habbal, S. R. and Withbroe, G. L.: 1981, Solar Phys. 69, 177.
Howard, R. A., Sheeley, N. R., Jr., Michels, D. J. and Kooman, M. J.:
    1986, this volume.
Huber, M. C. E. and Malinovsky-Arduini, M.: 1985, in Solar Magneto-
    hydrodynamics, ESA Spec. Publ. SP-220, ESA, Noordwijk, in press.
Hundhausen, A. J.: 1972, Coronal Expansion and the Solar Wind,
    Springer-Verlag, New York.
Kahler, S. W., Davis, J. M., Harvey, J. W.: 1983, Solar Phys., 87, 47.
Kohl, J. L., Munro, R. H., Weiser, H., Withbroe, G. L. and Zapata, C.
    A.: 1984, Bull. Amer. Astron. Soc. 16, 531.

Kohl, J. L. et al.: 1986, in this volume.

Kohl, J. L. et al.: 1980, Astrophys. J. **241**, L117.

Kohl, J. L. and Withbroe, G. L. 1982, Astrophys. J. **256**, 263.

Kohl, J. L., Withbroe, G. L., Zapata, C. A., and Noci, G.: 1983, in M. Neugebauer (ed.), Solar Wind Five, NASA, Washington D. C., p. 47.

Krieger, A. S., Timothy, A. F. and Roelof, E. C.: 1973, Solar Phys. **29**, 505.

Levine, R. H.: 1977, in J. B. Zirker (ed.), Coronal Holes and High Speed Solar Wind Streams, Colorado Assoc. Univ. Press, Boulder.

Levine, R. H.: 1982, Solar Phys. **79**, 203.

MacQueen, R. M.: 1980, Phil. Trans. Roy. Soc. **33**, 219.

Mullan, D. J. and Ahmad, I. A.: 1982, Solar Phys. **75**, 347.

Munro, R. H. and Jackson, B. V.: 1977, Astrophys. J. **213**, 874.

Nolte, J. T., Davis, J. M., Gerassimenko, M., Laxaras, A. T. and Sullivan, J. D.: 1977, Geophys. Res. Let. **4**, 291.

Orrall, F. Q., Rottman, G. J. and Klimchuk, J. A.: 1983, Astrophys. J. **266**, L65.

Rickett, B. J. and Coles, W. A.: 1983, in M. Neugebauer (ed.), Solar Wind Five, NASA, Washington D. C., p. 315.

Rottman, G. J., Orrall, F. Q. and Klimchuk, J. A.: 1981, Astrophys. J. **247**, L135.

Rottman, G. J., Orrall, F. Q. and Klimchuk, J. A.: 1982, Astrophys. J. **260**, 326.

Saito, K.: 1965, Pub. Astron. Soc. Japan **17**, 1.

Schwenn, R.: 1986, this volume.

Sheeley, N. R., Jr. and Harvey, J.: 1981, Solar Phys. **70**, 237.

Sheeley, N. R., Jr. et al.: 1982, Space Sci. Rev. **33**, 219.

Sheeley, N. R., Jr., Howard, R. A., Koomen, M. J. and Michels, D. J.: 1985, in Solar High-Resolution Astrophysics using the Pinhole/Occulter Facility, NASA, Marshall Space Flight Center, Huntsville, AL, in press.

Svestka, A., Solodyna, C. V., Howard, R. and Levine, R. H.: 1977, Solar Phys. **55**, 359.

Tandberg-Hanssen, E. A., Hudson, H. S., Dabbs, J. R. and Baity, W. A. (eds.): 1983, The Pinhole/Occulter Facility, NASA Technical Paper 2168, NASA, Washington, D. C.

Wagner, W. J.: 1984, Ann. Rev. Astron. Astrophys. **22**, 267.

Withbroe, G. L.: 1983, Astrophys. J. **267**, 825.

Withbroe, G. L. et al.: 1976, Astrophys. J. **203**, 528.

Withbroe, G. L., Kohl, J. L., Weiser, H. and Munro, R. H.: 1982a, Space Sci. Rev. **33**, 17.

Withbroe, G. L., Kohl, J. L., Weiser, H. and Munro, R. H.: 1985, Astrophys. J. **297**, in press.

Withbroe, G. L., Kohl, J. L., Weiser, H., Noci, G. and Munro, R. H.: 1982b, Astrophys. J. **254**, 361.

Zirker, J. B. (ed).: 1977, Coronal Holes and High Speed Solar Wind Streams, Colorado Assoc. Univ. Press, Boulder, CO.

Zirker, J. B.: 1981 in S. Jordan (ed.), The Sun as a Star, NASA SP-450, NASA, Washington D.C., p. 135.

THE HELIOSPHERIC ENERGY SOURCE

E.N. Parker
Laboratory for Astrophysics and Space Research
933 East 56 Street
Chicago, Illinois 60637
USA

ABSTRACT. The solar wind and the heliosphere exist as a consequence of the heat input to the corona, particularly the coronal holes. The necessary energy input to coronal holes has been estimated to be $10^6$ergs/ $cm^2$sec, requiring Alfven waves with rms fluid velocities of $10^2$km/sec. Observational upper limits on coronal fluid velocities are of the order of 25km/sec, which may not apply to the transparent coronal hole. Alternatively it has been suggested that coronal holes may be heated by agitation from neighboring active regions, suggesting that the vigor of a coronal hole depends upon its location. The Ulysses Mission will provide a direct comparison of the strength of the high speed wind from coronal holes at low latitude and coronal holes at high latitude, from which we should be in a better position to judge the nature of the presently unknown energy sources of the coronal holes and the resulting structure of the heliosphere. The question is fundamental to the dynamics of the windspheres of all stars.

1. INTRODUCTION

The solar wind and the heliosphere are a product of coronal expansion, driven by the heat input to the corona. The heat input may be transient when there is a coronal transient or flare, and otherwise more or less steady, as a consequence of the dissipation of small-scale motions and magnetic inhomogeneities. The heat input is qualitatively different in active regions and in coronal holes, and, at the present time, no theoretical scheme for heating either active regions or coronal holes has been confirmed by observation. Indeed, the contrary! The existing observations constrain the theoretical ideas on coronal heating to only one or two possibilities. That is to say, the basic cause of the heliosphere, punched out of the interstellar medium by the solar wind, is not clear. And until there is a firm theory we cannot understand the X-ray emission and wind sphere of the sun or other star.

To sort out the many pieces to the heliospheric puzzle, we begin with the fact (cf. Hundhausen, 1972; Krieger, Timothy, and Roelof, 1973; Zirker, 1977; Rottman, Orrall, and Klimchuk, 1982) that the high speed

33

*R. G. Marsden (ed.), The Sun and the Heliosphere in Three Dimensions, 33–38.*
© *1986 by D. Reidel Publishing Company.*

streams come from the coronal holes, covering perhaps a fifth or a tenth
of the surface of the sun. It is presumed that the slow wind issues from
the quiet corona around the periphery of the coronal hole. Thus, the
polar wind is fast and tenuous, issuing from the large polar coronal
holes, while the wind at low latitudes is a mixture of slow wind and
fast streams from middle latitudes (Wilcox, 1968; Svalgaard et al. 1975).
Except for an occasional blast from a large flare, it is not evident
that the active coronal regions contribute to the solar wind. The lines
of force in active regions are re-entrant, enclosing and confining that
part of the corona. Coronal expansion occurs only where the gas is able
to overcome the magnetic field and stretch the lines of force outward
into space. To state it the other way around, the solar wind issues only
from open fields. Hence it is the energy input to the coronal holes and
perhaps some of the quiet corona that causes the solar wind and the
heliosphere. The total integrated effect of the active regions contributes only a small part.

Now active coronal regions require an energy input of about
$1 \times 10^7$ergs/cm$^2$sec (Withbroe and Noyes, 1977) regardless of their scale L
($10^4$km for an X-ray bright point and $10^5$km for a normal active region).
The temperature is of the order of $2.5 \times 10^6$ °K, the density is close to
$10^{10}$atoms/cm$^3$, the field is about $10^2$gauss, and the Alfven speed $V_A$ is
approximately $2 \times 10^8$cm/sec. Hence the Alfven transit time along a coronal
loop of length 2L ranges from 10sec to $10^2$sec. There is no direct detection of waves in active coronal regions (but see Beckers, 1976;
Beckers and Schneeburger, 1977). Spicules may contribute substantially
to the heat input, but only at the bottom of the corona, whereas a more
uniformly distributed source is needed (cf. Rosner, Tucker, and Vaiana,
1978; Golub et al. 1980). There is an observational upper limit of
$\langle v^2 \rangle^{1/2} \lesssim 25$km/sec on unresolved fluid motions in the line of sight (Athay
and White, 1978; 1979a,b). It is allowable, then, to assume that the
25km/sec represents unresolved Alfven waves. In that case the net energy
flux $2\rho \langle v^2 \rangle V_A$ for two states of polarization cannot be in excess of
$4 \times 10^7$ergs/cm$^2$sec. Unfortunately, no one has been able to demonstrate how
waves of such small amplitude $(\Delta B/B \sim 10^{-2})$ can dissipate effectively,
and more or less uniformly, over any scale as small as $10^4$-$10^5$km, let
alone over all scales in the range $10^4$-$10^5$km. Hollweg (1984a,b) intro-
duces the assumption that the waves develop a Kolmogoroff spectrum and
decay by nonlinear cascade to large wave number.

There remains the possibility, then, that the dynamical nonequilibrium arising in the current sheets produced in the field by the random
shuffling of the footpoints is the principal source of energy for the
active corona (Parker, 1979,1982,1983a,b,1984,1985; Low, 1986).

## 2.   HEATING CORONAL HOLES

The coronal hole proves to be an entirely different situation. The field
is open, with only one end attached to the sun, so that the random shuffling of the footpoints produces Alfven waves which propagate away into
space. There is no accumulation of wrapping and winding of the lines of
force about their neighbors, and, hence, no reason to expect current

sheets with their internal dynamical nonequilibrium. The gas temperature is of the order of $1.5 \times 10^6$ °K, the density is perhaps $10^8$ atoms/cm$^3$, and the magnetic field is of the order of 10 gauss so that the Alfven speed $V_A$ is again about $2 \times 10^8$ cm/sec (Withbroe and Noyes, 1977; Withbroe et al. 1985). The required energy input is approximately $1 \times 10^6$ ergs/cm$^2$ sec, most of which goes into the expansion that produces the solar wind. It follows that if this energy is supplied by Alfven waves, then the rms fluid velocity $\langle v^2 \rangle^{1/2}$ must be of the order of 40km/sec in the lower part of a coronal hole. This is well above the observational upper limit of 25km/sec mentioned above, but perhaps it escapes observation because the low density of the gas in the coronal hole contributes so little to the integrated line profiles along the line of sight. Or perhaps there are no Alfven waves of such amplitude in the coronal hole and the heat source is some different, and unknown, process.

The fact is, then, that the heating of the coronal holes—the basic cause of the solar wind—is without substantiation. Indeed the possibilities are severely constrained by observations, to where there is little or no latitude for theoretical maneuvers.

To explore the situation in more detail, note that the velocity amplitude of Alfven waves evolves in proportion to $\rho^{-1/4}$ along a slowly varying field in a slowly varying fluid density $\rho$ . Suppose, then, that a wave amplitude of 0.4km/sec is produced at the photosphere by the convective motions below. (The Alfven speed in the photosphere where the number density is $10^{17}$ atoms/cm$^3$ is 0.07km/sec in a mean field of 10 gauss and 0.7km/sec in $10^2$ gauss. So the conjectured wave amplitude is not small.) The 0.4km/sec at the photosphere extrapolates to 22km/sec in the active corona (where the density is $10^{10}$ atoms/cm$^3$) and 70km/sec in the coronal hole ($10^8$ atoms/cm$^3$). So there is a consistent picture for the view that the wave amplitude at the base of a coronal hole may be sufficient to supply the necessary $10^6$ ergs/cm$^2$ sec. The picture is founded on a conjecture, of course, that the wave amplitude is 0.4km/sec at the photosphere. Ground based observations cannot check this conjecture, and something like the Solar Optical Telescope will be necessary to establish or deny its existence. The trouble is that there are no other observations that check either the 0.4km/sec at the photosphere or the 70km/sec at the base of the coronal hole. Thus, for instance, Withbroe et al. (1985) report an upper limit of $\langle v^2 \rangle^{1/2} \cong$ 70km/sec (based on the width of $L_\alpha$ ) in a coronal hole at $4R_\odot$ where the number density is of the order of $10^5$ atoms/cm$^3$. This upper limit extrapolates (according to the $\rho^{-1/4}$ law) to only 13km/sec at the base of the coronal hole (where the density is $10^8$ atoms/cm$^3$). Of course we may imagine that there are strong waves ($\langle v^2 \rangle^{1/2} \gtrsim$ 40km/sec) at the base of the hole which are somehow dissipated to 70km/sec at $r=4R_\odot$ . But the observed upper limit at $4R_\odot$ offers no suport, and supplies an additional constraint. We cannot yet assert that we know the cause (viz. the coronal heating) of the solar wind and the heliosphere. That means that we do not understand why any star has a thermally driven stellar wind.

Now, if it should be established ultimately that there are Alfven waves of sufficient amplitude in coronal holes, there remains the problem of their dissipation. The waves must be largely dissipated before reaching $r = 4R_\odot$ (i.e. in $2 \times 10^6$ km) in view of the observational upper

limit imposed by Withbroe et al. (1985). This is far less stringent than
the seemingly impossible requirements for an active region, of course,
where the dissipation must take place equally in $10^4$km in an ephemeral
active region and in $10^5$km in a normal active region. It would seem that
the characteristic damping length in the coronal hole must be of the
general order of $10^6$km, traversed in roughly $10^3$sec, or 10 wave periods
for waves of $10^2$sec periods. Both fast and slow mode waves, in an essentially collisionless plasma like the solar corona, are subject to
strong Landau damping (Barnes, 1969,1979; Barnes, Hartle, and Bredekamp,
1971), so that they are dissipated in distances of several wave lengths.
But it is Alfven waves that survive into the corona, and plane transverse
Alfven waves are not subject to significant damping. Nor, in view of
their small amplitude ($\Delta B/B \sim 0.04$) are they subject to strong nonlinear
effects. However, there is a geometrical effect, that seems to have been
ignored heretofore, which may introduce the necessary dissipation. We
presume that the Alfven waves are generated in the photosphere with
frontal widths of the order of $10^3$km, corresponding to the scale of the
granules that excite them. The wave lengths are less than 100km for periods of $10^2$sec, so that the waves are broad and nearly plane. However,
upon reaching the corona where the Alfven speed increases to $10^3$km/sec
or more, the wavelength increases to $10^5$km while the frontal width is
still of the order of $10^3$-$10^4$km. Insofar as the wave is not purely
torsional in form, it develops a strong longitudinal component and is,
therefore, subject to strong Landau damping (Barnes, Hartle, and Bredekamp, 1971; Habal and Leer, 1982). So the dissipation may take care of
itself.

    To consider other possibilities for the coronal hole, Fla et al.
(1984) have pointed out the interesting possibility that coronal holes
may be heated wholly, or in part, by disturbances that propagate into
them from the surrounding active coronal regions. Fast mode waves may
refract into the coronal holes and be dissipated sufficiently quickly
to deposit their energy (see also Davila, 1985). Insofar as coronal
holes are heated by disturbances from active regions, we would expect
the vigor of the wind from a coronal hole to depend upon the vigor and
proximity of active regions. There may be significant differences between the fast streams from the polar coronal holes (Orrall, Rottman,
and Klimchuck, 1983) and holes at low latitude. Specifically we need
measurements from Ulysses of the ratio of the mass flux to magnetic
field intensity $\rho v/B$ in space. This ratio can be extrapolated back to
the coronal hole at the sun, where a knowledge of the mean field strength
$B_c$ provides a direct figure for the mass flux $\rho_c v_c$ in the coronal hole.
The energy input to the wind in the coronal hole is $\rho_c v_c \mathcal{E}$ ergs/cm$^2$sec,
where $\mathcal{E}$ is the energy per unit mass, $\frac{1}{2} w^2 + GM_\odot/R_\odot$, necessary to lift the
wind away from the sun and provide it with a velocity $w$. Thus, for in-
stance, a wind velocity $w = 400$km/sec implies $\mathcal{E} = 2.7 \times 10^{15}$ergs/gm while
600km/sec implies $3.7 \times 10^{15}$ergs/gm. A typical mass flux of $1.7 \times 10^{-10}$gm/cm$^2$
sec in a coronal hole ($10^8$hydrogen atoms/cm$^3$ moving outward with a speed
of 10km/sec)implies an energy input to the wind of $0.5 \times 10^6$ergs/cm$^2$sec and
$0.6 \times 10^6$ergs/cm$^2$sec, respectively. What we need from the Ulysses Mission
are figures for $\rho v/B$ for the wind from coronal holes at different
latitudes, at widely varying distances from active regions. Perhaps the

most difficult task will be to obtain reliable numbers for the mean field in coronal holes at high latitudes, and to know the point on the sun from which originated the mass flux observed at a particular point in space.


3.   SUMMARY

To summarize, then, the energy source that creates the heliosphere is shrouded in uncertainty. A resolution of the difficulty requires observational input. Any detection or upper limit on wave motions and disturbances of any kind in coronal holes would help very much (cf. Bruner, 1978). The Solar Optical Telescope will play a crucial role in determining the precise nature of the small-scale motions that drive the field at the photosphere. It is these motions that produce the Alfven waves, and the wrapping and interweaving of the lines of force in the corona above. The Ulysses Mission will make a fundamental contribution in its direct study of the wind from the coronal polar holes, to show what, if any, distinctions can be made between coronal holes near and distant from active regions. It is to be hoped that the mystery of the energy of the solar wind can be cleared up in the next decade. It is a fundamental problem in the physics of stars.

This work was supported, in part, by the National Aeronautics and Space Administration under grant NGL-14-001-001.


REFERENCES

Athay, G. and White, O.R. 1978, Astrophys. J. **226**, 1135.
Athay, G. and White, O.R. 1979a, Astrophys. J. **229**, 1147.
Athay, G. and White, O.R. 1979b, Astrophys. J. Suppl. **39**, 333.
Barnes, A. 1969, Astrophys. J. **155**, 311.
Barnes, A. 1979, in Solar System Plasma Physics, Vol. **I**, ed. E.N. Parker, C.F. Kennel, and L.J. Lanzerotti (New York, North Holland Pub.Co.) pp. 249-319.
Barnes, A., Hartle, R.E., and Bredekamp, J.H. 1971, Astrophys. J. Letters, **166**, L53.
Beckers, J.M. 1976, Astrophys. J. **203**, 739.
Beckers, J.M. and Schneeburger, T.J. 1977, Astrophys. J. **215**, 356.
Bruner, E.C. 1978, Astrophys. J. **226**, 1140.
Davila, J.M. 1985, Astrophys. J. **291**, 328.
Fla, T., Habbal, S.R., Holzer, T.E., and Leer, E. 1984, Astrophys. J. **280**, 382.
Golub, L., Maxson, C., Rosner, R., Serio, S., and Vaiana, G.S. 1980, Astrophys. J. **238**, 343.
Habbal, S.R. and Leer, E, 1982, Astrophys. J. **253**, 318.
Hollweg, J.V. 1984, Astrophys. J. **277**, 392.
Hundhausen, A.J. 1972, Coronal Expansion and the Solar Wind, Springer-Verlag, New York, Chap. V.
Krieger, A.S., Timothy, A.F., and Roelof, E.C. 1973, Solar Phys. **29**, 505.
Low, B.C. 1986, Solar Phys. **100**, 309.

Orrall, F.Q., Rottman, G.J., and Klimchuk, J.A. 1983, Astrophys. J. Letters, **266**, L65.

Parker, E.N. 1979, Cosmical Magnetic Fields, Oxford, Clarendon Press, pp. 359-391.

Parker, E.N. 1982, Geophys. Astrophys. Fluid Dyn. **22**, 195.

Parker, E.N. 1983a, Geophys. Astrophys. Fluid Dyn. **23**, 85.

Parker, E.N. 1983b, Geophys. Astrophys. Fluid Dyn. **24**, 79.

Parker, E.N. 1984, in Proceedings of III Trieste Workshop, Sacramento Peak Observatory, 18-25 August; ed. by R. Stalio, J.B. Zirker.

Parker, E.N. 1985, Geophys. Astrophys. Fluid Dyn. (in press).

Rosner, R., Tucker, W.H., and Vaiana, G.S. 1978, Astrophys. J. **220**, 643.

Rottman, G.J., Orrall, F.Q., and Klimchuk, J.A. 1982, Astrophys. J. **260**, 326.

Sterling, A.C. and Hollweg, J.V. 1984, Astrophys. J. **285**, 843.

Svalgaard, L., Wilcox, J.M., Scherer, P.H., and Howard, R. 1975, Solar Phys. **45**, 83.

Wilcox, J.M. 1968, Space Sci. Rev. **8**, 258.

Withbroe, G.L. and Noyes, R.W. 1977, Annual Rev. Astron. Astrophys. **15**, 363.

Withbroe, G.L., Kohl, J.L., Weiser, H., and Munro, R.H. 1985, Astrophys. J. (Oct. 1).

Zirker, J.B. 1977, Coronal Holes and High Speed Wind Streams, Colorado Associated University Press, Boulder, Colorado; ed. by J.B. Zirker.

CORONAL SPECTROSCOPY AND IMAGING FROM SPARTAN DURING THE POLAR PASSAGE
OF ULYSSES

J.L. Kohl, H. Weiser, G.L. Withbroe
Harvard-Smithsonian Center for Astrophysics
60 Garden Street
Cambridge, MA  02138, U.S.A.

G. Noci
University of Florence
Institute of Astronomy, Florence, Italy

R.H. Munro
High Altitude Observatory
P.O. Box 3000
Boulder, CO 80307, U.S.A.

ABSTRACT.  A joint payload consisting of a UV Coronal Spectrometer and
a White Light Coronagraph is being developed for a series of flights on
NASA's Spartan carrier.  Spartan, which is deployed and retrieved by
Shuttle, is intended to provide at least 27 orbits of solar observations per mission.  The first flight of this payload is planned for
October 1986, and it is anticipated that a subsequent flight will occur
at each polar passage of Ulysses.  Coordinated measurements of solar
wind acceleration parameters in polar regions of the solar corona with
Spartan and "in situ" measurements of the polar wind from Ulysses will
provide a unique opportunity to establish empirical constraints on polar
wind models.  Spectroscopic diagnostics to be provided by Spartan include determinations of the electron and neutral hydrogen random velocities, outflow velocities in the 30 to 300 km/s range and electron
densities.

1.   INTRODUCTION

New ultraviolet instrumentation and plasma diagnostic techniques have
been developed and used during a series of sounding rocket flights to
initiate detailed spectroscopic studies of the extended solar corona
(i.e. heliospheric heights beyond about 1.5 solar radii from sun center). Results from the three flights have served to establish the
reliability of the instrument design (Kohl, Reeves & Kirkham, 1978) and
have also provided combined ultraviolet spectroscopic and white light
imaging measurements of the extended corona out to 3.5 solar radii.
These measurements have been used to develop a more complete empirical

*R. G. Marsden (ed.), The Sun and the Heliosphere in Three Dimensions, 39–43.*
© *1986 by D. Reidel Publishing Company.*

description of the extended solar corona.

Three polar regions have been observed. An observation of a polar coronal hole from April 13, 1979 has been discussed by Kohl et al., (1980) and a more detailed analysis is underway. An analysis (Withbroe et al., 1985) of a polar region observed on February 16, 1980, revealed subsonic outflow out to 3.5 solar radii. This result for a region which exhibits many of the characteristics of coronal holes is in sharp contrast to the supersonic outflow velocities inferred by Munro and Jackson (1977) from white light data for a broader and lower density polar hole observed in 1973. Their analysis was based on conservation of mass flux, but it was necessary for them to use equatorial rather than polar fluxes because of the unavailability of the latter. A preliminary analysis of combined ultraviolet and white light observations of a polar coronal hole from July 20, 1982 has provided evidence for supersonic outflow within 2 solar radii (Kohl et al., 1984). The measurements also can be used to determine the mass and energy fluxes in the solar wind near the sun. Coordinated observations with both Ulysses and Spartan will provide the first opportunity to utilize in situ values of polar fluxes together with spectroscopic diagnostics of the extended corona and white light determinations of electron density.


2.    INSTRUMENTATION

The White Light Coronagraph is similar to instruments described previously (MacQueen et al., 1974) and will only be discussed briefly here. It determines the polarization and absolute radiance of the broad band visible corona with about 25 arcsecond resolution simultaneously over most of the corona from 1.5 to 6 solar radii. Visible images are obtained with a CCD system and recorded on magnetic tape. This instrument's primary function is to determine the three dimensional electron density and its temporal properties over an extended field of view.

The UV Coronal Spectrometer is a relatively new instrument concept that consists of a reflecting telescope with internal and external occultation and a dual spectrometer. It measures the intensity and spectral line profile of H I Ly-$\alpha$, which consists of both resonantly and electron scattered components, and it also measures, simultaneously, the intensities of O VI $\lambda1032$ and $\lambda1037$. The occulted telescope is nearly identical to the sounding rocket system that has been described previously (Kohl et al., 1978). The field of view is illustrated on Figure 1. The instantaneous field of view corresponds to the projection of the spectrometer's three entrance slits onto the corona. The telescope mirror is split so that the two slits used to observe the intensity of the O VI resonance lines and the H I Ly-$\alpha$ resonance profile respectively, are projected onto the lower spatial element in Figure 1, and the slit used to observe the electron scattered component of H I Ly-$\alpha$ is projected onto the upper spatial element. Internal mirror motions are used to scan the corona in a radial direction from 1.5 to 3.5 solar radii. The entire Spartan is rolled about the sun line in order to view the full corona. The spatial resolution depends on the entrance slit dimensions and is 5' x 0.53', 5' x 2.6' and 5' x 4' for

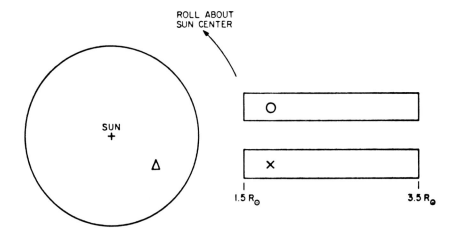

Figure 1.   Field-of-View of UVCS/Spartan 201.

observations of resonant H I Ly-$\alpha$, O VI, and electron scattered Ly-$\alpha$, respectively, where the longer dimension of the resolution element is always parallel to the limb-tangent.

The dual spectrometer consists of a 750 mm Fastie-Ebert section that is optimized for maximum efficiency at wavelengths near 1216 Å and has a spectral resolution of 0.3 Å for the resonantly scattered Ly-$\alpha$. The other section consists of a Rowland circle mount with a 600 mm radius of curvature and a spectral resolution of 2.5 Å at 1035 Å.  A discrete anode microchannel array detector is used to detect H I Ly-$\alpha$. Forty-eight anodes corresponding to 0.3 Å resolution elements are used to detect the resonance profile and forty-two anodes with 2 Å widths are used to detect the electron scattered profile and to isolate spectral lines located in the wing of H I Ly-$\alpha$.  The O VI lines are detected with channel electron multipliers with CsI photocathode coatings.

The Spartan itself consists of an evacuable instrument carrier and a service module.  The Spartan carrier provides the payload with battery power, a thermal shroud, a solar pointing control system, a preprogrammed sequencer that issues a series of coordinated commands to the payload and to the pointing system, and a magnetic tape recorder for data storage.  There is no telemetry link.


3.   SPECTROSCOPIC DIAGNOSTICS

Coronal plasma diagnostics to be used for the Spartan missions have been described by Kohl and Withbroe (1982) and by Withbroe et al. (1982). The physical processes involved are illustrated in Figure 2.  All diagnostics are based on single scattering events in an optically thin medium, and there is no appreciable absorption of the scattered radiation between the scatterer at the sun and the detector except in the center of the H I Ly-$\alpha$ line which is affected by exospheric hydrogen. Because of

SOLAR WIND PARAMETERS in POLAR SOURCE-REGIONS

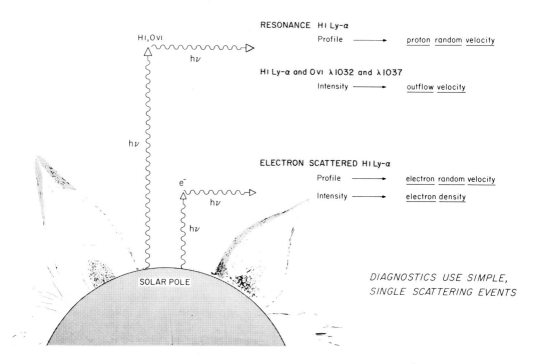

Figure 2.  An illustration of the physical processes being used for
           spectroscopic diagnostics of plasma parameters in the
           solar wind source region.

the low densities, the collision rates are small enough that the probability of the primary process being interrupted is negligibly small.
Hence, except for the physical scale, the spectroscopy of the extended
corona is similar, in many respects, to a controlled diagnostic measurement of an optically thin laboratory plasma.  Some information is
exceptionally straightforward to derive from the data.  For example, the
line-of-sight average of the random velocity distribution of neutral
hydrogen (the velocities of the protons and neutral hydrogen are believed to be identical due to their rapid charge transfer rate) is
determined directly by the H I Ly-α profile.  This information can be
used as a hard empirical constraint on theoretical models of the observed region.  Specification of local random velocities requires a
model of the line of sight.  The random velocities are due to both thermal and nonthermal processes and can be parameterized as a kinetic temperature.  The capability of the UV Coronal Spectrometer to perform this
diagnostic is now well established.
     In principle, the determination of the electron random velocities
is similarly straightforward, but in practice it is more difficult
because of the much smaller intensities.  This diagnostic will be

attempted, for the first time, during the Spartan flight planned for
October 1986.

The determination of outflow velocities is based on the Doppler
dimming of resonantly scattered radiation (cf. Withbroe et al., 1982).
In general, this diagnostic utilizes a comparison between the intensities of electron-scattered and resonantly scattered radiations. In the
absence of Doppler dimming (i.e. a static corona), the two intensities
would be expected to fall off with height in a similar fashion since
both would be approximately proportional to electron density, except for
small changes in the ionization balance due to electron temperature
variations. Outflow velocities shift the coronal absorption profile off
the incoming line radiation from the solar disk, and this results in a
decrease in the resonantly scattered density as a function of the outflow velocity. The sensitivity to velocity depends on the widths of the
spectral lines involved and hence on the mass of the scatterer. Rocket
observations of H I Ly-α, which is sensitive to velocitites in the 100
to 300 km/s range, have now been used to demonstrate the Doppler dimming
technique for that line. Spartan will also measure the intensities of
the O VI resonance lines which are sensitive to outflows in the 30 to
250 km/s range (Noci, Kohl and Withbroe, 1985).

REFERENCES

Kohl, J.L., Munro, R.H., Weiser, H., Withbroe, G.L. and Zapata, C.A.,
    1984, Bull. Amer. Astron. Soc. 16 531.
Kohl, J.L., Reeves, E.M. and Kirkham, B., 1978, in New Instrumentation
    for Space Astronomy, K. van der Hucht and G.S. Vaiana, eds. (New
    York: Pergamon Press) p. 91.
Kohl, J.L., Weiser, H., Withbroe, G.L., Noyes, R.W., Parkinson, W.H.,
    Reeves, E.M., Munro, R.H. and MacQueen, R.M., 1980, Astrophys. J.
    241, L117.
Kohl, J.L. and Withbroe, G.L., 1982, Astrophys. J. 256, 263.
MacQueen, R.M., Gosling, J.T., Hildner, E., Munro, R.H., Poland, A.J.
    and Ross, C.L., 1974, Proc. SPIE, 44, 207.
Munro, R.H. and Jackson, B.V., 1977, Astrophys. J. 213, 874.
Noci, G., Kohl, J.L. and Withbroe, G.L., in preparation.
Withbroe, G.L., Kohl, J.L., Weiser, H. and Munro, R.H., 1982, Space
    Sci. Rev. 33, 17.
Withbroe, G.L., Kohl, J.L., Weiser, H. and Munro, R.H., 1985, Astrophys.
    J. 297, in press.

# THE STRUCTURE AND ROTATION OF THE SOLAR CORONA: IMPLICATIONS FOR THE HELIOSPHERE.

D. G. Sime
High Altitude Observatory/NCAR
P.O. Box 3000
Boulder, Colorado 80307 USA

ABSTRACT. Observations of the solar corona and measurements of its rotation are reviewed in order to suggest what rotational characteristics of the solar wind might be observed from the Ulysses mission. Results from 20 years of data from the HAO K-coronameters on Mauna Loa, Hawaii, are described. They show that the corona rotates, on average, rather more rigidly than do any of the usual photospheric or chromospheric tracers on the sun, displaying an average synodic rotation rate of 14.2 degrees per day. The suggestion is made that a similar rotation character will be observed by Ulysses, and some evidence to support that from interplanetary observations is discussed.

## 1. INTRODUCTION

Since the corona forms the lower boundary for the solar wind, it provides a means by which to infer certain properties of the interplanetary medium to which we may not have direct access. This is particularly valuable in anticipation of the results which will be returned from the Ulysses spacecraft because, although the Ulysses mission will travel through high latitude regions of the solar wind which have never before been measured *in situ*, the corona itself has been measured routinely at all latitudes for some time. In particular, the rotation characteristics of the corona are well described as a function of latitude, and allow us to make an estimate of the solar wind rotational properties to be observed from Ulysses. In this paper, we review the results for the electron corona, postulate their extension into interplanetary space, and indicate to what extent evidence exists already to support the suggestion.

## 2. OBSERVATIONS OF CORONAL ROTATION

Daily observations of the solar white light (K-) corona have been made for the last 20 years at the High Altitude Observatory's Mauna Loa, Hawaii, facility. The data are a record of the distribution of polarized radiance of the corona (pB) which arises from Thompson scattering of photospheric light off the coronal electrons. As such, the observations yield a description of the large scale density structure of the corona, and by extension, some aspects of the structure of the heliosphere.

*R. G. Marsden (ed.), The Sun and the Heliosphere in Three Dimensions, 45–51.*
© *1986 by D. Reidel Publishing Company.*

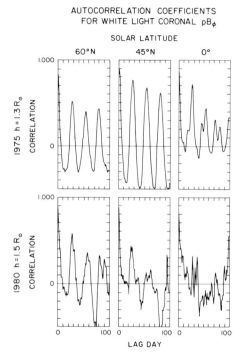

Figure 1. Examples of the autocorrelation coefficients for the 15° wide latitude bins centered at 60N, 45N and the equator are plotted for two years. The data near solar minimum (1975) are taken at a height of $1.3R_\odot$ and show a strong recurrence due to the coronal rotation. The observations from 1980, at solar maximum show a pronounced reduction in the peak at one rotation lag. The data for 1980 were taken at $1.5R_\odot$ .

This long data set has recently been analyzed to reveal the distribution and evolution of large scale (wave number < 6) features in the corona (Fisher and Sime 1984a). In particular, a correlation analysis has revealed the character of the coronal rotation (Fisher and Sime 1984b). From the daily series of limb observations, the data were formed into series representing 15° strips of latitude, and extending for about a calendar year. After the removal of a secular trend from the series, an auto-correlation analysis was performed, sample outputs from which are displayed in Figure 1. For each maximum near a lag of about 27 days, the position of the center of mass of the peak was estimated and interpreted as a measure of the period of recurrence of the coronal features. If the recurrence of a feature on the limb is inter-

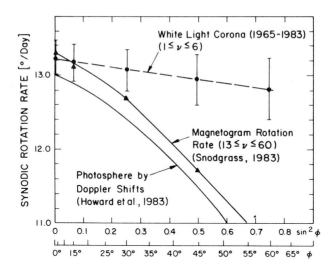

Figure 2. Average synodic rotation rate of the white light corona, shown as a function of latitude and compared with other measures of solar rotation. The middle curve is from Snodgrass (1983) and shows the rotational properties of magnetograms from the Mount Wilson observations, and closely resembles the result of Newton and Nunn (1951). The bottom curve represent values of the spectroscopic rotation rates from Mount Wilson data presented by Howard et al 1983.

preted as due purely to the rotation of the corona, the period or rate of rotation follows immediately. Samples of autocorrelation functions from 2 intervals during the period of interest, 1975 and 1980, are shown in Figure 1 for several latitudes. In both epochs, there is an identifiable peak at a lag of about 27 days. However, the amplitude of the recurrence signal is significantly lowered in the year near maximum, indicating that in any analysis such as this, it is important to verify that the basic assumption of stationarity is not violated by the actual evolution of the corona.

From autocorrelations such as the one in figure 1, average rotation periods can be estimated as a function of time or latitude. The overall average synodic rotation rate found in such a manner is 14.2°/day. Of particular interest is the fact that this rate does not vary widely as a function of latitude. Figure 2 show the average profile of rotation rate (or period) with latitude produced by averaging all years of data together. Also shown for comparison are 2 other measures, both photospheric, of solar rotation; plasma motion and small scale magnetic field motion. Clearly, although all the rotation rates displayed here are comparable at or near the equatorial region, the reduction of rotation rate with latitude in the corona is significantly less than that in the photospheric tracers. This suggests that the corona rotates in an almost rigid fashion and confirms the earlier result by Parker et al. (1980) that the white light corona displays less differential rotation than the photosphere does. However, an important point made in the Fisher and Sime analysis is that, although there is a marked contrast between the behavior of the small scale magnetic field field in the photosphere and that of the corona, the rotation rate of the large scale magnetic field, as expressed in the interplanetary neutral sheet (Hoeksema, et al., 1983), is very close to that of the corona. A further result reported by Fisher and Sime (1984b) was the existence of a cyclic variation in the individual yearly average rotation periods about the overall average period for the interval 1964-1983, with an amplitude of about 0.7 days. A change of a similar magnitude had been reported by Parker et al, but because of the duration and epoch of the data available to them, Parker and colleagues had not been able to distinguish whether the variation was secular or cyclic.

## 3. IMPLICATIONS FOR THE SOLAR WIND

Since it forms the lower boundary to the solar wind, the corona provides a basis from which to infer the organization of the heliosphere. A useful example is in the relationship between the long-lived high speed solar wind streams and the magnetically open rarefied regions in the corona, known as coronal holes. The mapping of high speed streams from 1AU to sources within coronal holes is well established empirically, and also quite well understood in basic physical terms (Holzer and Hundhausen (1982);

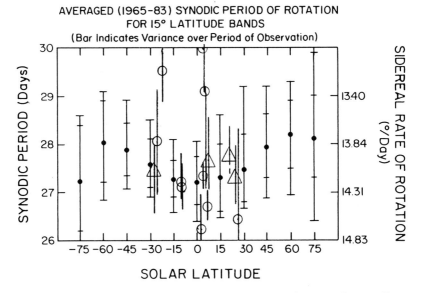

Figure 3. Time averaged values of the inferred coronal synodic rotation period derived for the period 1965-1983. The bars indicate the peak variation of the inferred yearly averages in each latitude bin. Also shown, as open circles, are values derived by Armstrong for the solar wind rotation rate derived from IPS observations, together with additional values derived from later observations (triangles). The bars on the IPS points indicate one standard deviation uncertainty.

Leer and Holzer (1985)). Consequently, the correspondence of features in the coronal density structure to to flow features in the solar wind suggests that the rotation properties of the solar wind will reflect those of the corona. That is, that the solar wind itself will display a rigid rotation similar to that of the corona.

If this is indeed the case, as the Ulysses spacecraft climbs to higher heliographic latitudes, it will measure only slightly lower rotation rates for the interplanetary medium than those in the ecliptic. The slowing of rotation would be much more marked if the solar wind rotation was tied to the photospheric rate, but this is not what we expect.

In a spherical solar wind, the magnetic field lines will deviate from radial in the characteristic spiral pattern more slowly at high latitudes than in the equatorial plane. Thus the radial distance at which a given spiral angle with respect to the radial is reached increases as $1/\cos(\theta)$ where $\theta$ is the latitude. Also, for a particular radial flow speed, the position at which the field direction deviates from the radial by a given amount is proportional to the rotation rate. Thus, increasing latitude and a slowing of rotation both would have the effect of reducing the wind up of the field lines and hence of reducing the development of some dynamical processes in the solar wind such as the development of corotating interaction regions. However, the implication of our speculation that the solar wind shares the almost rigid rotation of the corona is that any reduction in the development of the interplanetary spiral field lines to be detected from Ulysses will arise almost entirely from the purely geometrical effect of increasing latitude. In particular, the reduction in the development of the spiral will be significantly less than that which would be expected for the case in which the solar wind rotated in the same manner as the photosphere.

On the other hand, if there is a component of the solar wind which arises from, or is tied to, the photospheric plasma or magnetic fields, then it may be able to be distinguished from the background solar wind because of the different rate at which its rotational modulation will be detected.

## 4. OBSERVATIONS OF THE SOLAR WIND

Some clues already exist as to the rotation of the solar wind out of the ecliptic. High latitude measurements of the wind speed have been made for some time by the interplanetary scintillation (IPS) technique. In particular, measurements have been made from 1972 to the present, by the UCSD group (Coles, 1986) and from these, Armstrong (1975) has analyzed the recurrence of features in the velocity series. Armstrong's analysis, which is similar to the coronal analysis discussed above, gives results for the rotation period of the solar wind over a range of heliographic latitudes from approximately 40S to 30N. The periods are consistent with the almost rigid rotation of the solar wind at rates similar to those of the corona. Results from Armstrong are shown in Figure 3, augmented by some additional points derived from UCSD scintillation data by the author. The estimates based on IPS data are shown as squares and are plotted for comparison over the distribution of average coronal rotation rates with latitude. As can be seen in Figure 3, and although the data are sparse and somewhat uncertain, they do provide an indication that the large scale structure of the solar wind follows the rigid rotation of the corona, rather than the highly differential behavior with latitude seen in the photosphere.

Inference of the rotation properties of the solar wind can also be made from the results of an analysis of *in situ* measurements by Gosling et al.,(1976). In this analysis, the recurrence of features in the solar wind was once again measured, yielding in essentially the same manner as for the coronal work, a synodic rotation period for the interplanetary medium of 27.1 days. In particular, this result was measured when the wind was dominated by long lived high speed streams, corresponding as in the coronal study to intervals when the recurrence level was high and repeatable. This value corresponds quite well to the average equatorial rate of almost all tracers of solar rotation. However, the high speed streams discussed by Gosling et al. were unipolar, and are known now to have arisen from polar coronal holes which extended close to the equator. Their character was therefore organized at high latitude, and they must as a result show the rate of high latitude rotation in the interplanetary medium. Thus, once again, the indication is that the rotation of the solar wind at high latitudes is similar to that near the equator; the rotation of the solar wind is on average rigid.

## 5. SUMMARY

The overall average rotation of the density structure of the corona appears over the last two solar cycles to have been more rigid than the that of the photosphere. This is interpreted to indicate that the solar wind will also rotate almost rigidly, and there is fragmentary evidence from the IPS and some *in situ* measurements that this is indeed the case. As a result, the rotation rate of the solar wind at high latitudes may not be significantly different from that in the ecliptic. Any reduction which might therefore be expected in the dynamical effects in the solar wind due to slower winding up of field lines into the interplanetary spiral will be small. Any such effect will arise largely from the increase in latitude since there will be no reduction in rotation rate with latitude. If any component of the solar wind is tied to photospheric features it may be possible to distinguish or separate it from the background due to the difference in their rotation rates.

## ACKNOWLEDGEMENT

This manuscript has benefitted from a careful reading by R. R. Fisher, for which the author is grateful.

## REFERENCES

Armstrong, J. W. 1975 Ph.D. Thesis University of California at San Diego.

Coles, W. A. 1986 This proceedings.

Gosling, J. T., J. R. Abridge, S. J. Bame, and W. C. Feldman 1976 *J Geophys Res.,* **81,** 5061.

Fisher, R. and D. G. Sime 1984a *Astrophys. Jour.*, **285**, 354.

Fisher, R. and D. G. Sime 1984b *Astrophys. Jour.*, **287**, 959.

Gosling, J. T., J. R. Abridge, S. J. Bame, and W. C. Feldman 1976 *J Geophys Res.*, **81**, 5061.

Hoeksema, T., J. Wilcox, and P. Scherrer 1983 *J Geophys Res.*, **88**, 9910.

Howard, R., J. M. Adkins, J. E. Boyden, T. A. Cragg, T. S. Gregory, B. J. LaBonte, S. P. Padilla, and L. Webster 1983 *Solar Phys.*, **83**, 321.

Hundhausen, A. J. and T. E. Holzer 1980 *Phil. Trans. R. Soc. Lond.*, **A297**, 521.

Leer E. and T. E. Holzer 1985 **ESA** *SP-235, 3.*

Parker, G., R. Hansen, and S. Hansen 1980 *Solar Phys.*, **80**, 185.

Snodgrass, H. 1980 *Astrophys. Jour.*, **270**, 288.

# OVI DIAGNOSTICS OF SOLAR WIND GENERATION

Giancarlo Noci
Istituto di Astronomia dell' Universita' di Firenze
Largo E. Fermi 5 - 50125 Firenze, Italy

John L. Kohl, George L. Withbroe
Harvard-Smithsonian Center for Astrophysics
60 Garden st. - Cambridge MA 02138, U. S. A.

ABSTRACT. The OVI resonance doublet is partly collisionally and partly radiatively excited in the solar corona. In the solar wind the OVI ions can attain sufficient outflow speed to cause excitation of the $2P_{\frac{1}{2}}$ level by the chromospheric CII $\lambda 1037.0$ line. We show that this extends the diagnostic possibilities of the OVI resonance doublet. In particular, the determination of the intensity ratio of the doublet lines at several heights can be sufficient to yield the solar wind velocity at those heights and hence information on the mechanisms of solar wind acceleration.

## 1. INTRODUCTION

The possibility of using the Doppler dimming effect as a diagnostic tool to measure outflow velocities in the extended solar corona (up to a few solar radii of heliocentric distance) has been discussed in some recent papers (Beckers and Chipman 1974; Kohl and Withbroe 1982; Withbroe et al. 1982). Withbroe et al. have concluded that the OVI resonance doublet at 1031.9 and 1037.6 A is a promising tool for the investigation of the low speed (30-100 km/sec) solar wind.

In this paper we show that this velocity range is considerably extended as a result of the existence of a chromospheric CII line at 1037.0 A, which pumps coronal OVI ions to the upper level of the 1037.6 line.

The absolute intensity of the coronal lines depends, besides other parameters, on the abundance of the ion considered. To obtain this, one has to know other quantities, i.e. the abundance of the element and the ionization state, which are known with a considerable uncertainty. On the other hand, as pointed out by Kohl and Withbroe (1982), the intensity ratio of the resonance doublet of an ion of the Li

*R. G. Marsden (ed.), The Sun and the Heliosphere in Three Dimensions, 53–58.*
© *1986 by D. Reidel Publishing Company.*

isoelectronic sequence, like OVI, does not depend on the ion population and therefore the information contained in it is much more model independent than that arising from the absolute intensity of a line. Furthermore it is not affected by the error originating from the absolute calibration. Thus we will concentrate on the doublet intensity ratio.

## 2. THEORY

Both lines of the OVI resonance doublet have a radiative and a collisional component, which originate from transitions from the ground level. Therefore, since the solar corona is a low density medium, the intensity ratio is given by

$$\frac{I_{12}}{I_{13}} = \frac{\int_{-\infty}^{\infty} (h\nu_{12} \, C_{12} \, N_1/4\pi + j_{12}) \, dx}{\int_{-\infty}^{\infty} (h\nu_{13} \, C_{13} \, N_1/4\pi + j_{13}) \, dx} = \frac{\langle C_{12} N_1 + 4\pi j_{12}/h\nu_{12}\rangle}{\langle C_{13} N_1 + 4\pi j_{13}/h\nu_{13}\rangle} , \qquad (1)$$

where the suffix 12 refers to the $S_{1/2}$– $P_{1/2}$ transition ($\lambda$1037) and the suffix 13 to the $S_{1/2}$– $P_{3/2}$ one ($\lambda$1032). Here $x$ is a coordinate which runs along the line of sight, $C_{ik}$ is the collisional excitation rate, $j_{ik}$ the emissivity of the radiative component, $N_1$ the number density of OVI ions in the ground level and $\nu_{ik}$ the frequency of the ik transition ($\nu_{12} \simeq \nu_{13}$). Since both $j_{12}$ and $j_{13}$ are due to radiative excitations from the ground level and thus depend linearly on $N_1$ , the intensity ratio is independent of the OVI concentration, as already pointed out.

The region of maximum contribution to the integral at the numerator can be different from that to the integral at the denominator, depending on the density and velocity distributions in corona. A proper calculation of the intensity ratio, therefore, requires a coronal model. However, in order to gain an insight on the information contained in the intensity ratio we need to work out an analytical expression, even if approximate. The mean values in Equation (1) are equal to the actual values at some point along the line of sight inside the coronal feature observed: we make the approximation that this point is the same for numerator and denominator. In a spherically symmetric corona this is likely to be the point along the line of sight of minimum heliocentric distance, since the densities are maximum there. Therefore, for the sake of simplicity, we will refer, in the following, to this point and consider in it the ratio between the emissivities,

$$\rho = \frac{C_{12} + 4\pi j_{12}/N_1 h\nu_{12}}{C_{13} + 4\pi j_{13}/N_1 h\nu_{13}} , \qquad (2)$$

that we approximate as equal to the intensity ratio.

The radiative components of the coronal OVI doublet originate from photons emitted in the lower atmosphere. If $I_e(\lambda, \underline{n}')$ is the intensity of the exciting radiation at the wavelength $\lambda$ in the direction $\underline{n}'$, it can be shown (Noci et al. 1985) that the expression for j is

$$j(\underline{n}) = Bh\lambda N_1 \int_\Omega p(\varphi)d\omega \int_0^\infty I_e(\lambda - \delta\lambda, \underline{n}') \, \phi(\lambda - \lambda_0) \, d\lambda/4\pi, \qquad (3)$$

where h is Plank's constant, B the Einstein coefficient for absorption, $\underline{n}$ the unit vector parallel to the line of sight directed towards the observer, $\varphi$ the angle between $\underline{n}'$ and $\underline{n}$, $\Omega$ the solid angle subtended by the source of exciting radiation and $\phi$ the coronal absorption profile, whose width and shape (Gaussian) are due to the velocity spread of the OVI ions. $p(\varphi)$ gives the angular dependence of the scattering, $\delta\lambda$ is the Doppler shift associated to the solar wind outflow velocity w, which is assumed to be radial, and $\lambda_0$ the laboratory wavelength of the considered transition.

In this paper we will only discuss briefly the case $\delta\lambda \neq 0$, i.e. the case of coronal regions where the plasma is not confined by closed magnetic fields.

Following Kohl and Withbroe (1982), we define the dimensionless quantity

$$D(r,w) = \frac{\int_\Omega p(\varphi)d\omega \int_0^\infty I_e(\lambda - \delta\lambda, \underline{n}') \, \phi(\lambda - \lambda_0)d\lambda}{\int_\Omega p(\varphi)d\omega \int_0^\infty I_e(\lambda, \underline{n}') \, \phi(\lambda - \lambda_0)d\lambda}, \qquad (4)$$

where r is the heliocentric distance, which we plot in Figure 1 both for the 1032 and the 1037 line as a function of w for uniformly bright chromosphere and transition region and r = 2 solar radii.

The curves of Figure 1 are obtained by using, as values of the exciting intensity, the data of Vernazza and Reeves (1978). Figure 1 shows clearly that while the intensity of the coronal OVI line at 1032 A is sensitive to solar wind speeds in the interval 30 - 100 km/sec, the 1037 A line is sensitive to speeds up to 280 km/sec.

We now transform Equation (2) putting $C = N_e q(T_e)$, q, $N_e$ and $T_e$ being the electron impact excitation coefficient, the electron density and the electron temperature, respectively, and taking j from Equation (3).

Defining $\eta = \lambda_{13}^2 B_{13} r^2 \int_\Omega p(\varphi)d\omega \int_0^\infty I_e(\lambda) \, \phi(\lambda - \lambda_0)d\lambda/cqF$, where $F = N_e r^2 w$ is the electron flux in the solar wind, we get:

Kohl, J.L., and Withbroe, G.L. 1982, Ap. J., 256, 263.

Noci, G., Kohl, J.L., and Withbroe, G.L. 1985, in preparation.

Vernazza, J., and Reeves, E.M. 1978, Ap. J. Suppl., 37, 485.

Withbroe, G.L., Kohl, J.L., Weiser, H., and Munro, R. 1982, Space Sci. Rev., 33, 17.

# SOLAR WIND OBSERVATIONS NEAR THE SUN

J. W. Armstrong[1], W. A. Coles[2], M. Kojima[3] and B. J. Rickett[2]

1 Jet Propulsion Laboratory, Pasadena, CA 91193, U.S.A.
2 Univ. of California, San Diego, La Jolla, CA 92093, U.S.A.
3 The Research Institute of Atmospherics, Nagoya Univ.,
  Toyokawa 442, Japan.

ABSTRACT. The first observations of interplanetary scintillations with the VLA telescope are reported. The solar wind in the accelerating region from 3 to 12 solar radii was observed by scintillation of the radio source 3C279. The results obtained outside of 7 solar radii showed good agreement with previous work but observations between 3 and 4.5 solar radii were new and unexpected. Turbulence in the solar wind has a spatially anisotropic structure elongated in the radial direction, the flow direction being also in the radial direction. An abrupt change of both the velocity and the spatial anisotropy of turbulence was found at distances from 3 to 4.5 solar radii. There is a large random velocity component inside of 12 solar radii which is comparable to the bulk flow speed, and it has a spatially anisotropic probability distribution. From the measured cross-correlation functions we found evidence which may be related to the complex structure of the magnetic field.

## 1. Introduction

Interplanetary scintillations (IPS) can be used as a probe of the solar wind in regions where direct measurements are not available such as at high latitude and near the sun. The IPS technique can provide information on the velocity distribution (bulk velocity and random velocity) and the spatial spectrum of the plasma turbulence. Multi-station IPS observations have been made regularly at frequencies of 74 MHz (Armstrong and Coles, 1972), 69 MHz (Kakinuma et al., 1973) and 327 MHz (Kojima et al., 1982). These have probed the region outside 30 solar radii. The solar wind inside 30 solar radii is another inter-esting target of the IPS observations. In this region the energy of the magnetic field is larger than that of the flow, and plasma waves, such as Alfven modes, are expected to be contributing to the acceleration of the solar wind. However it has been difficult to produce a unique model of this acceleration region due to impossibility of in situ mea-surements of the solar wind plasma in this region. Remote sensing tech-niques can give useful information on the plasma in this region; however, single-frequency and single-antenna observations are inadequate

59

*R. G. Marsden (ed.), The Sun and the Heliosphere in Three Dimensions, 59–64.*
© *1986 by D. Reidel Publishing Company.*

to observe such active solar wind (Scott et al., 1983) and therefore
observations near the sun are hampered by the lack of suitable observatories.    Here we report the first IPS observations with the VLA telescope (operated by the U.S. National Radio Astronomy Observatory in New
Mexico). This telescope is almost ideal for  observations near the sun
as it has twenty seven antennas with a well-arranged base line configuration, it can be operated at frequencies of 1.4, 5 and 15 GHz (hereafter these frequencies are called L-, C- and U-band) and it can measure
both the complex field and the intensity scintillations simultaneously.
The freedom to choose the observation frequency provided by the VLA
allows us to select the proper strength of scintillation, so that the
base line length matches the scale size of a diffraction pattern. The
ability to measure the electricfield correlation is important in understanding the turbulence spectrum, because there is one-to-one relation
between the 'visibility' and the turbulence spectrum which is independent of the strength of scattering.
    Observations were carried out from Oct. 5th to 11th in 1983. On
5th the IPS source 3C279 was observed at about 11 solar radii to the
east of the sun; on 8th it was occulted; on 7th and 9th it made the
closest approach to the sun of 3.2 solar radii; on the last day it was
at about 11 solar radii to the west. The latitude was slightly north of
the equator, about 4 to 14 degrees.


2. Observational results

    The spatial anisotropy of the intensity diffraction pattern is

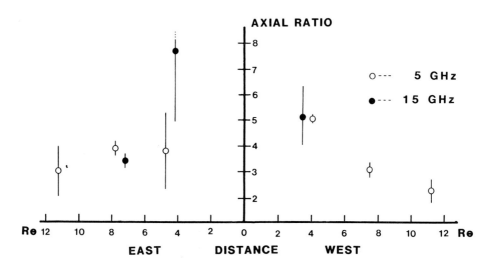

Figure 1.   The estimated axial ratio versus heliocentric distance. Each
data point is an average of observations for each day. The data point of
5 GHz is   shifted slightly in horizontal from that of 15 GHz to avoid
overlap.

shown in Figure 1. The spatial structure of an intensity diffraction
pattern was approximated with a characteristic ellipse and the anisotropy is represented by its axial ratio.  The axial ratios observed in C-
and U-band outside of 7 solar radii are consistent with earlier work
(Ekers and Little, 1971; Scott et al., 1983). As the C-band and U-band
scintillation observed outside of 7 solar radii was weak, the anisotropy
of the turbulent medium may be slightly more isotropic  than indicated
by the observations.  The axial ratio showed a rapid increase at the
nearest approach to the sun. The C-band observations at 4 solar radii
are probably underestimated by a factor of 1.5-2.0 due to the finite
receiver bandwidth.

   The mean elongation direction of the turbulence is in the radial
direction from the sun as shown in Figure 2. This figure shows a  marginally significant  northward deflection  at both sides of the sun. The
distribution function of the flow direction is also shown in Figure 2
and this is quite similar to that of the elongation direction of the
turbulence at both sides of the sun. This suggests that the bulk flow
and the magnetic field were both essentially radial between 3 and 12
solar radii.

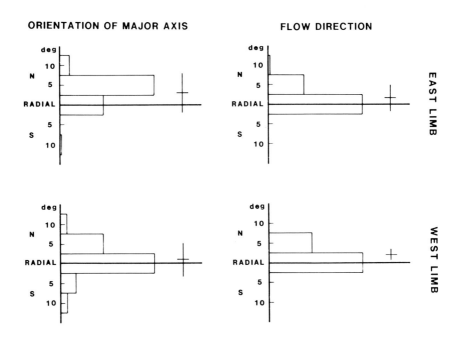

Figure 2.   The probability distribution of the elongation of the turbulence (left side) and the mean flow direction (right side). Spatial
structure of the turbulence is approximated with a characteristic
ellipse and the elongation direction is represented by orientation of a
major axis of the ellipse. Distribution at the east and the west side
are shown separately. The mean value of the distribution is shown by a
cross.

The estimated mean velocity is shown in Figure 3. The abscissa, of course, represents both space and time as the source moves about 4 solar radii per day. The data from 7 to 12 solar radii are consistent with earlier work as summarized by Scott et al. (1983). This earlier work showed a great deal of scatter and a general increase from about 200 km/s to 400 km/s between 10 and 20 solar radii. The most interesting feature of the new data is the rapid change between 3 and 4.5 solar radii. This could be an actual acceleration if the observation was in a region of rapidly diverging geometry (as shown for example by Munro and Jackson 1977). However the line of sight is not moving exactly along a streamline. It may be that the flow is rather filamentary and the line of sight crossed a thin high speed stream. Observations by spectroscopic techniques (Withbroe et al. 1982) and using conservation of flux arguments (Munro and Jackson 1977) suggest that rapid changes in velocity can be expected near 4 solar radii. However it is clear that such observations must be repeated and carefully compared with coronal observations to understand both the magnetic geometry and the characteristics of the underlying plasma.

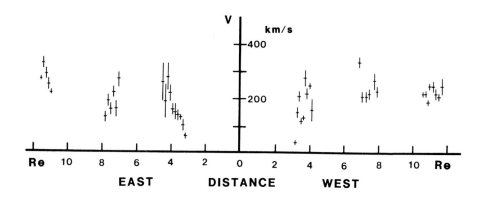

Figure 3.   The estimated solar wind mean flow speed versus heliocentric distance.   Typical observation time of each data point is 20 to 30 minutes.

All data taken inside of 12 solar radii showed evidence of a large random velocity component, in good agreement with the previous work which is  summarized by Scott et al. (1983). Six data samples are shown in Table I.  The velocity has an anisotropic probability distribution: the component parallel to the bulk flow is comparable to the bulk flow speed and the perpendicular component is not larger than 100 km/s. This velocity distribution could be a result of the line of sight passing through regions of different mean velocity. However it seems unlikely that the mean velocity could vary so widely. Hydromagnetic waves could also be the cause of random velocity. Theoretical calculations with Alfven waves (Hollweg, 1978; Leer et al., 1982) give a random velocity

around 200 km/s. This kind of random velocity can have an anisotropic
distribution.

Table I. Sample random velocity data

| DATE | TIME | VO | Svx | Svy |
|------|------|------|------|------|
|      | LST  | km/s | km/s | km/s |
| OCT10 | 09:24 | 235 | 250 | 100 |
| 10 | 10:38 | 192 | 200 | 100 |
| 10 | 11:53 | 254 | 300 | 0 |
| 10 | 13:08 | 286 | 200 | 100 |
| 11 | 10:37 | 310 | 200 | 100 |
| 11 | 12:27 | 268 | 300 | 50 |

VO is the bulk velocity, Svx is
the random velocity parallel to the
bulk flow and Svy is the perpendicular component.

Peculiar cross-correlation functions were found in some observations, namely functions with a double-peak. These were found under the
special conditions that the antenna pairs, between which the correlation
was made, had a base line component perpendicular to the bulk flow and
that scattering was weak. These correlation functions cannot be explained by special velocity distributions or a diffraction pattern with
a simple structure like a symmetrical ellipse, but by a superposed
pattern of variously-oriented ellipses. This can be thought of physically in several ways much like a 'random velocity'. First, the line of
sight could pass through a number of scattering media in which the
magnetic field has different configurations and the turbulence is elongated along it. Second, the turbulence may have a complex spatial spectrum like a butterfly. There is no particular reason that the spectrum
should be a symmetrical ellipse, especially when hydrodynamic waves are
present. Third, the orientation of anisotropy could change randomly with
time if the magnetic field is waving. Thus this peculiar correlation
function probably indicates the complex structure in the magnetic field.


Acknowledgement

We are particularly grateful to the VLA staff as such non-standard
observations require a great deal of assistance. The National Radio
Astronomy Observatory is operated by Associated Universities, Inc.,
under contract with the National Science Foundation.

## References

Armstrong, J.W. and Coles, W.A.: 1972, J. Geophys. Res. **77**, 4602
Ekers, R.D. and Little, L.T.: 1971, Astron. Astrophys. **10**, 310
Hollweg, J.V.: 1978, Rev. Geophys. Space Phys. **16**, 689
Kakinuma, T., Washimi, H. and Kojima, M.: 1973, Publi. Astron. Soc.
     Japan. **25**, 271
Kojima, M., Ishida, Y., Maruyama, K. and Kakinuma, T.: 1982, Proc. Res.
     Inst. Atmospherics, Nagoya Univ., **29**, 61
Leer, E., Holzer, T.E. and Fla, T: 1982, Space Sci. Rev. **33**, 161
Munro, R.H. and Jackson, B.V.: 1977, Astrophys. J. **213**, 874
Scott, S.L., Coles, W.A. and Bourgois, G.: 1983, Astron. Astrophys.
     **123**, 207
Withbroe, G.L., Kohl, J.L., Weiser, H. and Munro, R.H.: 1982, Space Sci.
     Rev. **33**,17

# THREE-DIMENSIONAL RECONNECTION AFTER A PROMINENCE ERUPTION

R. A. Kopp
Los Alamos National Laboratory
Los Alamos, NM 87545

G. Poletto
Osservatorio Astrofisico di Arcetri
50125 Firenze

ABSTRACT. A currently widely held explanation of the diverse phenomena following a prominence eruption (most clearly displayed during two-ribbon flares) is that these are the direct result of the ensuing reconnection of open magnetic field lines created by the eruption. Kopp and Pneuman (1976) and Kopp and Poletto (1984a,b) developed a 2-D model wherein an analytical time-dependent magnetic field geometry was used to account quantitatively for the major characteristics of many of these post-flare effects. In the present paper, after mentioning a few situations for which a 2-D treatment would nevertheless clearly be inadequate, we extend to three dimensions our earlier model. Assuming as before that reconnection causes the coronal field line system to relax to a nearly potential configuration (in contrast to the pre-eruptive field, which is probably highly sheared), a 3-D representation of the field may be constructed through the well-known technique of expansion of the magnetic scalar potential in spherical harmonics. This method can be applied to both real and model configurations. Here we present some preliminary results obtained for two hypothetical model situations, in both of which the opposite polarity regions of the surface field distribution are skewed with respect to one another.

## INTRODUCTION

Disrupting events, such as prominence eruptions, solar flares, and coronal transients, often create stressed open field configurations whose geometry has been modeled both analytically and numerically. Kopp and Pneuman (1976) and Kopp and Poletto (1984a,b) have developed an analytical time-dependent magnetic field model to represent the post-eruptive relaxation back to an equilibrium field configuration, in terms of which they were able to explain many features characterizing the decay phase of major two-ribbon (2-R) flares. This model hypothesizes that the excess energy vested in the open configuration is partially

65

*R. G. Marsden (ed.), The Sun and the Heliosphere in Three Dimensions, 65–70.*
© *1986 by D. Reidel Publishing Company.*

converted, via magnetic reconnection, into thermal and kinetic energy of
the plasma contained on newly formed closed field lines.

The magnetic configuration at any time was assumed to consist of an
axisymmetric field, potential between the solar surface ($r = R_o$) and a
source surface ($r = r_1 \geqslant R_o$), where it joins smoothly to a non-potential
radial field extending to greater heights.  The source surface height $r_1$
is allowed to increase with time to describe the occurrence of reconnection at progressively higher levels in the corona.  For simplicity the
normal component of the surface magnetic field was approximated by one
"lobe" of a single high-degree Legendre polynomial.

This model, being a two-dimensional one, can be applied only to active regions with very simple morphologies; it is inherently incapable
of describing features related to more complicated spatial structuring.
If we assume that the surface field distributions observed in 2-R flares
are representative of configurations where disruptions tend to occur, we
easily find cases where a 2-D model will either be unrealistic or
provide an incomplete description of the reconnection scenario.

For example, the large 2-R flare of 21 May, 1980 showed the classical behavior expected of magnetic reconnection – mass ejecta, a growing
loop system, and an increasing ribbon separation.  Concurrently,
however, the HXIS instrument aboard SMM imaged gigantic arch features
which do not lie in a plane normal to the solar surface (Švestka et al.,
1982) and which therefore cannot be represented by the 2-D geometry used
to describe the rising loops.  While the mechanism of origin of these
enormous arches is still unclear, their oblique orientation relative to
the flare loops they surmount (Švestka, 1984) requires an interpretation
in terms of a 3-D magnetic model.

More striking examples of complicated reconnecting geometries are
provided by multiple-ribbon flares (Tang, 1985).  In 3-R flares
the field lines from a single region of one polarity apparently connect
to two distinct regions of opposite polarity, while in 4-R flares up to
four discrete sets of loops may be formed, crisscrossing the corona in
different directions.  Obviously such exotic structures can only be
described in three dimensions.


THE 3-D MODEL

The basic hypothesis upon which the 3-D model is based remains the same
as for the 2-D model.  Namely, at any time t the field is assumed to
be potential between the solar surface ($r = R_o$) and an equipotential
source surface ($r = r_1$) and to extend radially outward beyond $r = r_1$.  A
magnetic field model that satisfies these requirements may be constructed through the well-known technique of expansion of the magnetic
scalar potential $\Psi$ in spherical harmonics.  The solution of Laplace's
equation in the domain $R_o \leqslant r \leqslant r_1$ is

$$\Psi(r,\theta,\phi) = R_o \sum_{n=1}^{\infty} \sum_{m=0}^{n} P_n^m(\theta) \left\{ \left[ c_n^m \left(\frac{r}{R_o}\right)^n + (1-c_n^m)\left(\frac{R_o}{r}\right)^{n+1} \right] g_n^m \cos(m\phi) + \right.$$

$$+ \left[ d_n^m \left(\frac{r}{R_o}\right)^n + (1-d_n^m)\left(\frac{R_o}{r}\right)^{n+1} \right] h_n^m \sin(m\phi) \Big\} ,$$

where $P_n^m(\theta)$ are the Legendre functions, $g_n^m$ and $h_n^m$ are constants to be determined from magnetic data and have dimensions of gauss, and $c_n^m$ and $d_n^m$ are constants that give the fraction of field contributed by external sources. The latter may be determined by requiring that the potential $\Psi(r,\theta,\phi)$ vanish at $r = r_1$, from which it follows that

$$c_n^m = d_n^m = - \left[ (r_1/R_o)^{2n+1} - 1 \right]^{-1} .$$

Once the potential $\Psi(r,\theta,\phi)$ is known, the magnetic field components at any point $(r,\theta,\phi)$ are given by

$$B_r = - \frac{\partial \Psi}{\partial r} , \quad B_\theta = - \frac{1}{r} \frac{\partial \Psi}{\partial \theta} , \quad B_\phi = - \frac{1}{r \sin \theta} \frac{\partial \Psi}{\partial \phi} ,$$

and from these a pictorial fieldline display can be constructed.

Evaluation of the constants $g_n^m$ and $h_n^m$ has been carried out through a least mean square fit to the line-of-sight field, minimizing the sum of the squares of the differences between the observed line-of-sight field and the calculated line-of-sight field. The reader may refer to the work of Altschuler and Newkirk (1969) for more complete details about this procedure. These authors, who were interested in the global structure of the corona, used synoptic magnetograph data to identify line-of-sight fields on individual surface elements. In the present case we wish to focus on the behavior of a single active region, and their technique is unsuitable because magnetic changes on a time scale of several days would sensibly affect the active region configuration. In the following we will therefore assume that the input data always refer to a single observation.


RESULTS

The method described above may be applied to either hypothetical model fields or real data, and the 3-D field structure at different times is obtained by allowing the source surface to assume different heights. A limitation of the method is comprised by the long computing time and large computer memory required whenever the reconstructed field must reproduce data of high spatial resolution. This problem can be overcome for model fields, however; these, even if defined on a coarse mesh of unrealistically large dimensions, provide an "enlarged" view of what happens in real fields when the same distribution of magnetic polarities is limited to smaller areas. We have analyzed the restructuring of two such ideal configurations with a resolution of 10 x 10 square degrees, where the positive and negative polarity regions run parallel to each other and to the solar equator. In the first one the two polarities are confined to the same longitude interval, but the maximum positive and negative fields are at different longitudes within this interval. In the second one, shown in Figures 1 and 2, the field maxima occur at the

same longitude, but the longitude intervals for the two polarities are
slightly shifted one relative to the other.   Input data have been
reconstructed by truncating the Legendre expansion at N = 16, and the
CPU time needed to evaluate the constants $g_n^m$ and $h_n^m$ was about 75 minutes
on a VAX 750.   These two configurations represent simplified examples of
"sheared" photospheric field distributions.

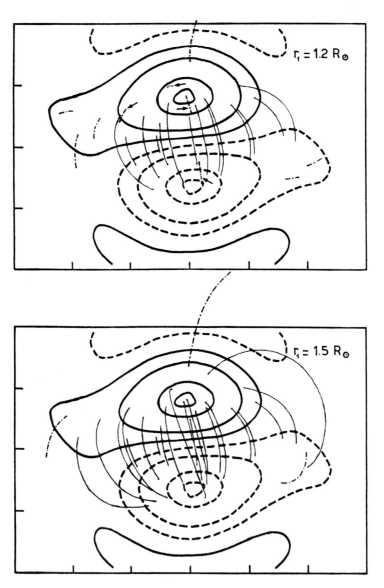

Figure 1.   Field topologies for the source surface at 1.2 and 1.5 $R_o$.
Open field lines are shown as dot-dash lines.   Inner isogauss levels are
for fields 3, 6, and 7 times larger than those on the outermost contour.

Figures 1 and 2 show typical field-line plots obtained when the source surface is at 1.2, 1.5, 2.0, and 2.5 solar radii. Field lines have been drawn sparsely all over the region, the only criterion for selection being to provide a good visual representation of the coronal magnetic geometry. These have been projected onto the surface as if viewed from directly above. Open field lines are more numerous when the source surface is at lower heights and appear to originate from both high- and low-intensity fields. Closed field lines, on the other hand,

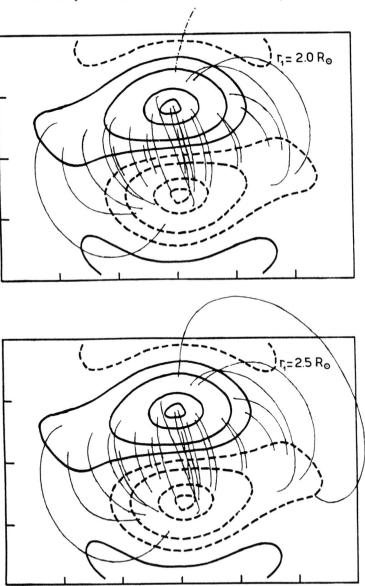

Figure 2. Same as Figure 1, but for the source surface at 2.0 and 2.5 R$_\odot$. Tick marks indicate 10$^\circ$ intervals of longitude and latitude.

shift but little from one map to another, because their maximum height decreases only slightly as the source surface rises.

Field lines bridging directly across the neutral line generally close at relatively low heights. The "peripheral" closed loops extending much higher, however, may be related to the gigantic arches recorded by HXIS. In the reconnection scenario, these would appear at later times when the source surface has risen to great heights, or they could possibly persist throughout a disrupting event which is limited to central areas of the active region.

CONCLUSIONS

The results presented here should be regarded as preliminary. Work is underway to improve the technique for displaying model fields, as well as to handle observed active region fields with reasonably high spatial resolution.

With regard to model fields one should keep in mind that the present displays are incomplete and miss some interesting structures. One way to avoid this would be to display at each source surface height, instead of always the same set of field lines, those which close just as they reach the source surface. Such maps would clearly emphasize the most recently closed field lines.

To simulate real fields it will be necessary to decrease significantly the computing time required per configuration. To do this we plan to restrict the calculation to a limited sector of the sun, replacing the surface field values outside this sector with a periodic repetition of those therein. Pilot calculations using this technique seem promising; for example, for a sector of longitudinal width $180^{\circ}$ (rather than $360^{\circ}$), the computing time is smaller by a factor of seven than that previously needed.

REFERENCES

Altschuler, M. D., and Newkirk, G. A. Jr.: 1969, Solar Phys. 9, 131.
Kopp, R. A. and Pneuman, G. W.: 1976, Solar Phys. 50, 85.
Kopp, R. A. and Poletto, G.: 1984a, Solar Phys. 93, 351.
Kopp, R. A. and Poletto, G.: 1984b, Proc. of IAU Coll, No. 86 (G. A. Doschek, ed.), p. 27.
Švestka, Z., Stewart, R., Hoyng, P., and 13 co-authors: 1982, Solar Phys., 75, 305.
Švestka, Z.: 1984, Solar Phys. 94, 171.
Tang, F.: 1985, Solar Phys., submitted.

# SECTION II:  SOLAR HARD X-RAYS

STEREOSCOPIC MEASUREMENTS OF HARD SOLAR X-RAYS, AND RELATED TOPICS

K. Hurley
Centre d'Etude Spatiale des Rayonnements
B.P. 4346
31029 Toulouse Cedex, France

ABSTRACT.   Solar  X-ray  observations  in  the  >20  keV  range  are
discussed,  with  emphasis  on  their  implications  for   electron
acceleration.  It is shown how measurements of  fine  time  structure,
directivity,  and  the  production  height  of  X-radiation  may  be
interpreted  in  the  contexts  of  thermal  and  non-thermal  electron
populations.  Recent  measurements  are  reviewed,  and  the  expected
contribution of Ulysses is discussed.

KEY WORDS: Solar flares, X-rays

1. INTRODUCTION

The solar flare, being a short time  constant  phenomenon,  is  perhaps
the most spectacular manifestation of solar  activity.    Nevertheless,
on a purely energetic scale, it is  relatively  insignificant:  even  a
large flare, such as that of August 4, 1972, liberates only about  2  x
$10^{32}$ erg over $10^3$ s, and perhaps 50% or more of this energy  is  in
electrons with energies >20 keV (Lin,  1982).    For  comparison,  the
steady state luminosity of the sun is 4 x $10^{33}$  erg/s,  and  most  of
this is of course in visible light.   Solar  flares,  however,  are  of
great  interest,  not  only  because  of  their  effects  on  the
interplanetary  medium  and  the  terrestrial  environment,  but  also
because we look  hopefully  to  the  sun  as  a  key  to  understanding
particle acceleration in general.  Solar flare hard X-radiation is,  in
turn, energetically a small part of the overall  solar  flare;  in  the
example of the August 4, 1972 event, some $10^{26}$ erg were  detected  in
X-rays above 25 keV (Hoyng et al., 1976).  But because it is  generally
accepted that this X-radiation is generated by  bremsstrahlung  by  the
energetic electrons near the acceleration site,  it  carries  with  it
information  which  is  essential  to  our  understanding  of  the
acceleration process. Phenomenologically, a flare may pass  through
three phases, identified as precursor, flash, and gradual (see,  e.g.,
the review by Brown et al., 1981).  During the flash  phase  there  may
be an impulsive phase, in which energy is rapidly released in the  form
of electromagnetic radiation, from radio to gamma rays, and  in
particles.  It is generally agreed that the source of  this  energy  is

73

R. G. Marsden (ed.), The Sun and the Heliosphere in Three Dimensions, 73–86.
© 1986 by D. Reidel Publishing Company.

magnetic, and located above the photosphere, possibly in the corona. Detailed reviews of flare theory and of the impulsive phase have been given by Sturrock (1980), Spicer and Brown (1981), and Kane et al. (1980a). The question which will be treated here is: what does the study of hard solar X-radiation tell us about the parent electron population? Specifically, what can we learn about the location of the electron acceleration region, the electron energy spectrum, and the electron velocity distribution function, and how will Ulysses contribute to these studies?

Broadly speaking, electron energization in the active region may take place via thermal or non-thermal processes. By thermal it is understood that the time required to thermalize the electrons is less than the duration of the impulsive phase; the plasma need not be isothermal, but temperatures in the range $10^8$ - $10^9$ K are required to explain the hard X-radiation. In nonthermal models, the electrons are accelerated to energies several hundred times greater than the temperatures prevailing before the impulsive phase. Beaming may result, which will be shown below to have important observational consequences. In the thermal models, hard X-radiation is produced by bremsstrahlung, and collisional energy losses are negligible. In the non-thermal models, however, the X-radiation is produced by thick- or thin-target bremsstrahlung between the energetic electrons and a much cooler plasma; energy losses due to Coulomb collisions are then typically $10^{4-5}$ times greater. One immediate and important consequence of the distinction between thermal and non-thermal models is therefore that, to generate a given X-ray flux, the non-thermal interpretation may require many times more electrons than the thermal interpretation. Indeed, in some cases a nonthermal interpretation implies that a large fraction of the energy dissipated in a flare go into electron acceleration, placing difficult requirements on the acceleration efficiency. Hence the attempt to distinguish between the two observationally and theoretically has assumed considerable importance.

A large number of models of electron energization and their subsequent hard X-ray emission have been elaborated based on the above considerations. They have been reviewed by Kane et al. (1980a), and a brief summary is provided here. It should be emphasized that it is difficult or even impossible to evaluate them without taking into account a large body of observational data which includes microwave, EUV and electron measurements; in addition, theoretical considerations may strongly favor aspects of one model or another. For conciseness, however, they are presented here in a manner which will emphasize the observational consequences as far as hard X-radiation is concerned.
a) Non-thermal, thin target models. Here, the electrons create X rays in a low density region ($<10^{10}$ ions/cm$^3$, or distances >2000 km above the photosphere), through which they pass, possibly by mirroring in the magnetic field of an active region, on their way to still lower density regions where they do not create X radiation. The X ray source may extend high into the corona, although this feature is not unique to this model. The variations in the observed X-ray flux directly reflect the variations in the electron injection or acceleration process.

b) Non-thermal, thick target, trap models. Here the X-radiation is produced by electrons trapped in a magnetic field; it is "thick-target" in the sense that the electrons lose a substantial part of their energy in the trap, but the ion density may be either low or high, with the production region either high or low, respectively. In some of the trap models, the injection of electrons takes place over a period much less than that of the impulsive phase.

c) Non-thermal, thick target, beamed. In this model, the electrons are accelerated in the corona with a relatively narrow pitch angle distribution. They travel to the low altitude, denser regions of the chromosphere, where they lose all of their energy in collisions and thick-target bremsstrahlung. Several observational consequences of this model, in addition to a low altitude X-ray source, are polarization, a directivity in the X-ray emission, and a tendency to detect steeper X-ray spectra near the solar limb than at the center (Langer and Petrosian, 1977).

d) Thermal models. Here, a plasma with high ion and electron temperatures is generally, but not always, confined at high altitudes. Contrary to what might be expected, it has been shown that the X-ray spectrum from such a source can mimic many spectral shapes, including a power law (Brown, 1974), and that mild degrees of polarization and directivity may also be found in the X-radiation (Emslie and Brown, 1980).

## 2. X-RAY MEASUREMENTS IN GENERAL

It is now possible in principle to obtain polarization, time history, spectral, and imaging or spatial information over a wide range of X-ray energies. In practice however, all techniques have not yet been applied at all energies, and this discussion will be limited to measurements or techniques for which the ULYSSES solar X-ray/cosmic gamma ray burst experiment (Cotin et al., 1983) can make contributions similar or comparable to those made to date.

a) X-Ray Time Histories. All X-ray detectors provide measurements of the time histories, or light curves of solar flare X-radiation. Over the past few years, the tendency has been to increase the time resolution to the 10 ms level. Emslie (1983) and Kiplinger et al. (1984) have discussed the constraints which observations of fine time structure impose on non-thermal models: for an instantaneous electron injection into a thick target, the electron collisional energy loss time cannot be greater than the time constant of the observed X-ray structure. This fact leads to limits on the electron pitch angle distribution, the density in the target, and the distance between the acceration site and the target. Also, the fact that X-ray and microwave observation timescales are nearly compatible can be exploited to make detailed comparisons (e.g. Kaufmann et al., 1984); this has led to the concept of "quasi-quantization" of energy release in the impulsive phase (Sturrock et al., 1984) via elementary bursts.

b) Energy Spectra. Due to the fact, noted above, that multitemperature thermal sources may produce a wide variety of X-ray

spectra, the measurement of <u>individual</u> solar flare X-ray spectra
with low time and/or energy resolution, over small dynamic ranges, is
generally of little use in distinguishing between thermal and
non-thermal sources. However, several types of spectral information
are indeed useful for constraining theory. The first is the
measurement of a large number (say >100) of energy spectra by a single
instrument (to eliminate intercalibration effects) over a wide range

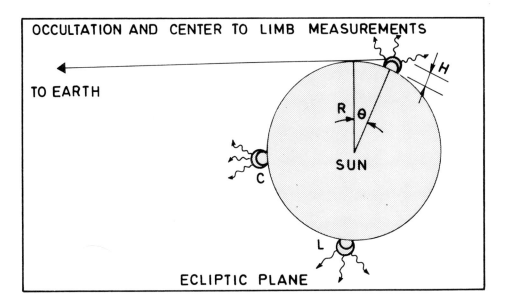

Figure 1. The principle of occultation and center-to-limb
measurements of solar X rays. In the first case, solar
flares observed from the earth are viewed in partial
occultation (i.e., occulted up to a height H) if they occur
at an angle θ behind the limb. For H=50000 km,
θ≅21°. However, the Hα region is not
visible from earth, and its position must be inferred by
extrapolation. In the center-to-limb measurements, a large
number of events are observed over a wide range of heliographic
longitudes, resulting in the exploration of a correspondingly
wide range of flare/sun center/spacecraft angles. For the
center flare C, the angle is 0°, while for the limb flare L
it is 90°.

of heliographic longitudes. These center-to-limb measurements (Figure
1) provide the basis for evaluating directivity in the X-ray emission,
and inferring the amount of beaming in the electron acceleration
process. The inherent assumption is that by observing a large number
of events, the differences in the geometries of individual active
regions can be averaged out, and the dominant effect will be that of
the changing spacecraft/active region/sun angle.
     Spectral measurements of individual flares are useful provided

that they have high energy resolution ($\sim$1 keV) and/or time resolution ($\sim$250 ms). In the first case, this has led to the discovery of a hitherto unsuspected hot isothermal component ($T \sim 3 \times 10^7$K) in flares (Lin et al., 1981); in the second, it may provide additional information on the nature of the electron distribution and X-ray production region (Kiplinger et al., 1984).

c) Spatial Information. Information on the spatial distribution of solar X-rays may now be obtained by a wide variety of techniques; two dimensional images with 5-8" resolution up to 30-40 keV (van Beek et al., 1980; Makishima, 1982) have been obtained on the SMM and Hinotori spacecraft. At higher energies it is possible to obtain the integral height distribution of hard X rays near the solar limb by making use of the fact that the dense regions in the lower chromosphere absorb and/or scatter X-rays efficiently (Fig. 1). This "occultation" technique has provided a number of interesting results, discussed below, although it should be noted that, when practiced from a near-earth spacecraft, it relies on the assumption that the flare location, invisible from earth, may be accurately determined, e.g., by extrapolating the position of an active region which has just rotated out of sight.

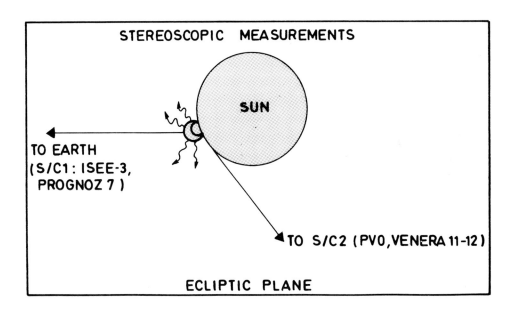

Figure 2. The principle of stereoscopic measurements of solar hard X-radiation. One spacecraft, taken arbitrarily here to be an earth satellite, observes a flare in its entirety. A second spacecraft, widely separated in heliocentric coordinates from the first, also observes the flare. If observed unocculted by the second spacecraft, a directivity measurement results; if occulted, a measurement of the height distribution of the X-radiation.

d) Multispacecraft Observations.     Observations of hard X-radiation
from a given active region by two  or  more  spacecraft  (Fig.  2)  may
fulfill a wide variety of objectives, depending upon the conditions  of
observation.  The detection of fine time structure in the  X-ray  light
curve by two spacecraft not only adds statistical significance  to  the
observation, but also provides the  basis  for  an  imaging  technique,
"triangulation", which can in principle achieve 10" resolution  in  1-3
dimensions at any energy (Hurley et al., 1983).  If the spacecraft  are
widely separated in  heliocentric  coordinates,  the  measurements  of
spectra and time histories can be used to determine the directivity  of
the emission (Kane et al., 1980b).  If the active  region  is  observed
in partial occultation from one spacecraft, but viewed in its  entirety
from  another,  unique  information  may  be  obtained  on  the  height
distribution of the X-radiation (Kane et al. 1982), sometimes  without
the disadvantage noted above, namely that the event is  invisible  from
earth.

## 3. SOME RECENT RESULTS

     This section will  be  devoted  to  a  brief  review  of  results
obtained on hard solar X-rays using the techniques  explained  above.
Table 1 lists some of the observations of fine time  structure  in  the
X-ray light curves of solar flares.  The  instrumental  resolution  has
increased by over an order of magnitude in recent years, and with  this
increase has come the discovery of finer and  finer  time  structure.
Structure at the 20-50 ms level is apparently quite  rare,  and  indeed
its reality has been questioned by Brown  et  al.  (1985).   If  real,
however, it  imposes  significant  constraints  on  nonthermal,  beamed
models, as noted above; hence it will  be  important  to  continue  the
examination  of  x  ray  time  histories  for  more  examples  of  such
structure.
     Table 2 presents the results of three studies  of  center-to-limb
variations in solar flare X-rays (Fig. 1).   The  fact  that  no
variations were found in the first two studies was initially  taken  as
evidence against beamed, thick target  models.   However,  Langer  and
Petrosian (1977) and others have noted that  in  the  energy  range  of
these studies (<100 keV), Compton scattering of the X radiation in  the
target can wash out any initial anisotropy due to beaming.   The  SMM
measurements (Vestrand et al., 1985) at  >300  keV  provide  the  first
experimental  evidence  for  electron  beaming  and  subsequent  X-ray
anisotropy.  At these energies, Compton scattering  is  less  efficient
and the angular dependence of the  bremsstrahlung  cross  section  is
stronger, leading to observable anisotropy.  The  SMM  observations  are
consistent  with  a  downward-directed  electron  beam  producing  X
radiation by thick target emission.
     Table 3 summarizes four  sets  of  stereoscopic  observations  of
hard solar X-radiation.  Brown et al. (1983) and  Koul  et  al.  (1985)
have shown that the observations of Kane et  al.  (1982)  may  also  be
interpreted as a nonthermal, beamed electron population radiating in  a
thick target at low altitudes.  Note that in  this  interpretation,  as

## TABLE 1. OBSERVATIONS OF FINE TIME STRUCTURE IN HARD SOLAR X RAYS

| VEHICLE | X RAY ENERGIES, KEV | NUMBER OF FLARES | TIME STRUCTURE MS. | REFERENCE |
|---|---|---|---|---|
| Balloon | >30 | 1 | ~1800 | Hurley & Duprat, 1977 |
| Balloon | 20-150 | 1 | ~120 | Frontera & Fuligni, 1979 |
| Prognoz-7, Venera 11&12 | >100 | 2 | 55 | Hurley et al., 1983 |
| SMM | 25-400 | Several | 20 | Kiplinger et al., 1983 |

## TABLE 2. SOME CENTER-TO-LIMB OBSERVATIONS OF HARD SOLAR X RAYS

| SPACECRAFT | X-RAY ENERGIES, KEV | NUMBER OF FLARES | VARIATIONS IN FREQUENCY OF OCCURRENCE | SPECTRAL VARIATIONS | INTENSITY VARIATIONS | REFERENCE |
|---|---|---|---|---|---|---|
| OGO-5 | 10-100 | ~300 | None | ---- | ---- | Kane, 1974 |
| OSO-7 | ~20 | 148 | None | None | <40% | Datlowe et al., 1977 |
| SMM | >300 | 121 | 43% More at limb | Harder at limb | ---- | Vestrand et al., 1985 |

in that of the SMM measurements, the source of the electrons is coronal.

Table 4 displays some of the measurements of the height distribution of X-rays achieved by both occultation and imaging. The occultation measurements, from the OSO and OGO spacecraft, clearly indicate the production of X-rays at coronal altitudes. However, as Kane et al. (1982) have demonstrated, this may still be quite consistent with thick target production at low altitudes: the coronal source may represent only a small fraction (~10%) of the total X-ray emission. The two dimensional images of limb flares provided by SMM and Hinotori, however, give clear evidence of purely coronal X-ray sources. Tsuneta (1983) has proposed that there are three basic flare types: A) a hot thermal flare, with a low altitude, small hard X-ray source in a small coronal loop, B) a nonthermal flare in which an electron beam produces thick target emission in the chromosphere in the impulsive stage, and C) an event occurring high in the corona, with nonthermal electrons in a thick target trap.

## 4. MEASUREMENTS IN THE ULYSSES ERA

The Ulysses spacecraft, currently scheduled for a May 1986 launch, contains a solar X-ray/cosmic gamma ray burst experiment built by the Centre d'Etude Spatiale des Rayonnements (Toulouse), Max-Planck Institute (Garching), and the Space Research Institute (Utrecht). Briefly, it consists of 4 sensors in two packages: two Si surface barrier detectors, each $500\mu m$ thick by 0.5 $cm^2$ in area, for measuring 5-20 keV X rays in 4 channels, and two hemispherical CsI(Tl) scintillation crystals, each 2 mm thick by 4.6 cm diameter, for measuring 15-150 keV X-rays in 16 channels. The hard X-ray detector has numerous operating modes, which provide 8 or 32 ms resolution time history data and 1-16 s spectral data. A more detailed description of the instrument has appeared in Cotin et al. (1983). It was initially planned to have a virtually identical instrument aboard the second spacecraft to provide detailed stereoscopic measurements; following the abandonment of this program, a proposal was made and accepted within the framework of the Franco-Soviet collaboration to incorporate similar, although not identical, detectors aboard the two Soviet Phobos spacecraft, to be launched in 1988. The current plan is to place two sensors aboard each spacecraft, consisting of 5.1 cm diameter by 3 cm thick NaI(Tl) crystals; these will record time histories with resolution of 8 ms and up, and 128 channel energy spectra in the 5 keV-1 MeV range.

Over the Ulysses mission, the heliocentric distance will range from 1 to 5 AU, and the effective instrumental sensitivity to solar X-ray bursts will vary by over an order of magnitude. The instrument should, however, observe many hundreds of hard X-ray bursts with good time resolution, and the examination of these time histories might be expected to produce perhaps ten examples of very fast time structure, if it indeed exists. All of the bursts observed on the earth-facing solar hemisphere (i.e. which can be localized by associating them with Hα regions) can also be used to search for evidence of

TABLE 3. STEREOSCOPIC OBSERVATIONS OF HARD SOLAR X RAYS

| SPACECRAFT | X RAY ENERGIES, KEV | NUMBER OF FLARES | DIRECTIVITY OF X RADIATION | IMPLIED HEIGHT OF X RAY PRODUCTION, KM | REFERENCE |
|---|---|---|---|---|---|
| ISEE-3/PVO | >50 | 1 | ---- | <<25000 | Kane et al., 1979 |
| ISEE-3/PVO | 50-100 | 8 | None | ---- | Kane et al., 1980b |
| ISEE-3/PVO | 100-500 | 3 | ---- | <2500 | Kane et al., 1982 |
| Prognoz-7/ Venera 11&12 | >100 | 2 | ---- | <96000 | Hurley et al., 1983 |

TABLE 4. SOME HEIGHT MEASUREMENTS OF HARD SOLAR X RAYS BY OCCULTATION AND IMAGING

| SPACECRAFT | X RAY ENERGIES, KEV | NUMBER OF FLARES | IMPLIED HEIGHT OF X RAY PRODUCTION, KM | REFERENCE |
|---|---|---|---|---|
| OSO-5, OGO-5 | 15-250 | 1 | >25000 | Frost & Dennis, 1971; Kane & Donnelly, 1971 |
| OSO-7 | 30-64 | 8 | 0-18000 | McKenzie, 1975 |
| OSO-7 | 10-200 | 1 | >66000 | Hudson, 1978 |
| SMM | 3.5-30 | 1 | Up to 30000 | Van Beek et al., 1981 |
| OSO-7 | 5-200 | 1 | >45000 | Hudson et al., 1982 |
| Hinotori | 17-50 | 1 | 14000-30000 | Takakura et al., 1983 |
| Hinotori | 14-38 | 1 | 40000 | Tsuneta et al., 1984 |

anisotropic emission through center-to-limb statistical studies, although detectable effects are unlikely to appear below about 100 keV for the reasons noted above.

A unique feature of the experiment results from the Ulysses orbit, as illustrated in Figure 3. It is well known that solar

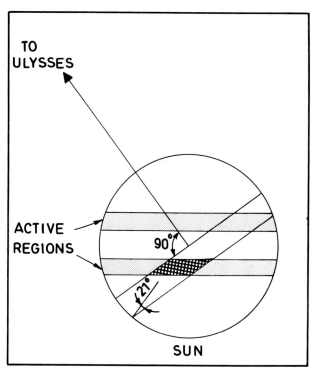

Figure 3. An illustration of the unique geometry for solar X-ray measurements provided by Ulysses. The sun's active regions are concentrated in two bands near the solar equator. The 21° wide band in which Ulysses views X-rays in partial occultation up to 50000 km intersects these bands. For Ulysses heliographic latitudes of 80° or more, 100% of one band is viewed in partial occultation.

activity is concentrated in two narrow bands of latitude. Thus as the heliographic latitude of the spacecraft increases, the fraction of a single band which is viewed in partial occultation by the experiment increases, reaching 100% for heliographic latitudes above about 80°.

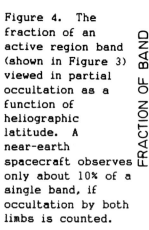

Figure 4. The fraction of an active region band (shown in Figure 3) viewed in partial occultation as a function of heliographic latitude. A near-earth spacecraft observes only about 10% of a single band, if occultation by both limbs is counted.

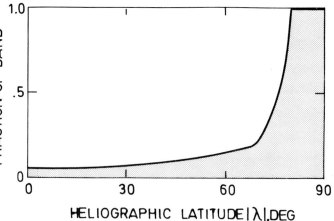

The fraction as a function of heliographic latitude is shown in  Figure
4, assuming  occultation  heights  of  50000  km  or  less,  and  solar
activity concentrated in the 8-15° latitude  band.    This  unusually
favorable geometry leads to two major improvements in occultation and

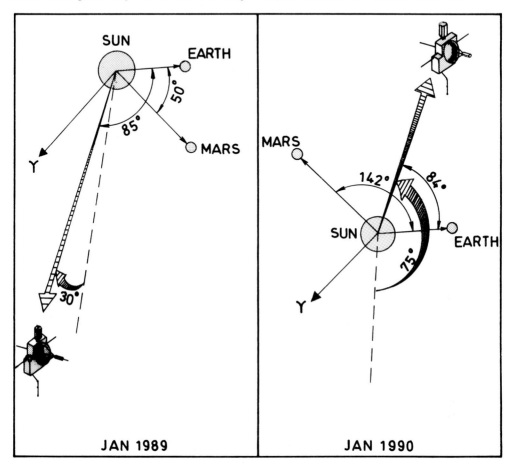

Figure 5.   Two examples of the   geometry  (particularly   the
Ulysses/sun/second spacecraft angles) which will result   for
stereoscopic observations using Ulysses in conjunction  with
the Soviet Phobos spacecraft and/or near-earth satellites.
The line to the first   point  of  Aries,   the   dashed   line,
Mars, and the Earth lie in the ecliptic plane.    In   January
1989 Ulysses is about 30° out of this plane, and one   year
later, about 75° out.

stereoscopic measurements.  First, on the order of 100 events  will  be
viewed  in  partial  occultation  by  Ulysses  during  its  two  polar
passages,  counting  only  those  events  on  the  earth  facing  solar
hemisphere.  This represents  not  only  a  major  improvement  in  the
number of such observations, but also in the accuracy of measuring  the

occultation height, since the Hα flares will be visible from earth and thus well-localized. Second, as illustrated in Figure 5, a wide range of spacecraft/sun/spacecraft angles will become available for stereoscopic measurements using Ulysses and Phobos or Ulysses and an earth satellite. Both occultation and directivity measurements can be performed using these observations. Tables 1-4 demonstrate that, with the exception of the center-to-limb measurements, the statistics of many of the hard solar X-ray observations bearing on the question of electron acceleration processes are still quite poor. The long lifetime and unique geometry of the Ulysses mission should improve this situation considerably.

## 5. ACKNOWLEDGEMENTS

On the French side, the Ulysses solar X-ray/cosmic gamma-ray burst experiment was supported by CNES Contracts 78-212 through 84-212. The author wishes to thank Sharad Kane for a critical reading of this manuscript.

## 6. REFERENCES

Brown, J. 1974, in Coronal Disturbances, Ed. G. Newkirk, Jr., Reidel Publishing Co. (Dordrecht, Holland) p. 395

Brown, J., Smith, D., and Spicer, D. 1981, in The Sun As a Star, Ed. S. Jordan, NASA SP 450, p. 181

Brown, J., Loran, J., and MacKinnon, A. 1985, Astron. Astrophys. (submitted)

Brown, J., Carlaw, V., Cromwell, D., and Kane, S. 1983, Solar Phys. 88, 281

Cotin, F. et al. 1983, in The International Solar Polar Mission-Its Scientific Investigations, Ed. K.-P. Wenzel, R. Marsden, and B. Battrick, ESA SP-1050, p.209

Datlowe, D., O'Dell, S., Peterson, L., and Elcan, M. 1977, Ap. J. 212, 561

Emslie, A. and Brown, J. 1980, Ap. J. 237, 1015

Emslie, A. 1983, Ap. J. 271, 367

Frontera,F. and Fuligni,F. 1979, Ap. J. 232, 590

Frost, K. and Dennis, B. 1971, Ap. J. 165, 655

Hoyng, P., Brown, J., and van Beek, H. 1976, Solar Phys. 48, 197

Hudson, H. 1978, Ap. J. 224, 235

Hudson, H., Lin, R., and Stewart, R. 1982, Solar Phys. 75, 245

Hurley, K. and Duprat, G. 1977, Solar Phys. 52, 107

Hurley, K., Niel, M., Talon, R., Estulin, I., and Dolidze, V. 1983, Ap. J. 265, 1076

Kane, S. and Donnelly, R. 1971, Ap. J. 164, 151

Kane, S. 1974, in Coronal Disturbances, Ed. G. Newkirk, Jr. (Dordrecht:Reidel), p. 105

Kane, S., Anderson, K., Evans, W., Klebesadel, R., and Laros, J. 1979, Ap. J. 233, L151*

Kane, S. et al. 1980a, in Solar Flares: A Monograph from Skylab Solar Workshop II, Ed. P. Sturrock, Colorado Associated University Press, Chapter 5

Kane, S., Anderson, K., Evans, W., Klebesadel, R., and Laros, J. 1980b, Ap. J. Lett. 239, L85

Kane, S., Fenimore, E., Klebesadel, R., and Laros, J. 1982, Ap. J. Lett. 254, L53

Kaufmann, P., Correia, E., Costa, J., Dennis, B., Hurford, G., and Brown, J. 1984, Solar Phys. 91, 359

Kiplinger, A., Dennis, B., Emslie, G., Frost, K., and Orwig, L. 1983, Ap. J. Lett. 265, L99

Kiplinger, A., Dennis, B., Frost, K., and Orwig, L. 1984, Ap. J. Lett. 287, L105

Koul, P., Moza, K., Khosa, P., and Rausaria, R. 1985, Ap. J. 292, 725

Langer, S. and Petrosian, V. 1977, Ap. J. 215, 666

Lin, R., Schwartz, R., Pelling, R., and Hurley, K. 1981, Ap. J. Lett. 251, L109

Lin, R. 1982, in Gamma Ray Transients and Related Astrophysical Phenomena, Eds. R. Lingenfelter, H. Hudson, and D. Worrall, AIP Conference Proceedings No. 77, p. 419, AIP Press, New York

Makishima, K. 1982, in Proceedings of the Hinotori Symposium on Solar Flares, ed Y. Tanaka et al. (Tokyo: Institute of Space and Astronautical Science), p. 120

McKenzie, D. 1975, Solar Phys. 40, 183

Spicer, D. and Brown, J. 1981, in The Sun as a Star, Ed.  S.  Jordan, NASA SP 450, p, 413

Sturrock, P. 1980, in Solar Flares: A  Monograph  from  Skylab  Solar Workshop II, Ed. P. Sturrock, Colorado Associated  University  Press, Chapter 9

Sturrock, P., Kaufmann, P., Moore, R., and Smith, D. 1984, Solar  Phys. 94, 341

Takakura, T. et al. 1983, Ap. J. 270, L83

Tsuneta, S. 1983, in  Proceedings  of  the  Japan-France  Seminar  on Active Phenomena in the Outer Atmosphere of the Sun and  Stars,  eds. J.-C. Pecker and Y. Uchida, Meudon Observatory, Paris, p. 243

Tsuneta, S. et al. 1984, Ap. J. 284, 827

van Beek, H., Hoyng, P., Lafleur,  B.,  and  Simnett,  G.  1980,  Solar Phys. 65, 39

van Beek, H. et al. 1981, Ap. J. Lett. 244, L157

Vestrand, W. Forrest, D., Chupp. E., Rieger, E., and  Share,  G.  1985, preprint to be submitted to the Astrophysical Journal

TEMPORAL EVOLUTION OF AN ENERGETIC ELECTRON POPULATION
IN AN INHOMOGENEOUS MEDIUM. APPLICATION TO X-RAY BURSTS.

N. Vilmer*, A.L. MacKinnon**, G. Trottet*
* U.A. 324, Observatoire de Paris, Section d'Astrophysique de
  Meudon, D.A.S.O.P., F92195 MEUDON CEDEX.
**Department of Astronomy, University of Glasgow, GLASGOW,
  G128QQ, U.K.

ABSTRACT. The interpretation of spatially resolved hard X-ray data
implies the use of models where the spectral, spatial and temporal
evolution of the X-ray flux can be studied. As a first step to construct
detailed hard X-ray source models, we present here time-dependent
solutions of the electron continuity equation in an inhomogeneous medium
and we illustrate the evolution of the electron population and of the X-
ray flux for simple situations. Applications of the results to
stereoscopic X-ray observations is briefly discussed.

## 1. INTRODUCTION

Spatially resolved hard X-ray data have been obtained recently either
through direct imaging as for example aboard S.M.M. (Van Beek et al, 1980)
or aboard Hinotori, (e.g. Tsuneta et al, 1983) or through stereoscopic
observations of partially occulted X-ray flares (Kane, 1983). Models
predicting the spatial, temporal and energetic behaviour of the X-ray
flux are necessary to interpret such data. In the case of the non-thermal
hard X-ray emission, the instantaneous X-ray flux produced at one point of
the atmosphere is indeed directly related to the instantaneous fast
electron spectrum at that point. The evolution of the fast electron
population is described in a proper way by a Fokker-Planck equation
(Rosenbluth et al, 1957), which has been solved at least in the solar
physics literature for steady-state or impulsive ($\delta$ function) electron
injections (e.g. Leach and Petrosian, 1981 ; Kovalev and Korolev, 1981).
But its application in cases where the electron collisional lifetime and
injection durations are of similar magnitudes has not yet been
considered, although such situations are relevant for long duration hard
X-ray events where electrons must be continuously injected during a
finite period. In such cases, a simpler mathematical approach is the use
of a first order, non dispersive continuity equation in phase space taking
into account mean rates of change of the variables. This equation relates
the number of electrons of a specified energy and mean pitch angle at a
given point to the arbitrary source function. Such an equation cannot be
used to describe the exact angular distribution of the fast electrons, but

87

R. G. Marsden (ed.), The Sun and the Heliosphere in Three Dimensions, 87–92.
© 1986 by D. Reidel Publishing Company.

describes its mean behaviour. However, pitch-angle scattering is adequately included for purposes of hard X-ray spatial distribution calculations. Such a treatment has moreover the advantage of providing analytic, time dependent solutions for arbitrary source functions and ambient density structures and it then allows to study the temporal and spatial evolution of the X-ray spectrum.

## 2.   BASIC CHARACTERISTICS OF THE MODEL

Energetic electrons are injected into an inhomogeneous plane parallel atmosphere defined by a density $n(z)$ and a magnetic field $B(z)$ where $z$ is the depth from some arbitrary point in the atmosphere. $B(z)$ is assumed to be slowly varying with $z$ and to remain essentially in the $z$ direction. Under this condition, the electron pitch-angle $\theta$ can be approximated as the angle between the z-axis and the electron velocity ($\mu = \cos \theta$ is positive, resp. negative for an electron moving downwards, resp. upwards). Electrons are injected at a rate $q(E,t,z,\mu)$ for $t \geqslant 0$. The electron population $N(E,t,z,\mu)$ evolves then in the medium through different processes such as Coulomb collision energy losses (mean rate $dE/dt$) and scattering due to collisions or magnetic field gradients (mean rate $d\mu/dt$). The continuity equation is then given by :

$$\frac{\partial N(E,t,z,\mu)}{\partial t} + \frac{\partial}{\partial E}\left[ N(E,t,z,\mu)\,\frac{dE}{dt} \right] + \frac{\partial}{\partial z}\left[ N(E,t,z,\mu)\,\frac{dz}{dt} \right] +$$

$$\frac{\partial}{\partial \mu}\left[ N(E,t,z,\mu)\,\frac{d\mu}{dt} \right] = q(E,t,z,\mu) \tag{1}$$

where $dE/dt$ is the energy loss rate through electron-electron Coulomb collisions (Bai and Ramaty, 1979) and $d\mu/dt$ is the mean change of pitch angle under the combined action of Coulomb collisions (Brown, 1972) and the magnetic field gradient (adiabatic invariance of the magnetic moment). The solutions, their physical interpretation and their illustration in simple cases are discussed in Vilmer et al. (1985) and Craig et al. (1985). The approximation which consists in omitting the velocity dispersion due to Coulomb collisions is also discussed in Vilmer et al (1985). For mildly relativistic electrons (initial energy below 200 keV), the mean behaviour of the electrons is fairly well described by the analytic treatment.

## 3.   EVOLUTION WITH HEIGHT AND TIME OF THE ENERGETIC ELECTRON POPULATION

The two extreme cases where either pitch-angle scattering due to the magnetic field gradient or to Coulomb collisions is negligible have been examined. In both cases, a non-thermal electron distribution is continuously injected at an arbitrary depth $z = 0$ in a stratified medium with a density scale height $H$ [$n(z) = n_0 e^{z/H}$] and a magnetic field $B(z)$. We then compute :

$$N(E,t,z) = \int_{-1}^{1} d\mu\, N(E,t,z,\mu) \qquad \text{for } E < 160 \text{ keV} \tag{2}$$

For simplicity, $q(E,t,z,\mu)$ is chosen as :

$$q(E,t,z,\mu) = S_o\, E^{-\gamma}\, F(t)\, G(z)\, H(\mu)$$

where : 

$\begin{cases} F(t) = t\,(2t_o - t) & \text{for}\quad 0 \leqslant t \leqslant 2t_o \\[2mm] \quad\;\; = 0 & \text{elsewhere} \end{cases}$  (3)

$G(z) = \delta(z)$ \qquad where $\delta(z)$ is the Dirac delta function

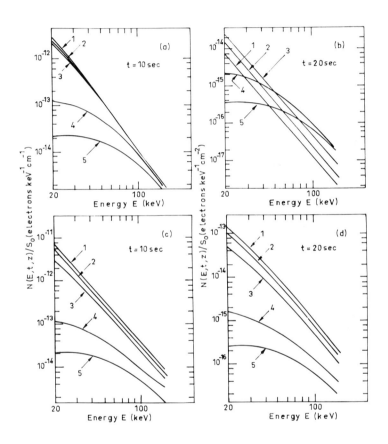

Figure 1. Evolution with depth and time of $N(E,t,z)/S_o$ when the pitch angle scattering is due to a varying magnetic field with $R_m = 2$ and $z_T = 3$ $10^9$cm. Figures 1a and 1b correspond to a beamed injection ($\mu_o = 0.5$) while Figures 1c and 1d correspond to an isotropic one. Curves 1, 2, 3, 4 and 5 correspond respectively to $z = 5\ 10^7$cm, $10^8$cm, $2\ 10^8$cm, $5\ 10^8$cm and $6$ $10^8$cm. The chosen parameters are : $\gamma = 3$, $t_o = 10$ sec, $n_o = 10^{10}$ cm$^{-3}$ at the injection point ($z = 0$), $H = 10^8$cm.

Two extreme cases are considered for $H(\mu)$ : a beam distribution [$\delta$ ($\mu$ − $\mu_O$)] and an isotropic one.

As an example, Figure 1 shows electron spectra as a function of depth at $t = t_O$ (maximum of the injection) and at $t = 2t_O$ (end of the injection) for both a beamed injection and an isotropic one when the magnetic field is given by (e.g. Bai, 1982) :

$$
B(z) = \begin{cases} B_O \left[ 1 + \dfrac{z^2}{z_T^2} (R_m - 1) \right] & z < z_T \\[3ex] B_O\, R_m & z \geqslant z_T \end{cases}
\qquad (4)
$$

where Rm is the mirror ratio at $z = z_T$. The general behaviour of electron spectra with depth is a progressive hardening. However, there are

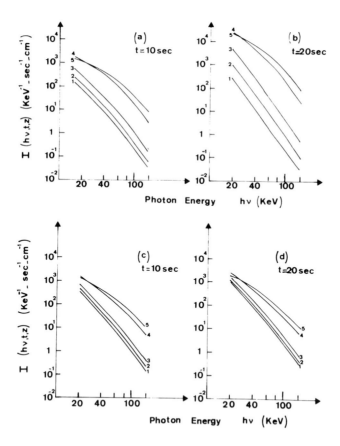

Figure 2. Evolution with depth and time of the X-ray flux $I(h\nu, t, z)$ produced by the energetic electron population shown on Figure 1.

differences between a beamed and an isotropic injection as well as between
the different scattering processes which have been discussed in detail in
Vilmer et al. (1985). For a similar injection spectrum, harder spectra
at low depths are expected when the scattering is due to the magnetic
field gradient. Moreover, as expected, it can be noticed that the electron
spectrum at a given depth, as well as the hardness difference between
given depths, strongly depend on the injection height.

## 4.  EVOLUTION WITH HEIGHT AND TIME OF THE X-RAY SPECTRUM

Figure 2 shows the X-ray flux $I(h\nu,t,z)$ produced at energy $h\nu$ and depth $z$
by the electron population shown on Figure 1. The X-ray flux $I(h\nu,t,z)$
produced at depth $z$ generally increases and continuously hardens with
depth. However, the hardening of the X-ray emission with depth depends on
the initial pitch-angle distribution of the injected electrons, the
hardening being less important for an isotropic injection. As shown on
Figure 3, in the case of an isotropic injection, the variations of the X-
ray spectral hardness between given depths depend moreover on the
injection depth. All the above conclusions are also valid for the
integrated hard X-ray flux emitted above a given depth. The continuous
increase and spectral hardening of this flux with depth is then consistent
with the general results of the stereoscopic observations (Kane, 1983).

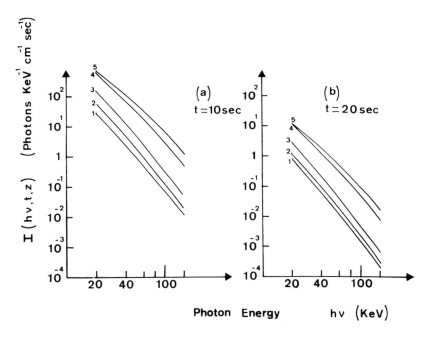

Figure 3.  Evolution with depth and time of the X-ray flux $I(h\nu,t,z)$
produced by an isotropic electron injection at $z_O = 1.5\ 10^8$cm and $n_O = 2$
$10^9$cm $[n(z_O) = 10^{10}$ cm$^{-3}]$. The other parameters are the same as for Figure
1.

## 5.  DISCUSSION

As shown above, various evolutions with depth of hard X-ray spectra may be predicted with the present model, depending on the characteristics of the electron injection, of the ambient medium and of the injection depth. Such evolutions of the X-ray flux with depth have indeed been observed with stereoscopic observations of partially occulted X-ray flares (Kane, 1983). The occulted X-ray flux (observed by the instrument which saw only the higher part of the flaring source) is smaller than the unocculted one and the occulted spectrum is softer. However, for coronally occulted events (occultation height ⩾ 25000 km above the photosphere), in spite of a similar occulting height, the flux ratio and its energy dependence are not the same for all events. Moreover, occultation ratios at different energies may present different temporal evolutions. Such variations of the flux ratio and of its energy dependence from one event to another are easily understood with the present model when the characteristics of the ambient medium and of the electron injection vary from one flare to the other. Moreover, different temporal evolutions of the occulted and unocculted X-ray fluxes may also be expected. At present, stereoscopic observations are still rare but this situation should improve in the near future. Indeed, the comparison of high time resolution X-ray measurements in the 15-150 keV range which will be made aboard Ulysses (Cotin et al, 1983) with the ones obtained in the ecliptic plane will greatly favour the observations of partially occulted X-ray flares.

REFERENCES

Bai. T. and Ramaty, R. : 1979, Astrophys. J. **227**, 1072.
Bai. T. : 1982, Astrophys. J. **259**, 341.
Brown, J.C. : 1972, Solar Phys. **26**, 441.
Cotin, F., de Jager, C., Henoux, J.C., Heise, J., Hilhorst, M., Hurley, K., Niel, M., Paschmann, G., Sommer, M., Van Rooijen, J. and Vedrenne, G. : 1983, ESA Special Publication, **SP 1050**, "The International Solar Polar Mission", 211.
Craig, I.J.D., MacKinnon, A.L. and Vilmer, N. : 1985, Astrophys. and Space Science, **116**, 377.
Kane, S.R. : 1983, Solar Phys. **86**, 355.
Kovalev, V.A. and Korolev, O.S. : 1981, Sov. Astron. **25**, 215.
Leach, J. and Petrosian, V. : 1981, Astrophys. J. **251**, 781.
Rosenbluth, M.N., MacDonald, W.M., Judd, D.L. : 1957, Phys. Rev. **107**, 1.
Tsuneta, S., Takakura, T., Nitta, N., Ohki, K., Makishima, K., Murakami, T., Oda, M. and Ogawara, Y. : 1983, Solar Phys. **86**, 313.
Van Beek, H.F., Hoyng, P., Lafleur, B. and Simnett, G.M. : 1980, Solar Phys. **65**, 39.
Vilmer, N., Trottet, G. and MacKinnon, A.L. : 1985, Astron. Astrophys. in press.

INFLUENCE OF SOLAR FLARES ON THE X-RAY CORONA

D. M. Rust and D. A. Batchelor
The Johns Hopkins University Applied Physics Laboratory
Johns Hopkins Road
Laurel, Maryland 20707 USA

ABSTRACT.  Sequences of X-ray images of solar flares, obtained with the
Hard X-ray Imaging Spectrometer (HXIS) on the SMM spacecraft, reveal
many dynamical phenomena.  Movies of 20 flares recorded with 6-sec time
resolution were examined.  We present here a preliminary analysis of the
events as a group, and discuss some new aspects of the well-studied 1980
May 21 flare and a 1980 November 6 flare.

1.  INTRODUCTION

   Solar flares influence the corona by local heating, by thermal
conduction to the surroundings, by electron beam heating, and by
restructuring of the magnetic field.  Depending on the details of the
magnetic fields nearby, the heat generated in flares may be confined to
a few small loops or it may be conducted over distances of a solar
radius or more.  Heating is detected by enhanced soft X-ray emission.
The velocity of the thermal fronts is > 1000 km/sec at flare onset.  The
existence of these fast-moving thermal fronts was inferred first from X-
ray images provided by the HXIS experiment on SMM (Rust et al., 1985).

2.  PRELIMINARY RESULTS ON GROWTH OF SOFT X-RAY SOURCES

   The Hard X-ray Imaging Spectrometer (HXIS) on SMM observed some 300
flares.  Observations with spatial resolution of 8 and 32 arc sec were
made, in 6 energy bands covering the range 3.5-30 keV.  The source
structures are highly dynamic, and the HXIS images reveal many apparent
motions and changes of shape.  There are overall brightenings of faint
loop structures, brightness contour changes that suggest propagation of
disturbances along loops, evidence of successive heating of neighboring
loops, and filling of loops by luminous material in a manner suggestive
of chromospheric evaporation.  Even transverse motions of loops have
been noted.

R. G. Marsden (ed.), The Sun and the Heliosphere in Three Dimensions, 93–98.
© 1986 by D. Reidel Publishing Company.

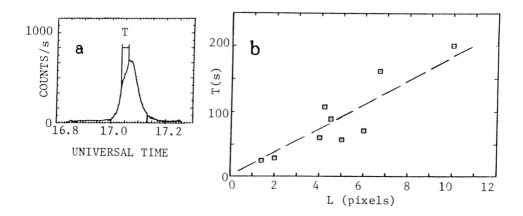

Figure 1a.  Typical flare hard X-ray burst profile, illustrating
            how T is defined.

      1b.  Time (T) from half-peak to full-peak emission vs.
            soft X-ray flare diameter (2) for 9 flares.

Lateral motions of contours of constant brightness in HXIS images have
been discussed in Rust et al. (1985), who interpreted them as indicative
of expanding thermal conduction fronts. Velocities at flare onset in
the range 300-1600 km/sec were inferred.

Here, we take another approach to determine whether soft X-ray contour
changes represent moving disturbances. As an example, Figure 1a shows
the time history of soft X-ray emission from a flare (3.5-16 keV),
summed over the entire 304-pixel HXIS fine field-of-view. We examined
the array of counts that represent the corresponding soft X-ray source
at its time of peak emission, and found the region including all pixels
containing counts of at least 25% of the maximum count level. The
length of this region in pixels we called L. (The 8 arcsec length of a
pixel corresponds to 5600 km on the Sun.) The time from half of peak to
full peak emission in the time history was designated T. Figure 1b is a
plot of T versus L for 9 flares, chosen without regard to T. The nearly
linear relationship of L and T suggests that a common growth velocity
characterizes the soft X-ray sources. If the growth is from a central
point toward both ends, this speed is about 80 km/sec; if the spreading
is from one end to the other, a speed twice this size is inferred, about
160 km/sec. (This method characterizes the bursts at a time near the
soft X-ray peak rather than the very early phase with faster apparent
motions studied by Rust et al.) The inferred velocities are not very
different from flow velocities inferred from Doppler shifts of the Ca
XIX line (Antonucci et al., 1984). Therefore, our results of 160 km/sec
in this case are suggestive that the inferred motion may be due to
chromospheric evaporation.


3.   SOFT AND HARD X-RAY SOURCES OF THE 1980 MAY 21 FLARE

The flare of 1980 May 21 at 2055 UT produced hard X-ray emission
in the 16-30 keV band from footpoints of a coronal arch (Regions A and
B, discussed by Hoyng et al., 1981). In Figure 2, we show images of the
flare in the soft and hard ranges, accumulated for four time intervals
during the early impulsive phase: gradual emission prior to the
sharpest, largest rise of hard X-rays; the largest rise and peak of hard
X rays; the decline after this peak; and the second largest peak. A
large fraction of the impulsive hard X-rays, ~ 50% at burst maximum,
actually came from the arch, between the footpoints, rather than regions
A and B shown by Hoyng et al. The radiation is therefore interpreted in
terms of the thermal conduction-front model rather than the nonthermal
thick-target model alone (see Batchelor et al., 1985 and refs.
therein). A conduction front, traveling at a few times the ion-sound
speed, 2000 km/s, could have traversed the 50,000 km long arch in the
rise time of the burst (~ 11 s).

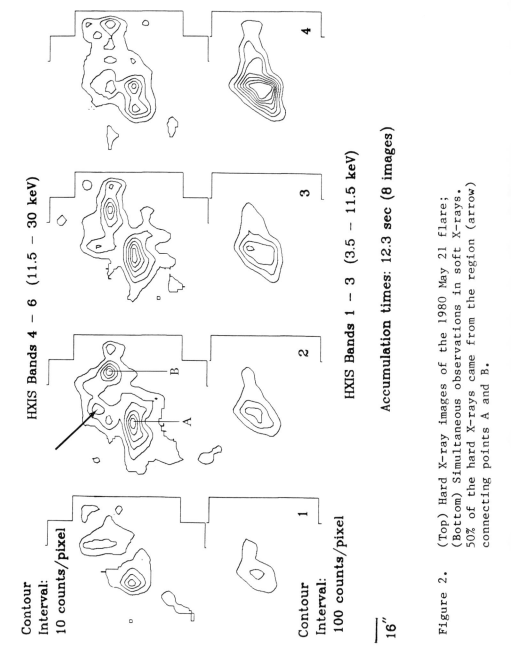

Figure 2.  (Top) Hard X-ray images of the 1980 May 21 flare; (Bottom) Simultaneous observations in soft X-rays. 50% of the hard X-rays came from the region (arrow) connecting points A and B.

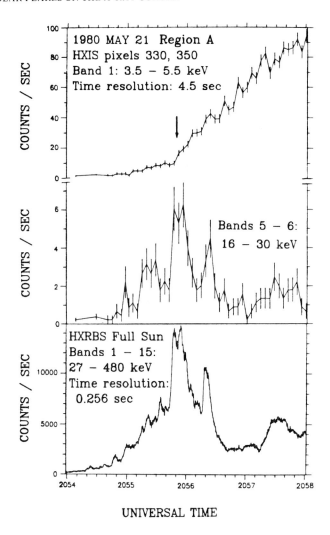

Figure 3.    (Top) Count rate in selected pixels vs. time;
(Bottom) Simultaneous 27–480 keV hard X-ray burst
for the 1980 May 21 flare.

In Figure 3 we show the soft and hard X-ray emission early in the impulsive phase of the flare, and hard X-rays observed with HXRBS. Antonucci et al. (1984) presented evidence for chromospheric evaporation in the region of HXIS pixels 330 and 350 (Region A of Hoyng et al.) The sudden rise in soft X-rays from those pixels, shown in the top of Fig. 3, tends to corroborate that claim.

## 4. THE 1980 NOVEMBER 6 FLARE AT 1725 UT

In this flare, a coronal loop 130,000 km long was observed in soft and hard X rays. The points at the extreme ends of the loop showed the brightest hard X-ray emission, as in the May 21 flare. A measurement of density in the loop was made before the hard X-ray burst, yielding values from 1.6 to 3 x $10^{10}$ $cm^{-3}$. At these densities, a 20-keV electron would be stopped in 13,000 to 25,000 km by Coulomb collisions. Thus the thick-target model can be ruled out as the origin of the hard X-rays because the required electron beam would lose its energy while in the loop before it reaches the chromosphere. Most of the hard X-rays from an electron beam would then come from the loop, contrary to the observations.

This work was supported by NSF Grant ATM-8312720 and NASA Grant NSG 7055.

REFERENCES

Antonucci, E., Gabriel, A. H., and Dennis, B. R. (1984)., Astrophys. J., 287, 917.
Batchelor, D. A., Crannell, C. J., Wiehl, H. J., and Magun, A. (1985), Astrophys. J., 295, 258.
Hoyng, P., et al. (1981), Astrophys. J., 246, L155.
Rust, D. M., Simnett, G. M., and Smith, D. F. (1985), Astrophys. J., 288, 401.

# SECTION III:  CORONAL TRANSIENTS AND MASS EJECTIONS

# CORONAL TRANSIENTS AT HIGH HELIOGRAPHIC LATITUDES

D. G. Sime
High Altitude Observatory/NCAR
P.O. Box 3000
Boulder, Colorado 80307 USA

ABSTRACT. One of the few properties of coronal transients to have been demonstrated to show a variation which may depend on the phase of the solar cycle is their distribution with latitude. Observations from Skylab (1973-74) showed a distribution which was restricted largely to the region between 45°N and 45°S. Data from the the Solar Maximum Mission, on the other hand show transients over a much wider range of latitudes. A spacecraft at high heliographic latitude near the time of solar maximum is thus likely to encounter coronal transients directed towards it and because of its trajectory, offers a more favorable opportunity to identify the interplanetary signature of these events than is found near earth or in the ecliptic.

## 1. INTRODUCTION

The Ulysses spacecraft will enter a region of the interplanetary medium which has never previously been explored with *in-situ* measurements, namely regions of high heliographic latitude. Since observations are already available of the corona at all latitudes, and because the corona and the interplanetary medium are so closely coupled it is possible to remove some of the uncertainty about what the mission will encounter. In this paper, some aspects of our knowledge of the properties of coronal transients (mass ejections) will be reviewed in order to anticipate some of the opportunities for the study of coronal transients which may come from the Ulysses observations.

## 2. OBSERVATIONS OF TRANSIENTS

Observations of coronal mass ejection transients are now available from the epochs near solar minimum (Skylab, 1973-74) and solar maximum (Solar Maximum Mission, 1980). When they are observed with comparable instruments and interpreted with similar techniques, there appear to be no significant differences between individual transients at the two epochs. Masses, velocities, shapes, and extents are reported to be substantially the same for the two intervals.

The principal differences to have been established to date are ones which relate to the global distributions of these events. In particular, Hundhausen et al. (1984) have shown that the latitude distribution of the occurrence of transients is

101

*R. G. Marsden (ed.), The Sun and the Heliosphere in Three Dimensions, 101–106.*
© *1986 by D. Reidel Publishing Company.*

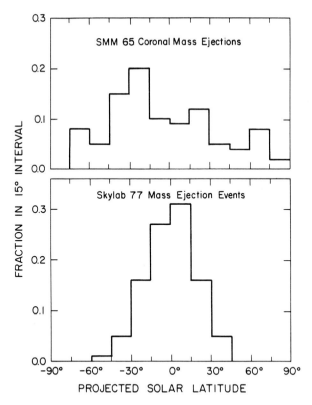

Figure 1. The projected heliographic latitudes of coronal transients. The top panel shows the distribution of projected latitudes for the 65 SMM mass ejections. The bottom panel show the distribution for the 77 events observed during the Skylab Mission (Hundhausen et al., 1983).

significantly different, near the last solar maximum, from the distribution observed during the Skylab mission. Figure 1 displays their result. In the upper panel, the fraction of the set of 65 events observed is shown as a function of latitude in $15^\circ$ wide bins, while the same quantity is shown for the Skylab data (77 events) in the lower panel. The Skylab epoch transients are restricted almost entirely to the interval between $45^\circ$N and $45^\circ$S latitude. The data from SMM, on the other hand, show events at almost all latitudes. Thus, at the time of polar passage of the Ulysses Mission, ie. near solar maximum, it should be expected that coronal transients will occur at almost all latitudes and in particular, at latitudes similar to those of the spacecraft.

A global property which does not appear to change when examined in a consistent way, is the rate of occurrence of mass ejection transients. Hundhausen et al. (1984) showed that the rate of transients observed by the coronagraph/polarimeter on SMM, when corrected for the duty cycle of the instrument, was 0.9 per day. This is not significantly different from the rate of mass ejections (0.75 per day) which the Skylab observations yield when they are identified in a consistent manner and similarly corrected for duty cycle. Such a constancy of rate was not expected. From an

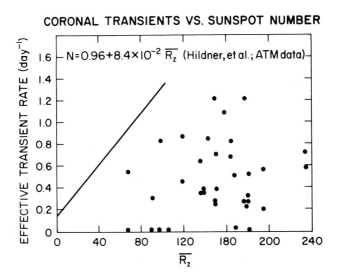

Figure 2. Effective rate of coronal transients during SMM for 90° sectors compared with the sunspot number averaged over the appropriate sector. For comparison, the linear relation found by Hildner et al. (1976) from the Skylab data is shown.

analysis of the Skylab data, Hildner et al. (1976) had found a linear relationship between the rate of transients occurring in a sector of longitude and the averaged sunspot number for that sector. Since the average sunspot number at solar maximum was expected to be some 3 times that at the epoch of Skylab, it was expected that the rate of transients would similarly be increased.

Not only is the overall constancy of rate at odds with the prediction, but application of the same analysis to the SMM data as was applied to the Skylab observations yields the result shown in Figure 2. In this figure, the rate of transients observed in each 90° sector of longitude during SMM is plotted against the averaged sunspot number for that sector, as first done for the Skylab data by Hildner et al. (1976). Clearly, the relation found for the 1973-74 data does not hold. Further, apparently no non-trivial linear relation can be fit significantly to these data. We interpret this as indicating that the occurrence of transients shows no clear dependency on sunspot number in 1980. Since the principal difference in the organization of sunspots between the 2 epochs was their relative uniformity in longitude during SMM, we suggest that the result from the Skylab data reflects the large scale organization of the corona at that time. Rather than indicating a relationship with sunspot number *per se* the Hildner et al. result comes from the dominance in 2 sectors of large coronal holes from which no transients arose. Thus, if the global conditions at the time of polar passage are similar to those of the last maximum, then about 0.4 transients per day will be ejected uniformly in longitude at latitudes of more than 30°.

Examination reveals that no systematic differences occur between the character of transients at high latitudes and others. Of the 14 events occurring above 45° latitude during SMM, for which associated activity could be identified, 7 were associated with flares and 7 with prominence eruptions. These events adhered to the general result that their morphology in the corona tends to depend on the nature of the chromospheric activity which can be associated with them, in common with all the Skylab and SMM transients (Sime 1985). Similarly, the events above 45° show no peculiarities in their apparent speeds, or masses.

If any striking difference has been suggested between observations of transients near solar minimum and those at the maximum of the cycle, it may be the single observation in the SMM data of one event which may be consistent with a

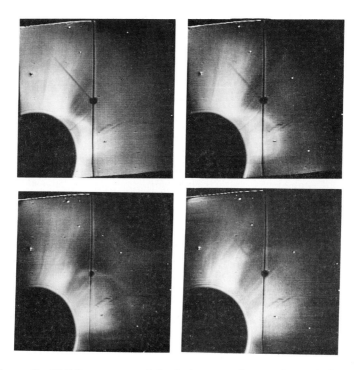

Figure 3. SMM coronagraph/polarimeter observations of the coronal transient of day 188 1980. The images are of the West sector of the corona; North is to the top left, south to the bottom right. The time spanned by the four images is 31 minutes. A faint loop is seen to appear and expand rapidly, interacting with the streamer to the north. At the point of interaction, the loop kinks, and the streamer is not disrupted until the loop front has completely penetrated it.

propagating shock within the corona. This transient, which is strikingly different in appearance and evolution from the typical loop transients (Sime, MacQueen and Hundhausen, 1984) has been identified in the SMM data by Sime and Hundhausen (1984) and is illustrated in Figure 3. Its velocity, shape and interaction with the surrounding corona, together with signs of possible refraction, all separate it from the events previously and usually observed and are consistent with the behavior of a propagating shock wave. It is not known whether its detection during SMM is due to instrumental properties and cadence, or whether it is of a kind which occurs only near solar maximum.

It is difficult to say what the Ulysses spacecraft will encounter during its mission, because the interplanetary manifestations of coronal mass ejection transients, if any, are not well understood. This arises in part because the physical nature of the transients themselves is poorly understood, and in part because the identification of an interplanetary feature with a coronal event is indirect and difficult. Most attempts to do so, thus far, rely on statistical association, often inconsistent with physical intuition. Several suggestions stand as to the nature of the interplanetary manifestations of corona transients, but most are offered on an *ad hoc* basis, with little physical explanation. Suggestions exist that the signature of a transient is a magnetic bubble ( Klein and Burlaga, 1982), a helium enhancement (Borrini et al., 1982), a non-compressive density enhancement (Gosling et al., 1977), and a shock (Sheeley et al., 1985). Although for most of these suggestions, individual events have been studied at least to the point that a plausible association can be made, the justification for them relies heavily on similarity in statistics, such as occurrence rates, between the features and coronal transients.

A further uncertainty lies in the lack of knowledge of the propagation of these effects in interplanetary space. Although early studies indicated that even a very narrow coronal ejection could produce a broad front near 1AU, (DeYoung and Hundhausen 1971), there are now suggestions that only transients which are seen to be near the equator in the corona can be detected in the ecliptic at a fraction of an AU (Sheeley et al 1984). However, the Ulysses Mission will provide several advantages in identifying the interplanetary signature of coronal transients. The first arises quite simply from the trajectory, which reaches high latitudes, and provides the opportunity to detect any effects which propagate from the transient at a large latitudinal separation from it. Second, at high latitudes, because the spiral wrapping of the magnetic field lines is reduced, the interplanetary signature of transients will be freer of dynamical interactions with the interplanetary medium than it would be at comparable distances in the ecliptic. Finally, the trajectory of the Ulysses spacecraft is such that at certain phases of the mission, observations of the corona from the earth and its vicinity will be able to identify transients ejected from the corona in the direction of the spacecraft, and permit a much closer identification of individual mass ejections with a particular interplanetary effect than has been achieved previously.

## 3. SUMMARY

On the basis of previous observations of the corona, the Ulysses spacecraft will be expected to encounter the interplanetary signatures of transients at high latitudes. If the global structure of the corona is the same near the time of polar passage as it was in 1980, these transients will be distributed uniformly in longitude and occur at a rate of 0.4 per day above $30^{\circ}$ latitude. The high latitude events are not expected to be different in average properties from their ecliptic counterparts, but the Ulysses mission because of its trajectory, may be better suited to distinguishing the interplanetary signatures of these events than any previous one.

ACKNOWLEDGEMENT It is a pleasure to thanks R. M. E. Iliing for helpful comments on the manuscript.

# REFERENCES

DeYoung, D. S. and A. J. Hundhausen, 1971 *J Geophys Res.*, **76**, 2245.

Hildner, E., J. T. Gosling, R. M. MacQueen, R. H. Munro, A. I. Poland, and C. L. Ross 1976 *Solar Phys.*, **48**, 127.

Gosling, J. T., E. Hildner, J. R. Abridge, S. J. Bame, and W. C. Feldman 1977 *J Geophys Res.*, **82**, 505.

Klein, L. and L. Burlaga 1982 *J Geophys Res.*, **87**, 8763.

Hundhausen, A. J., C. Sawyer, L. House, R. M. E. Illing, and W. J. Wagner 1983 *J Geophys Res.*, **86**, 2639.

Borrini, G., J. T. Gosling, S. J. Bame, and W. C. Feldman 1983 *Solar Phys.*, **83**, 367.

Sime, D. G., R. M. MacQueen, and A. J. Hundhausen 1984. *J Geophys Res.*, **89**, 2113.

Sheeley, N. R. Jr., R. A. Howard, M. J. Koomen, D. J. Michels, R. Schwenn, K. H. Muhlhauser, and H. Rosenbauer 1985 *J Geophys Res.*, **90**, 163.

Sime, D. G. and A. J. Hundhausen 1985 *EOS, Trans. Am. Geophys. Union,* **65**, 1070.

Sime, D. G. 1985 Submitted to *Solar Phys.*

# THE SOLAR CYCLE DEPENDENCE OF CORONAL MASS EJECTIONS

R. A. Howard, N. R. Sheeley, Jr., D. J. Michels
Naval Research Laboratory
Washington, D.C. 20375   USA

M. J. Koomen
Sachs/Freeman Associates, Inc.
Bowie, MD USA

ABSTRACT. The Solwind white light coronagraph on P78-1 has been making routine observations of the solar corona since 28 March 1979.   Data from the 1984/1985 time period has just been analyzed.   During this interval, a period of low solar activity, coronal mass ejections (CMEs) occurred at the rate of 0.2- 0.4/day, in contrast to the rate of 1.8/day during the period around solar maximum, 1979-1981.  The rate of equatorial CMEs also dropped by the same amount during this period.  A class of CMEs, "streamer blowouts", occurred at the same rate during the two epochs.  Many of the parameters associated with CMEs, their type, their angular span, central latitude, mass, speed, and energy, have changed from their distributions at solar maximum.   During the 1984/1985 period, CMEs are confined to low latitudes, rarely reaching the high latitudes seen during the maximum years.  They are smaller, slower, less massive, and less energetic.

## 1.    INTRODUCTION

Since the discovery of the coronal mass ejection in 1971, the question of the rate of CMEs has been one of extreme interest.  Each of the four orbiting white light coronagraphs to date have measured the rate of CMEs.  From the OSO-7 data, Tousey et al (1974) estimated that the sun produced about 14 transients per month for the period November, 1971 through May, 1973.  From the Skylab data, Hildner et al (1976) estimated that the sun produced 30 transients per month for the period May, 1973 through February, 1974. Additionally, Hildner et al discovered an apparent fluctuation in the rate with sunspot number, which led them to predict that the rate during sunspot maximum could be as high as 3 transients per day. Hundhausen et al (1984) in analyzing the SMM data for 1980, found that the rate was only about 0.9/day, which only reprsented a small increase over the Skylab rate during the declining phase of the previous solar cycle.  Finally, Howard et al (1985) have found that the frequency of CMEs observed with the Solwind coronagraph was 1.8 CMEs/day during the period 1979-1981, and that no

*R. G. Marsden (ed.), The Sun and the Heliosphere in Three Dimensions, 107–111.*
© *1986 by D. Reidel Publishing Company.*

obvious and persistent correlation with sunspot number could be found
for that period.

We have now analyzed data from June 1, 1984 – May 1, 1985.  We
find that the sun is producing far fewer CMEs than during the 1979–1981
period and that their properties are quite different.


## 2.    OCCURRENCE RATE OF CMEs

In an analysis of CMEs observed during 1979–1981, Howard et al (1985)
identified approximately 1000 CMEs which corresponded to a corrected
rate of occurrence of 1.8 CME/day.  If only major CMEs were considered'
they obtained 0.9 CME/day.  Thus about half of the CMEs observed were
minor, but definite, mass ejections.  They further found that the
fluctuations in the CME rate did not match the fluctuations in the
sunspot number.  We have processed 11 months of recent data from the
Solwind coronagraph using the same techniques as were used in Howard et
al (1985).  A total of 59 CMEs were detected, which is equivalent to
about 5 CMEs per month.  This can be compared to a rate of
approximately 34 per month during the 1979–1981 interval.  The
instrumental duty cycle was comparable between the two periods, so that
the uncorrected rates can be compared directly.

In Figure 1 we compare the average yearly CME rates (corrected for
duty cycle) as a function of sunspot number.  We are not certain how to
correct the current data for instrumental duty cycle, so that both the
uncorrrected rate and the rate corrected in the same manner as during
sunspot maximum are plotted.  As can be seen, the 1984 rate is a factor
of 5 to 10 lower than the rate Howard et al (1985) obtained for the
period around sunspot maximum.  The 1982 rate is a preliminary result
from about 6 months of data.

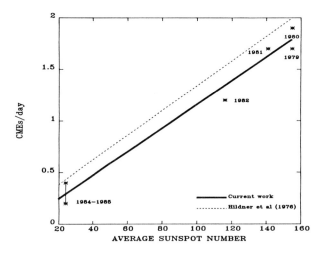

Figure 1.   Variation of Rate of Coronal Mass Ejections (1979–1985)

The line through the points is a best fit approximation to the individual yearly averages.  For comparison we show the regression line of Hildner et al (1976) for the Skylab data.  In view of the scatter of the individual points in both the Skylab data and our data the agreement is surprisingly good, both for the slope and for the absolute rate.  The agreement of the absolute rate perhaps is fortuitous, but the variation with solar cycle is probably valid.


3.   EQUATORIAL OCCURRENCE RATE

Can the difference in rates between the two epochs be accounted for by a lack of CMEs at high latitudes, with the rate at low latitudes being approximately constant?  To answer this question we have examined the rates at low latitudes by counting the number of events whose edges came within $45^{\circ}$ of the solar equator for both the east and west limbs. Table 1 shows that the rate of equatorial events is also lower by a factor of 2-5 in 1984-1985 than in prior years.  Similar results are obtained if the latitude is restricted to $1^{\circ}$ from the equator rather than the $45^{\circ}$ band used above.  Also, excluding the minor events does not change the trends shown above.

Table 1.  Equatorial Occurrence Rate of CMEs for East and West Limbs

| Year | Number of days | Number of CMEs on E Limb | Corrected E Limb Rate | Number of CMEs on W Limb | Corrected W Limb Rate |
|------|------|------|------|------|------|
| 1979 | 212 | 131 | .76 | 112 | .65 |
| 1980 | 268 | 95 | .61 | 131 | .84 |
| 1981 | 360 | 175 | .72 | 199 | .82 |
| 1982 | 194 | 58 | .46 | 81 | .64 |
| 1984/5 | 299 | 15 | .12 | 43 | .33 |

In a study of the relation of CMEs to interplanetary shocks, Sheeley et al (1985) found that the shock-associated CMEs were at least $45^{\circ}$ in angular span and crossed the heliographic equator.  Only 1 of the 33 CMEs observed during the 1984-1985 period analyzed to date satisfied the same criteria as before.  We have not yet tried to extend the interplanetary shock/CME study to this phase of the solar cycle, but if the same criteria are valid, we would not expect to find many IP shocks during this period.


4.   PROPERTIES OF CMEs

Howard et al (1985) have presented histograms of various CME proper-ties.  A similar presentation for the 59 events during the 1984-1985 period is shown in Figure 2.  The differences between the two epochs is striking.  The events are smaller, slower, at lower latitudes, and less energetic (Table 2).

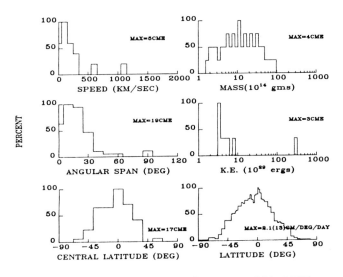

Figure 2. Properties of CMEs: 1984-1985

Table 2.    Average Properties of CMEs

|                      | 1979-1981 | 1984 |
|----------------------|-----------|------|
| Speed (km/s)         | 472       | 208  |
| Span (deg)           | 45        | 24   |
| Mass ($10^{15}$ g)   | 4.1       | 2.1  |
| K.E. ($10^{30}$ erg) | 3.5       | 0.3  |

Howard et al (1985) found it convenient to divide the various CMEs into 9 different structural classes depending upon their visual appearance. We have extended that classification to the current epoch. In Table 3 we compare the relative rates of each structural class during the two epochs.

Table 3.   Relative Rates of CME Structural Classes

|                   | 1979-1981 | 1984-1985 |
|-------------------|-----------|-----------|
| Spike             | 22%       | 24%       |
| Double Spike      | 12        | 17        |
| Multiple Spike    | 19        | 3         |
| Curved Front      | 15        | 14        |
| Loop              | 1         | 0         |
| Halo              | 2         | 0         |
| Complex           | 5         | 0         |
| Streamer Blowout  | 5         | 27        |
| Diffuse Fan       | 10        | 5         |
| Other             | 9         | 10        |

Note that the streamer blowout classification has increased by a
factor of 5. This matches the reduction in the overall rate, so that
the absolute rate of streamer blowouts is constant over the solar
cycle. In Figure 3 we present a comparison of the rate of streamer
blowouts as a function of sunspot number. This apparent constancy of
streamer blowouts throughout the solar cycle may be a result of the
natural evolutionary process of coronal streamers, which of course are
always present. Since streamer blowouts are occurring at lower
latitudes during the declining phase, it will be interesting to see of
they have abnormal interplanetary signatures.

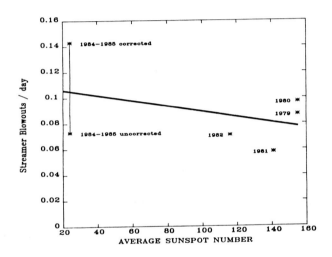

Figure 3. Rate of Streamer Blowout CMEs During 1979-1985

## 5. REFERENCES

Hildner, E., Gosling, J.T., MacQueen, R.M., Munro, R.H., Poland, A.I.
    and Ross, C.L.: 1976, Solar Phys., 48, 127.
Howard, R.A., Sheeley, Jr., N.R., Koomen, M.J. and Michels, D.J.: 1985,
    J. Geophys. Res., 90, 8173.
Hundhausen, A.J., Sawyer, C.B., House, L., Illing, R.M.E. and
    Wagner, W.J.: 1984, J. Geophys. Res., 89, 2639.
Sheeley, Jr., N.R., Howard, R.A., Koomen, M.J., Michels, D.J., Schwenn,
    R., Muhlhauser, K.H. and Rosenbauer, H.: 1985, J. Geophys. Res.,
    90, 163.
Tousey, R., Howard, R.A., Koomen, M.J.: 1974, Bull. Amer. Astron. Soc.,
    6, 295.

HELIOS IMAGES OF CORONAL MASS EJECTIONS

B. V. Jackson
University of California at San Diego
La Jolla,
California 92093
USA

ABSTRACT. The zodiacal light photometers on board the HELIOS spacecraft can be used to form images of coronal mass ejections in the interplanetary medium. Several aspects of these data are unique: they trace coronal mass ejections using Thomson scattering techniques to distances from the Sun greater than 0.5 AU; their perspective from the HELIOS orbits allow information to be gained about the three-dimensional shapes of specific mass ejections viewed both by coronagraphs and HELIOS; the global view afforded by the spacecraft photometers can image the mass ejection from within and thus relate in situ measurements to the shape of the whole structure. To date, the HELIOS photometers have been used to study coronagraph-observed mass ejections including those which originated at the Sun on 7 May, 24 May and 27 November 1979, and 21 May, 18 June, 29 June and 6 November 1980.

1. Introduction

Coronal mass ejections represent a great concentration of mass and energy input into the lower corona. It is estimated (Howard et al., 1985) that more than 5% of the solar wind in the ecliptic plane could be composed of mass from these discrete ejections of material. Coronal mass ejections can be observed near the solar surface by Earth-orbiting coronagraphs (MacQueen et al., 1974; Koomen et al., 1975). Spacecraft nearer the Sun not in Earth orbit have given in situ evidence of the progress of coronal mass ejections as they move outward from the Sun (e.g., Burlaga et al., 1982; Sheeley et al., 1984). However, the true interplanetary extent of these ejections both in and out of the ecliptic plane has here-to-fore only been guessed by extrapolating from coronagraph images. The three-dimensional shapes and positions of coronal mass ejections are uncertain from coronagraph observations alone. Coronagraph polarization measurements give some information of how distant the material of a mass ejection is from the plane of the sky. However, the shapes of mass ejections are not easily determined from these Earth-based observations.

R. G. Marsden (ed.), The Sun and the Heliosphere in Three Dimensions, 113–118.

## 2. HELIOS Observations

The HELIOS zodiacal light photometers (Leinert et al., 1975; 1981) were originally intended to measure the distribution of dust in the interplanetary medium between Sun and Earth. However, they can also be used to image the variations of brightness produced by large-scale changes in the interplanetary electron content. Because the HELIOS spacecraft are not Earth-orbiting, they can view mass ejections from an entirely different perspective than Earth. Thus, the HELIOS spacecraft give information on mass ejection three dimensional structures using stereoscopic techniques. Polarization data from the photometers further the ability of HELIOS spacecraft observations to disentangle the components of large mass ejections. The HELIOS zodiacal light sensors consist of three photometers fixed on the spacecraft and rotating with it on an axis perpendicular to the plane of the ecliptic. The photometers of HELIOS B point 16°, 31° and 90° north of the ecliptic plane with the 16° and 31° photometers clocking data into 32 longitude sectors at constant ecliptic latitude around the sky. The sixteen sectors nearest the Sun are lengths of 5.6° in ecliptic longitude; sector lengths of 11.2° and 22.4° are formed for sectors at greater solar elongations. The 16°, 31° and 90° photometers have apertures of 1°, 2° and 3° respectively. Helios A photometers point south of the ecliptic plane. The data are compiled over an 8.6-minute period from a set of ultraviolet, blue and visual filters, and polaroids oriented in three directions and each photometer, and renewed in sequence nominally every 5.2 hours.

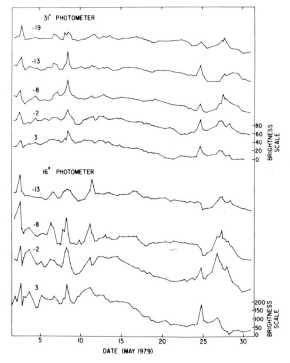

**Figure 1. Time series data for the month of May 1979, for selected HELIOS B photometer positions presented as in Richter et al. (1982) in units of tenth magnitude star brightness normalized to 1AU for the V (visual) photometric band. The large slowly-varying component from zodiacal light has been largely removed from the data. Sixteen degree photometer sectors with centers at ecliptic longitudes relative to the sun of 3°, -2°, -8°, and -13° and thirty-one degree photometer sectors at the same relative solar ecliptic longitudes plus an additional sector at -19° are presented. A brightness scale (number of 10th magnitude stars per square degree) is given for each set of photometers. Evident in the data are the brightness peaks during May 8-9 that are imaged as a mass ejection.**

Figure 1 presents the time series of selected individual photometer sectors as in Richter et al (1982) or Jackson and Leinert (1982) for May, 1979. The HELIOS photometers are shown to be stable over month-long periods; rapid daily variations from the zodiacal light component are negligible even at times when known meteor streams were in view. Measurement to measurement variations in brightness from sources other than electron Thomson scattering would generally be interpreted as noise in the data. Photometer brightness variations where the electron density variations are expected to be small at large elongations and at the greatest solar distances, show that random brightness variations of less than a tenth magnitude star per square degree are typical from one photometer measurement to another. Variations of the signal above this level are taken as significant and assumed due solely to changes in the amount of scattered light from electrons.

The HELIOS photometer data show brightness variations with time that occur later in the 31° photometers than in the 16° photometers and thus are outward propagating disturbances. Richter et al. (1982) give velocities for several of these outward-propagating "plasma clouds" and show that their speeds are consistent with or somewhat higher than solar wind speeds. Jackson et al. (1985) and Jackson and Leinert (1985) use these data to produce elongation-time diagrams of the onsets and brightest portions of mass ejections observed in the lower corona by both SOLWIND and Solar Maximum Mission coronagraphs. In Jackson (1985) both the photometers of HELIOS A and HELIOS B are used to view the same ejection that is also observed by the SOLWIND coronagraph on 27 November 1979. The coronal mass ejection observed by the SOLWIND coronagraph on 7 May 1979, is used as a demonstration in the following section.

3. Images of a Coronal Mass Ejection

To form images of coronal mass ejections from the HELIOS spacecraft photometer data, it is necessary to combine the observations from each sector into a meaningful spatial representation at a single instant in time. To do this, the time series from one photometer position for an eight day period is displayed as in Figure 1. A straight line baseline placed through the time series gives an estimate of the excess brightness at each data transmission time for the period in question. With this procedure, these data are then linearly interpolated to give brightness values at any instant in time. Electron numbers can be related to brightness from Thomson scattering as in Billings (1966). Without knowing the position of the electron condensation along the line of sight, an assumption that the electrons are at the point closest the Sun gives a lower limit estimate to the total number of electrons represented by the brightness increase. If brightnesses are interpreted as columnar density, it is possible to combine data from individual photometer sectors much as has been done from coronagraph image pixels (eg., Hildner et al., 1975; Jackson and Hildner, 1978; Poland et al., 1981). The contoured surface determined in terms of excess density from Helios photometer data is essentially an image of the density distribution around the spacecraft.

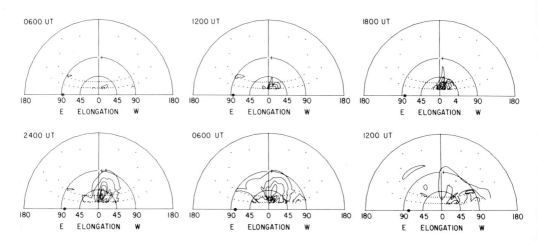

Figure 2. HELIOS contour plots for the 7 May 1979, ejection as it moves outward from the Sun over a period from 0600UT 08 May to 1200UT 09 May. In this presentation, the Sun is centered and various solar elongations labeled on the abscissa form semi-circles above the ecliptic plane (represented by the horizontal line). The vertical line is the great circle to the north of the spacecraft. The position of the Earth is marked as the '⊕' near east 90° and the solar north pole tilt indicated by the short line segment crossing 90° elongation. Positions of the sector centers are marked by dots. Electron columnar density is contoured in levels of $3 \times 10^{14}$ $cm^{-2}$ beginning at $3 \times 10^{14}$ $cm^{-2}$. The larger elongations are generally the lowest level contoured.

Figure 2 presents HELIOS B images of the ejection of 7 May 1979, at several stages contoured in quantities of excess columnar density. The contour plotting program interpolates and extrapolates from sparsely placed data points by determining directional derivatives of the surface from adjacent positions. In the presentation of Figure 2, all material travelling radially outward from the Sun moves in a straight line away from zero degrees elongation. In these "fisheye" lens-type views, the antisolar point maps onto a semi-circle at 180° elongation. Because the interpolation is large above 31° ecliptic latitude, high accuracy is not claimed for the contours presented there although they are shown for completeness.

As viewed in SOLWIND the ejection of 7 May 1979 is seen from Earth travelling to the solar northwest, reaching the outer extent of the coronagraph at 10 $R_{\odot}$ at approximately 2200 UT 7 May. SOLWIND coronagraph observers describe this ejection as three-pronged and moving at a speed of 150 km $s^{-1}$ (Poland et al., 1981). In SOLWIND data, the coronal mass ejection is confined to approximately 70° of position angle. The ejection on 7 May 1979, (as in Jackson and Leinert, 1985) depicted in Figure 2 is viewed by HELIOS B nearly 90° from the Sun-Earth line as it travels out to and above the spacecraft situated at 0.3 AU. The HELIOS B spacecraft first observed the outermost portion of the ejection at approximately 0100 UT 8

May in the innermost photometers (15R$_0$). As seen from HELIOS B the ejection has two distinct prongs and is confined to 80° of position angle in its initial stages. Further information on the three-dimensional structure of this mass ejection can be derived from the HELIOS polarization information which shows that the portion of the ejection to the north of the Sun is not as highly polarized as the material to the solar northwest. This implies that the northern portion of the ejection is heading toward and above the HELIOS B spacecraft. By midday 9 May, all of the spacecraft photometer positions show some evidence of a surrounding increase in electron density.

4. Conclusions

The resolution of the zodiacal light photometer imaging is adequate to show the larger features of coronal mass ejections. Excess electron density is well-calibrated by stars in the photometer fields of view. The difficulty of determining the distance along the line of sight to the excess density (a problem that concerns most measurements by coronagraphs) can be partially overcome by comparison with Earth perspective views as in the case of the 7 May 1979 ejection. It should be possible to use the HELIOS photometers to image other features observed in the solar corona that are expected to extend into the interplanetary medium. These features include coronal streamers and the density increases behind shocks as well as coronal mass ejections. One unique feature of these data is the ability to remotely measure the density of structures that are measured in situ at the same time. Thus, three-dimensional shapes of the density increases observed in situ may become more clear.

Data from the HELIOS photometers extend over nearly a complete solar cycle. To date, less than 2% of the available data has been used to form images. HELIOS A, launched December, 1974, has operated successfully until the present although various failures and limited power currently decrease the usefulness of the data. HELIOS B, launched January 1976, operated through the beginning of 1980. The photometer experiments were generally turned off when the spacecraft were at maximum distance from the Sun; they normally sent photometer data for only four of their six-month orbital periods. Nevertheless, the interplanetary medium from nearly a complete solar cycle is available to be viewed by this technique.

Acknowledgements

I appreciate many helpful discussions with my colleagues about this report, especially those with Ch. Leinert who has graciously supported my analysis of the HELIOS photometer data. Special appreciation is due R. Howard and N. Sheeley, Jr. of the United States Naval Research Laboratory, who originally led me to the HELIOS photometer data. Particular thanks is due UCSD students Steve Hurlbut and Ray Ng whose assistance and analysis of the data have made presentations of the HELIOS images possible. The work of B. Jackson is supported by Air Force contract F19628-85-K-0037 and National Science Foundation grant ATM84-06487 with the University of California, San Diego.

References:

Billings, D.E.: 1966, *A Guide to the Solar Corona*, New York: Academic Press, 150.

Burlaga, L.G., Klein, L., Sheeley, N.R., Jr., Michels, D.J., Howard, R.A., Koomen, M.J., Schwenn, R., and Rosenbauer, H.: 1982, *Geophys. Res. Lett.* **9**, 1317.

Hildner, E., Gosling, J.T.,MacQueen, R.M., Munro, R.H., Poland, A.I., and Ross, C.L.: 1975, *Solar Phys.* **42**, 163.

Howard, R.A., Michels, D.J., Sheeley, N.R., Jr. and Koomen, M.J.: 1982, *Astrophys. J. Lett.* **263**, L101.

Howard R.A., Sheeley, N.R., Jr., Koomen, M.J., and Michels, D.J.: 1985, *J. Geophys. Res.*, in press.

Jackson, B.V. and Hildner, E.: 1978, *Solar Phys.* **60**, 155.

Jackson, B.V.: 1985, *Solar Phys.* **95**, 363.

Jackson, B.V., and Leinert, Ch.: 1985, *J. Geophys. Res.*, in press.

Jackson, B.V., Howard, R.A., Sheeley, N.R., Jr., Michels, D.J., Koomen M.J., and Illing, R.M.E.: 1985, *J. Geophys. Res.*, **90**, 5075.

Koomen, M.J., Detwiler, C.R., Bruecker, G.E., Cooper, H.W., and Tousey, R.: 1975, *Applied Optics* **14**, 743.

Leinert, C., Link, H., Pitz, E., Salm, N., and Kluppelberg, D.: 1975, *Raumfahrtforschung* **19**, 264.

Leinert, C., Pitz, E., Link H., and N. Salm, N.: 1981, *J. Space Sci. Instr.* **5**, 257.

MacQueen, R.M., Eddy, J.A., Gosling, J.T., Hildner, E., Munro, R.H., Newkirk, G.A., Jr., Poland, A.I., and Ross, C.L.: 1974, *Astrophys. J. Lett.* **187**, L85.

Poland, A.I., Howard, R.A., Koomen, M.J., Michels, D.J., and Sheeley, N.R., Jr.: 1981, *Solar Phys.* **69**, 169.

Richter, I., Leinert, C., and Planc, B.: 1982, *Astron. Astrophys.* **110**, 115.

Sheeley, N.R., Jr., Howard, R.A., Koomen, M.J., Michels, D.J., Schwenn, R., Mulhauser, K.H., and Rosenbauer, H.: 1984, *J. Geophys. Res.* **90**, 163.

# RELATIONSHIP OF CORONAL TRANSIENTS TO INTERPLANETARY SHOCKS:   3D ASPECTS

Rainer Schwenn*
Earth and Space Sciences Division
Los Alamos National Laboratory
Los Alamos, NM 87544

More than 1000 coronal mass ejections (CMEs) caused by different types of coronal transients have been analyzed up to now, based on the images from white-light coronagraphs on board the OSO 7, Skylab, P78-1, and SMM spacecraft. In many cases, the CME images give the impression of loop-like, more planar structures, similar to those of prominence strutures often seen in $H_\alpha$ pictures. There is increasing evidence, though, for a three-dimensional bubble- or cloud-like structure of CMEs.

Many CMEs generate interplanetary shock waves which can be detected in situ using space probes. The associations between CMEs and shocks have been analyzed recently in much detail. The state of our present knowledge on the nature of CMEs and their interplanetary effects can be summarized in the following points:

1) The real cause for the onset of a CME is still unclear.

2) In many cases the CME takes off well before the impulsive phase of an associated $H_\alpha$-flare (if there is one at all), thus indicating a common cause in the corona for both the CME and the flare.

3) The energy for the CMEs stems from the magnetic field reservoir.

4) The evolution of a CME probably involves magnetic reconnection. Its contribution to the driving forces is still unknown.

5) Due to some shearing of the original coronal field lines (the top of which may form a prominence arch) the reconnection process may lead to nested magnetic helices instead of closed planar loops, thus forming a plasmoid with a truly three-dimensional shape.

6) The plasmoid may contain "cold" chromospheric prominence material, which does not always reach coronal ionization state on the way through the corona, or "hot" flare heated coronal material, or both.

*permanent address: Max Planck Institute für Aeronomie
                    D3411 Katlenburg-Lindau, W. Germany

*R. G. Marsden (ed.), The Sun and the Heliosphere in Three Dimensions, 119–121.*
© *1986 by D. Reidel Publishing Company.*

7) The CME drives a pressure wave into the ambient medium. Eventually, a shock wave may form, provided the injection speed relative to the local medium exceeds the medium's characteristic speed (400 to 500 km s$^{-1}$ in the lower corona).

8) There may exist at least two different types of CMEs

   a) the fast "flare type" CME, producing strong interplanetary shocks which on occasion may extend nearly around the sun and propagate beyond the present observational limits at about 30 AU.

   b) the slow "eruptive prominence type" CME, post-accelerated for several hours, producing weak shocks or, rarely, NCDEs (Non-compressive density enhancements).

9) The "cold" prominence material sometimes visible in big "loop-like" CMEs follows the "hot" CME loop top by a distance. There is evidence that both the cool and the hot material are being acted upon by the same magnetic forces.

10) The pressure wave (or shock wave) travels ahead of the main CME loop and may be associated with the so-called "forerunner" which is a very faint feature sometimes discernible ahead of the CME.

11) It is still unclear how coronal shocks can give rise to type II radio emission which, in some cases, apparently originates from the CMEs' legs rather than from their tops.

12) Detached plasmoids or "magnetic clouds" can on occasion be identified both in optical observations of the corona and in situ measurements in interplanetary space.

13) Any CME faster than $\sim$ 400 km s$^{-1}$ produces an interplanetary shock wave observable at any point within the angular extent of the CME. Some shocks, though, can have significantly larger angular extents than their associated CMEs.

14) The driver gas which is often enclosed in a magnetic cloud is probably identical to that of the CME plasmoid. Typical signatures, are: strongly enhanced helium abundance, low electron and proton temperatures, high magnetic field strength with low variance, bidirectional streaming of supra-thermal electrons and energetic particles and, at times, unusual ionization states.

15) For only about half of all shocks can some signatures of a driver gas be found. This probably means that the angular extension of the shock front is generally larger than that of the driver gas. The possible existence of real blast waves without ejecta following them cannot be ruled out.

16) On the average, shock fronts at 1 AU have an approximately circular appearance with respect to longitude. There is evidence for some flattening with respect to latitude.

17) Individual shock fronts may be distorted significantly locally, owing to the highly structured ambient solar wind.

18) The number of major CMEs (loops, spikes, etc.) appears to increase dramatically with solar activity although this is not unanimously agreed upon by all abservers. The range of projected latitudes also increases, resulting in a significant number of CMEs apparently ejected over the sun's poles.

19) The number of interplanetary shocks follows the solar activity cycle very closely.

Based on these conclusions we can now point out some of the problems to be attacked next. There seems to be a series of open questions concerning the very origin of the different types of CMEs and the detailed sequence of events in their early phase. Obviously we are not yet able to discern in all respects the causes from the effects. Furthermore, we do not know much about shock evolution in the highly structured solar wind. To improve this situation, theoretical modelling including three-dimensional shock propagation in a realistic environment is required. Furthermore, we need concerted and contiguous observations from several points of view distributed all around the sun. There may be new techniques to be developed in order to bridge the gap between solar observations (now ending at 10 $R_\odot$) and in situ measurements (now starting at 60 $R_\odot$) and to cover the largely unexplored regions out of the plane of the ecliptic. The long desired Ulysses mission will probably provide us many new observational facts on the three-dimensional structure of shocks. There is an excellent chance that Ulysses will encounter many shocks caused by both low- and high-latitude CMEs, when it finally starts its journey over the sun's poles in 1989, i.e. close to the maximum of the next sunspot cycle.

( A comprehensive review by the same author of the subject matter dealt with in this paper has been accepted for publication in Space Science Reviews - Ed.)

# THREE-DIMENSIONAL CONFIGURATIONS OF INTERPLANETARY DISTURBANCES ASSOCIATED WITH CORONAL MASS EJECTIONS

Takashi Watanabe, Takakiyo Kakinuma, and Masayoshi Kojima
The Research Institute of Atmospherics, Nagoya University
Toyokawa 442
Japan

ABSTRACT.  Three-dimensional propagation properties of interplanetary consequences of coronal mass ejections are studied on the basis of interplanetary scintillation (IPS) and spacecraft observations in 1978 – 1981.  Bubble-like interplanetary disturbances were associated with some white-light coronal mass ejections.

## 1.  INTRODUCTION

Since white-light coronal mass ejections (CMEs) have been observed in projection on the sky, their three-dimensional (3-D) configurations are not fully understood (see reviews given by e.g., Wagner, 1984; Michels et al., 1984).  An approach to investigate three-dimensional properties of CMEs will be to observe their interplanetary (IP) consequences at many points in IP space.  The interplanetary scintillation (IPS) technique will be a promising means for our purpose because many observation points can be obtained even in the region away from the ecliptic (e.g., Watanabe and Kakinuma, 1984).  In this paper, we discuss several examples of "CMEs in IP space" on the basis of IPS and spacecraft observations.

## 2.  IPS OBSERVATIONS OF INTERPLANETARY DISTURBANCES

IPS measurements observe an extended region along the line-of-sight of a radio source.  Because of the rapid radial fall off in electron density fluctuations, the region observed is confined to the vicinity of the closest approach point to the Sun on the line-of-sight.  Since IPS measurement of a radio source is made for 1 – 2 hours at meridian transit of the radio source, the chance to detect the arrival of the shock front at the line-of-sight of the radio source will be small.  Although the identification of the shock front is quite difficult, we may identify an IP disturbance by transient increase in the flow speed and, in many cases, in the level of scintillation  due to the turbulent, high-speed post-shock region (a shocked gas layer and a postshocked gas

R. G. Marsden (ed.), The Sun and the Heliosphere in Three Dimensions, 123–128.
© 1986 by D. Reidel Publishing Company.

layer).    According to a statistical analysis of IP shock wave
disturbances at 1 AU, which was made by Borrini et al. (1982), the
radial thickness of the shocked gas layer is typically 12 hours (i.e., $\sim$
0.14 AU), and the postshocked gas layer (or the shock driver) with high
helium abundance has the typical thickness of 1.5 days (i.e., $\sim$0.4 AU).
Thus many IP disturbances will be detected by the IPS measurement in
spite of its poor time coverage, although some high-speed IP
disturbances with  thin  post-shock region will escape from detection.
The observed flow speed is a weighted average of the flow speeds in the
post-shock region.  Since the radial thickness of the shock driver is
thicker than that of the shocked gas layer, observed flow speed of the
post-shock region may be attributed to that of the shock driver (Chao,
1984), although the shocked gas layer will be observed on occasions.

     Since the solar wind density is not obtainable from the IPS
measurement, we cannot use the conservation law of the mass flux to
estimate the hydrodynamical shock speed.  Watanabe and Kakinuma (1984)
proposed a simple method to estimate the shock speed using the observed
post-shock flow speed.  The theory of piston driven shocks (e.g.,
Liepmann and Roshko, 1957) shows that, in a strong shock approximation,
the shock speed in an ideal gas is about 4/3 times as high as the speed
of the shock driver or the flow speed of the shocked gas.  In-situ
observations show that the multiplication factor is 1.1 - 1.2.  Thus the
approximate shock speed may be estimated from IPS measurements in such a
way that the observed post-shock flow speed is multiplied by 1.2
(Watanabe and Kakinuma, 1984).  To construct the directional diagram of
the shock speed, we normalize the local shock speed to that at a
specific heliocentric distance assuming that the shock speed $V_s$ is given
by a formula, $V_s \propto R^{-n}$ where R is the heliocentric distance.  We
estimate 'n' on the basis of spacecraft observations for each event by
the method given by Gosling et al. (1975).

     The position of an "observation point" on a line-of-sight has a
spatial ambiguity because the typical angular spread of the principal
scattering zone seen from the Sun is $50^{\circ}$ (Rickett, 1975).  Thus small
scale structures of the IP disturbance will be smeared out.


3.   EDGE-ON CMEs IN 1979 - 1981

Sheeley et al. (1985) made a comparison between Solwind observations of
CMEs and Helios observations of IP shocks during 1979 - 1982.  The
comparison was made when one of the Helios spacecraft was situated
roughly above the east or the west solar limb, where CMEs are most
visible from the Earth.  IPS measurements can fill gaps between Helios
spacecraft and near-Earth spacecraft observations.  Among the examples
of CMEs, the directional diagrams of the shock speeds (normalized to
those at the heliocentric distance of 0.5 AU) of well-identified nine IP
disturbances are shown in Figure 1.  When an IP disturbance was detected
by near-Earth spacecraft, we also plot the shock speed which is
normalized to that at 0.5 AU.  The power-law deceleration coefficient is
estimated from Helios observations (Sheeley et al., 1985).  It is seen
that some high speed edge-on CMEs propagated in IP space as bubble-like

IP disturbances and that a part of them could have hit the Earth.

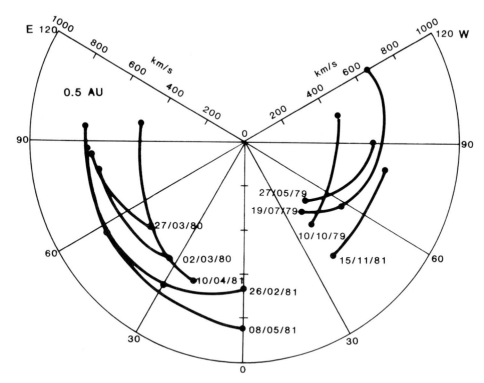

Figure 1.    Directional diagrams of the shock speeds (normalized to those at 0.5 AU) for the interplanetary disturbances associated with nine edge-on CMEs in 1979 - 1981 (Sheeley et al., 1985).    The date of each CME event is represented as DAY/MONTH/YEAR.

## 4.   A HALO CME OF NOVEMBER 27, 1979

A halo CME or a line-of-sight CME is a $360^\circ$ coronal density enhancement encircling the occulting disk.    A typical example was obtained with the Solwind coronagraph on November 27, 1979 (Howard et al., 1982).    The solar source of the CME has been attributed by Howard et al. (1982) to a north-south aligned disappearing solar filament near disk center (N05W03) accompanied by a 1N solar flare (N14E05) at 0647 UT, November 27, 1979.    The estimated frontal speed of the CME was 1160 km/s (Howard et al., 1982).    This CME, which moved outward along the Sun-Earth line, was also detected by the Helios photometers (Jackson, 1985). A 3-D representation of the angular distribution of the shock speed of the IP disturbance associated with the halo CME is shown in Figure 2.    It is concluded that the IP manifestation of the halo CME was a bubble-like IP disturbance.    Somewhat slower shock speeds were obtained around the normal of the center of the disappearing filament.    Detailed description on the event with a preliminary directional diagram of the shock speed is given elsewhere (Watanabe, 1985).

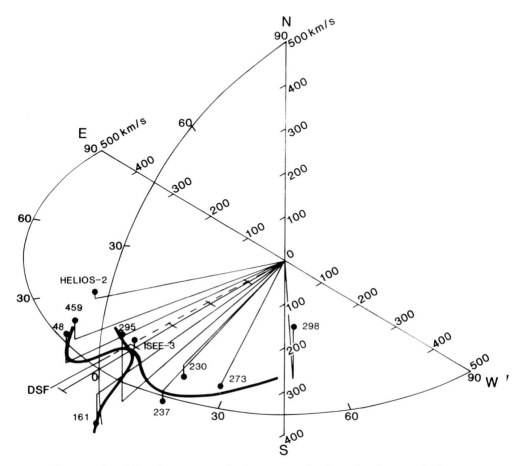

Figure 2.   The directional diagram of the shock speed for an interplanetary disturbance associated with the halo CME of November 27, 1979.   Each dot represents the shock speed normalized to that at 1 AU. For IPS observations, the 3C numbers of observed radio sources are given.   The shock speed at ISEE-3 is taken from Russell et al. (1983).   The flow speed of the enhanced-density solar wind which was observed at Helios 2 at 1600 UT, November 28 (Jackson, 1985) is also plotted.   The normal of the center of disappearing solar filament of November 27, 1979, which was observed immediately before the first detection of the CME, is indicated by "DSF".

## 5.   DISAPPEARING SOLAR FILAMENT OF AUGUST 23, 1978

Trottet and MacQueen (1980) proposed a strong correlation between loop like CMEs and north-south aligned disappearing solar filaments. Disappearance of a north-south aligned solar filament took place near the solar disk center (N15E03) at 1100 ± 0300 UT on August 23, 1978 (Joselyn and Bryson, 1980). No relevant solar-flare activity was reported. A 3-D directional diagram of the shock speed for the IP disturbance associated with the disappearing filament is shown in Figure 3. It is seen that the IP consequence of the disappearing filament of August 23, 1978 was a bubble-like IP disturbance. Since rather slower flow speeds were observed in the region apart from the solar equatorial plane (3C295, 3C161), the 3-D configuration of the IP disturbance is approximated by an oblate sphere having an axial ratio of 1.7 (Watanabe and Marubashi, 1985). Tappin et al. (1983) also discussed this event on the basis of IPS observations at Cambridge.

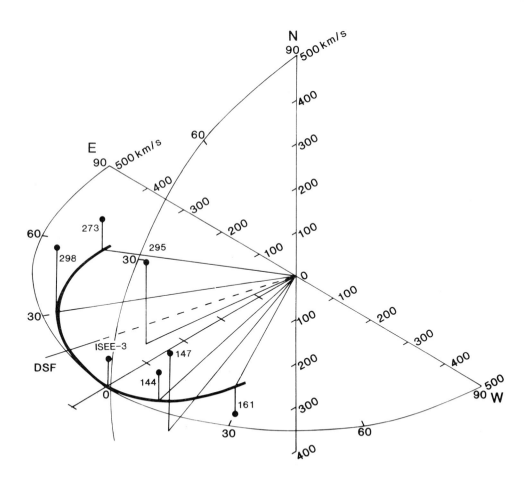

Figure 3. Same as Figure 2 but for the interplanetary disturbance associated with the disappearing solar filament of August 23, 1978. The shock speed at ISEE-3 is taken from Ogilvie et al. (1982).

## 6.   CONCLUDING REMARKS

We have studied IP consequences of edge-on and halo-type CMEs, and an IP
disturbance associated with a north-south aligned disappearing filament,
which may be attributed to the source of a loop-like CME (Trottet and
MacQueen, 1980).   It is concluded that IP consequences of some CMEs are
bubble-like IP disturbances.   Since IPS measurements can be used in the
investigation of the 3-D propagation properties of CMEs in IP space,
cooperation between the IPS network and the Ulysses spacecraft will be
valuable.

## REFERENCES

Chao, J. K.: 1984, Adv. Space Res., **4**, 327.
Gosling, J. T., Hildner, E., MacQueen R. M., Munro, R. H., Poland, A.
    I., and  Ross, C. L.: 1975, Solar Phys., **40**, 439.
Howard, R. A., Michels, D. J., Sheeley, Jr., N. R., and Koomen, M. J.:
    1982, Astrophys. J., **263**, L101.
Jackson, B. V.: 1985, Solar Phys., **95**, 363.
Joselyn, J., and  Bryson, Jr., J. F.: 1980, in Solar and Interplanetary
    Dynamics, eds., M. Dryer and E. Tandberg-Hanssen, D. Reidel,
    Dordrecht, Holland, p. 413.
Liepmann, H. W., and Roshko, A.: 1957, Elements of Gasdynamics, John
    Wiley & Sons, Inc., New York, p. 62.
Michels, D. J., Sheeley, Jr., N. R., Howard, R. A., Koomen, M. J.,
    Schwenn, R., Mühlhäuser, K. H., and Rosenbauer, H.: 1984, Adv. Space
    Res., **4**, 311.
Ogilvie, K. W., Coplan, M. A., and Zwickl, R. D.: 1982, J. Geophys.
    Res., **87**, 7363.
Rickett, B. J.: 1975, Solar Phys., **43**, 237.
Russell, C. T., Smith, E. J., Tsurutani, B. T., Gosling, J. T., and
    Bame, S. J.: 1983, in M. Neugebauer (ed.), Solar Wind Five, NASA CP-
    2280, p. 385.
Sheeley, Jr., N. R., Howard, R. A., Koomen, M. J., Michels, D. J.,
    Schwenn, R., Mühlhäuser, K. H., and Rosenbauer, H.: 1985, J. Geophys.
    Res., **90**, 163.
Tappin, S. J., Hewish, A., and Gapper, G. R.: 1983, Planet. Space Sci.,
    **31**, 1171.
Trottet, G., and MacQueen, R. M.: 1980, Solar Phys., **68**, 177.
Wagner, W. J.: 1984, Ann. Rev. Astron. Astrophys., **22**, 267.
Watanabe, T.: 1985, Proc. Res. Inst. Atmospherics, Nagoya Univ., **32**, 11.
Watanabe, T., and Kakinuma, T.: 1984, Adv. Space Res., **4**, 331.
Watanabe, T., and Marubashi, K.: 1985, J. Geomag. Geoelectr., in press.

THE THREE-DIMENSIONAL EXTENT AT 1 AU OF INTERPLANETARY DISTURBANCES

ASSOCIATED WITH DISAPPEARING SOLAR FILAMENTS

C. S. Wright
IPS Radio and Space Services
Department of Science
P.O. Box 702
Darlinghurst NSW 2010
Australia

ABSTRACT. The average three-dimensional extent at 1AU of interplanetary (IP) disturbances caused by disappearing filaments, is inferred from the levels of geomagnetic activity following the disappearance of filaments with many different orientations and locations on the sun. In this sense, the magnetic field of the earth is used as a 'probe' to determine the 'directivity pattern' of the IP disturbance. The results suggest that IP disturbances have preferred directivity patterns with respect to the underlying parent filaments and their associated magnetic fields. The most geoeffective parts of the disturbances propagate at angles of about 50 degrees from the radial direction.

1. INTRODUCTION

Several methods have been employed to study the extent of IP distur-bances arising from large solar events such as flares and filament disappearances. Near the sun (within about 10 solar radii) the distur-bances are observed directly with white-light coronagraphs. Subtracted coronagraph images show the two-dimensional electron density distribu-tion against the plane of the sky (Howard et al, 1985). Further from the sun, 0.3-1.2AU, IP disturbances have been investigated by their scintillation effects on signals from occulted radio sources (Gapper et al, 1982; Watanabe and Kakinuma, 1984). The interplanetary scintilla-tion (IPS) technique determines the two-dimensional distribution in the plane of the sky of fluctuations in the electron density, but not the electron density itself. A third technique employs satellite-borne pho-tometer observations of zodiacal light (Jackson, 1985) to determine the two-dimensional electron density distribution in IP disturbances from near the sun to beyond 1AU.

This paper describes a technique that can provide additional infor-mation about the extent, averaged over many events, of IP disturbances associated with the disappearance of solar filaments.

R. G. Marsden (ed.), The Sun and the Heliosphere in Three Dimensions, 129–134.

## 2.  METHOD

The technique relies upon the established connection between the disappearance of large solar filaments and the occurrence of geomagnetic
activity (Joselyn and McIntosh, 1981; McNamara and Wright, 1982; Wright
and McNamara, 1983).  The most pronounced geomagnetic effects were found
to occur with delays of about 4 days.  McNamara and Wright (1982) argued
that the geomagnetic activity was the magnetospheric response to the
passage of the IP disturbances.  If this interpretation is correct, the
average angular extent of the IP disturbances can be determined by examining the geomagnetic response following each of many disappearances
from widely separated heliographic locations.
     In this sense the earth is used as a'probe' to detect the 'geoef-
fective' components of the IP disturbances.  To be successful it is
necessary to define the 'response function' (Section 2.1) of the geomagnetic field in terms of the solar wind parameters near the earth.  A
further complication arises because filaments are greatly elongated
structures and, unlike flares, cannot be considered as 'point sources'.
Specification of a filament's position must also include its orientation.  A solution to this problem is to classify the filaments according
to their perspective, or orientation, as seen from the earth (Section
2.2).

### 2.1  Response Function

The disturbance level of the geomagnetic field is related to the solar
wind momentum flux, $nv^2$, and the flux of magnetic inductance, $vB$, by the
empirical relationship (Svalgaard, 1977),

$$am \sim vB(nv^2)^{1/3} \tag{1}$$

where B, v and n are the magnetic inductance, velocity and density
respectively near the earth and am is a three-hourly index of geomagnetic activity. (A fourth parameter that is important in this context is
the orientation of the IMF with respect to the dipole axis of the earth.
However, for present purposes, its role is secondary to that of the
IMF.) Equation 1 is a statistical relationship derived from a large
number of data  and therefore the parameters in this expression are
averages.  Approximate ranges over which they vary are (King, 1983):

$$1 < n < 100 \text{ cm}^{-3}, \quad 250 < v < 750 \text{ km sec}^{-1}, \quad 1 < B < 40 \text{ nT}.$$

It follows that the magnitudes of geomagnetic disturbances that follow
filament disappearances with fixed delays (for which the velocity should
be approximately constant) will be more sensitive to the IMF than to the
density. Thus an increase in geomagnetic activity may be interpreted as
a density increase or, more probably, a magnetic field increase.

## 2.2  Filament-centred Reference Frame

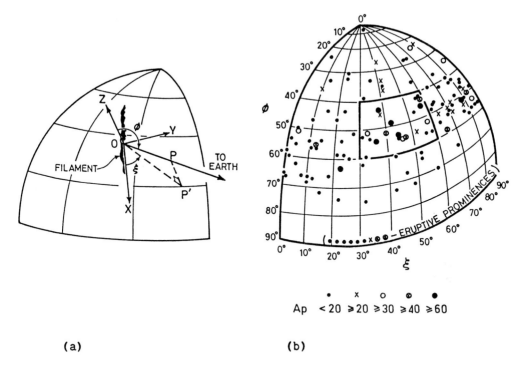

            (a)                                    (b)

Figure 1.  (a) The direction of the earth with respect to a solar filament.  OX and OY are perpendicular and lie in the plane tangential to
the solar surface at the midpoint, O, of the filament. Z indicates the
zenith.  P lies on the line-of-sight to the earth and P' is the perpendicular projection of P in the plane XOY. The direction to the earth is
defined by the angles ZOP, ∅, and XOP, ξ. (b) The distribution of the
earth's direction for 125 disappearing filaments. The level of geomagnetic activity on the fourth day after the disappearance of each filament is shown by the various symbols. The roving window is indicated by
the rectangular outline. Here, it delineates the 'geoeffective domain'
(see text).

For an observer on the solar surface at the mid-point of a filament, the
line-of-sight to the earth may be uniquely specified (Figure 1(a)) in
terms of a 'zenith' angle, ∅, measured from the vertical and an
'azimuth' angle, ξ, measured from the filament axis to the projection of
the line-of-sight in the plane of the solar surface.  Hence ∅ is also
the angular distance from disc centre to the mid-point of the filament
and therefore varies from 0 to 90 degrees. On the other hand, ξ varies
from 0 to 360 degrees.  However if a filament is assumed to have two-
fold symmetry (about its axis and the line perpendicular to its axis)
then the range of ξ also becomes 0 to 90 degrees.

## 2.3  Analysis

The directions to the earth in terms of $\xi$ and $\emptyset$ were determined (Figure
1(b)) for all 125 large filaments that disappeared between 1974 and
1980.  To determine the dependence of geomagnetic activity upon $\xi$ and $\emptyset$,
a roving 'direction window' of arbitrary size was positioned over the
$\xi-\emptyset$ domain (Figure 1(b)).  The level of geomagnetic activity associated
with the filaments that 'viewed' the earth through this window was then
determined by two methods.  First, a superposed epoch analysis was performed to determine the average value of the geomagnetic index, Ap, on
days following the disappearances. Secondly, the proportion of filaments
which were accompanied on the fourth day (when the average disturbance
was greatest) by Ap$\geq$30 was calculated, and the significance of the
results determined using the binomial theorem.


## 3.  RESULTS

Since few large filaments are observed at latitudes less than 20
degrees, there are few points in the region defined by $\emptyset$<20 degrees,
corresponding to the region near disc centre (Figure 1(b)). Similarly,
the relative absence of points in the region $\emptyset$>70 degrees reflects the
aspect sensitivity of filament observations near the limbs.
        Each point in Figure 1(b), in addition to showing the direction to
the earth with respect to a filament, also indicates the level of
geomagnetic activity that occurred (on the fourth day) following the
disappearance of that filament. Large disturbances are clustered in the
domain 40<$\emptyset$<60 degrees. Of the 20 filaments that disappeared within 40
degrees of disc centre, only two were followed by Ap greater than 30.
This contrasts strongly with the pattern of geomagnetic activity for
flare-associated disturbances which shows a strong preference for flares
near disc centre.  The most noteworthy feature of the distribution in
Figure 1(b) is the highly significant clustering of large disturbances
within the indicated window.  This is made more apparent by Figure 2(a).
        The upper three and lower three panels of Figure 2(a) summarise the
geomagnetic response when the earth lies inside and outside the window,
respectively.  Within the window, the level of geomagnetic activity that
occurred on the fourth and fifth days was high.  Also, 11 of the 21
filaments were accompanied by Ap$\geq$30 on the fourth day whereas only two
such instances were expected by chance. The probability of this result
arising by chance can be calculated from the binomial distribution to be
<$10^{-6}$.  In contrast, outside the window, average levels of geomagnetic
activity prevailed on the fourth and fifth days.  Only 11 of the 116
filaments were accompanied by Ap$\geq$30 on the fourth day, a result not significantly different from the chance value of 9. (The significant effect
with a delay of 8-10 days (Wright, 1983) has no direct relevance here.)
        In summary, significant geomagnetic effects were observed only when
the earth lay within a tightly constrained range of directions with
respect to the filaments. This 'geoeffective domain' is defined by
40<$\emptyset$<60 degrees and 30<$\xi$<60 degrees.

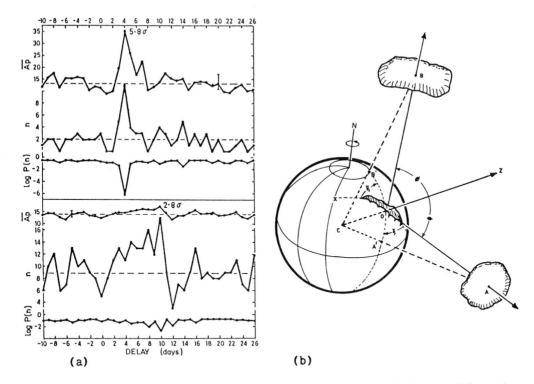

(a)                                                    (b)

Figure 2.   (a) Summary of the geomagnetic response of the 21 filaments
when the earth is inside the geoeffective domain - upper three panels -
and the 104 filaments and prominences when the earth is on or outside
the domain - lower three panels. The uppermost graph in each set shows
the average Ap in an epoch surrounding the filament disappearances, the
central graph shows the number of filaments, n, followed on each day of
the epoch by Ap$\geq$30 and the lowermost graph shows the probability, P(n),
of the number, n, arising by chance. (b) Schematic illustration showing
the orientation of an IP disturbance with respect to the underlying
filament.  XO indicates the filament axis and C is the solar centre. A
and B indicate the centres  of two geoeffective components, while A' and
B' are the respective projections on the solar surface.


4.  DISCUSSION

     It is not possible from these results to discriminate between the
effects of density and magnetic field within the IP disturbance.  For
example, it is conceivable that matter is concentrated radially and that
the geomagnetic effects arise from non-radial concentrations of magnetic
field.  In this sense the results are not directly comparable with those
derived from other techniques, which apply only to density or to density
fluctuations.  It seems clear, however, that the most geoeffective portions of these IP disturbances propagate in directions determined by the
underlying configuration of the parent filaments and their associated
magnetic fields.

It should be emphasised that Figure 1(b) is constructed upon the assumption that filaments have two-fold symmetry, thereby restricting the range of ξ from 360 to 90 degrees. Each filament may therefore be considered to have four geoeffective domains since no distinction is made between quadrants. The results are consistent with a filament having at least two geoeffective domains since otherwise it would be highly improbable that so many of the 21 filaments would be followed by major geomagnetic disturbances when the earth lay within the window shown in Figure 1(b). Figure 2(b) provides a schematic illustration of a configuration in which a filament disappearance is assumed to give rise to an IP disturbance with two geoeffective components. It is intended as a pictorial aid to illustrate the geometry involved. At 1AU, the geoeffective portions of the IP disturbance subtend angles of about 20-30 degrees at the sun. The overall extent of the IP disturbance may be as large as 120 degrees.


5.  CONCLUSIONS

The main conclusions derived from this study are:

1.  The directions of propagation of filament-associated IP distur-
    bances are controlled by the underlying filament and magnetic field
    configuration.
2.  The geoeffective components of such IP disturbances are concen-
    trated at angles of abʋut 50 degrees from the radial direction and
    about 45 degrees from the axis of the parent filament.

REFERENCES

Gapper, G. R., A. Hewish, A. Purvis, and P. J. Duffett-Smith, 1982,
    Nature, **296,** 633-636.
Howard, R. A., N. R. Sheeley, Jr., M. J. Koomen and D. J. Michels, 1985,
    to appear in J. Geophys. Res..
Jackson, B. V., 1985, to appear in Solar Phys..
Joselyn, J. A. and P. S. McIntosh, 1981, J. Geophys. Res., **86,** 4555-
    4564.
McNamara, L. F. and Wright, C. S., 1982, Nature, **299,** 537-538.
Watanabe, T. and T. Kakinuma, 1984, Adv. Space Res., **4,** 331-341.
Wright, C. S., 1983, Proc. Astron. Soc. Aus., **5,** 198-202.
Wright, C. S. and L. F. McNamara, 1983, Solar Phys., **87,** 401-417.

THREE-DIMENSIONAL, TIME-DEPENDENT, MHD MODEL OF A SOLAR
FLARE-GENERATED INTERPLANETARY SHOCK WAVE

M. Dryer
NOAA Space Environment Laboratory (R/E/SE)
Boulder, Colorado  80303, USA

S.T. Wu
Department of Mechanical Engineering
University of Alabama in Huntsville
Huntsville, Alabama  35899, USA

and

S.M. Han
Department of Mechanical Engineering
Tennessee Technological University
Cookeville, Tennessee  38505

ABSTRACT. Three-dimensional model of the propagation of an
interplanetary shock wave into a representative ambient three-
dimensional heliospheric solar wind is demonstrated. The numerical MHD
simulation is initialized by assuming a peak shock velocity of 1000 km
sec$^{-1}$ at the center of a right circular cone of $18^{\circ}$ included angle at 18
solar radii. Examination of the shocked plasma and IMF parameters
reveals several fundamental results that differ from our earlier 2D and
2-1/2D simulations under similar input conditions that were confined to
the ecliptic plane. The differences include: (i) diminution of the
solar wind peak velocity occurs in the 3D example because of flow
divergence introduced in the heliolatitudinal direction; (ii)
concentration of the peak density at each radius in an annular region
with the minimum density at the central axis due, again, to flow
divergence; (iii) similar behavior of the IMF magnitude that starts
(near the sun) with peaks at high latitudes and to the west of the
simulated flare's central meridian; and (iv) twisted, helical-like, IMF
rotation due to a large amplitude, non-linear Alfven wave in the shocked
plasma. The amplitude is highest along the central axis and decays to a
minimum at the sides of the MHD shock where the fast-mode jump
conditions prevail.

1.    INTRODUCTION

It has been known for a number of years that flare-generated shock waves

*R. G. Marsden (ed.), The Sun and the Heliosphere in Three Dimensions, 135–140.*
© *1986 by D. Reidel Publishing Company.*

can achieve high latitudinal extents.  This was dramatically
demonstrated by the Culgoora radioheliograph in March 1969 (Smerd, 1970)
via the type II plasma emission mechanism from shock-excitation of
coronal electrons.  More recent interplanetary scintillation (Watanabe
and Kakinuma, 1984; and Hewish, 1984) and white light coronagraph
observations have repeatedly confirmed the large latitudinal extent of
both flare-generated and coronal hole-generated disturbances in the
solar wind.  The need to obtain in situ observations of such
disturbances by ULYSSES is clearly indicated.  It is also desirable that
numerical simulations be used to aide in the interpretation and
understanding of the ground-based IPS observations, the ecliptic
baseline studies by spacecraft (WIND, SOHO, PHOBUS, etc.), as well as by
ULYSSES during the latter´s excursion to high heliolatitudes.  The
purpose of this paper is to illustrate the capability of a 3D MHD time-
dependent computer code to simulate the solar wind response near the sun
to a solar flare-generated shock wave.


2.   BACKGROUND


Previous time-dependent MHD modeling has been limited to the ecliptic
plane (Wu et al., 1983; Gislason et al., 1984; and Dryer et al., 1984)
in the 2-1/2D (or non-planar) formulation that ignored latitudinal
variation of all dependent variables.  The present time-dependent, fully
3D, code requires, as its starting point, a steady-state solution as a
physical basis. Insight in this regard has been provided by the steady-
state, quasi-radial, analytical study of Nerney and Suess (1975) and the
co-rotating, time-independent, numerical work of Pizzo (1982).  A full
description of the mathematical formulation and numerical techniques is
given by Han et al. (1984) for the steady-state and time-dependent
aspects of the present example of an initial boundary value problem.

     Briefly, the 3D steady-state solar wind used here consists of a
spiralling IMF embedded within the supersonic, superalfvenic solar wind.
At t = 0 hr, a shock wave is used to initialize the time-dependent
computation at the lower boundary, $r = 18R_o$, of a domain contained
within the boundaries: $18 \leq r \leq 222R_o$, $\pi/4 \leq \theta \leq 3\pi/4$; and $0 \leq \phi \leq
\pi/2$ where r, $\theta$, $\phi$ are the helio-radial, latitudinal, and -longitudinal
coordinates, respectively.  (As more computer resources become available
to us, the latitudinal and longitudinal boundaries will be extended.)
The shock velocity is assumed to be 1000 km sec$^{-1}$ at the equatorial mid-
point of this domain and to decay sinusoidally to zero at the periphery
of a 18$^o$ included solid angle centered at the sun.  The Rankine-Hugoniot
jump conditions are input for 1-1/2 hr at the center of this "bulls-eye"
at 18R$_o$, decaying to zero strength at its periphery as just noted.  The
temporal input was chosen to represent a somewhat-average, piston-driven
shock as suggested by a short duration, thermal phase of a flare as
measured in the soft X-ray bandwidth range (say, 1-4 Å).  The spatial
diminution of the shock was chosen to represent a rather highly-
collimated, non-spherical shock wave that could be expected to "fit"
within the computational domain as it moved outwardly to the outer
boundary of ~1.1 AU.

## 3.    RESULTS

Some of the three-dimensional results for the plasma velocity and IMF magnitude at r = 32R$_0$ are shown in Figure 1 at t = 5 hr. The top panels show the shock-accelerated solar wind's radial velocity component in the expected, collimated shape. The right side of the figure shows the magnitudes on the spherical, " $\phi$ - $\theta$ ", surface of the velocity components and |$\underline{B}$|. The middle panels show the transverse velocity with the null value exactly at the center of the disturbed solar wind. The

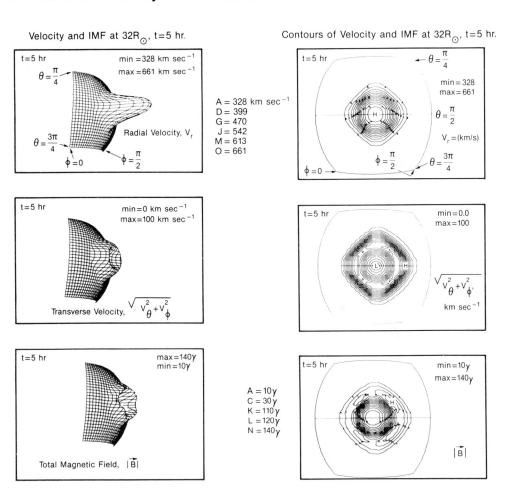

Figure 1.      Velocity components and total IMF magnitude at t = 5 hr, r = 32R$_0$. Left side: oblique view of non-spherical shock parameters. Right side: contours of radial velocity, transverse velocity, and |$\underline{B}$| within shocked volume. Initial shock velocity: 1000 km sec$^{-1}$.

lower panels show the total IMF magnitude. It is seen that the peak
magnetic field magnitudes develop first at the mid-latitudes within the
western (right side, when viewed toward the sun) portion of the shocked
plasma. This is due to the poleward spiralling IMF topology wherein the
largest values of the shock's quasi-perpendicularity, $\theta_{nB}$, occur first
at the mid-latitudes. At t = 20 hr., however, after the shock passes
the r = 72$R_o$ position (not shown here), the spiral IMF in the equatorial
plane becomes effective on the western side, and the two "high"
contours-previously separate at the 32$R_o$ location-now join across the
equatorial plane.

    The plasma density contours, in Figure 2, also show an asymmetrical
response to the shock. Close to the sun at t = 5 hr (32$R_o$, as seen in
the upper left panel), the nearly wholly-gasdynamic behaviour dominates
because of the essentially high beta plasma; that is, we can detect a
plasma snow-plow effect. As a result of this fact, plus the nature of
the "bulls-eye" input, the plasma is pushed outward from the center,
creating a high density annular ring. At the larger distances, an
asymmetry develops, biased toward the eastern side of the shocked
volume. At t = 30 hr (lower right panel in Figure 2), the density in
the re-developing annular ring is twice (69 cm$^{-3}$ at 72$R_o$) the value (33
cm$^{-3}$) at the center of the expanding volume.

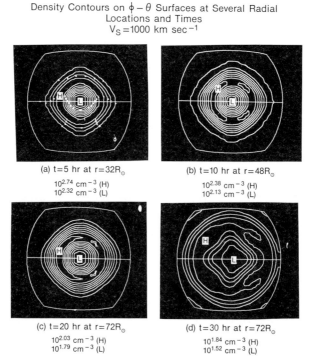

Density Contours on $\phi - \theta$ Surfaces at Several Radial
Locations and Times
$V_S = 1000$ km sec$^{-1}$

(a) t=5 hr at r=32$R_\odot$
$10^{2.74}$ cm$^{-3}$ (H)
$10^{2.32}$ cm$^{-3}$ (L)

(b) t=10 hr at r=48$R_\odot$
$10^{2.38}$ cm$^{-3}$ (H)
$10^{2.13}$ cm$^{-3}$ (L)

(c) t=20 hr at r=72$R_\odot$
$10^{2.03}$ cm$^{-3}$ (H)
$10^{1.79}$ cm$^{-3}$ (L)

(d) t=30 hr at r=72$R_\odot$
$10^{1.84}$ cm$^{-3}$ (H)
$10^{1.52}$ cm$^{-3}$ (L)

Figure 2.      Density contours on   $\phi - \theta$   surfaces at several radial
locations and at several times as noted.

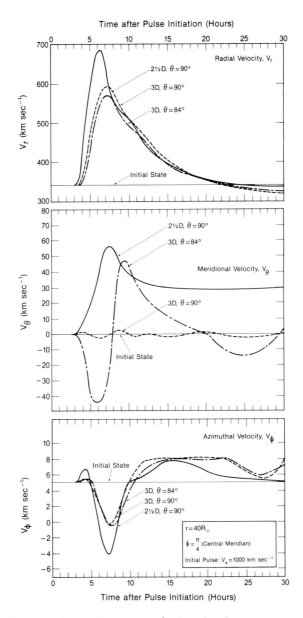

Figure 3.    Time series of solar wind velocity components at a point, $r = 40R_\odot$, along the central axis of the simulated flare.    The initial shock is input at $18R_\odot$ as described in the text with no computations from $1R_\odot$ to $18R_\odot$.

Finally we demonstrate another aspect of the difference between a fully 3D and a non-planar, 2-1/2D, computation. Figure 3 shows a time series of the three velocity components at $r = 40R_0$ as the shock arrives (at t = 3 hr), then recedes beyond this position. The upper panel for the radial component, $V_r$, shows that a very significant decrease in the peak value occurs as a direct result of the additional divergence in the meridional direction. This result is also reflected in the heliolongitudinal direction (bottom panel). The middle panel also demonstrates in 3D that a large amplitude MHD wave passes the spacecraft with, first, a 45 km sec$^{-1}$ motion away from the equatorial plane (at $\theta$ = 84°), followed by an equal swing around to and pointed toward the opposite pole. This motion (confirmed by large scale IMF rotation not shown here) would have been impossible to predict from the 2-1/2D computation.

On the basis of this representative flare-generated shock simulation we suggest that ULYSSES will continue to observe flare shocks whose strengths will diminish as the higher latitudes are reached. However, large scale IMF rotations may be expected to be observed as the spacecraft approaches the latitudes of flares that are generated in the southern hemisphere.


4.   ACKNOWLEDGMENT

The authors wish to thank the following organizations for their support during the preparation of this work: NOAA, via Contract NA84RAA05519 (SMH) and Contract No. NA84RAC05132 (STW); NASA, via Interagency Order No. W-15,361 (Mod. 2) (MD); NASA/NAGW-9 (STW); USAF/AFGL, via Project Order ESD 5-618 (Adm. No. 1) (MD); and the National Center for Atmospheric Research for the use of their CRAY-1 computer facility.


5.   REFERENCES

Dryer, M., S.T. Wu, G. Gislason, S.M. Han, Z.K. Smith, J.S. Wang, D.F. Smart, and M.A. Shea, 1984, Astrophys. Space Sci., 105, 187.
Gislason, G., M. Dryer, Z.K. Smith, S.T. Wu, and S.M. Han, 1984 Astrophys. Space Sci., 98, 149.
Han, S.M., S. Pantichob, S.T. Wu, and M. Dryer, 1984 in Proceedings of the XII Southeastern Conference on Theoretical and Applied Mechanics, Vol. II, Auburn University, Engineering Extension Service, Auburn, Alabama 36849, USA, May 10-11, 1984, 39.
Hewish, A. 1984, Private Communication.
Howard, R.A., D.J. Michels, N.R. Sheeley, Jr. 1982, Astrophys. J., 263, L101.
Nerney, S.F. and S.T. Suess 1975, Solar Phys., 45, 255.
Pizzo, V.J. 1982, J. Geophys. Res., 87, 4374.
Smerd, S.F. 1970, Proc. Astron. Soc. Austral., 1, 305.
Watanabe, T. and T. Kakinuma 1984, Adv. Space Res., 4, 331.
Wu, S.T., M. Dryer, and S.M. Han 1983, Solar Phys., 84, 395.

# SECTION IV:  SOLAR WIND

INTERPLANETARY SCINTILLATION OBSERVATIONS OF THE SOLAR WIND AT HIGH
LATITUDES

W. A. Coles and B. J. Rickett
Electrical Engineering and Computer Sciences
University of California at San Diego
La Jolla, California  92093

ABSTRACT.  The ULYSSES spacecraft will provide the first direct
measurements of the polar solar wind.  Thus our present knowledge of
this region is derived from various indirect methods.  These include
radio propagation observations, cometary observations and optical
observations (such as doppler dimming of spectra - see Withbroe this
report).  The purpose of this paper is to review the radio
propagation measurements, particularly their solar cycle dependence;
and to discuss the analysis of these observations, the underlying
assumptions in the models used, and their potential weaknesses.

1.  INTRODUCTION

There are various types of propagation observation depending on the
probing signal and receiving system available.  If one has a
coherent signal source, such as a spacecraft beacon or a radar echo,
then phase sensitive measurements can be made.  From these one can
determine the mean electron density and the large scale fluctuations
in density.  One can also measure the spectral broadening which
provides a good estimate of the small scale fluctuations.
Unfortunately spectral broadening is also dependent on anisotropy so
observations inside of 10 solar radii, where there may be large
anisotropy, should be interpreted with caution.  Of course phase
coherent observations have been restricted to the ecliptic because
the spacecraft and the radar targets are in the ecliptic.
    Incoherent probing signals can be used in several ways.  The
simplest is to measure the angular scattering with an
interferometer.  This provides very good information on the
fluctuations with scales of the order of the interferometer
baseline.  Interference between various components of the angular
spectrum causes intensity scintillations.  These form a spatial
diffraction pattern which drifts past the receiver(s) with the solar
wind velocity.  If several receiving antennas are available the
velocity can be measured.  In this case the turbulence is acting
solely as a tracer for velocity measurement.  One can also measure
the spectrum of the intensity variations which gives good

143

R. G. Marsden (ed.), The Sun and the Heliosphere in Three Dimensions, 143–151.

information about fluctuations on scales larger than the radius of
the first Fresnel zone.  For the purpose of comparison with ULYSSES
data and prediction of the ULYSSES environment, intensity
scintillations are the more appropriate technique because they
provide greatest sensitivity at distances of the order of 1 AU.

A common problem with indirect methods is the use of an
oversimplified model to interpret the observations.  For example the
early intensity scintillation observations were interpreted under
the assumption that the turbulence spectrum was gaussian.  It is now
known to be power law and the analysis has been revised in several
important respects.  An effect of this error was that rms electron
densities of a few % were estimated when, in fact, the rms density
is not a useful measure for a power law process.  For such a process
the rms depends on the averaging time; for a very long time it would
approach 100%.  A similar problem will be caused by the recent
discovery (Kojima et al. this report) that the turbulence near the
sun is highly anisotropic.  This will substantially change the
interpretation of most propagation observations taken inside about
10 solar radii.

A line of sight integration is inherent in all propagation
measurements and, indeed, in all remotely sensed observations.  This
is a major cause of bias or distortion in the resulting data.  When
intensity scintillations (IPS) are used to estimate the velocity the
following problem can occur.  The velocity estimate is,
approximately, a weighted average of the velocities perpendicular to
the line of sight.  However the weighting function is the "level of
turbulence", another variable.  Normally the observation is
dominated by the region of the line of sight which is closest to the
sun because the level of turbulence falls like the 4th power of the
distance.  However an intensely turbulent region can dominate the
observations even when it is not near the closest point of
approach.  If this region happens to have a relatively low velocity
it may obscure a high velocity stream which happens to be at the
closest point of approach.

## 2.  CALIBRATION

In view of the potential errors discussed above it is
particularly valuable to compare indirect with direct measurements,
where they are available.  Such a comparison can uncover such
biases, improve the models, reduce the necessary assumptions and, in
effect, calibrate the propagation methods.  An important side
benefit of the ULYSSES mission will be to extend such comparisons to
high latitudes.

A sketch of the line of sight showing the relative contribution
to intensity scintillation from different positions (assuming
spherical symmetry) is given in figure 1.  The pattern velocity
reflects only that component of the flow velocity perpendicular to
the line of sight.  If one assumes that the entire contribution
comes from the point P where the weighting function for spherical

symmetry peaks and that there are no temporal variations other than
simple rotation; then a comparison can be made between spacecraft
and IPS observations (of ecliptic sources). This comparison is
shown for two periods in figure 2. The first period in 1974 shows
quite good agreement in that most high speed streams appear in both
data sets. During this period the solar wind was very stable and
the corotation assumption was very good. The discrepancies have
been shown to be caused by domination of the weighting region by a
slow but turbulent stream (Coles et al., 1978). The second period
in 1979 showed general agreement but there are more discrepancies.
This is anticipated because, in addition to the problems of the 1974
period, there are frequent transients which do not map according to
a simple rotation.

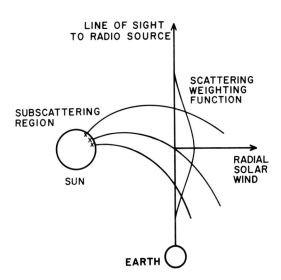

Figure 1.   Line of sight geometry for an IPS observation.

## 3.   SYNOPTIC MAPS

The foregoing discussion suggests that IPS observations are
well suited to study of the average behavior of the solar wind but
that interpretation on any given day may not be very reliable. As
multi-station IPS observatories are necessarily small they are
limited by sensitivity to a small number of sources. For example

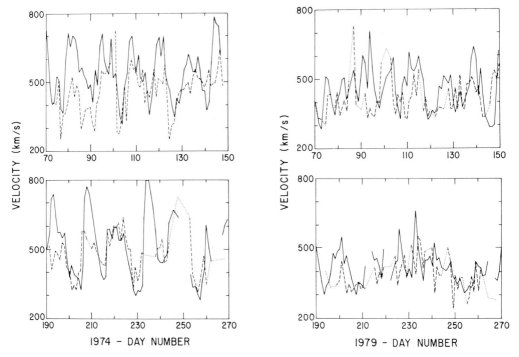

Figure 2.  Comparison of IPS derived velocity and spacecraft
observations mapped to the point of maximum scattering.  The
spacecraft data are drawn with a solid line and the IPS data are
connected with a dashed line.  When the IPS observations are
separated by more than one day the connection is shown dotted.

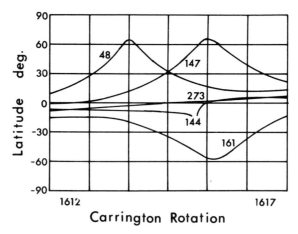

Figure 3.  Track of the "subscattering point" on the solar
surface.  Five strong sources used at UCSD are shown for six solar
rotations in 1974.  This track is repeated in the first half of
every calendar year.

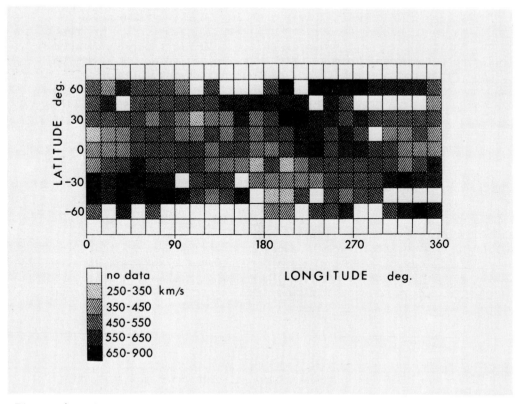

Figure 4.    Average solar wind speed mapped to the solar surface for the first six solar rotations of 1974.

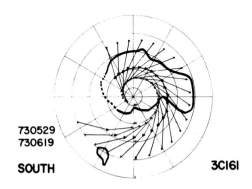

Figure 5.    Track of the "subscattering point" on the solar surface for one source in one solar rotation. The south pole is shown. The heavy line is an estimated coronal hole boundary. The extent of the half power weighting region on the IPS line of sight is indicated by the light lines.

the UCSD system regularly observes 20 sources of which 6 to 10 are
useful at any given time (whereas the Cambridge single antenna
system is large enough to observe several 1000 sources). The track
of the "scattering region" mapped to the solar surface is sketched
in figure 3 for those sources which were useful to the UCSD system
for the first half of the year. One can see that to obtain a
reasonable coverage of the sphere several rotations must be averaged
together. Such an average for 1974 is shown in figure 4 from Coles
et al. (1980). The solar wind was quite stable during this period
and a clear map was obtained.

     These maps are effectively smoothed by line of sight averaging
so one cannot determine the sharpness of the stream boundaries
directly. A good idea of the extent of the weighting region at
higher latitudes (assuming spherical symmetry) is given by figure 5
from Rickett et al. (1976). Here a polar view of the sun is given
with a sketch of a coronal hole and the IPS weighting regions for
each day projected back to the solar surface. The peak of the
weighting region and the two 50% points are linked by a light line
and there is one such line plotted for each day. Kakinuma et al.
(1982) have used model fitting to improve the resolution of such
maps. Their results show that the data are consistent with abrupt
stream boundaries.


4.   SOLAR CYCLE CHANGES

     There are two types of solar observation which are particularly
useful for comparison: white light coronagraph and surface
magnetograph data. The magnetograph data must be extrapolated
(using a potential field model) to some "source surface" in the
solar wind. Of particular interest here is the "neutral sheet"
which separates regions of opposite magnetic field polarity. The
white light data are primarily useful, in this comparison, for the
ability to outline the cool low-density regions of open magnetic
field geometry called "coronal holes". Averaged maps in the format
of figure 4 are shown for the IPS velocity, the HAO Mark II white
light, and the Stanford Magnetic Field in figure 6.

     The 1974 map is typical 1972-1975. A comparison of the IPS
velocity and the white light intensity established quite clearly
that high speed polar streams emanate from polar coronal holes (at
least during solar minimum). During this period the polar stream
and the corresponding hole were centered 20 or 30 degrees off the
rotation axis. The 1976 and 1977 maps are very similar; there is a
well defined polar stream and coronal hole, but they are centered on
the rotation axis. Here the magnetic field neutral line, shown
light in the rightmost frame, corresponds closely with a low-
velocity high-density region. The maps for 1978 and 1980 are
typical of 1978-1981. The equivalent comparison is much less
obvious but a careful study shows that both the polar streams and
the polar coronal hole have disappeared. Similar comparisons are
displayed by Rickett and Coles (1982) for the period 1972 to 1982.

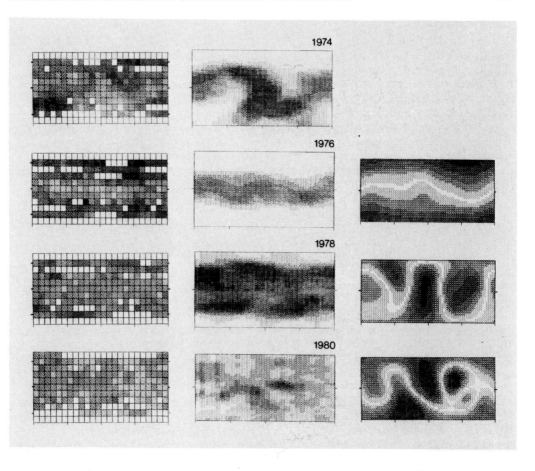

Figure 6. Synoptic maps of IPS velocity, polarized brightness, and magnetic field strength in the solar wind (left to right) in the same format as figure 4.

The magnetic field configuration suggests a different view of the evolution discussed above. One is led to consider the view that the neutral lines defines a region of low velocity and high density (and necessarily closed magnetic configuration). At solar minimum it follows a great circle tilted say 20 or 30 degrees from the rotational equator. This leaves high velocity regions of open magnetic field configuration over the poles. However at solar maximum the neutral line wrinkles up into a complex structure that reaches into the polar regions. This eliminates the polar stream (and hole) or at least reduces it below the detection threshold. In the maps for 1980 the regions of high complexity appear to have the lowest velocity, but the velocity is low everywhere.


## 5. CONCLUSIONS

ULYSSES will encounter Jupiter in July 1987 thus it will probably see a simple polar stream/hole structure only in the first year or so after the encounter. As ULYSSES passes over the poles solar activity should be high; the coronal magnetic field configuration is likely to be complex; and the overall solar wind velocity is likely to be low. However in making predictions based on IPS measurements it must be remembered that significant averaging occurs on the line of sight. It is possible, for example, that the core of a polar stream is much higher in speed than the 700 km/s that we observe and it is possible that narrow high speed streams exist even at solar maximum. It should also be remembered that the "calibration" against spacecraft data was done in the ecliptic. The polar solar wind must have quite different characteristics, for example, stream-stream interactions must be less important. It is not clear how this will affect IPS observations although the agreement should be improved if the polar solar wind is indeed more homogeneous.

At present multiple antenna IPS observations are made regularly by the Research Institute for Atmospherics of Nagoya University in Japan and by the University of California at San Diego. A new system is almost operational at the Physical Research Laboratory in Ahmedabad, India. Single antenna observations with much greater sensitivity are made irregularly by the Cavendish Laboratory of Cambridge University and by the Lebedev Institute at Puchino.


ACKNOWLEDGEMENTS

This work has been supported from the beginning by the Atmospheric Sciences Division of the U. S. National Science Foundation and since 1977 by the U. S. Airforce Office of Scientific Research. We thank all our colleagues at UCSD who have helped make the observations and analyze the data.

REFERENCES

Coles, W.A., Harmon, J.K., Lazarus, A.J. and Sullivan, J.D.: 1978,
    J. Geophys. Res. **83**, 3337.
Coles, W.A., Rickett, B.J., Rumsey, V.H., Turley, D.G.,
    Ananthakrishnan, S., Armstrong, J.W., Harmon, .K., Scott, S.L.
    and Sime, D.G.: 1980, Nature **286**, 239.
Kakinuma, T., Washimi, H., Kojima, M.: 1982, U. S. Airforce Report
    AFGL-TR-0398, "Proceedings of STIP Symposium on Solar Radio
    Astronomy, IPS and Coordination with Spacecraft", Shea, Smart,
    McLean and Nelson eds., pg. 153.
Rickett, B,.J., Sime, D.G., Sheeley, N.R.Jr., Crockett, W.R. and
    Tousey, R.: 1976, J. Geophys. Res. **81**, 3845.
Rickett, B.J. and Coles, W.A.: 1982, NASA Conference Pub. 2280,
    "Solar Wind Five", M. Neugebauer ed., pg. 315.

# EXPLORATION OF HELIOSPHERE BY INTERPLANETARY SCINTILLATION

R.V. Bhonsle, S.K. Alurkar, S.S. Degaonkar,
H.O. Vats and A.K. Sharma
Solar and Plasma Astrophysics Area
Physical Research Laboratory
Ahmedabad-380009, India

ABSTRACT. The Physical Research Laboratory has been
engaged in setting up a three station radio observatory
operating at a frequency of 103 MHz in Western India for
exploration of the heliosphere by interplanetary scintillation (IPS) technique. The three radio telescopes are
situated at Thaltej (Ahmedabad), Rajkot and Surat and are
separated by about 200 km from each other. These are being
operated as "transit" telescopes in correlation interferometer mode with excellent relative time accuracy for interplanetary scintillation observations of compact radio
sources.
    IPS observations of different scintillating radio
sources are made over a wide range of their solar elongations. Such IPS observations enable the determination of
not only the solar wind velocity at different heliolatitudes and other travelling interplanetary disturbances but
also the structure of heliosphere upto a distance of about
1.0 AU.

## 1. INTRODUCTION

Compact radio sources such as quasars or radio galaxies show
rapid fluctuations of intensity when they are observed by
sensitive radio telescopes. These fluctuations of intensity
or interplanetary scintillations (IPS) are caused as a result
of scattering of the radio waves by the plasma density irregularities in the interplanetary medium (Hewish et al 1964).
The power spectrum of intensity fluctuations gives information about the scale sizes of plasma density irregularities.
The scintillation index, which is the rms intensity variations over the mean source intensity, is a measure of the
turbulence in the solar plasma. Under weak scattering conditions, the pattern velocity measured at three IPS antennae
separated by about 100-200 km can be related to the solar

153

R. G. Marsden (ed.), The Sun and the Heliosphere in Three Dimensions, 153–158.
© 1986 by D. Reidel Publishing Company.

wind velocity (Dennison & Hewish, 1967).

Realizing the importance and versatility of the IPS technique, the Physical Research Laboratory decided to set up three radio telescopes at 103 MHz for IPS work separated by about 200 km from one another in Western India. This paper describes the radio telescope designed and developed for IPS work at 103 MHz located at Thaltej (72° 30'E; 23° 02'N) near Ahmedabad, the other two IPS telescopes situated at Rajkot (70° 44'E; 22° 17'N) and Surat (72° 47'E; 21° 09'N) being of similar design. At present, the telescopes at Thaltej and Rajkot are being operated on daily basis. The Surat telescope has been installed and its regular operation is expected to start from October 1985.


## 2. IPS RADIO TELESCOPE AT 103 MHz

The IPS radio telescope is operated as a correlation inter-ferometer at 103 MHz set up at Thaltej near Ahmedabad (Alurkar et al 1982).

The radio telescopes at Rajkot and Surat are of the same design except that the antenna apertures at these stations are 5000 $m^2$ as against 10,000 $m^2$ at Thaltej. The time synchronization at all the three stations is accompli-shed by adjusting local crystal clocks with ATA time signals transmitted from New Delhi, on 10 MHz. Post detector time constant 0.1 second enables intensity fluctuations upto 10 per second to be recorded. The ionospheric scintillat-ions are usually slower than 0.1 Hz, which are digitally filtered out with a low-frequency cut-off at 0.1 Hz before computing IPS spectra. The ionospheric scintillation at this latitude is not usually observed.

When a scintillating radio source is observed by a phase-switching interferometer, there appear three main components in the output. They are:
(a) A broad band system noise voltage covering a frequency range from zero to the higher cut-off determined by the receiver time constant.
(b) A slowly-varying voltage due to the passage of source through the E-W Antenna pattern of the telescope.
(c) A desirable rapidly changing voltage is due to the scintillations. It has a noise-like character and its frequency spectrum depends on the existing conditions in the solar wind, solar elongation, the apparent angular diameter, the receiver bandwidth etc.

Presently radio sky survey at 103 MHz is being carried out using radio telescope at Thaltej as a result of which few hundred radio sources, both scintillating and non-scintillating ones, have been observed. The minimum detect-able scintillating flux by this telescope is around 1 Jy rms.

## 3. PRELIMINARY OBSERVATIONS

Table I gives the list of scintillating sources which are observed for IPS work by means of radio telescope at Thaltej and Rajkot. These sources fall in the declination range of ± 30 of the zenith and are unambiguously identifiable. The

Table I. List of the strongly scintillating sources observed by Thaltej IPS telescope

| Source | Code | Beam | Declination (deg.) | Flux (Jy) |
|---|---|---|---|---|
| 3C   2 | 01 | 12L | +1 | 28 |
| 48 | 12 | 5B | +33 | 61 |
| 67 | 19 | 2R | +28 | 22 |
| 74 | 1D | 33L | +19 | 23 |
| 119 | 2C | 9R | +41 | 21 |
| 123 | 2E | 3R | +29 | 289 |
| 125 | 30 | 8R | +39 | 25 |
| 144 | 37 | 1L | +22 | 1623 |
| 147 | 38 | 13R | +50 | 52 |
| 161 | 3E | 14L | -06 | 76 |
| 186 | 46 | 7R | +37 | 29 |
| 196 | 4B | 12R | +48 | 115 |
| 237 | 64 | 8L | +08 | 36 |
| 254 | 6C | 9R | +41 | 48 |
| 263.1 | 6F | 1L | +22 | 29 |
| 265 | 71 | 4R | +31 | 32 |
| 267 | 73 | 5L | +12 | 25 |
| 270.1 | 77 | 6R | +34 | 27 |
| 273 | 79 | 11L | +2 | 142 |
| 274 | 7A | 6L | +12 | 1552 |
| 286 | 82 | 4R | +31 | 29 |
| 298 | 8C | 9L | +07 | 81 |
| 318 | 96 | 2L | +20 | 21 |
| 324 | 99 | 2L | +21 | 23 |
| 368 | AC | 7L | +11 | 32 |
| 380 | AE | 12R | +48 | 112 |
| 409 | B1 | 1R/1L | +23 | 144 |
| 454 | C0 | 3L | +18 | 20 |
| 459 | C4 | 10L | +04 | 44 |
| CTA 21 | CC | 4L | +16 | 12 |

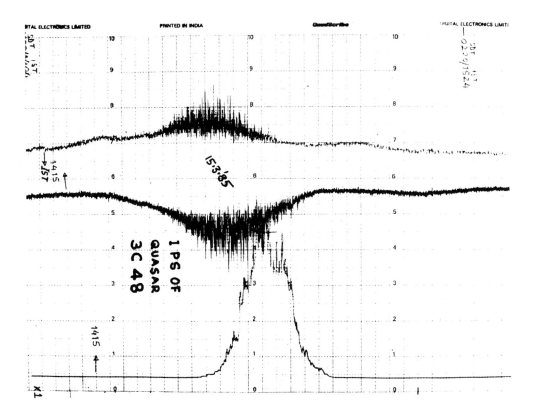

Figure 1.   A sample IPS recording of 3C 48   made on
15-3-1985 at Thaltej.   From top to bottom, Cosine, Sine
and Scintillometer outputs are shown.


radio telescope at Surat has been installed but will be
commissioned for regular operation from October 1985.
Figure 1 shows a sample IPS recording of radio source
3C 48 as recorded on Sine, Cosine and scintillometer
channels.   On this day the scintillation index of 3C 48
was 0.35 for the solar elongation of $44°.9$ heliolongitude
and heliolatitude of the point of the closest approach of
the line of sight to the source from the sun being $40°.6$
and $21°.4$ respectively.   It has been found from our IPS
observations that  m  is a highly variable parameter on
day to day basis.   But on long term basis, it does vary
significantly as a function of solar elongation.   It is
possible to plot  m  against solar elongation for all the

sources listed in Table I.  Since the solar elongation,
heliolatitude and heliolongitude of the points of closest
approach of the line of sight from the sun varies slowly
from day to day, 'm' and its daily variation can be related
to fluctuations in electron density of heliosphere within
the radial distance from 0.5 to 1.0 AU applying the theory
of weak scattering.

It can be shown that, far from the thin diffraction
screen in the Fraunhofer region (i.e. $Z \gg a^2/\lambda$) where
a = scale length of irregularities, Z = distance of the
screen from the earth and $\lambda$ = wavelength.

$$m = \sqrt{2}\,\phi_0 = \sqrt{2}\,\pi^{\frac{1}{4}}\, r_e \lambda \langle N^2 \rangle^{\frac{1}{2}} \, (aL)^{\frac{1}{2}} \quad \text{where}$$

$\phi_0$ = phase deviation, and the scale of the diffraction
pattern is the same as that of irregularities in the IPM.
Hence the observation of the scintillation index permits
an estimate of $\langle \Delta N^2 \rangle^{\frac{1}{2}}$ to be made, if a and L can be
determined.  It has been shown that, in practice, for a
radial law $\Delta N \propto r^{-2.5}$ the expression

$$\phi_0^2(p) = \pi^{\frac{1}{2}}\, r_e^2 \langle \Delta N^2 \rangle (p)\, a(p)\, p/\nu^2$$

is adequate to determine $\langle \Delta N^2 \rangle^{\frac{1}{2}}$ from measurements of $\phi_0$,
where  p  is distance of the point of closest approach of
a line of sight to the sun [Readhead (1971) and Little
(1976)] .

4.  CONCLUSION

The IPS technique at meter wavelengths constitutes a powerful method for monitoring plasma turbulence in the interplanetary medium from about 0.5 AU to 1.0 AU, which is a
small fraction of the heliosphere.  Still it is important
to be able to measure scintillation index reliably which
can be related to $\langle \Delta N \rangle^{\frac{1}{2}}$.  Furthermore, it is possible
to choose a grid of scintillating radio sources around the
sun in order to keep watch on the scintillation activity in
three dimensions and its evolution on day to day basis.  It
has been demonstrated by Russian and Cambridge radio astronomers that the clouds of excess plasma turbulence and their
daily movements across the sky can be followed and thus keep
track of "interplanetary weather".  There will be an excellent opportunity for the next few years when the solar activity is low, to study the interplanetary plasma processes

associated with polar coronal holes, high speed coronal
streams etc. simultaneously with the International Solar
Polar Mission "Ulysses".


ACKNOWLEDGEMENT

We thank Prof. S.P. Pandya, Director, Physical Research
Laboratory, for his keen interest in this work.  We also
thank our colleagues in the Radio Astronomy group for
their help in design, development and operation of IPS
telescopes.  The IPS project was financed by the Department of Science and Technology and Department of Space,
Government of India.


REFERENCES

Alurkar, S.K., Bhonsle, R.V., Sharma, R., Bobra, A.D.,
Sohan Lal, Nirman, N.S., Venat, P. and Sethia, G. 1982,
J. Instn. Electronic & Telecom. Engrs. 28, 577.
Dennison, P.A. & Hewish, A. 1967, Nature, 213, 343.
Hewish, A., Scott, P.F & Wills, D. 1964, Nature, 203, 1214.
Little, L.T. 1976, Methods of Experimental Physics, Ed. Meeks,
M.L., Academic Press, 12, 118.
Readhead, A.C.S. 1971, Mon. Not. Roy. Astron. Soc. 155, 185.

EVOLUTION OF TURBULENCE AND WAVES IN THE SOLAR WIND IN RADIUS AND
LATITUDE

Aaron Barnes
Theoretical Studies Branch
NASA-Ames Research Center
Moffett Field, CA, USA

ABSTRACT.  Fluctuations of hydromagnetic scale ($\gtrsim$ the ion gyroradius)
have long been studied in interplanetary plasma and magnetic field data
near the orbit of Earth.  These fluctuations have been interpreted
variously as hydromagnetic waves and turbulence, and potentially are a
key to advancing our understanding of nonlinear phenomena in cosmic and
laboratory plasmas.  Moreover, it is likely that hydromagnetic waves
and/or turbulence are a major factor in the acceleration of the solar
wind.  Observations near 1 AU have provided considerable insight, but
there are a number of unresolved fundamental issues.  We review what is
known about the evolution of the hydromagnetic fluctuations with
heliocentric distance, and discuss how these observations bear on the
questions of the nature and origin of the fluctuations.  With the
exception of limited and tentative indirect information related to the
outer corona, observations to date have been limited to near the
ecliptic plane.  One of many important advances to be expected from the
Ulysses mission is determination of the variation, if any, of intensity
and nature of the interplanetary fluctuations with heliographic
latitude, which may further elucidate the character of the fluctuations
and their role in accelerating the solar wind.

INTRODUCTION

Large-amplitude plasma and magnetic fluctuations on the hydromagnetic
scale are found throughout the interplanetary medium at all times.  Most
theoretical studies of these fluctuations have been based on the theory
of hydromagnetic waves, but much of the recent work has adopted the
viewpoint that it may be more appropriate to regard the hydromagnetic
fluctuations as a turbulent phenomenon.  The character of the
fluctuations varies from place to place in the solar wind.  Near stream
interaction regions they are very complex and difficult to interpret.
The situation seems to be simpler in stream rarefaction regions and at
the trailing edges of high-speed streams (Belcher and Davis, 1971),
where the fluctuations are characterized by roughly constant density and
magnetic field strength, but with large variations in the directions of

159

*R. G. Marsden (ed.), The Sun and the Heliosphere in Three Dimensions, 159–166.*
© *1986 by D. Reidel Publishing Company.*

the magnetic field and plasma velocity.  The global variation of the
character of these "Alfvenic" fluctuations must be accounted for by an
adequate theory of interplanetary fluctuations; hence the variation of
the Alfvenic fluctuations with heliocentric distance and heliographic
latitude has a direct bearing on the resolution of the waves vs.
turbulence issue.  Direct in situ observations so far have been limited
to near the ecliptic plane.  In this paper we review the current state
of knowledge about the variations of the Alfvenic fluctuations with
heliocentric distance, and discuss these results in the context of the
waves-turbulence dialogue.  In addition, we discuss in a tentative way
the variations in heliographic latitude that might be expected on
theoretical grounds.

## ALFVENIC FLUCTUATIONS: WAVE LANGUAGE

The simplest idealization of interplanetary microfluctuations is that
they may be described as a field of Alfven waves, i.e., fluctuations in
magnetic field direction, but not field strength, at constant pressure
and density.  The associated fluctuation in flow velocity, $\Delta V$, is
related to the magnetic fluctuation, $\Delta B$, by

$$\pm \Delta V = -[(\Delta B)/\{4\pi\rho\}^{1/2}]\Phi. \tag{1}$$

Here $\rho$ is the local mass density of the plasma, and
$\Phi = \sqrt{\{1+4\pi(P_t - P_b)/B^2\}}$, where B is the magnetic field strength, and the
P's refer, respectively, to the thermal pressures transverse and
parallel to the magnetic field.  The choice of sign in eq. (1) is
determined by the  propagation direction in the plasma rest frame, the
upper sign corresponding to propagation more nearly parallel than
antiparallel to the mean magnetic field direction.  The wave field is
not restricted to plane waves (e.g., Goldstein et al., 1974; Barnes,
1976), but may be tangled and quite disordered.

     In this simplest model one considers the large-scale variation of
the wave field to be describable by the WKB approximation (Weinberg,
1962; Bazer and Hurley, 1963).  In this approximation the fluctuations
are assumed to have a length scale that is short compared to the
macroscopic scale of the plasma.  The WKB theory provides a means of
computing the wave amplitude throughout the system, which, at least for
Alfven waves, is not limited to small wave amplitude (Barnes and
Hollweg, 1974) and does not require the waves to be locally planar
(Hollweg, 1974).  Idealizing the solar wind as a steady, azimuthally
symmetric flow in a reference frame corotating with the sun, the WKB
theory gives the following expression for the variation of the amplitude
of an outward-propagating Alfven wave as it propagates along a magnetic
field line (Barnes, 1979, eq. 6.21):

$$\langle \Delta B^2 \rangle = \text{const. } \rho^{1/2} \Phi^{-3} / \{M_A + 1\}^2 \tag{2}$$

Here $M_A = (4\pi\rho)^{1/2} \Phi^{-1} \{V_{sw}^2 + r^2 \Omega^2 \sin^2\theta\}^{1/2}/B$, and $V_{sw}$, r, $\Omega$, and $\theta$ are,
respectively, the radial solar wind flow speed, heliocentric distance,

angular velocity of the coronal base, and colatitude.  At the coronal
base of the field line $\Phi \approx 1$ and $M_A \ll 1$, so that eq. (2) can be written in
the form

$$\langle _\Delta B^2 \rangle = (4\pi)^{3/2} \{F_0 / B_0\} \rho^{1/2} \Phi^{-3} / \{M_A + 1\}^2 \tag{3}$$

where $B_0$ is the magnetic field strength at the foot of the field line,
and $F_0 (=\{\langle _\Delta B^2 \rangle / 4\pi\} B_0 / (4\pi\rho_0)^{1/2}$ is the Alfven-wave energy flux at the
base.  Thus, eq. (3) states that the variance of Alfvenic fluctuations
is proportional to $F_0 / B_0$ times a function of measurable quantities.
Along a given corotating field line $M_A$ varies inversely as $\Phi\sqrt{\rho}$, so that
when $M_A$ is very large

$$\langle _\Delta B^2 \rangle \sim \rho^{3/2} / \Phi. \qquad (M_A^2 \gg 1) \tag{4}$$

Because $\Phi \sim 0(1)$, large gradients in $\langle _\Delta B^2 \rangle$ reflect either large gradients
in $\rho$ or large gradients in source conditions (cf. eq. (3)).  In
particular, this WKB model gives $\langle _\Delta B^2 \rangle \sim r^{-3}$ at large heliocentric
distances (and fixed latitude) and any significant latitude gradient in
$\langle _\Delta B^2 \rangle$ represents a gradient in coronal conditions, specifically, in
$F_0 / B_0$.

     A recent study of Pioneer 10-11 magnetic data by Bavassano and
Smith (1985) finds that the power in Alfvenic fluctuations declines as
$r^{-3.5}$ between 1 and 5 AU, somewhat more rapidly than the $r^{-3}$ of the
simplest WKB model.  However, these authors note that the WKB
calculation for an expanding high-speed stream (Hollweg, 1975) predicts
a somewhat steeper decrease than $r^{-3}$; this is so because of the greater
density gradient in a stream rarefaction region (cf. eq. (4)).  It
should also be noted that if $\Phi$ increases with r, $\langle _\Delta B^2 \rangle$ must increase
faster than $r^{-3}$.  Altogether, the Bavassano and Smith study indicates
consistency of the Pioneer observations with the WKB model over the
range 1-5 AU.  Most earlier studies, over a narrower range of
heliocentric distance, are in agreement with this conclusion (a summary
of most of the earlier work is given in the review of Barnes, 1979, sec.
6).

     The situation at distances inside 1 AU is more complicated,
however.  Studies based on Helios 1-2 data (Denskat and Neubauer, 1982;
Bavassano et al, 1982) over the range 0.3-1.0 AU indicate a steeper
variation of power, especially at (spacecraft-frame) frequencies above
$\sim 0.01$ Hz.  In particular, Bavassano et al find that at low frequencies
$(2.8 \times 10^{-4} - 2.5 \times 10^{-3}$ Hz) the average power density varies with distance
in proportion to $r^{-3.2}$, roughly consistent with WKB, while at the higher
frequencies the variation is $r^{-4.2}$, indicating substantial dissipation
for wavelengths larger than a few times $10^4$ km.

     There are other theoretical and observational reasons to question
the relevance of the WKB theory to interplanetary fluctuations (e.g.,
see Barnes, 1979, secs. 6 and 8).  In particular, near 1 AU one
typically finds $_\Delta B \sim 0(B)$, suggesting that some nonlinear decay process
may limit the amplitudes.  Hollweg (1973) argued that such nonlinear
decay would result in plasma heating large enough to be important in
solar wind acceleration.  Hollweg modeled this heating by the assumption

that the wave amplitude saturates at $_\Delta B \gtrsim B/\sqrt{2}$.  In regions where such saturation controls the wave amplitude, we would anticipate that $\langle _\Delta B^2 \rangle$ is strongly correlated with $B^2$, but not with $\rho$ as eq. (4) indicates. The saturation model would then give $\langle _\Delta B^2 \rangle$ proportional to $r^{-4}$ in the inner heliosphere and $r^{-2}$ in the outer heliosphere.  The near-equatorial observational studies mentioned above would seem to favor the WKB model, except for the shorter wavelength Alfvenic fluctuations well inside 1 AU.  Now consider latitudinal variations in the saturation model.  At a given heliocentric distance r, the ratio of the power on an equatorial field line to that on a polar field line would be

$$\langle _\Delta B^2 \rangle_e / \langle _\Delta B^2 \rangle_p = (B_{0e}/B_{0p})^2 \{1+(r\Omega/V)^2\} \tag{5}$$

At r=2 AU this ratio would be about 5 $(B_{0e}/B_{0p})^2$.


## ALFVENIC FLUCTUATIONS: TURBULENCE LANGUAGE

The WKB model of Alfvenic fluctuations discussed above appears to be consistent with observed variations of fluctuation amplitude in heliocentric distance.  We reiterate that this theory (Eq. 3) does not require the assumption of plane waves.  In fact, there is a long-standing debate about whether or to what degree the observed interplanetary fluctuations can be described as plane waves (Denskat and Burlaga, 1977; Barnes, 1979, 1981; Dobrowolny et al, 1980a,b; Herbert et al, 1984).  In particular, the WKB theory predicts that plane wave normals should be directed radially outward.  But the wave normals should then coincide with the direction of minimum magnetic variance (a direct consequence of the solenoidal character of $\underline{B}$), whereas in fact the observed minimum variance directions are strongly aligned with the local mean field $\langle \underline{B} \rangle$  (Daily, 1973, and many subsequent corroborations).
        Moreover, interplanetary magnetic power spectra, and in particular the Alfvenic regions, typically show a power-law dependence on $k_r$, generally consistent with $k_r^{-5/3}$, extending over two or more decades in $k_r$.  This power-law behavior would be consistent with isotropic fluctuations with power spectrum proportional to $k^{-5/3}$ associated with the inertial (Kolmogorov-Obukhov) range of the energy cascade familiar from the theory of fluid turbulence (e.g., see Batchelor, 1960).
        These turbulence-like power spectra, together with the inconsistency of the observed minimum-variance directions and the WKB refraction theory, indicate that a wave-language description may be misleading, and that a description using the language of turbulence theory may be more appropriate.  However, the application of the turbulence concept to the interplanetary Alfvenic fluctuations raises new questions.  For example, even though the observed minimum-variance directions are inconsistent with the WKB theory, there usually is a well-defined minimum variance direction, a result not generally expected for a turbulent system.  However, the alignment of the minimum-variance direction and $\langle \underline{B} \rangle$ is a straightforward consequence of stochastic variations in a field of constant magnitude (Barnes, 1981); in other words, the stochastic nature of the fluctuations in combination with the

constancy of field strength may account for the observed orientations of
the minimum-variance directions.

A second and more basic question raised in the application of the
turbulence concept to interplanetary fluctuations has to do with the
fact that in fluid turbulence the transfer of energy from longer to
shorter scale eddies is associated with the Reynolds stress $\rho \langle \Delta V \Delta V \rangle$.
In MHD turbulence the corresponding energy transfer is due to the sum of
Reynolds and Maxwell stresses $\rho \langle \Delta V \Delta V \rangle - \langle \Delta B \Delta B / 8\pi \rangle$. It is readily
verified from eq. (1) that this sum vanishes*, so that for pure Alfvenic
fluctuations there can be no energy cascade (Dobrowolny et al, 1980a,b;
Barnes, 1981).Accounting for the observed power law spectrum is thus a
fundamental problem that must be faced by any theory that treating
Alfvenic fluctuations as a turbulent phenomenon.

A separate issue is the consistent sign of the dynamical alignment
of $\Delta V$ and $\Delta B$ (cf. eq. (1)), which corresponds to propagation of the
fluctuations in the antisunward direction (as viewed in the plasma frame
[Coleman, 1967; Belcher and Davis, 1971; and a number of subsequent
authors]). This 'dynamical alignment' has been taken as very strong
evidence that the fluctuations are of solar origin (more precisely,
produced in a region where the solar wind flow speed is smaller than the
Alfven speed, probably at $r \lesssim 20 \ R_0$). A theory that treats the Alfvenic
fluctuations as a more local phenomenon must account not only for the
development of the dynamical alignment, but also for the consistency of
its sign.

Theoretical studies of hydromagnetic turbulence are at an early
stage of development, and essentially all computational studies of the
topic have confined themselves to turbulence in an incompressible
medium. Therefore, when applying results of this work to the Alfvenic
fluctuations, it is important to keep in mind that the theory prohibits
variations in density. On the other hand, the Alfvenic fluctuation (eq.
(1) et seq.) is an exact solution of the compressive MHD equations: this
exact solution happens to give constant density, but the constancy of
density in no way prohibits other compressive solutions to these same
MHD equations. In particular, the fact that the interplanetary Alfvenic
fluctuations exhibit only small density variations does not by itself
justify the direct application of theoretical work based on the
assumption that the medium is incompressible. In fact, substantial
fluctuations of density do occur outside the Alfvenic regions. Hence,
the application of results of incompressible turbulence theory to the
Alfvenic fluctuation problem will always raise the question of the
consequences of the neglect of compressibility.

Altogether, there are four fundamental issues that must be
answered if the language of turbulence is to be the key to our
understanding of the Alfvenic fluctuations. (1) The constancy of $(\Delta B)^2$
must be explained. (2) The formation of the power-law form of the

------------------------------------------------------------------

*If there is thermal anisotropy ($\Phi \neq 0$) the resulting nonzero contribution
from the pressure stress tensor must be included, but the sum still
vanishes.

Alfvenic power spectrum must be explained.  (3)  The evolution of the
dynamical alignment (eq. (1)) and the consistency of its sign with
outward Alfvenic alignment must be explained.  (4)  If a theory based on
imcompressible MHD is used, a convincing argument must be given that a
compressible treatment would give at least qualitatively similar
results.

     Having stated the strengths and weaknesses of the wave approach,
and having given caveats about the challenges faced by a turbulence-
theory approach, let us now consider some of the attempts that have been
made to apply the language of turbulence to the interplanetary
fluctuations.  Remembering that pure Alfvenic fluctuations admit no
energy cascade, consider the possibility that the primary turbulent
region is fairly near the sun (say inside $r=20$ $R_0$).  Indeed, radio
scintillation observations suggest that local velocity variations are
much larger in this part of the solar wind than elsewhere (Ekers and
Little, 1971; Scott, 1978; Armstrong and Woo, 1981).  If this turbulence
develops fully and establishes an inertial spectral range, a $-5/3$ power
law could be characteristic of velocity and magnetic fluctuations.  If
the turbulence decays leaving only "fossil" Alfvenic fluctuations, the
energy cascade would subside, but the $-5/3$ power law might remain.  As
long as the decay takes place in the region of sub-Alfvenic flow, only
outward-propagating fluctuations could pass into the super-Alfvenic
region.  Such fossil Alfvenic fluctuations would automatically satisfy
criteria (1) to (4) given above.

     However, this scenario begs the question of why the near-solar
turbulence region should decay into Alfvenic fluctuations.  One possible
resolution is that the near-solar turbulence could involve large
compression in magnetic field strength.  Compressive fluctuations are
subject to strong resonant interaction with thermal ions or electrons,
which may cause strong dissipation (Barnes, 1966, 1967); being
noncompressive, Alfvenic fluctuations are not damped in this way.
Dobrowolny et al (1980a,b) have suggested an alternative way to produce
the Alfvenic fluctuations, arguing that (at least in an incompressible
medium) an initial spectrum consisting of Alfvenic fluctuations of mixed
alignment (in the sense of eq.(1)) and their nonlinear interactions will
decay nonlinearly to a state dominated by one or the other sign for
alignment.  If all this happens inside the region of sub-Alfvenic flow,
only the outward-propagating fossil fluctuations would be seen in the
interplanetary medium.  Significant developments in observation and/or
theory will be required to determine whether either (or both) of these
ideas is a reasonable description of the true situation.

     If the Alfvenic fluctuations are not of near-solar origin, then
they must be generated dynamically in the flow of the interplanetary
medium.  The solar wind consists of streams of various speeds that
interact with each other, and the interaction regions tend to be
extremely turbulent and exhibit compressive fluctuations.  If one views
a large region of the solar wind as a turbulent system, perhaps the
"stirring of the system" is most apparent in the stream interaction
regions, with Alfvenic fluctuations as the product that naturally
appears in the portions of the streams farthest removed from stream
interfaces.  Again, the power-law spectrum and the existence of dynamic

alignment might be understood as the end product of the decay of
compressive fluctuations or of nonlinear interaction of oppositely-
aligned Alfvenic components.  However, this scenario still leaves the
dilemma that the alignment is found to correspond to outward
propagation.

A possible resolution of this last difficulty was recently
suggested by Matthaeus et al (1983).  This argument was based on
numerical simulations using the incompressible MHD equations, focussing
on the evolution of the cross-helicity, defined as

$$H_c = \int \underline{V} \cdot \underline{B} \; d^3 x \qquad\qquad\qquad (6)$$

where the integration is taken over all space.  Clearly the consistency
of the dynamical alignment of the Alfvenic fluctuations can be thought
of in terms of the cross-helicity in the Alfvenic portion of the
spectrum.  The calculations of Matthaeus et al show that, under some
conditions, a system whose largest-scale eddies have large (i.e., nearly
dynamically aligned) cross helicity initially generates cross-helicity
of the opposite sign at shorter wavelengths.  They argue that the
largest scale cross-helicity in the solar wind always has sign opposite
to outward Alfvenic propagation, which might account for the outward
propagation of shorter-wavelength Alfvenic component.  Their
calculations further indicate that after a long enough time, the
spectrum evolves to the point where there should be a roughly equal
mixture of positive and negative cross-helicity at the shorter
wavelengths.  Applied to the solar wind, this scenario suggests that at
r=5-10 AU a noticeable component of inward-propagating Alfvenic
fluctuations might be expected.  Matthaeus and Goldstein (1982) have in
fact reported one example of mixed sign for cross helicity in the
inertial range near 5 AU.

Matthaeus and Goldstein (1982) also developed a method for recovering
the spectrum of another of the rugged invariants of MHD turbulence
theory, the magnetic helicity

$$H_m = \int \underline{A} \cdot \underline{B} \; d^3 x \qquad\qquad\qquad (7)$$

where $\underline{A}$ is the magnetic vector potential.  The conservation of this
quantity holds for any choice of gauge, but the Coulomb gauge ($\nabla \cdot \underline{A}=0$) is
convenient for spectral studies.  The sign of $H_m$ indicates the
"handedness" of magnetic structures ($H_m>0$ indicates right-handedness).
Matthaeus and Goldstein found that the magnetic helicity spectrum
alternates sign over the whole spectral range, indicating that
right- and left-handed structures are equally likely.  They also found
that the net magnetic helicity normally lies at scale lengths much
larger than the magnetic scale length.  The latter result is consistent
with the idea of an inverse cascade of magnetic helicity (Frisch et al,
1975) or freely decaying selective decay (Matthaeus and Montgomery,
1980).

It is fair to say that the attempts to apply turbulence language to
interplanetary fluctuations have produced some intriguing results, but

the results to date are not conclusive.  The theory is still in a rather
primitive state, and the relevant observations must be made throughout
the entire heliosphere.  Further work in this field is needed and should
be encouraged.


## REFERENCES

Armstrong, J.W. and Woo, R.: 1981, Astron. Astrophys., 103, 415.
Barnes, A.: 1966, Phys. Fluids. 9, 1483.
Barnes, A.: 1967, Phys. Fluids 10, 2427.
Barnes, A.: 1976, J. Geophys. Res., 81, 281.
Barnes, A.: 1979, in Solar System Plasma Physics, vol. 1, edited by E.N.
Parker, C.F. Kennel, and L.J. Lanzerotti, pp. 249-319, North Holland,
Amsterdam, 1979.
Barnes, A.: 1981, J. Geophys. Res., 86, 7498.
Barnes, A., and Hollweg, J.V.: 1974, J. Geophys. Res., 79, 2302.
Batchelor, G.K.: 1960, The Theory of Homogeneous Turbulence, Cambridge.
Bavassano, B. and Smith, E.J.: 1985, EOS Trans. Am. Geophys. Un., 66,
334.
Bavassano, B., Dobrowolny, M., Mariani, F. and Ness, N.F.: 1982,
J. Geophys. Res., 87, 3617.
Bazer, J. and Hurley, J.: 1963, J. Geophys. Res., 68, 147.
Belcher, J.W. and Davis, L., Jr.: 1971, J. Geophys. Res., 76, 3534.
Coleman, P.J., Jr.: 1967, Planet. Space Sci., 15, 953.
Daily, W.D.: 1973, J. Geophys. Res., 78, 2043.
Denskat, K.U. and Burlaga, L.F.: 1977, J. Geophys. Res., 82, 2693.
Denskat, K.U. and Neubauer, F.M.: 1982, J. Geophys. Res., 87, 2215.
Dobrowolny, M., Mangeney, A. and Veltri, P.: 1980a, Phys. Rev. Lett.,
45, 144.
Dobrowolny, M., Mangeney, A. and Veltri, P.: 1980b, Astron. Astrophys.,
83, 26.
Ekers, R.D. and Little, L.T.: 1971, Astron. Astrophys., 10, 310.
Frisch, U., Pouquet, A., Leorat, J. and Mazure, A.: 1975, J. Fluid
Mech., 68, 769.
Goldstein, M.L., Klimas, A.J. and Barish, F.D.: 1974, in Solar Wind
Three, edited by C.T. Russell, pp. 385-387, University of California
Press, Los Angeles.
Herbert, F., Smith, L.D. and Sonett, C.P.: 1984, Geophys. Res. Lett.,
11, 492.
Hollweg, J.V.: 1973, Astrophys. J., 181, 547.
Hollweg, J.V.: 1974, J. Geophys. Res., 79, 1539.
Hollweg, J.V.: 1975, J. Geophys. Res., 80, 908.
Matthaeus, W.H. and Montgomery, D.C.: 1980, Ann. N.Y. Acad. Sci., 357,
203.
Matthaeus, W.H. and Goldstein, M.L.: 1982, J. Geophys. Res., 87, 6011.
Matthaeus, W.H., Goldstein, M.L. and Montgomery, D.C.: 1983, Phys. Rev.
Lett., 51, 1484.
Scott, S.L.: 1978, Ph.D. Thesis, University of California, San Diego.
Weinberg, S.: 1962, Phys. Rev., 126, 1899.

# A COMPUTER SIMULATION STUDY OF THE MICROSCOPIC STRUCTURE OF A TYPICAL CURRENT SHEET IN THE SOLAR WIND

M. Roth
Institute for Space Aeronomy
3, Avenue Circulaire
B-1180 Brussels
Belgium

## 1. INTRODUCTION

Kinetic theories of tangential discontinuities (TD) in space plasmas have been discussed by a number of authors (e.g., Alpers, 1969; Roth, 1978; 1979; 1980; and 1983; Lee and Kan, 1979). These theories consider unidimensional plane current layers and the determination of their microscopic structure is based on both Vlasov and Maxwell's equations for plasma and fields. It is outside the scope of this paper to give a detailed account of these theories, but the interested reader can refer to Roth (1980, 1983) for the theoretical aspect sustaining the numerical results given in this note.

In section 2 we will give numerical results for the internal structure of a typical current sheet in the solar wind. Finally, the time resolution of plasma measurements across such a layer will be discussed in section 3 in the frame of a Interdisciplinary Study of Directional Discontinuities in the Solar Wind with the Ulysses Mission (Lemaire et al., 1983).

## 2. THE INTERNAL STRUCTURE OF A TYPICAL CURRENT SHEET

In a cartesian coordinate system, the plane of a tangential discontinuity is parallel to the (Y,Z) plane and all the variables are assumed to depend on the X coordinate, normal to the discontinuity.

Plasma boundary conditions for a typical current sheet in the solar wind are given in table I. On the left hand side of the transition, or side 1, i.e. for large negative values of X, there is a "hot" plasma of hydrogen and helium, while on the right hand side or side 2, i.e. for large positive values of X, there is a "cold" plasma of hydrogen and helium. Notice that on side 1, each plasma component has a distinct mean velocity. Note also that the "hot" plasma species have a vanishing number density on side 2 while the "cold" ones have a vanishing number density on side 1. The transition itself will be a region where these two plasmas of different characteristics are interpenetrated. The plasma in the transition will therefore

167

R. G. Marsden (ed.), The Sun and the Heliosphere in Three Dimensions, 167–171.

| | $N_1$ | $N_2$ | $T_1$ | $T_2$ | $V_{y_1}$ | $V_{y_2}$ | $V_{z_1}$ | $V_{z_2}$ |
|---|---|---|---|---|---|---|---|---|
| e | 5 | 3 | 15 | 10 | 219 | 175 | -311 | -285 |
| $H^+$ | 4.5 | 2.7 | 6 | 4 | 223 | 175 | -315 | -285 |
| $H_e^{++}$ | 0.25 | 0.15 | 16 | 12 | 180 | 175 | -272 | -285 |
| | $cm^{-3}$ | | eV | | km/s | | km/s | |

Table I. The indices 1 and 2 refer respectively to sides 1 and 2, i.e., to large negative and large positive values of X.

be considered as a three-components plasma ($e^-$, $H^+$ and $He^{++}$, an admixture of the two adjacent plasmas. The magnetic field on side 1 is assumed to be uniform ($B_z = - B_y = 3.5$ nT). Its intensity on side 2 is, of course, predetermined from the pressure balance condition. On each side of the transition, the plasma boundary values satisfy the charge neutrality and the zero current density conditions.
    The first three panels of figure 1 illustrate the number densities. The second panel displays the total number density of electrons ($e^-$), protons ($H^+$) and helium particles ($He^{++}$), i.e. the sum of the number densities of the cold and hot particles, illustrated on the first and third panels. Also in figure 1 are displayed the temperatures: 5th panel for the "hot" plasma species, 7th panel for the "cold" plasma species and 6th panel for the average temperatures of the admixture. Notice that the cold and hot electron temperatures do not change throughout the transition. This is a consequence of a particular choice for the parameters of the electron velocity distribution functions. Indeed, for the two electron species, it has been assumed that the velocity distribution functions remain isotropic about their mean velocity. Computation of the temperature (and also of moments of non-zero order) becomes meaningless, of course, when the corresponding number density becomes vanishingly small, as shown in the 5th and the 7th panels. Panels 4 and 8 of figure 1 illustrate, respectively, the y and z components of the energy flow in the admixture. The thickness of the transition in unit of the proton gyroradius, $R(H^+)$, can be deduced from the scale shown in the upper part of the panels. It can be seen that the variation in the $H^+$ number density occurs in about 20 $R(H^+)$, i.e., in about 800 km.
    The 12 panels of figure 2 illustrate the electric potential (1st panel), the electric field (5th panel), the relative charge density (9th panel), the current density components (2nd panel), the magnetic field components (6th panel) and hodogram (10th panel), the contribution of each plasma species to the current density (3rd, 4th, 7th and 8th panels) and to the plasma kinetic pressure (11th panel) and finally (12th panel) the pressure balance condition.
    Figure 3 shows how each plasma species contributes to the bulk velocity of the plasma. Panels 1-2-3, 5-6-7 and 9-10-11 illustrate

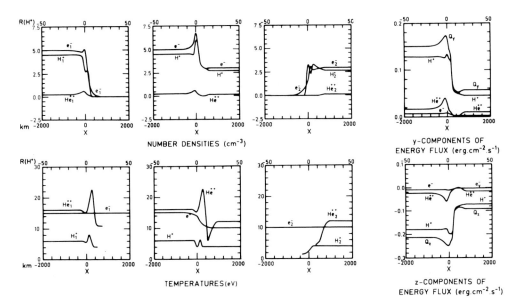

Figure 1. Plasma characteristics across a TD whose boundary conditions are given in table I.

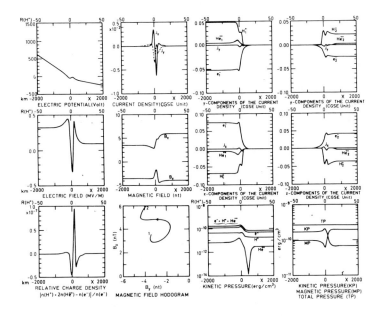

Figure 2. Fields and current characteristics across a TD whose boundary conditions are given in table I.

respectively the mean velocity of electron, proton and helium species.
The hodograms illustrated in panels 8 and 12 show that the protons
carry most of the bulk velocity of the plasma, as can also be seen
from panels 4 and 6.

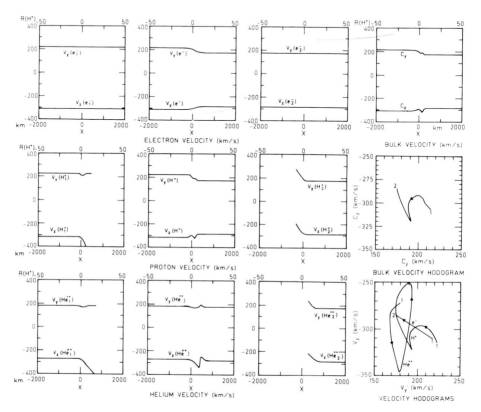

Figure 3. Flow characteristics across a TD whose boundary conditions
are given in table I.

## 3. CONCLUSIONS

The results shown in this paper indicate how to model a tangential
discontinuity in the solar wind. This modeling of current sheets
is one aspect of the Interdisciplinary Study of Directional Discontinuities in the Solar Wind with the Ulysses Mission (Lemaire et al.,
1983). One of the objective of this Interdisciplinary Study is a
detailed comparison of theoretical calculations with magnetic-field
and particle-flux measurements. As shown in this paper, the theoretical
thickness for current sheets in the solar wind is expected to be of
the order of 20 $R(H^+)$ or less. The time resolution of present-day
magnetic-field measurements in space is usually high enough to de-

termine the fine structure of such sharp magnetic-field "discontinuities", but the best time resolution for direct plasma measurements
is generally much lower. A time of at least 10 s is required to sample
particles in all energy ranges and in all velocity directions. Over
such a long period, a spacecraft has travelled a distance of 4000 km
in the frame of the supersonic solar wind plasma. Since the actual
time resolution of solar wind plasma instruments will be much larger
than the time (1-5 s) required for a interplanetary vehicle to pass
through a thin current sheet, the particle fluxes measured in successive energy channels and successive solid angles must be compared
directly with the corresponding values deduced from theoretical velocity distributions calculated at different depths in the current sheet.

REFERENCES

ALPERS, W., 1969, Astrophys. Space Sci., **5**, 425-437.
LEE, L.C. and J.R. KAN, 1979, J. Geophys. Res., **84**, 6417-6426.
LEMAIRE, J., M. ROTH, M. SCHERER and M. SCHULZ, 1983, ESA SP-1050,
     265-271.
ROTH, M., 1978, J. Atmos. Terr. Phys., **40**, 323-329.
ROTH, M., 1979, ESA-SP-148, 295-309.
ROTH, M., 1980, Aeronomica Acta A n° 221, also in 1984, Acad. R.
     Belg., Mém. Cl. Sci., Collection in 8e-2e série, **44**(7) 222 pp.
ROTH, M., 1983, pp. 139-147, in : W. Bötticher, H. Wenk and E.
     Schulz-Gulde (Eds.), Proceedings of the XVI International
     Conference on Phenomena in Ionized Gases, Düsseldorf.

# SOLAR WIND COMPOSITION AND WHAT WE EXPECT TO LEARN FROM OUT-OF-ECLIPTIC MEASUREMENTS

J. Geiss and P. Bochsler
Physikalisches Institut
University of Bern
Sidlerstrasse 5
CH-3012 Bern / Switzerland

ABSTRACT. Elemental abundances in the solar wind are fractionated relative to the solar abundances by atom-ion separation in the upper chromosphere and by ion-ion separation in the corona, where also the charge states of the ions are frozen-in. Thus solar wind composition and charge states of the elements can be used to study conditions and processes at the solar surface and in the corona. The velocity distributions of individual ion species reflect wave-particle interactions, collisions and stream-stream interactions in interplanetary space. The SWICS and SWPE experiments carried by Ulysses are well equipped for measurements of the abundances of a number of elements and their charge and velocity distributions. The scan over virtually all solar latitudes effected by Ulysses ought to give new insight into processes and conditions in the solar wind source regions, and it will provide data for a 3-dimensional picture of solar wind expansion and heliospheric processes, which is also important for studying the interactions of the galactic cosmic rays and the interstellar gas with the heliosphere.

## 1. GENERAL

After twenty years of solar wind research and an even longer period of optical corona studies, the most fundamental questions concerning the expanding solar atmosphere are only partially answered:

(1) The way in which solar matter is continuously fed to the corona is not clearly identified,

(2) There is no consensus about the mechanism of heating this matter to a temperature of $10^6$K, and

(3) The process of acceleration and the source of momentum driving coronal material to supersonic velocity are not uniquely determined.

This impasse can only be overcome by more data and new methods of observation, most likely in three areas: (a) Remote sensing of the corona to determine its composition as well as the velocities and temperatures of some constituents as a function of altitude and in relation to solar surface characteristics, (b) investigation of the global three-dimensional properties of the solar wind and their relationship to coronal features, and (c) a more complete and continuous record of the abundances and the charge state distributions of elements (and isotopes) in the solar wind. It is hoped that significant progress in the second and third areas will result from the ensuing Ulysses mission.

Solar wind composition studies can play a crucial role in answering the three fundamental questions mentioned above, because abundances

R. G. Marsden (ed.), The Sun and the Heliosphere in Three Dimensions, 173–186.
© 1986 by D. Reidel Publishing Company.

of elements and isotopes in the solar wind and their charge state distributions provide direct information on processes in the corona and
the chromosphere, since the mixture of solar wind ions remains unalter-
ed on the way from the outer corona to the region in interplanetary
space where the solar wind is measured. Furthermore, ion composition
measurements are particularly sensitive indicators for processes in the
source region, because the circumstances under which these processes
occur must be marginal in three respects which are summarized in
Table I.

TABLE I:  Physical processes in the solar wind source region.

---

Acceleration
Conditions are in between solar exospheric escape ($\rightarrow$ $H^+$ only) and
dense wind ($\rightarrow$ no fractionation).
Consequence:  Strong compositional changes, e.g. He/H = 0.001-0.3
indicate changing conditions in the acceleration region.

Incomplete Ionization
Heavier elements are not fully stripped of electrons.
Consequence:  Charge state distributions of ions reflect the tem-
perature structure of the corona.

Ion-Atom Separation
The coronal plasma is apparently fed from a partially ionized gas.
Consequence:  Overabundance of elements with low first ionization
potential.

---

Although we have already a considerable set of data on the solar
wind ion composition, these data do not provide a good coverage over
the changing solar wind conditions. Nevertheless, a clear relation between ion abundances and solar wind characteristics begins to emerge.
In Table II, we give abundance ratios of elements or isotopes in three
solar wind regimes, and in Table III, we summarize the findings on
freezing-in temperatures (cf. Hundhausen, 1972), which are derived from
the charge state distributions of the elements (data from references in
Table II, and from Ipavich et al., 1983 and Galvin et al., 1984). It is
evident that composition and charge states are significantly different
in the fast streams, the average low speed solar wind, and the driver
plasma. These three types of solar wind originate from parts of the
corona with very different physical characteristics.

The bulk properties of the solar wind as well as the abundances,
charge states, velocities and temperatures of individual ion species
are determined by a sequence of physical mechanisms. Observations and
theoretical work lead us to distinguish three types of processes occurring in three different radial domains:

(1) Elements with low first ionization potential (FIP) are enriched, relative to the solar surface, in corona, solar wind and solar
flare particle populations. It is thought that this enrichment is
caused by an atom-ion separation process operating in the upper chromosphere and leading to a supply of fractionated gases into the corona.

(2) The solar wind is accelerated out of the corona, driven thermally and by (nonresonant) momentum transfer from waves. The accelera-

tion process can lead to further <u>fractionation</u> <u>of</u> <u>elements</u> <u>and</u> <u>iso-</u>
<u>topes</u>. The <u>charge</u> <u>states</u> of heavier elements <u>freeze-in</u> deep in the co-
rona, probably below the transsonic point and remain virtually un-
changed from thereon.

(3) It is observed in interplanetary space that a) helium and
heavier ions travel faster than hydrogen (with velocity increments that
are limited by the Alfvén speed), and b) helium and heavier ions have
temperatures which are proportional to mass. These two observations are
interpreted as being caused by <u>resonant</u> <u>wave-particle</u> <u>interaction</u> in
the <u>outer</u> <u>corona</u> and in <u>interplanetary</u> <u>space</u>.

TABLE II:  Abundance ratios in different solar wind regimes.

| | Low Speed (Interstream) | | High Speed (Coronal Holes) | | Driver Plasma | |
|---|---|---|---|---|---|---|
| He/H | 0.04 | (1,2) | 0.05 | (3) | >0.15 | (4,5) |
| $^4$He/$^3$He | 2100 | (6,7) | 2150 | (7) | – | |
| He/O | 80 | (8) | 60 | (8) | 200 | (9) |
| Ne/O | 0.17 | (8) | – | | – | |
| Fe/H | $5 \cdot 10^{-5}$ | (9) | – | | $9.1 \cdot 10^{-5}$ | (9) |
| Fe/He | $\leq 2 \cdot 10^{-3}$ | (10) | $3 \cdot 10^{-3}$ | (10) | $8 \cdot 10^{-4}$ | (9) |

(1) Ogilvie, 1972                    (6) Coplan et al., 1984
(2) Neugebauer, 1981                 (7) Bochsler, 1984
(3) Bame et al., 1977                (8) Bochsler et al., 1985
(4) Hirshberg et al., 1972           (9) Bame et al., 1979
(5) Borrini et al., 1982            (10) Schmid, 1985 (preliminary
                                          result)

TABLE III:  Approximate ranges of freezing-in temperatures (in $10^6$K) in different solar wind flows.

| Low Speed (Interstream) | High Speed (Coronal Holes) | Driver Plasma |
|---|---|---|
| 1.6 - 2.1 | 1.1 - 1.5 | >2 |

General Systematics:  T(iron)   < T(oxygen)
                      T(carbon) < T(oxygen)

The relevant observations and the picture that emerges of the processes in these three regions were recently reviewed in an ESA publication by Marsch (1985) and by Geiss (1985). Thus, in what follows we
shall give only a brief account of the findings obtained so far and
discuss what can be learned from out-of-ecliptic measurements of ion
abundances in the solar wind.

## 2.  GLOBAL CHARACTERISTICS OF THE SOLAR WIND

The solar cycle is accompanied by a systematic variation in the global
structure of the corona. Under minimum conditions, the dipole component
of the solar magnetic field is dominant, i.e. there are large areas at
the poles with open field lines that manifest themselves as coronal
holes (cf. Figure 1, left), from which fast solar wind streams emanate.
Thus during solar minimum, we would encounter a relatively regular solar wind pattern channeled by a B-field configuration similar to the
one depicted in Figure 2a. The Ulysses spacecraft is expected to carry
out its pass through the middle and high solar latitudes mainly during

Figure 1.  The solar corona photographed with a radial filter. Left: 30 June 1973, late
descending phase of solar cycle 20. Right: 16 February 1980, maximum of solar cycle 21.
After Sime (1985); courtesy High Altitude Observatory, National Center for Atmospheric
Research, Boulder, CO.

the years of solar maximum (Volland et al., 1983), i.e. when the magnetic field of the sun often has a complicated global structure and
when solar activity is strong. The 3D solar wind expansion to be scanned by Ulysses will reflect the coronal structures typical for solar
maximum (cf. Figure 1, right). Near the sun, domains with open and with
closed field lines alternate frequently, as sketched in Figure 2b. This
pattern will, however, undergo rapid changes - globally as a result of
adjustments caused by the reversal of the global B-field and locally as
a result of coronal transients. In the maximum phase of the solar cycle, the solar wind properties are obviously more difficult to understand than during solar minimum. The fine structure in the corona during the maximum phase makes it difficult to trace a solar wind parcel
back to a coronal feature, and modelling of the solar wind expansion is
complicated by the geometrical structure in the corona and by time variations. Models would probably remain arbitrary if they could be checked only against some bulk properties of the solar wind and not against
the set of composition and charge state data that Ulysses is expected
to return.

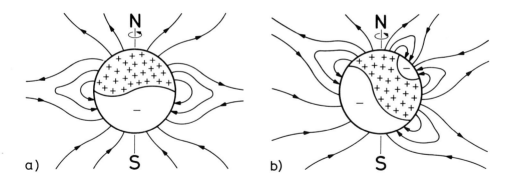

Figure 2. Sketches of solar magnetic field configurations: (a) The dipole component dominates during the minimum phase of solar activity. High speed solar wind is generated in the large open field line region (coronal holes). The low latitude current sheet gives rise to a belt of streamers which in turn are sources of low speed solar wind. (b) During solar maximum conditions the B-field structure is more complicated (cf. Figure 1, right) and more rapidly changing.

## 3. EXPERIMENTS

The Solar Wind Plasma Experiment SWPE and the Solar Wind Ion Composition Spectrometer SWICS on board the Ulysses spacecraft complement each other ideally.

SWPE (Bame et al., 1983) has 7 electron and 16 ion detectors and will precisely measure the velocity distribution functions of the electrons and ions in three dimensions as a function of energy per charge. In this way, velocities, temperatures and deviations from Maxwellians can be determined for individual solar wind species with a time resolution of 4 minutes. The energy/charge (E/Q) analyzers employed in this experiment will allow to measure not only protons and $\alpha$–particles, but also heavier ion species if the solar wind temperatures are sufficiently low. The potential of E/Q analyzers for composition studies has been proven in the past; an example is given in Figure 3.

SWICS (Gloeckler et al., 1983) is a new type of instrument. It combines an E/Q analysis in an electrostatic deflection system, a time of flight measurement, and a determination of the total energy in a solid state detector. These properties of SWICS will make for the first time a two-parameter analysis of the ion composition in the solar wind, i.e. mass/charge (M/Q) spectra as well as a low resolution mass analysis of the ions will be obtained. This will eliminate many degeneracies that occur in solar wind M/Q spectra, e.g. SWICS will separate $He^{2+}$ from $C^{6+}$ and $C^{5+}$ from $Mg^{10+}$, and thus add C and Mg to the list of elements which can be determined in the solar wind. It should also substantially increase the number of charge states of other elements that can be identified. The coincidence techniques employed by SWICS lead to a very low background, enabling measurements of rare solar wind constituents, shock accelerated ions, or particles of other sources. The performance of the instrument and in particular its ability to measure simultaneously mass and M/Q ratio is demonstrated in Figure 4, where calibration count rates are plotted as a function of M/Q and M.

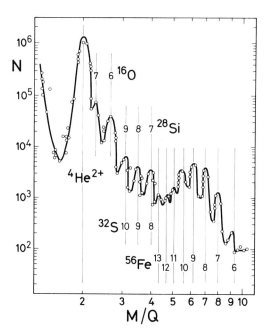

Figure 3. Solar wind ener-
gy/charge spectrum obtained
by Zastenker et al. (1985)
with the Prognoz-7 space-
craft. The authors converted the E/Q scale to an M/Q
scale. The figure demon-
strates that high resolu-
tion M/Q spectra can be obtained with electrostatic
instruments at times of low
kinetic temperature.

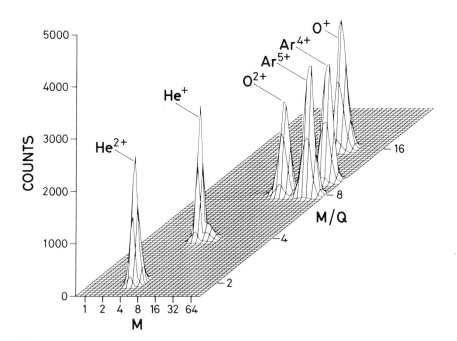

Figure 4. A 3D presentation of calibration data of the SWICS experiment to be flown
with Ulysses (Galvin and Rettenmund, 1985). Note the distinction between ions with identical M/Q ratio such as the pair O(2+) and Ar(5+). Atmospheric Ar-40 was used. Ion energies (1-6) × $10^4$ eV/q.

The CHEM instrument carried by the AMPTE CCE spacecraft is very similar to SWICS. When the magnetosphere is compressed, the CCE spacecraft enters into the magnetosheath analyzing the heated solar wind. Gloeckler et al. (1985) obtained spectra which clearly separate $C^{6+}$ from $He^{2+}$, and this enabled them to give for the first time unequivocal carbon abundance data in solar wind matter.

## 4.  ION-ATOM SEPARATION IN THE UPPER CHROMOSPHERE

There is ample evidence that elements with low first ionization potential are overabundant in parts of the corona (Veck and Parkinson, 1981), in solar flare particle populations (Cook et al., 1984; Fan et al., 1984; Meyer, 1985; Breneman and Stone, 1985) and in the solar wind (Geiss and Bochsler, 1984; Geiss, 1985). These observations are summarized in Figure 5.

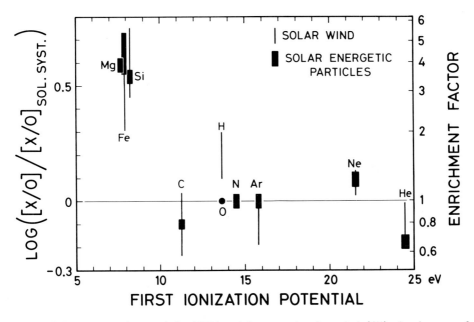

Figure 5.  Solar energetic particle (SEP) and low speed solar wind (SW) abundances relative to oxygen and normalized to the solar system values. Uncertainties in the solar system abundances are not included in the error estimates. SEP data were taken from Cook et al. (1984). The SW data are from the sources quoted in Table II. In addition, the following data are included: He, Ne, and Ar abundances from Geiss et al. (1972), and the preliminary C/O abundance ratio obtained by Gloeckler et al. (1985).

The conditions under which an ion-atom separation process could operate can be delimited to number densities $<10^{10}$ cm$^{-3}$ and temperatures of the order of $10^4$K which points to the upper chromosphere as the likely location (Geiss, 1982). As a first step, for studying this separation process, Geiss and Bochsler (1984) investigated the ioniza-

tion of nine important elements as a function of time under solar surface conditions, taking into account exitation and ionization by photons and electrons as well as recombination. The resulting ionization
curves are reproduced in Figure 6 for a total hydrogen density of
$10^{10}$ cm$^{-3}$ and a constant temperature of $10^4$K. As expected, ionization
rate and first ionization potential are generally well related, but it
appears that in some respects the ionization times reflect the observed
abundances even better than the ionization potential (cf. Geiss, 1985).
Figure 6 implies that the separation process operates on a time scale
of the order of 10-100 seconds.

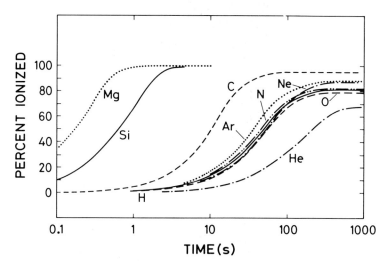

Figure 6. Ionization of various elements as a function of time in an optically thin
layer at the solar surface. A constant temperature of $10^4$K and a constant gas density of
$n(H) + n(H^+) = 10^{10}/cm^3$ were assumed (Geiss and Bochsler, 1984).

The atom-ion separation will depend on the ratio ionization time
to diffusion time for different elements, regardless whether diffusion
is driven by pressure gradients, thermal forces, or electrical forces
resulting from plasma processes. The simplest mechanism - studied by
von Steiger and Geiss (1985) - envisages an upward flow of solar matter
in a relatively narrow flux tube. Atoms can diffuse across the magnetic
field lines and are lost from the upward flowing plasma. In this case,
the governing parameter for ion-atom separation is

$$L = 4.81 \sqrt{D \, t_{ion}}$$

where $t_{ion}$ is the characteristic ionization time, D is the diffusion
constant of the atomic species, and 4.81 is the appropriate scale parameter for a cylindrical geometry. Since the diffusion constants change
with time because of ionization and diffusive losses, quantitative models involve numerical integration. However, a simple plot of the SW
and SEP abundances as a function of L gives an indication of the gener-

al trend. Figure 7 shows such a plot for T = $10^4$K, a total hydrogen
density of $10^9$ cm$^{-3}$, and 50 percent hydrogen ionization. The diffusion
constants are rather similar for the atomic species considered, with
the exception of H and O which have large resonant charge exchange
cross sections with protons, and therefore diffuse more slowly than
other atoms. Three groups of elements can be distinguished in Figure 7:
(a) Mg, Si, Fe which - like other metals - ionize very fast, (b) an
intermediate group with O, C, H, Ar, N, Ne, and (c) the slowly ionizing
He. The L-scale in Figure 7 indicates that the ion-atom separation process considered here would have to work on a fine lateral scale.

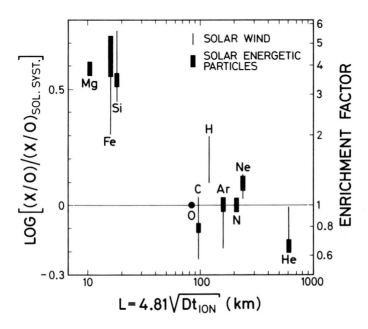

Figure 7. The abundances of elements in SW and SEP populations given in Figure 5 are
plotted against the characteristic length that atoms can diffuse across B at the solar
surface before getting ionized (see text).

    The Ulysses mission ought to contribute relevant data for even-
tually pinpointing the geometry and the mechanism of the ion-atom separation process that the SW and SEP data so strongly imply. Properties
of the chromosphere and the transition zone show systematic changes on
the solar surface. For instance, the density and temperature gradients
in the transition zone are lower inside coronal holes than in other
solar regions (cf. Figure 8). Thus, a scan over the whole solar surface
could reveal a relation between the degree of ion-atom separation and
some solar surface property.

    Many characteristics of an atom are directly or indirectly related to its first
ionization potential. Thus, a variety of processes can lead to FIP dependent separation. For instance, in a nebula of cosmic composition, elements with low FIP are found
in the grains, whereas those with high FIP remain largely in the gas phase. If condensed

planetary matter is added to the corona by comets disintegrating near the sun, or by the Poynting-Robertson effect acting on interplanetary grains, it would be low-FIP elements that are enriched, just as observed in the SW and SEP populations.

To be sure, the observational evidence we have so far does not at all indicate that in the present epoch external sources affect the coronal composition. On the contrary, there are three observations suggesting that this effect is small: (a) From interplanetary meteoroid data and zodiacal light observations source and sink terms for the interplanetary dust can be estimated (Grün et al., 1985). The throughput derived by these

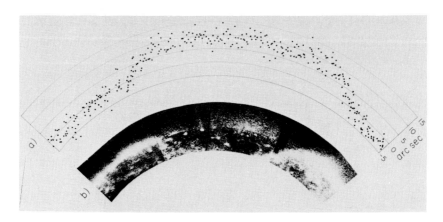

**Figure 8.** Transition zone over south polar hole. (a) Height of the Ne VII (46.5 nm) limb relative to the Lyman continuum limb. (b) Mg X (62.5 nm), delineating the coronal hole. The rise of the Ne VII limb by 10 arc sec inside the coronal hole demonstrates the strong widening of the transition zone (after Huber et al., 1974).

authors is small. Only a comparatively large and continuous deposition of grains directly into the corona (possibly by small comets) could have a significant effect; (b) The volatilities from grains of low-FIP elements differ appreciably. Depending on the mechanism of grain deposition in the corona, successive evaporation could lead in the low corona to stronger enhancement for the refractory elements Al, Ca, or Ti than for the alkalis or Zn. Such a trend is not apparent in the UV data of Veck and Parkinson (1981) or in the recent multi-element SEP data of Breneman and Stone (1985); (c) Whereas the region covered by the UV measurements of Veck and Parkinson (1981) and the SEP particle populations contain only external matter which has penetrated deep into the corona, the solar wind will pick up external ions at higher altitudes as well, thus further increasing the abundance of elements like Mg, Si, or Fe. A difference between SW and SEP in the enhancement of low-FIP elements is, however, not apparent (cf. Figures 5 and 7).

The arguments against external contamination of the corona carry considerable weight, but they are not yet conclusive. The remaining doubt might be removed by the Ulysses mission. Asteroids, short-period comets and the dust derived from them are concentrated near the ecliptic, only the long-period comets have evenly distributed orbits. Comparison of the latitudinal variations of solar wind composition and dust populations might reveal whether there is any relation between the two.

Recently, Russell et al. (1985) have interpreted a peculiar signature in the interplanetary B-field as being due to small, invisible comets. If this identification is confirmed, registering these events as well as observations of comets hitting the sun (Michels et al. 1982) could help to obtain an estimate of the flux of small comets into the innermost region of the solar system.


## 5.   THREE DIMENSIONAL SOLAR WIND EXPANSION

The solar activity will be in its maximum half-cycle during most of Ulysses' out-of-ecliptic journey. We expect, therefore, to encounter

coronal structures and solar wind expansion patterns similar to those
shown in Figures 1 (right) and 2b. Observations made during Carrington
rotation 1679 (Figure 9) which occurred near solar maximum (March 1979)
ought to give us a general idea of what to expect during Ulysses' latitudinal scan. The solar magnetic map and the polarity of the interplanetary B-field in the ecliptic plane shown in the bottom panel are from
Hoeksema et al. (1983). The data on solar wind helium velocity, helium
kinetic temperature and He/O abundance ratio were obtained with the
ISEE-3 Ion Composition Instrument. Solar wind data were mapped back to
the solar surface with the solar wind velocities observed at 1 AU,
roughly to account for the transit time delay.

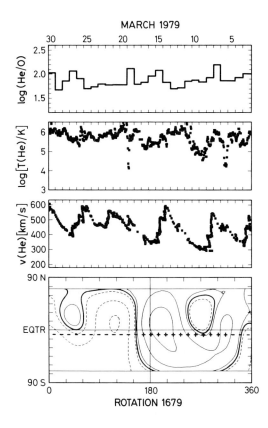

Figure 9. Comparison of
magnetic surface features
(from Hoeksema et al.,
1983) with He speed, He
kinetic temperature and
He/O ratio observed at 1
AU during Carrington rotation 1679. In the bottom
panel, the contour levels
are at 0, 1, 5, and 10
microteslas, negative contours (field pointing to-
ward the sun) are dashed.

Figure 9 shows the well established correlation between solar magnetic field structure and solar wind speed (cf. Bame et al., 1977).
Also the generally observed trend of increasing kinetic temperature
with increasing speed (Burlaga and Ogilvie, 1970) is apparent. However,
due to the complicated structure of the magnetic field near solar maximum, stream-stream interactions can severely mask the otherwise obvious
correlation between speed and temperature and will blur possible signatures in composition data. Furthermore, - even without stream-stream

interactions considered – it is difficult to unambiguously correlate
surface features with signatures in the interplanetary plasma at distances of the order of 1 AU, because of the relatively small scaled
structures in the source region.

The He/O data give an indication of the potential for, but also of
the difficulties with relating solar wind abundance data to coronal
structures and eventually to the expansion geometry and dynamics in the
solar wind source region. In Figure 9 there is a hint that during Carrington rotation 1679, high He/O ratios occurred at times of elevated
solar wind velocity. It is evident, however, that it will be difficult
to establish a significant relation, even if the analyses were extended
to a large data set.

The situation would be much improved with out-of-ecliptic solar
wind measurements, provided that spacecraft near the ecliptic would
return data at the same time. If the longitudinal separation between
Ulysses and a near-ecliptic spacecraft is not too large, say <$90^{\circ}$, then
we would obtain two parallel scans, instead of one, through magnetic
structures as the one shown in Figure 9. The two-spacecraft measurements would greatly improve the chances of correctly relating solar
wind measurements with corona observations. NASA and ESA are considering to set up a programme – which might appropriately be called Pene-
lope – for providing in-ecliptic data during Ulysses' out-of-ecliptic
journey by reactivating some spacecraft. The problem discussed in this
section shows the importance of such measurements not only while Ulysses is at high latitude, but during its entire latitude scan.

During the maximum phase of solar activity, non-steady state phenomena such as flares, rapid changes in field configuration, or different types of transients will dominate solar wind expansion. Ion composition data ought to be particularly valuable for attempts to identify
the causes and source of transient phenomena, because the abundances
and in particular the charge states of the elements place severe constraints on expansion models. A telling example is the unexpected discovery of high $He^{+}$ abundances sometimes following shocks (Schwenn et
al., 1980; Gosling et al., 1980), proving that solar matter occasionally passes the corona without being accommodated. A systematic investigation of such $He^{+}$ events including low temperature ions of other elements might allow to delineate this phenomenon.

## 6. HELIOSPHERIC PROCESSES

Solar wind propagation and processes in the heliosphere are covered by
other articles in this volume. We mention here just two of the areas
where investigations with the solar wind instruments on board Ulysses
ought to make a vital contribution.

Ions produced in the heliosphere from interstellar atoms are probably partly accommodated into the solar wind and partly shock-accelerated. For studying both these processes, measurements of abundances and
velocity distributions of ion species are an obvious requirement. The
Ulysses payload will provide composition data in a continuous energy
range from $\sim 10^2$ to $\sim 10^9$ eV. This capability is invaluable for studying

the physics of collisionless shocks including acceleration mechanisms.
Investigation of the entry of the galactic cosmic radiation via
high solar latitude was one of the foremost motives for undertaking the
Ulysses mission. For interpreting the 3D cosmic ray data, a quantitative description of the fields and waves out to large solar distances
is required. The velocity distributions of major and minor constituents
in the solar wind reveal a strong interplay between fields and particles (Marsch et al., 1982a, b; Isenberg and Hollweg, 1983). Thus only
in combination with measured particle velocity distributions can the
fields and waves observed by Ulysses be extrapolated with some confi-·
dence to large solar distances, as is required for studying the cosmic
ray entry.

## ACKNOWLEDGEMENTS

We wish to thank W.I. Axford, A. Bürgi, G. Gloeckler, E. Grün, M.C.E. Huber, K.W.
Ogilvie, D.G. Sime, and R. von Steiger for discussions and A.G. Galvin, G. Gloeckler,
F.M. Ipavich, U. Rettenmund, and J. Schmid for making available to us unpublished results. The camera-ready manuscript was efficiently prepared by Mrs. G. Troxler, and the
drawings by K. Bratschi. Support from the Swiss National Science Foundation is acknowledged.

## REFERENCES

Bame, S.J., Asbridge, J.R., Feldman, W.C., and Gosling, J.T.: 1977, J.
     Geophys. Res. 82, 1487.
Bame, S.J., Asbridge, J.R., Feldman, W.C., Fenimore, E.E., and Gosling,
     J.T.: 1979, Sol. Phys. 62, 179.
Bame, S.J., Glore, J.P., McComas, D.J., Moore, K.R., Chavez, J.C.,
     Ellis, T.J., Peterson, G.R., Temple, J.H., and Wymer, F.J.: 1983,
     ESA Spec. Publ. SP-1050, 49.
Bochsler, P.: 1984, Habilitation Thesis, University of Bern.
Bochsler, P., Geiss, J., and Kunz, S.: 1985, Sol. Phys., in press.
Borrini, G., Gosling, J.T., Bame, S.J., and Feldman, W.C.: 1982, J.
     Geophys. Res. 87, 7370.
Breneman, H. and Stone, E.C.: 1985, Proc. 19th Int. Cosmic Ray Conf. 4,
     213.
Burlaga, L.F. and Ogilvie, K.W.: 1970, Astrophys. J. 159, 659.
Cook, W.R., Stone, E.C., and Vogt, R.E.: 1984, Astrophys. J. 279, 827.
Coplan, M.A., Ogilvie, K.W., Bochsler, P., and Geiss, J.: 1984, Sol.
     Phys. 93, 415.
Fan, C.Y., Gloeckler, G., and Hovestadt, D.: 1984, Space Sci. Rev. 38,
     143.
Galvin, A.B. and Rettenmund, U.: 1985, unpublished.
Galvin, A.B., Ipavich, F.M., Gloeckler, G., Hovestadt, D., Klecker, B.,
     and Scholer, M.: 1984, J. Geophys. Res. 89, 2655.
Geiss, J.: 1982, Space Sci. Rev. 33, 201.
Geiss, J.: 1985, ESA Spec. Publ. SP-235, 37.
Geiss, J. and Bochsler, P.: 1984, Int. Conf. "Isotopic ratios in the
     solar system", Paris, June 1984 (Proc. to be published by CNES).

Geiss, J., Bühler, F., Cerutti, H., Eberhardt, P., and Filleux, C.:
    1972, NASA Spec. Publ. SP-315, 14.1.
Gloeckler, G., Geiss, J., Balsiger, H., Fisk, L.A., Gliem, F., Ipavich,
    F.M., Ogilvie, K.W., Stüdemann, W., and Wilken, B.: 1983, ESA
    Spec. Publ. SP-1050, 77.
Gloeckler, G., Ipavich, F.M., Hamilton, D.C., Wilken, B., Stüdemann,
    W., and Kremser, G.: 1985, contribution 04.05.06 IAGA 5th General
    Assembly, Prag, 1985.
Gosling, J.T., Asbridge, J.R., Bame, S.J., Feldman, W.C., and Zwickl,
    R.D.: 1980, J. Geophys. Res. 85, 3431.
Grün, E., Zook, H.A., Fechtig, H., and Giese, R.H.: 1985, Icarus 62,
    244.
Hirshberg, J., Asbridge, J.R., and Robbins, D.E.: 1972, J. Geophys.
    Res. 77, 3583.
Hoeksema, J.T., Wilcox, J.M., and Scherrer, P.H., 1983, J. Geophys.
    Res. 88, 9910.
Huber, M.C.E., Foukal, P.V., Noyes, R.W., Reeves, E.M., Schmahl, E.J.,
    Timothy, J.G., Vernazza, J.E., and Withbroe, G.L.: 1974,
    Astrophys. J. 194, L115.
Hundhausen, A.J.: 1972, Coronal Expansion and Solar Wind, Springer,
    Berlin, Heidelberg, New York.
Ipavich, F.M., Galvin, A.B., Gloeckler, G., Hovestadt, D., Klecker, B.,
    and Scholer, M.: 1983, Solar Wind Five (M. Neugebauer, ed.), NASA
    Conf. Publ. 2280, 597.
Isenberg, P.A. and Hollweg, J.V.: 1983, J. Geophys. Res. 88, 3923.
Marsch, E.: 1985, ESA Spec. Publ. SP-235, 11.
Marsch, E., Mühlhäuser, K.H., Schwenn, R., Rosenbauer, H., Pilipp, W.,
    and Neubauer, F.M.: 1982a, J. Geophys. Res. 87, 52.
Marsch, E., Goertz, C.K., and Richter, K.: 1982b, J. Geophys. Res. 87,
    5030.
Meyer, J.-P.: 1985, Astrophys. J. Suppl. Ser. 57, 173.
Michels, D.J., Sheeley, N.R. Jr., Howard, R.A., and Koomen, M.J.: 1982,
    Science 215, 1097.
Neugebauer, M.: 1981, Fundamentals of Cosmic Physics 7, 131.
Ogilvie, K.W.: 1972, J. Geophys. Res. 77, 4227.
Russell, C.T., Phillips, J.L., Saunders, M.A., and Fedder, J.A.: 1985,
    contribution 04.01.10 IAGA 5th General Assembly, Prag, 1985.
Schmid, J.: 1985, contribution 04.05.02 IAGA 5th General Assembly,
    Prag, 1985.
Schwenn, R., Rosenbauer, H., and Mühlhäuser, K.H.: 1980, Geophys. Res.
    Lett. 7, 201.
Sime, D.G.: 1985, ESA Spec. Publ. SP-235, 23.
von Steiger, R. and Geiss, J.: 1985, unpublished.
Veck, N.J. and Parkinson, J.H.: 1981, Mon. Not. R. Astr. Soc. 197, 41.
Volland, H., Bird, M.K., and Edenhofer, P.: 1983, ESA Spec. Publ.
    SP-1050, 245.
Zastenker, G.N., Yermolaev, Yu I., Vaisberg, O.L., Omelchenko, A.N.,
    Borodkova, N.L., Avanov, L.A., Skalsi, A.A., Nemecek, Z., and
    Safrankova, J.: 1985, contribution 04.05.05 IAGA 5th General
    Assembly, Prag, 1985.

COMETS AND THREE-DIMENSIONAL SOLAR WIND STRUCTURE

John C. Brandt
Laboratory for Astronomy and Solar Physics
NASA/Goddard Space Flight Center
Greenbelt, Maryland, 20771, USA

ABSTRACT.  The principal ways in which the study of comets can
contribute to knowledge of three-dimensional solar wind structure are
briefly reviewed.  Calibration of the cometary results should be
considered an important scientific goal of space probes sent to high
heliocentric latitudes.

Cometary studies can contribute to the understanding of
solar-wind structure in three different ways:  (1) orientations of
plasma tails contain information concerning the solar-wind velocity
vector; (2) contours of the Lyman-alpha emission from the hydrogen
cloud contain information concerning the solar-wind flux; and (3)
periodic "disconnection events" (DEs), where the entire plasma tail
is detached from the comet, contain information on the location of
magnetic sector boundaries.
    The orientations of plasma tails have been used to study solar
wind velocity for many years (Biermann 1951).  The analysis rests on
the assumption that the plasma tails act as wind socks, that is, they
point in the direction of the momentum flow as seen by an observer
riding with the comet.  At present, there is no reason to doubt the
validity of this assumption (Biermann 1983).  For an assumed
parametric model, the position angles of plasma tails on the plane of
the sky can be calculated and compared with the measured values.
Standard least-squares procedures can be used to adjust the input
parameters and hence to determine a global, solar-wind model.  Models
based on approximately 700 to 800 individual observations (e.g.,
Brandt, Roosen, and Harrington 1972, Brandt and Mendis 1979)
consistently produce solar-wind radial speeds in the range 400-420
km/sec, azimuthal speeds of 5-7 km/sec, and very small polar speeds,
in agreement with essentially all other evidence.
    In principle, the astrometric approach can be used to test any
systematic variation in solar-wind velocity properties that can be
represented parametrically.  This technique was used to attempt to
verify the poleward increase in radial speed found in the radio
scintillation studies (e.g., Coles et al. 1980).  Unfortunately, no

187

*R. G. Marsden (ed.), The Sun and the Heliosphere in Three Dimensions, 187–189.*
© *1986 by D. Reidel Publishing Company.*

evidence for the poleward increase was evident in the comet data
(Brandt, Harrington, and Roosen 1975). We note that a constant
poleward gradient was assumed. Because the radio scintillation
results show periods (Coles et al. 1980) that closely resemble the
cometary results (i.e., little or no gradient), the most likely
interpretation is that the discrepancy results from selection effects
in sampling a strongly varying situation or that the last few solar
cycles studied by radio scintillations are significantly different
from the 75 years in the cometary sample. A reanalysis of the
cometary data with a poleward gradient varying with solar cycle may
resolve the discrepancy.

Additional evidence for poleward changes comes from studies of
the extent of the hydrogen cloud around comet Bennett (1969i), which
is sensitive to the solar wind flux. A higher flux is indicated for
solar latitudes greater than 45° (Bertaux, Blamont, and Festou
1973). This type of analysis should be repeated with additional
comets at high solar latitudes.

The astrometric technique has been extended to specific events by
applying the wind-sock assumption to the entire length of the tail
(Niedner, Rothe, and Brandt 1978; Niedner, Brandt, Zwickl, and Bame
1982). Full exploitation requires simultaneous cometary and
three-dimensional solar wind velocity data.

The final cometary approach utilizes the disconnection events
(DEs) of plasma tails (Niedner 1982). Basically, Niedner and Brandt
(1978, 1979) have proposed that a reversal of the polarity of the
interplanetary field being pressed into the cometary ionosphere would
cause magnetic reconnection to occur on the sunward side. This
process could sever the field lines connecting the plasma tail to the
head region. Clearly, sector boundaries are a prime location for
this scenario to occur, and DEs could be produced. If these events
are produced as described, they can be used to map the extent of the
sector boundaries in solar latitude. While there are certainly other
interpretations of DEs, the magnetic reconnection model is the best
at present (Niedner 1984).

The principal conclusions from the study of DEs are that the
sector structure can extend to solar latitudes $\gtrsim 45°$ for much of a
solar cycle and that the tilt angles (the inclination of the sector
boundary with respect to the solar equator) could be quite high. The
sector structure inferred from DEs was also found to be variable with
the solar cycle, although the exact nature of this variability is not
well-established from the DE data set.

Much of the initial resistance to these results on sector
boundaries (and by implication on the magnetic reconnection
interpretation itself) was due to general acceptance of the
"ballerina skirt" model with a small warp. The situation has clearly
changed over the past few years. For example, Hoeksema (1984) and
Hoeksema et al. (1982, 1983) have computed coronal magnetic field
models which appear to confirm Niedner's results.

The cometary results offer considerable insight into the
three-dimensional structure of the solar wind. When taken together
with the radio scintillation results, even stronger inferences can

probably be made about the properties of interplanetary space out of the ecliptic. Thus, one of the major contributions of space probes sent to high solar latitudes could be to calibrate the cometary and radio scintillation results so that we can infer the structure at other times.

## REFERENCES

Bertaux, J. L., Blamont, J. E., and Festou, M. 1973, Astr. and Ap., 25, 415.

Biermann, L. 1951, Zs. f. Astrophysik, 29, 274.

Biermann, L. 1983, in Topics in Plasma-, Astro- and Space Physics, G. Haerendel and B. Battrick, Eds, (Max-Planck-Institut für Physik und Astrophysik, Garching), p. 85.

Brandt, J. C., Harrington, R. S., and Roosen, R. G. 1975, Ap. J., 196, 877.

Brandt, J. C., and Mendis, D. A. 1979, in Solar System Plasma Physics Volume II, C. F. Kennel, L. J. Lanzerotti, and E. N. Parker, Eds. (North-Holland Publ. Co.), p. 253.

Brandt, J. C., Roosen, R. G., and Harrington, R. S. 1972, Ap. J., 177, 277.

Coles, W. A., Rickett, B. J., Rumsey, V. H., Kaufman, J. J., Turley, D. G., Ananthakrishnan, S., Armstrong, J. W., Harmon, J. K., Scott, S. L., and Sime, D. G., 1980, Nature, 286, 239.

Hoeksema, J. T. 1984, Thesis, Stanford University.

Hoeksema, J. T., Wilcox, J. M., and Scherrer, P. H. 1982, J. Geophys. Res., 87, 10,331.

Hoeksema, J. T., Wilcox, J. M., and Scherrer, P. H. 1983, J. Geophys. Res., 88, 9910.

Niedner, M. B. 1982, Ap. J. Suppl. 48, 1.

Niedner, M. B., 1984, in Magnetic Reconnection in Space and Laboratory Plasmas, Geophysical Monograph 30, p. 79.

Niedner, M. B., and Brandt, J. C. 1978, Ap. J., 223, 655.

Niedner, M. B., and Brandt, J. C. 1979, Ap. J., 234, 723.

Niedner, M. B., Brandt, J. C., Zwickl, R. D. and Bame, S. J. 1983, in Solar Wind Five, M. Neugebauer, Ed. (NAS CP-2280), p. 737.

Niedner, M. B., Rothe, E. D., and Brandt, J. C. 1978, Ap. J., 221, 1014.

# STRUCTURE AND DYNAMICS OF COROTATING AND TRANSIENT STREAMS IN THREE DIMENSIONS

L. F. Burlaga
NASA/Goddard Space Flight Center
Laboratory for Extraterrestrial Physics
Greenbelt, MD   20771

ABSTRACT.   This paper reviews current ideas concerning the sources of interplanetary flows and their dynamical evolution, with emphasis on results that are relevant to measurements which will be made by Ulysses.

## 1.   INTRODUCTION

The solar wind is a mapping from the sun to the heliosphere. Beyond a few solar radii, the mapping is essentially radial from a spherical surface near the sun to a 3-dimensional domain dominated by the sun. The rotation of the sun assigns an orientation to the space around it, and it maps azimuthal gradients into radial gradients. The mapping is complicated by the intrinsic variability of the sun on many scales. The basic parameters of interest for solar wind dynamics are the velocity V, density N, magnetic field $\vec{B}$, and temperature T.

The equation of motion describes spatial and temporal changes in velocity. These changes are perturbations on a constant velocity solution described by Parker (1963). Daily averages of Voyager 1 observations of B(R) between 1 AU and 5 AU are shown at

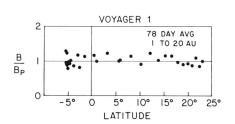

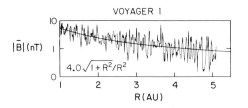

Fig. 1.   Voyager observations of the magnetic field strength versus latitude (top) and versus distance from the sun (bottom).

R. G. Marsden (ed.), The Sun and the Heliosphere in Three Dimensions, 191–204.
© 1986 by D. Reidel Publishing Company.

the bottom of Figure 1, from Burlaga et al. (1982) together
with the theoretical curve. One can see that Parker's model
provides a very good approximation to observations made near
the ecliptic. The fluctuations of the data with respect to
the theoretical curve for V=constant are the quantities of
interest in dynamical studies. Most of the high-field
regions are interaction regions ahead of streams and behind
shocks, but some magnetic clouds and "CME's" (Burlaga and
King, 1979; Burlaga et al., 1981) are included. Most of the
low-field regions are rarefaction regions in high speed
streams.

Parker's constant speed model predicts a latitudinal
variation of $\vec{B}$. Voyager 1 made observations at heliographic
latitudes from -7.5° to 23° as it moved from 1 AU to 20 AU.
Seventy-eight day averages of B are shown at the top of
Figure 1 as a function of heliographic latitude, normalized
with respect to the "spiral field strength". It is clear
that Parker's model provides a good approximation to the
magnetic field over a range of 30° in latitude. The
remainder of this paper is concerned with perturbations
about the values given by Parker's model.

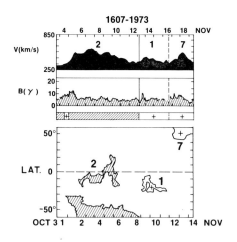

Fig. 2. Relations among co-
rotating interplanetary
streams, the magnetic field
strength and polarity, and
coronal holes.

## 2. COROTATING STREAMS

The principal type of flow in
the solar wind, particularly
at quiet times but also at
active times, is the "co-
rotating stream." Many
studies (Hundhausen; 1977)
have shown that coronal holes
are the sources of corotating
streams. Figure 2 from
Burlaga et al. (1978a) shows
three streams measured by the
Los Alamos instrument on Imp
7 and Imp 8 together with
magnetic field strengths and
polarity measurements from
the GSFC instrument on the
same spacecraft, and X-ray
measurements from the AS&E
experiment on Skylab. The
largest stream (#2) is
related to an equatorial
coronal hole, and the
polarity of the magnetic
field in the streams is the
same as that at the sun under
the coronal hole. A smaller

stream (#1) is associated with a smaller coronal hole which lies 5°-25° below the solar equator. A third stream (#7) is a polar coronal hole which does not extend below 45°. It has been established that flows from coronal holes spread out in latitude and longitude in the corona (see, e. g., Hundhausen, 1977, and Burlaga, et al., 1978), but little is known about the dynamics of flows in the corona. Note that an interaction region precedes the stream from an equatorial coronal hole, but an interaction region is not so clearly developed ahead of the stream from a polar coronal hole.

The dynamical evolution of a corotating stream between the corona and 1 AU is illustrated in Figure 3 from Pizzo (1981), based on a stationary, 2D MHD model. Very little is known about V, N, B, and T near the sun (this is one of the most outstanding problems for future work), so Pizzo took the inner boundary for his calculation at 0.3 AU, where observations from the experiment of Rosenbauer on Helios provided accurate input data. The speed profile is "mesa-like" with steep (but not discontinuous) sides and a flat top. The density in the stream is low, because it originated in a coronal hole. The temperature in the stream is high, presumably as a result of some heating process. Note that the leading edge of the stream is marked by an "interface" across which the density decreases, the temperature increases, and the speed increases. This signature is a reliable indication of a corotating stream. The computed profile at 1 AU which is shown on the right of Figure 3 is an accurate description of corotating flows at 1 AU. Comparison of the input and output profiles in Figure 3 shows a surprisingly large nonlinear evolution over as little as 0.7 AU. The trailing edge of the stream broadens kinematically as fast plasma in the stream moves away from slower plasma behind it. A rarefaction is thereby produced, further depressing the density,

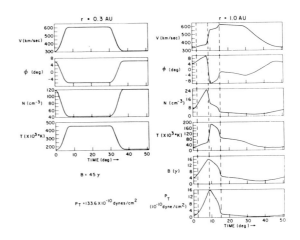

Fig. 3. Corotating streams profile at 0.3 AU (left) used as input to a 2-D stationary MHD model which produced the theoretical profile at 1 AU shown at the right.

but the density in the stream at 1 AU is low principally
because of the low density at the source.  As fast plasma at
the front of the stream overtakes and collides with slower
plasma ahead of it, matter and magnetic fields are compressed, the plasma is heated, and a pressure wave is
produced.  The pressure wave expands, accelerating the slow
plasma ahead by its positive gradient and decelerating fast
plasma in the stream by its negative gradient.  Shock waves
are beginning to develop at the front and rear of the pulse
in Figure 3, as indicated by vertical dashed lines. The
interface is clearly be seen at 1 AU (Burlaga, 1974), and
both observations (Schwenn et al., 1981) and theory
(Hundhausen and Burlaga, 1975) indicate that the interface
is thinner at 1 AU than at 0.3 AU.  Some slipping occurs
across the interface, as indicated by the change in the
azimuthal flow angle in Figure 3, which relieves the stress
and delays the formation of shocks.
     The evolution of corotating streams at high latitudes
should be qualitatively the same as that discussed above,
but the effects of compression, which are ultimately
associated with rotation of the sun, are smaller at higher
latitudes and they are absent over the poles.  Thus, the
enhancements in N, T, B, and P will be smaller at high
latitudes.  Since the interfaces observed near the ecliptic
at 1 AU appear to be produced by the overtaking of slow
plasma by fast plasma, which is related to solar rotation,
one expects the interfaces to be broader at higher
latitudes.  The jump in speed across the interface should be
larger at higher latitudes, where pressure gradients are
less effective in altering the speed profile.
     Having described the radial evolution of a single
corotating stream between the sun and 1 AU, let us now
consider the evolution of two streams between ≈1 AU and 5
AU.  Again, the evolution is best described by the results
of an MHD model.  In this case the input data are
observations made by Helios between 0.6 AU and 1 AU, and the
output is from the model of Whang (1984).  The results
(Whang and Burlaga, 1985) are summarized in Figure 4, where
the two corotating streams and their respective interaction
regions at the inner boundary are labeled B and C.  Each of
the streams produces a forward shock (F) and a reverse shock
(R) at the front and rear of its interaction region, respectively.  As shown in the lower panel of Figure 4, the
forward and reverse shocks move apart, and near 2 AU (point
$\alpha$) the reverse shock RB from stream B interacts with the
forward shock FC from stream C.  At this point, the separate
interaction regions coalesce to form a single "merged
interaction region", and a "secondary interaction region" is
formed between FC and FB.  The evolution of the interaction

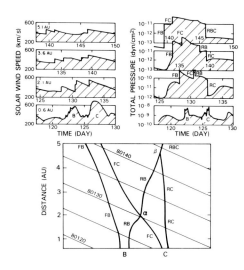

Fig. 4. Radial evolution of streams (top left), inter-action regions (top right) and shocks (bottom) resulting from the interaction of a fast stream overtaking a slower stream.

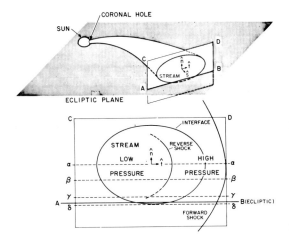

Fig. 5. Sketch of the 3-dimensional configuration of a corotating stream and inter-action region.

regions with increasing distance is shown by the pressure profiles at the top right of Figure 4, and the erosion of the streams is shown by the speed profiles at the top left of Figure 4.

Consider a co-rotating stream issuing from a compact near-equatorial coronal hole just above the ecliptic plane (Figure 5, from Burlaga et al., 1978). The stream will be confined to a spiral-shaped tube, and we con-jecture that it will be bounded by a thin inter-face interface on its leading edge (DB) at 1 AU, by a broad boundary on its trailing portion (CA) and by a layer of intermediate thickness at the top (CD) and bottom (AB). Schwenn et al. (1978) and Burlaga et al. (1978b) have shown that streams may be sharply bounded in latitude between 0.3 AU and 1 AU, and between the corona and 0.3 AU, respectively. A high pressure interaction region will form ahead of the stream, and its latitudinal extent will be greater than that of the stream.

A 3-D gas-dynamic model of a corotating stream from an extension of a polar coronal hole has been constructed by Pizzo (1977, 1980) for the region between 0.16 AU and 1 AU. The speed

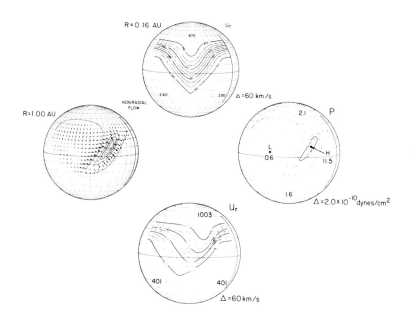

Fig. 6.    Theoretical  profiles  of  the  evolution of a co-
rotating stream from a polar coronal hole.

profile which was assumed for the computation is shown at
the top of Figure 6.   Fast plasma overtakes slow plasma en
route to 1 AU, resulting in a crowding of the lines of equal
speed on the leading side of the stream (higher longitudinal
speed gradients) and spreading of the lines on the trailing
side (smaller longitudinal speed gradients), as shown at the
bottom of Figure 6.   The overtaking produces a pressure
pulse, which is shown on the right of Figure 6.   Note that
although the speed gradient extends from the equator to 40°
N latitude, the pressure wave is concentrated at low latitudes, with a peak pressure at a latitude of $\approx 12°$.   The
pressure gradients induce a secondary flow, which is shown
at the left of Figure 6.   Although there is a measurable
meridional flow, which is a useful diagnostic of the stream
geometry, the associated mass transport is negligibly small.
      Beyond approximately 2 AU in the ecliptic (farther at
higher latitudes) a forward shock and a reverse shock will
form (Smith and Wolfe, 1976).   It is expected that the
forward shock will extend over a wide range of latitudes
than the reverse shock.   Forward shocks will probably be
seen at high latitudes by Ulysses, but reverse shocks are
less probable there.

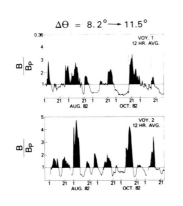

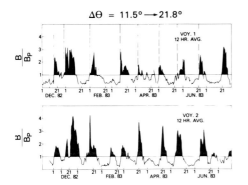

Fig. 7. Bottom: Merged interaction regions associated with stationary, recurrent lows. Top: Interaction regions associated with transient flows.

Some measurements of latitudinal variations of the magnetic field strength in interaction regions have been made by Voyager 1 and 2, beyond 10 AU. The magnetic field strength profiles from Voyager 1 and Voyager 2 for an 8-month interval in which corotating merged interaction regions were dominant and during which the latitudinal separation increased from 11.5° to 21.8° are shown the bottom of Figure 7 (Burlaga et al., 1985a). Allowance for the radial separation of the spacecraft has been made by shifting the time scales. There is a good correspondence between the merged interaction regions observed at Voyager 1 and those observed by Voyager 2. The maximum field strengths in the interaction regions at Voyager 2, which was near the ecliptic, were significantly-larger than those at Voyager 1, which was 11.5°-21.8° higher in latitude. This is consistent with the expectation that the compression will be smaller at higher latitudes, where the effects of solar rotation are diminished.

The observations just described refer to an interval in which corotating merged interaction regions were dominant. It is perhaps of interest to digress briefly to show similar measurements for an interval in which transient flows were dominant. These measurements (Burlaga et al., 1985) are shown at the top of Figure 6. Although Voyagers 1 and 2

are separated by only 8.2°-11.5° in this case, there is
little correlation between the two magnetic field profiles.
Again the largest peaks are at lower latitudes (Voyager 2),
although we cannot speak of latitude gradients in individual
interaction regions in this case.


## 3.  TRANSIENT SHOCKS

The shape of a shock depends on:  1) whether it is driven'
(Hundhausen, 1972) or detached (Burlaga et al., 1981; Smart
and Shea, 1985); 2) the ambient speed and pressure profiles
(Hirshberg et al., 1974; Heinemann and Siscoe, 1974; Burlaga
and Scudder, 1975); and 3) small-scale waves, discontinuities and turbulence (Burlaga et al., 1980).  A bewildering
variety of shock shapes can occur (Pinter, 1982), but for
the purpose of estimating what will be observed out-of-the-
ecliptic it is reasonable to use an average shape derived
from observations, rather than observation of a single shock
by any number of spacecraft.  Average shock shapes derived
from single spacecraft measurements and best guesses for the
sources were published by Hirshberg (1968), Ogilvie and
Burlaga (1969), and Chao and Lepping (1974), who estimated a
spherical shape with a radius of curvature of $\lesssim$ 1 AU at 1
AU and extending over
180° in longitude.
Recently radio observa-
tions by ISEE-3 have
made it possible to
identify the sources of
many shocks associated
with interplanetary
Type II bursts (Cane,
1985).   Unfortunately
only the fastest shocks
are selected in this
way, so the average
shock shape derived for
these shocks is not
necessarily representa-
tive of all shocks.
The average shock shape
derived by Cane (1985)
is shown at the top of
Figure 8.  The shocks
extend over nearly 180°
in longitude and the
average shock is nearly
spherical with a radius
of curvature of ≈1

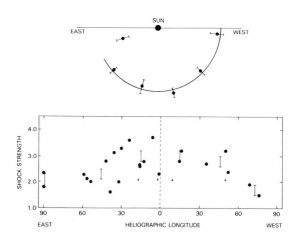

Fig. 8.  Top:  Average shock shape
derived from ISEE-3 observations.
Bottom:  Shock strengths versus
longitude.

AU. However, there is an east-west asymmetry indicating significant distortions from spherical symmetry near the eastern limb. The shock strength (measured by the jump in density) was determined as a function of longitude for the shocks considered by Cane (Figure 8, bottom). For sources near central meridian, the shocks are of moderate strength, and they are weaker at larger longitudes from the sun.

Let us now extrapolate the observations discussed above to estimate what will be observed out of the ecliptic. Most of the shocks observed by Cane (1985), as well as most of those discussed by Chao and Lepping (1974), were associated with solar flares. Flares are observed only at latitudes between +45° and -45°, the average latitude being ≈15°, depending on the phase of the solar cycle. To first approximation we can say that the flares are near the equator, and we can replace E-W in Figure 8 by S-N. Thus, we expect that shocks will be observed near the poles by Ulysses, although their strength will be weak, $n_2/n_1 < 2$. Polar shocks will be difficult to identify in magnetic field data alone, because many of them will be parallel shocks with no change in $|\vec{B}|$.

It is known that many shocks are associated with magnetic clouds (see, e.g., Burlaga et al., 1981; Klein and Burlaga 1982; and the references in Burlaga, 1984) and with coronal mass ejections (see the review by Schwenn in this volume). Coronal mass ejections (CME's) can occur at all latitudes, even over the poles (Sheeley et al., 1980). Hundhausen et al. (1984) have shown that there are more CMEs at high latitudes when the sun is more active. These results suggest that Ulysses will observe magnetic clouds and other ejecta associated with CMEs at high latitudes, and a significant fraction of the clouds should be preceded by shocks. Many of these shocks may be observed near the "nose" of the driver, hence they may be of moderate strength. As Ulysses moves from the earth to the sun's polar regions, it will observe shocks with increasing frequency. This is suggested by Helios observations of the frequency of shocks versus time (Figure 9) for the interval from 1975 to 1982 (Volkmer and Neubauer, 1984; Sheeley et al., 1985). The frequency increased from about one shock every other solar rotation in 1976 near solar minimum to approximately 5 shocks per solar rotation in 1981 near solar maximum. One can anticipate a similar increase in the rate of shocks during the first five years of the journey of Ulysses. Obviously, the state of the interplanetary medium will be very different when Ulysses moves over the poles than when it leaves the earth. Transient flows may be less regular and have smaller length scales than corotating flows (Goldstein et al., 1984), but systems of transients may have a large-scale "order" in the sense of being turbulent (Burlaga and Goldstein, 1984).

The configuration of shock waves may provide a more useful basis for conceptual organization of the structure of the solar wind near solar maximum than the flows and interaction regions. Figure 10 shows a possible scenario involving the production three shocks within 7 days. Shock $S_1$ is assumed to originate in an active region A at time $t_1$, and shock $S_3$ is assumed to originate in the same active region 7 days later at time $t_3 = t_1 + 7$. A different active region is assumed to produce shock $S_2$ at an intermediate time $t_2$, $t_1 < t_2 < t_3$. Figure 10 shows the situation at the time when shock $S_1$ is at 5 AU. Figure 10 is intended to show the intersection of the shocks with the ecliptic, but if the shocks are spherical and extend over 180° one can imagine the shocks as hemispheres extending out of the plane of Figure 10 and over the poles of the sun. Thus, the sun is surrounded by a "shell" of shocks".

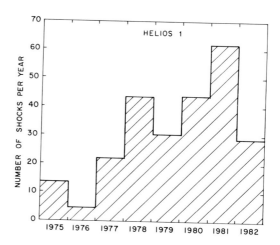

Fig. 9.  Number  of shocks per year versus time.

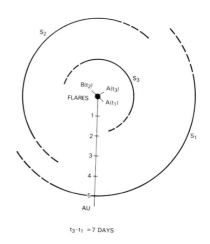

Fig. 10.  Hypothetical "shell" of shocks.

Based on our earlier discussion, the shocks might have a curvature which is as little as half that showing in Figure 10.  They may be bent on an intermediate scale owing to passage through streams and interaction regions; they may be rippled owing to inhomogeneities and fluctuations; and the shapes may be related to the driver speeds and configurations. Thus, the real configuration may be a distorted image of Figure 10, but the basic "shell" topology will remain. Burlaga et al., (1985b) have shown that the magnetic fields behind shocks may be turbulent on a scale of 0.1 AU near 1 AU, and Burlaga and Goldstein (1984) and Burlaga et al. (1985a) have shown that the turbulence may extend over large scales at larger distances. One can imagine a layer of turbulence behind each of the shocks in Figure 10 (Burlaga et al., 1985b) in which case the sun is surrounded by a patchy shell of turbulence. Such a shell is an effective barrier to cosmic rays near the ecliptic, and it is a dominant factor in long-term modulation of cosmic rays (Burlaga et al., 1985b). Since flare-associated shocks are weaker at higher latitudes, the turbulence and long-term modulation effects might also be weaker at higher latitudes.

## 4. CONCLUDING REMARKS

We have discussed a number of effects which one expects to observe out-of-the-ecliptic by the historic expedition of Ulysses, based on extrapolation of our present knowledge and experience. Of course, the most interesting results will be those which are not in agreement with our expectations. Ulysses will certainly answer many questions that have been raised, and Ulysses might offer surprises--entire new phenomena. It might be found that near the poles, where the field lines are radial and the field strength is low, the distribution functions are non-Maxwellian. In this case, the fluid approximation, which is the basis of all that was said in this review, may not be valid and all of our inferences are without foundation!

## 5. REFERENCES

Bell, B.: 1963, Smithsonian Contributions to Astrophysics, 5, No. 15.

Burlaga, L. F.: 1974, J. Geophys. Res., 79, 3717.

Burlaga, L.: 1984, Space Sci. Rev., 39, 255.

Burlaga, L. F., and Goldstein, M. L.: 1984, J. Geophys. Res., 89, 6813.

Burlaga, L. F., and King, J.:  1979, J. Geophys Res., 84, 6633.

Burlaga, L. F., and Scudder, J. D.:  1975, J. Geophys. Res., 80, 4004.

Burlaga, L. F., Behannon, K. W., Hansen, S. F., Pneuman, G. W., and Feldman, W. C.:  1978a, J. Geophys. Res., 83, 4177.

Burlaga, L. F., Ness, N. F., Mariani, F., Bavassano, B., Villante, U., Rosenbauer, H., Schwenn, R., and Harvey, J.: 1978b, J. Geophys Res., 83, 5167.

Burlaga, L. F., Lepping, R. P., Weber, R., Armstrong, T., Goodrich, C., Sullivan, J., Gurnett, D., Kellogg, P., Keppler, E., Mariani, F., Neubauer, F., Rosenbauer, H., and Schwenn, R.:  1980, J. Geophys. Res., 85, 2227.

Burlaga, L. F., Sittler, E., Mariani, F., and Schwenn, R.: 1981, J. Geophys. Res., 86, 6673.

Burlaga, L. F., Lepping, R. P., Behannon, K. W., Klein, L. W., and Neubauer, F. M.:  1982, J. Geophys. Res., 87, 4345.

Burlaga, L. F., McDonald, F. M., Goldstein, M. L., and Lazarus, A. J.:  1985a, submitted to J. Geophys Res.

Burlaga, L. F., McDonald, F. M., Goldstein, M. L., Lazarus, A. J., Mariani, F., Neubauer, F. M., and Schwenn, R.: 1985b, NASA TM-86213, submitted to J. Geophys. Res.

Cane, H. V.:  1985, J. Geophys. Res., 90.

Chao, J. K., and Lepping, R.:  1974, shocks, and solar activity, J. Geophys. Res., 79, 1799.

Goldstein, M. L., Burlaga, L. F., and Matthaeus, W. H.: 1984, J. Geophys. Res., 89, 3747.

Heineman, M. A., and Siscoe, G.:  1974, J. Geophys Res., 79, 1349.

Hirshberg, J.:  1968, Planetary Space Sci., 16, 309.

Hirshberg, J., Nakagawa, Y., and Wellck, R. E.:  1974, J. Geophys. Res., 79, 3726.

Hundhausen, A. J.:  1972, Coronal Expansion and Solar Wind, Springer-Verlag, New York.

Hundhausen, A. J., and Burlaga, L. F.:   1975, J. Geophys Res., 80, 1845.

Hundhausen, A. J.:   1977, in Coronal Holes and High Speed Wind Streams, p. 225, J. B. Zirker, Ed., Colorado Associated University Press.

Hundhausen, A. J., Sawyer, C. B., House, L., Illing, R. M. E., and Wagner, W. J.:   1984, J. Geophys Res., 89, 2639.

Klein, L. W., and Burlaga, L. F.:   1982, J. Geophys Res., 87, 613.

Ogilvie, K. W., and Burlaga, L. F.:   1969, Solar Phys., 8, 422.

Parker, E. N.:   1963, Interplanetary Dynamical Processes, Interscience Publishers, New York.

Pinter, S.:   1982, Space Sci. Rev., 32, 145.

Pizzo, V. J.:   1977, A three-dimensional model of high-speed streams in the solar wind, NCAR Coop. Thesis 43, Univ. of Colo. and Nat. Center for Atmos. Res., Boulder.

Pizzo, V. J.:   1980, J. Geophys. Res., 85, 727.

Pizzo, V. J.:   1981, Sol. Wind Proc. Conf., 4th, 153.

Schwenn, R., Montgomery, M. D., Rosenbauer, H., Miggenrieder, H., Mulhauser, K. H., Bame, S. J., Feldman, W. C., and Hansen, R. T.:   1978, J. Geophys. Res., 1011.

Schwenn, R., Mulhauser, K. H., and Rosenbauer, H.:   1981, Solar Wind Four, p. 118, ed. H. Rosenbauer, MPAE-W-100-81-31.

Sheeley, Jr., N. P., Howard, R. A., Koomen, M. J., Michels, D. J., and Poland, A. I.:   1980, Astrophys. J. Lett., 238, L161.

Sheeley, Jr., N. P., Howard, R. A., Koomen, M. J., Michels, D. J., Schwenn, R., Mulhauser, K. H., and Rosenbauer, H.:   1985, J. Geophys. Res., 90, 163.

Smart, D. F., and Shea, M. A.:   1985, J. Geophys. Res., 90, 183.

Smith, E. J., and Wolfe, J. H.:   1976, Geophys. Res. Lett., 3, 137.

Volkmer, P. M., and Neubauer, F. M.:   1984, submitted to
_Annals Geophysical_.

Whang. Y. C.:   1984, _J. Geophys. Res._, _89_.

Whang, Y. C., and Burlaga, L. F.:   1985, _J. Geophys. Res._,
_90_, 221.

PROPAGATION OF SOLAR WIND FEATURES: A MODEL COMPARISON USING
VOYAGER DATA

N. I. Kömle, H. I. M. Lichtenegger and H. O. Rucker
Space Research Institute of the
Austrian Academy of Sciences
Lustbühel Observatory
A-8042   Graz
Austria

ABSTRACT. For a three-dimensional description of the heliosphere it is
necessary to know the conditions and time variations of the solar wind
at remote positions not directly measured by in-situ observations from
a spacecraft. In the present paper we compare the predictions of
ballistic and hydrodynamic models with the actually observed solar wind
parameters. This was possible by the nearly radial alignment of both
Voyager spacecraft during the Jupiter pre-encounter period in 1978/79,
when the azimuthal separation was less than 4 degrees.
     For a selected period of time the propagation of solar wind
features was calculated both ballistically and hydrodynamically over a
distance of about 0.35 AU. A correlation analysis between the parameter
profiles from the model and the Voyager 1 observations shows that the
hydrodynamic propagation model yields essentially better results than
the ballistic propagation model.

1.   INTRODUCTION

A global understanding of the solar wind properties within the heliosphere requires the knowledge of how the solar wind changes its parameters as it moves radially outward from the sun. This is particularly
relevant if one wants to know the solar wind conditions not only along
the trajectory of a spacecraft but also at remote positions not always
accessible to direct measurement. Numerical models describing the
propagation of special HD and MHD disturbances in the solar wind have
been discussed extensively in the literature (see   e.g. Dryer and
Steinolfson, 1976; Wu, 1984 and references therein).
     As shocks and other discontinuities are common phenomena in the
solar wind it is imperative for a correct overall description of the
dynamics to use a numerical integration code which represents these
structures in an optimum way (no post-shock oscillations and no smearing).
     In the present paper we apply such a hydrodynamic model to the
propagation of the solar wind. As model input we used Voyager 2 (V2)
data during the Jupiter pre-encounter period in 1978.
     Fortunately we are able to check the level of reliability of the

*R. G. Marsden (ed.), The Sun and the Heliosphere in Three Dimensions, 205–210.*
© *1986 by D. Reidel Publishing Company.*

propagation model by comparing the calculated solar wind parameter
profiles with observations made by Voyager 1 (V1). The special
trajectories of both spacecraft enabled this comparison: after having
passed by V2 in early 1978 the V1 spacecraft steadily increased the
distance to V2, whereby both V1 and V2 were almost radially aligned
with respect to the sun. (The deviation was less than 4 degrees during
the period in question).

    After a brief description of the propagation models we analyse a
selected data sample. By means of a correlation analysis we are able to
determine the level of confidence inherent in the propagation models,
which will finally be discussed in detail.

## 2.  BALLISTIC AND HYDRODYNAMIC PROPAGATION OF SOLAR WIND FEATURES

Beyond the solar wind acceleration region the bulk flow speed of the
solar wind is roughly an order of magnitude higher than the propagation
speed of waves and shocks in the solar wind frame (Hundhausen, 1972).
Therefore in principle one can reproduce a solar wind feature at a
larger distance from the sun by advecting it outward with the local
solar wind speed. This approach takes into account only the radial and
azimuthal velocities (due to the bulk flow and the solar rotation,
respectively) and is known as the so-called "ballistic" propagation
(Desch and Rucker, 1983). As long as interactions between fast and slow
solar wind streams are of less importance this propagation model is
certainly useful to some extent. But if stream-stream interaction becomes important on the space and time scale to be investigated, a
hydrodynamic model must be used.

    In the present investigation we have applied the "Piecewise Para-
bolic Method" (PPM, Colella and Woodward, 1984) to the propagation of
solar wind features. PPM is an advanced hydrocode which is especially
well suited to treat properly discontinuities and shocks. Radial
symmetry is assumed in the model, i.e. all parameters depend only on the
distance r from the sun. An adiabatic energy equation with the adiabatic
index $\gamma = 5/3$ is used. For details of the code we refer to Kömle et al.
(1984) and Kömle and Lichtenegger (1984).

    As initial conditions we employ a simple steady state model with
a constant velocity and an inverse square law for density and pressure.
On the sunward boundary of the computational domain (position of V2) the
values for the bulk flow speed, the density and the pressure measured
by V2 are applied as boundary conditions. As time integration proceeds,
the structures move into the computational grid (consisting of 80 equally
spaced cells) and the resulting values at the outer boundary (position
of V1) are stored.

## 3.  SELECTED DATA SAMPLE

The efficiency of the developed hydrodynamic solar wind propagation
model is favourably checked at a high speed stream event including
several shocks. Such an occasion was offered in early December 1978 with

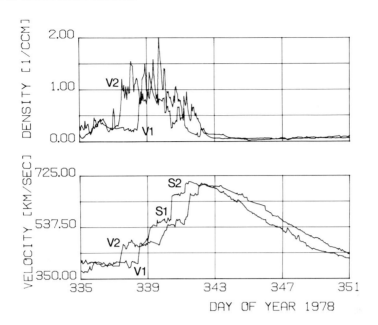

Figure 1. Velocity and density profiles as measured by Voyager 1 (V1) and Voyager 2 (V2).

a simultaneous V1 and V2 observation. In the present work we have used the interval from Dec. 1 to 16, 1978 (day 335.0 to 351.0, 1978). During this period V2 had a radial distance of about 4.3 AU from the sun and V1 was positioned further outward at 4.65 AU. Thus the solar wind structures passing V2 had to move over a distance of 0.35 AU before being detected by V1. Fig.1 shows the velocity and density data as measured by both Voyager spacecraft for the period in question. The data points are 1 hour averages. The V2 data needed as input for the model are fairly complete and contain only a few minor gaps which had to be interpolated for the numerical simulation. The data coverage of the subsequent V1 observation within the same period reached only 70%, but larger data gaps could be interpolated rather confidently.

      As can be seen from Fig.1 the interval covers a prominent high speed stream with a peak bulk speed of about 700 km s$^{-1}$ embedded in low speed solar wind. In the speed profiles the correspondence between the V1 and V2 observations is obvious, which is not the case for the density profiles.


4.   RESULTS AND DISCUSSION

The results of the ballistic and the hydrodynamic simulation are shown in comparison with the V1 data in Fig.2. The plots only cover the sub-interval from day 338 until 344, 1978, including the main features in the   respective profiles.

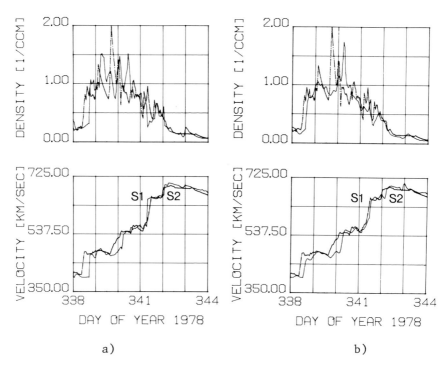

a)                                           b)

Figure 2. Comparison of measured V1 profile (dashed dotted line) with
a) ballistically propagated V2 profile (full line) and
b) hydrodynamically calculated V2 profile (full line).


As already seen in Fig.1 most features in the V2 profile can be
identified in the V1 data. Close inspection of the shock positions
reveals, however, that subsequent shocks change their relative positions
indicating an individual propagation speed in the solar wind frame,
which may be either forward or backward. The two shocks marked by S1
and S2 in Fig.1 obviously move towards each other, i.e. S1 moves backward in the solar wind while S2 moves forward. The existence of backward and forward propagating shocks in the solar wind is well known
(Hundhausen, 1972) and is reflected in our hydrodynamic model as can
be seen e.g. around day 342, 1978 in Fig. 2b. These special characteristic
inherent to hydrodynamic fluids obviously cannot be reproduced by a
ballistic propagation model, as demonstrated in Fig.2a.
        When simulating the solar wind propagation, a much more delicate
problem than the velocity is the correct prediction of the plasma density
profile. This is due to the fact that the temporal variations of the
density are much stronger than the corresponding variations of the bulk
speed. As already demonstrated in Fig.1 there is less good correspondence
between the actually measured V2/V1 density profiles than between the
respective velocity profiles. This fact is confirmed by a correlation
analysis as shown below. In the hydrodynamic simulation we observed a
density "spike melting" during the solar wind propagation from V2 to V1

which finally resulted in a single, but broader peak as marked in
Fig.2b. It is not yet clear and will be subject to further investigations
whether this spike melting is a numerical effect (possibly due to
numerical diffusion of the integration scheme)or a real physical interaction. In the actually-measured profile the spikes seem not to be
"lost", although their shape and height change considerably.
    The quality of the simulation of the solar wind propagation can be
determined by a correlation analysis. For this purpose we cross-
correlated the solar wind parameter profiles resulting from the
ballistic and hydrodynamic propagation with the V1 observations and
compared these findings also with the direct V2/V1 correlation. These
results are summarized in Fig.3. In the V2/V1 velocity correlation the
maximum correlation coefficient reaches $3\sigma$ ($\sigma$ = standard deviation) and
occurs at a position which coincides with the expected lag time of

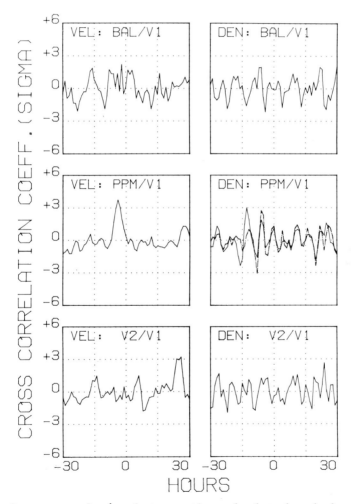

Figure 3. Cross correlation between the calculated and observed profiles
and between the V2/V1 data directly.

about 25 hours. The PPM/V1 velocity correlation reveals a coefficient that is somewhat higher than the direct V2/V1 correlation and occurs at a lag time around zero, because the PPM profile is already propagated to the position of V1. A surprising result is produced by the correlation between the ballistically propagated speed and the V1 velocity data. At the relevant lag time position the correlation coefficient remains insignificant.

The maximum density correlation for V2/V1 shows up at the expected position with a correlation coefficient close to $3\sigma$. Again the results from the ballistic model are worse. PPM also gives a low correlation (full line in Fig.3, DEN:PPM/V1). The reason for this behaviour lies in the spike melting discussed above. If those spikes that are melted in the PPM simulation (but separated in the V1 data) are excluded from the correlation, a density correlation coefficient close to $3\sigma$ occurs at the expected position (dotted-dashed line in Fig.3, DEN:PPM/V1).


## 5.   SUMMARY AND CONCLUSIONS

In this paper we have analysed the solar wind propagation by performing a case study. The speed and density profiles as measured by both Voyager spacecraft were used in a ballistic and a hydrodynamic propagation model.

Over the investigated period the relevant solar wind features remain clearly recognizable. This can be said especially for the speed profiles while the amplitude of the corresponding density spikes may vary considerably. The cross correlation analysis shows that the hydrodynamic model coincides considerably better with the observations than the ballistic one. Thus the present investigation makes evident the limitations of the ballistic solar wind propagation model and clearly demonstrates the improvement by a hydrodynamic model.


## ACKNOWLEDGEMENTS

We are indebted to H.S.Bridge, Voyager Plasma Science team principal investigator, for providing the Voyager plasma data and to M.D.Desch (NASA/GSFC) for the Voyager spacecraft ephemeries.


## REFERENCES

Colella, P. and P.R. Woodward, 1984: J.Comp.Phys., 54, 174
Desch, M.D. and H.O. Rucker, 1983: J.Geophys.Res., 88, 8999
Dryer, M. and R.S. Steinolfson, 1976: J.Geophys., 81, 5413
Hundhausen, A.J., 1972: Physics and Chemistry in Space, Vol 5, ed.
      J.G.Roederer. Springer Berlin-Heidelberg-New York
Kömle, N.I. and H.I.M. Lichtenegger, 1984: Comp.Phys.Comm. 34, 47
Kömle, N.I., H.I.M. Lichtenegger and K. Schwingenschuh, 1984: Report of
      the Space Res. Inst., Austrian Academy of Sciences, 31
Wu, S.T., 1984: Computer simulation of space plasmas, H.Matsumuto and
      T.Sato (eds.), 179, Reidel

# SECTION V:  HELIOSPHERIC STRUCTURE

3-D CORONAL AND HELIOSPHERIC STRUCTURE FROM RADIO OBSERVATIONS.

J.L. Bougeret, S. Hoang and J.L. Steinberg.
Département de Recherches Spatiales, U.A. CNRS No 264.
Observatoire de Meudon, 92195 Meudon Principal Cedex.France.

ABSTRACT. We review radio observations of the heliospheric 3-D
structure from the mid-corona to the Earth orbit, made from the Earth
and from space, using thermal emission, tracers in the form of radio
bursts or interplanetary scintillation of radio sources.

1. INTRODUCTION

In this review, we have concentrated on the structure of the mid- and
upper corona and the heliosphere in their steady state. There are good
recent reviews of solar radio astronomy by Kundu (1982) and Dulk (1985).
Solar transients and shock waves were covered in a recent Gordon
Conference (Stone and Tsurutani, 1985; Tsurutani and Stone, 1985).

1.1. Radiation mechanisms

Radio emissions are produced by the thermal motions of charged particles
of the plasma (thermal radiation) or triggered by suprathermal particles
(non-thermal mechanisms). In both cases, the emission takes place near
some critical level in the corona and heliosphere.
    Thermal emission depends upon the electron temperature of the
plasma and its coefficient of absorption (Kirchoff's law) which is an
increasing function of the density. Therefore, thermal emission comes
from the densest regions where the radiation can escape from; that is,
neglecting the effects of the magnetic field, regions where $f \gtrsim f_p$ ($f_p =$
$9 N^{1/2}$ is the plasma frequency in kHz; N is the particle density in
$cm^{-3}$). We note that a hot region 1 of low density can be seen colder
than another region 2 which is colder but denser than region 1.
    When the magnetic field is strong enough electrons on spiral
trajectories can radiate by the gyro-synchrotron mechanism at the
gyrofrequency and its harmonics. If the electrons are energetic enough,
we see a continuum of synchrotron radiation.
    Non-thermal emissions involve excitation, by some external agent,

213

R. G. Marsden (ed.), The Sun and the Heliosphere in Three Dimensions, 213–228.

of resonant modes of the medium; such a mode is found at the plasma
frequency $f_p$. If the excitation is strong enough, radiation at $2f_p$ is
produced by scattering of the plasma waves set in the medium.

The required conditions are not always met for the emitted radio
waves to leave the region where they are produced. Large scale
refraction and scattering on various inhomogeneities are observed. They
distort the images we see from the Earth.

The electron density decreases with the radial distance to the Sun.
Therefore, while observing at decreasing frequencies, we explore regions
located at increasing radial distances from the Sun. Meter waves are
emitted between 1 and 2 $R_o$ ($R_o$ is the solar radius), decameter waves
from 2 to 10 $R_o$. At frequencies lower than about 10 MHz, observations
must be carried out from space because of the ionospheric cut-off. On
the low frequency side, the limit is ultimately set by the cut-off
frequency of the interplanetary plasma, about 20 kHz at Earth.

Since we shall be reviewing results of the radio probing of the
corona and interplanetary medium, we shall mostly consider radio waves
whose wavelength is larger than about 1 meter.

1.2. Observable structures

The active regions are seen as bright regions since the corona is hotter
and denser above active centers than above quiet regions. There exist
other coronal regions which are relatively hot but less dense than the
surrounding medium: these are the coronal holes which have also been
observed by radio instruments.

Other structures are the magnetic loops or arches which sometimes
appear hotter than the surrounding corona, and the coronal streamers
which are denser than the surrounding medium. The contrast of these
structures – active regions, loops and streamers – as compared to the
rest of the corona is a function of frequency so that multifrequency
observations should lead to detailed models which could be compared to
those built on optical, EUV and X-ray observations.

The lines of force which support the observed structures guide
subrelativistic electrons which are sometimes injected onto them from
active centers and produce radio bursts. Tracking these radio sources at
different frequencies yields maps of the magnetic field at different
levels in the corona and interplanetary medium. For instance, when the
electrons travel over closed lines of force, they may produce the so-
called inverted-U burst or Type U burst which is a mere visualization of
a coronal arch; when they are injected onto open field lines of force,
they emit Type III bursts which can be observed as far as the Earth's
orbit and beyond.

1.3. Methods of 3-D observations

1.3.1. 2-D images and density models. Since the observing frequency
determines the heliocentric distance of the region we are looking at,
2-D observations can be used to map 3-D structures, providing we have a
density model. Thus we need images of quasi-permanent structures at
several frequencies which are obtained using radio heliographs (Clark

Lake, Culgoora, Nançay). For transient phenomena, such as bursts of
various types, the dynamic spectrum which shows gray-shaded or contour
plots of the intensity versus time and frequency contains information
about the structures through which the exciters (electrons) travel (see
Figure 3). In short, the exploration of frequency contains information
collected over different heliocentric distances or shells.

Density models can be obtained in a variety of ways: center-limb
observations (limited by refractive or scattering effects), optical, EUV
and X-ray observations (limited by the effects of integration over the
line of sight through an optically thin medium) or in-situ measurements
in the interplanetary medium.

1.3.2. Corotation of stable structures or sources. If the object is
stable enough, we can measure its apparent motion when the Sun rotates
over periods of days from which its radial distance can be deduced from
purely geometrical considerations.

1.3.3. Multispacecraft observations. The ideal method would be to obtain
stereoscopic pairs of radio images from widely separated sites. To
obtain radio images requires large radio heliographs. At the present
time we can only use ground-based images and space-based measurements of
parameters such as the intensity of radio bursts versus time. Intensity
ratios can be strongly influenced by nearby coronal structures.
Differences in arrival times yield the third dimension. At low
frequencies the ISEE-3 spacecraft yields a direction (two angles) so
that only 2 widely separated spacecraft are necessary to get a 3-D
position.

1.3.4. Scintillation of radio sources. Observations of the scintillation
of many radio sources can yield 2-D maps of interplanetary disturbances.
A rather tight grid is necessary (up to 1000 radio sources have been
used by Hewish) to get angular resolution. One can also study the
scintillation of a space probe telemetry link. Whether artificial or
natural radio sources are observed, there are effects of integration
over the line of sight to be subtracted from the data.


2. OBSERVATIONS

2.1. Active centers and coronal holes

As early as 1970, two-element interferometers operating on meter waves
recorded the signature of hot solar regions, superimposed on the "quiet
Sun" scans. These features could only be detected in the absence of
sporadic activity which appears in the form of short bursts.

Then multi-element interferometers were developed which provided
East-West scans with beam widths of a few minutes of arc. The time was
ripe for identifying radio regions with optical features. The effects of
the active regions and of the sporadic bursts could be removed and the
"quiet Sun" East-West profile was defined as the lower envelope of many
E-W scans taken over long periods of time. Hence all features with an

angular diameter smaller than the quiet Sun were considered as bright features due to some kind of hot region.

Leblanc (1970) and Axisa et al. (1971) published observations from the E-W Nançay radio heliograph and measured physical parameters of some coronal regions which were brighter than the quiet Sun. Then coronal holes and arched regions were discovered in UV and X-rays. The first radio identification of a coronal hole was obtained by Dulk and Sheridan

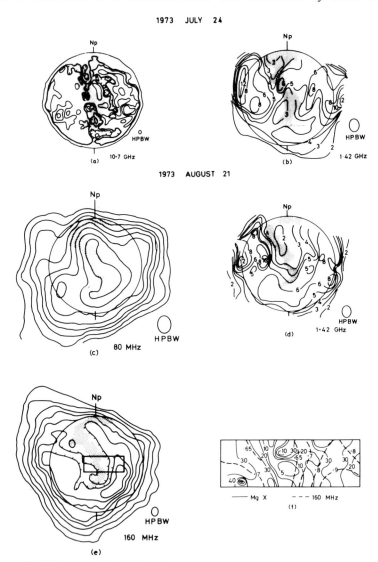

Figure 1. Maps of the Sun obtained at various radio frequencies and the MgX image, showing a coronal hole (shaded area) (After Dulk et al., 1977).

(1974) with the Culgoora 2-D radio heliograph operated at 80 and 160

MHz.These authors compared their maps to FeXV (284 A) images and unambiguously detected one coronal hole whose shape was roughly similar at radio and EUV wavelengths. They also derived the hole temperature (0.8 $10^6$ K) from the radio observations and that of the ambient corona ($10^6$ K); the hole density was about 1/4 of that of the corona. Looking at the limbs, they found, at fixed height, a radio brightness which correlated very well with that measured at the same height with a coronameter.

Lantos and Avignon (1975) analyzed the E-W Nançay array data using UV data as a guide to evaluate the N-S distribution of radiation in their E-W beam which had no N-S resolution. They concluded that the "quiet Sun level", obtained from lower envelope determinations was actually that of the coronal holes radiation, with a brightness temperature of 0.75 $10^6$ K. They did not find any solar cycle variations in this lower envelope.

Dulk et al. (1977) compared EUV images from Skylab with observations obtained at several radio frequencies: 80 and 160 MHz (Culgoora), 1.42 GHz (Fleurs) and 10.7 GHz (Bonn) (Figure 1). They deduced a model of the transition region and the corona which could not fit both radio and optical observations at the same time.

Recently, Alissandrakis et al. (1985) used the Nançay 2-D radio heliograph and aperture synthesis to produce 2-D maps of the slowly varying Sun with resolutions of 1.5' E-W and 4.2' N-S. This instrument had the high sensitivity and the fast response which were necessary to remove the effects of bursts: coronal holes were clearly visible, but several other features were also observed which could be identified with arches, streamers or the continua associated with Type I storm centers (Figure 2).

Our conclusion is that the only coronal structures which are clearly observed by meter wave radio heliographs are coronal holes and streamers. Other structures that might be seen as well will be discussed in the next paragraph. A very useful improvement would be to make simultaneous observations at several frequencies, larger than 169 MHz, in the hope that the deeper we look, the easier it will be to find optical features corresponding to the radio ones.

2.2. Coronal loops and arches in the low corona

Coronal magnetic structures are traced by both the thermal emission and the radio bursts.

2.2.1. Thermal emission. We have already mentioned the analysis of high resolution maps of the Sun by Alissandrakis et al. (1985). Using maps obtained during relatively quiet periods, and for several consecutive days, these authors carefully eliminated the sporadic contributions from Type I storms. The remaining features were usually weaker and did not coincide with either quiescent filaments or active regions. On the other hand, they were associated with neutral lines of the photospheric magnetic fields. Hence, Alissandrakis et al. suggested that these bright features, corresponding to denser or hotter regions, could be systems of magnetic arches, similar to those that can be seen in X-ray images,

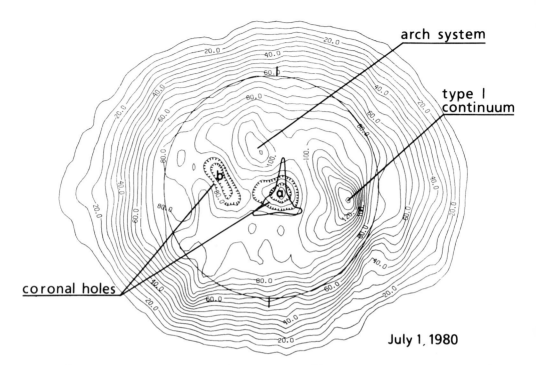

Figure 2. 2-D map of the Sun obtained at 169 MHz with the Nançay radio heliograph (After Alissandrakis et al., 1985).

but higher up in the corona (0.15 $R_0$ above the photosphere). This new result will have to be checked on a large number of periods and compared to other observations including X-rays, lower radio frequency maps, and coronagraphic observations. Alissandrakis et al. (1985) also suggested that shallow brightness depressions on the maps, which could not be identified with coronal holes, could be arch regions with low electron temperature and/or radio emission.

2.2.2. Radio bursts: type U. Recent analyses have revived the interest in type U bursts as tracers of coronal loops. Chains of U-bursts were found by Leblanc et al. (1983) to present a regular drift in the turning frequency - the top of the loops. The deduced velocities were typical of coronal transients, which showed that chains of U-bursts preferentially occurred in transient loops (Figure 3).

      On the other hand, the observation of storms of U-bursts by Leblanc and Hoyos (1985) in the frequency range 25-75 MHz (distance range from 1.8 to 3 $R_0$ from Sun center) implied that large magnetic arches can be stable for as long as a few days.

      In both chains and storms, the shape of the individual U-bursts suggested that the arches were diverging, a condition that seemed to be required for the production of radio emission in closed loops.

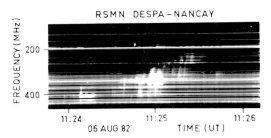

Figure 3. Dynamic spectrum of drifting U bursts obtained with the radio spectrograph in Nançay. This is a radio image of an expanding arch.

2.2.3. <u>Radio bursts: type I</u>. Other radio evidence for large scale magnetic arches was provided by the type I storm activity. It is currently accepted that this long lasting, highly polarized activity is generated in or near magnetic structures and thus traces those structures. In the meter wavelength range, it is usually associated with the feet of arch systems also seen in X-rays (Stewart and Vorpahl, 1977), but in some cases it obviously traces long distance magnetic links between active regions (Lantos-Jarry, 1970), demonstrating the existence of large scale magnetic arches.

Brueckner (1983) compared EUV observations of the spatial fine

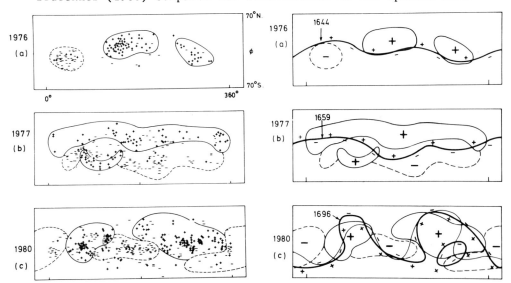

Figure 4. Left, yearly synoptic maps of the reduced positions and polarities (deduced from the sense of the radio polarization) of daily mean noise storm sources. Right, the zones of identical storm polarities compared with simulated heliospheric current sheet. The 4 digit numbers refer to Carrington rotations (After Stewart, 1985)

structure of the corona during the SKYLAB period with type I storm activity. He deduced that, in all cases, reconnection of coronal magnetic fields over large distances was the cause of type I storms, rather than changes of magnetic fields within an active region.

Stewart (1985) made a synoptic study of the occurrence and polarization of 160 MHz noise storms recorded at Culgoora during the current solar cycle. Such analyses can help understand the relationship between the coronal activity and the sector structure observed in the interplanetary medium. Indeed, the radio observation showed that strong magnetic fields in the low corona (where the type I storms are produced) tended to form a sector structure similar to that observed in the interplanetary magnetic field (Figure 4).

Combining the coronagraph/polarimeter observations on board the Solar Maximum Mission satellite and Nançay radio heliograph measurements, Kerdraon et al. (1983) found that type I storm onsets or enhancements were systematically associated with the appearance of additional material in the corona.

## 2.3. Streamers in the corona and interplanetary medium

2.3.1. Corona. Several works have used limb observations with the purpose of solving an old problem: are Type III bursts produced in overdense structures ? Type III radial distances at a given observing frequency f are found larger than that of the region where $f = f_p$ or $f = 2f_p$; and this observation can be interpreted in two ways: either some propagation effect makes these events appear higher than they actually are; or they take place in structures which are denser than the surrounding medium, of which we only know the average density taken over the line of sight. This old problem is probably not yet settled for good.

Since 1970 or so, radio observations of limb streamers have been combined with optical observations with coronameters. Bright streamers were detected both optically and with radio waves by Jackson et al. (1979) who found impossible to reconcile the models derived from the two sets of observations.

Pick et al. (1979) studied narrow streamers seen on the limb in 1974. These structures were about 10 times denser than the surrounding corona. This was roughly what was required to explain the height of radio bursts above the limb. More recently, Trottet et al. (1982) analyzed similar structures seen with the SMM C/P while Type III bursts were recorded with the Nançay radio heliograph. Type IIIs were seen along the extrapolated axis of narrow overdense structures; but this did not permit to conclude that the Type III electrons travel inside these structures (Figure 5).

Kundu et al. (1983) made a similar attempt at lower frequencies, comparing Clark Lake radio images to NRL–P78–1 corona images. Both the location and the estimated streamer electron density were consistent with the radio source being located within the dense regions of the streamer. But two possibilities were left, a classical dilemma in interpreting radio observations: a) the emission occurred at the second harmonic, and the actual position of the burst coincided with its apparent position, or b) the emission took place at the fundamental plasma frequency, but was ducted outwards, as suggested for example by Duncan (1979).

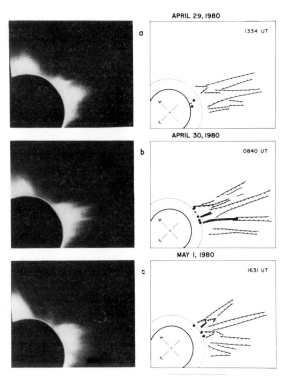

Figure 5. White-light coronal images (left) and a drawing of the main features (right) where are shown (black dots) the 169 MHz Type III positions measured at the same time (After Pick et al., 1979).

It was recognized from early radio heliograph observations, both at Nançay and Culgoora, that the position of sources of associated type III bursts could be widely scattered (Wild, 1970; McLean, 1970; Mercier, 1975). Observations by Mercier (1973) and Kai and Sheridan (1974) confirmed the idea that the type III bursts revealed the presence of coronal magnetic neutral sheets. The acceleration regions are thus preferentially distributed near plage filaments.

Mercier (1975) analyzed in detail high space-time resolution observations of several groups of type III bursts. His result strongly suggested that the magnetic field lines were widely diverging from the acceleration region. From his results, it can be estimated that the path of the exciters could be scattered in a cone as large as 60°. Sheridan et al. (1983) reached the same conclusion using observations at 4 frequencies with the Culgoora radio heliograph.

Lantos et al. (1984) observed simultaneously Type III bursts at 6 cm with the VLA and at 169 MHz with the Nançay radio heliograph. Changes of the 6 cm position by 10" induced changes of 0.5 $R_0$ of the meter wave position, thus demonstrating that the field lines issued from active centers are widely divergent.

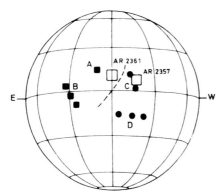

Figure 6. 10 arcsecond changes in the position of 6 cm bursts observed
in the active region 2357 with the VLA are simultaneous with shifts of
0.5 $R_o$ of the 169 MHz Type III positions (black dots) observed in Nançay
(After Lantos et al., 1984).

2.3.2. <u>Interplanetary medium</u>. Bougeret et al. (1984a, b and 1985), using
ISEE-3 observations, have studied several Type III storms; each of these
storms consisted of thousands of individual Type IIIs and persisted for
several days. This allowed the use of the apparent rotation rate method
to measure directly the radial distance at which a given frequency was
radiated. This work confirmed and extended earlier work (Fainberg and
Stone, 1970) with IMP-6 data: the emission took place in structures
which were overdense as compared with the average interplanetary medium.
The average IP medium density was determined using in-situ Helios data
(Bougeret et al., 1984). The density of the overdense structures was
observed to fall off as a function of the distance R from the Sun faster
than $R^{-2}$ while the mean density in the IP medium varies nearly exactly
as $R^{-2}$. At about 0.3 AU, the overdensity factor varied from storm to
storm: the higher it is, the faster the density falls off. At the Earth
orbit, the overdensity factor is only slightly larger than 1 (Figure 7).
Since ISEE-3 can measure not only the elongation of a source, but also
its elevation referred to the ecliptic, the work of Bougeret et al.
should soon be expanded to include the elevations, thus yielding a 3-D
mapping of these overdense regions.

2.4. Corotating turbulent or dense regions

In the last 15-20 years, the interplanetary medium has been studied
using radio source interplanetary scintillation (IPS), using larger and
larger antenna collecting areas and therefore an increasingly tight grid
of sources.
      The theory of these observations has been developed so that
characteristics of the medium producing the scintillations can be
obtained from the data, assuming, in general, spherical symmetry. At any
frequency, sources too close to the Sun cannot be used because the
scintillation regime changes in such a way that the scintillation index
(ratio of the rms intensity fluctuations to the average intensity)
decreases sharply. On the other hand, the way the intensity

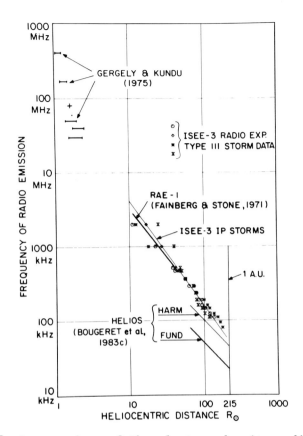

Figure 7. A comparison of the electron density radial variation in
streamers, obtained from kilometric noise storm positions measured from
ISEE-3, and in the "average" interplanetary medium obtained from in-situ
measurements (After Bougeret et al., 1984b).

scintillations are built up along the line of sight depends upon the
elongation of that line of sight and of the angular size and structure
of the radio source which is observed.

The pioneering works of Dennison and Wiseman (1968), Burnell
(1969), Watanabe and Kakinuma (1972), Houminer and Hewish (1972) and
Vlasov (1979) demonstrated the existence of corotating scattering
structures which were associated with stream–stream interaction regions.

The most comprehensive study was made by the Cambridge group
(Gapper et al., 1982; Tappin et al., 1983; Hewish et al., 1985). These
authors used a grid of about 900 sources observed daily at meridian
transit on 81.5 MHz. Maps of the reduced fluctuation of source intensity
$g = \Delta S / \overline{\Delta S}$ are shown in Figure 8. On these maps, interplanetary
travelling disturbances are seen, the shape and velocity of which can be
measured.

Interplanetary scintillation observations should yield a wealth of
information on the heliospheric structure, provided that they are
carried out regularly over long periods. Up to now, results have been

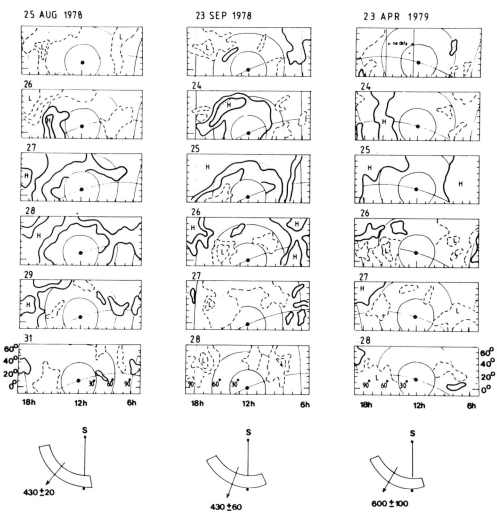

Figure 8. Sky maps of scintillation enhancements due to three
interplanetary transients. Contours are continuous or dotted for
scintillation index larger or smaller respectively than average. The
lower panel shows the approximate shapes and speeds of the density
transients as derived from the maps shown (After Hewish et al., 1985)

published using data acquired over periods of a few months only. There
is little doubt that this method is essentially sensitive to variations
of the plasma density although the theory shows that it should also be a
function of the distance from the observer to the scattering region and
the characteristic size (if any) of the density irregularities. These
(secondary ?) effects seem well understood and, in any case, the use of
many radio sources in widely different directions helps removing the
effects of the width of the weighting functions over the line of sight.
        Strong scattering effects are expected on interplanetary Type III
bursts, the frequency of which is much closer to the local plasma
frequency than the IPS observing frequencies.

## 2.5. The interplanetary magnetic field topology

Interplanetary Type III bursts, whether isolated or occurring in storms can be tracked by properly instrumented satellites; one can map the magnetic field lines which guide the electrons producing the radio emission. This can be done either using direction finding alone or with differences in the times of flight from the source to different spacecraft.

Several observations of the first method were reported in the literature (Baumback et al., 1976; Weber et al., 1977; Fitzenreiter et al., 1977; Gurnett et al., 1978). In most of these analyses, only a few bursts were observed with spacecraft whose difference in heliocentric longitude did not exceed 40° or so. They essentially confirmed that Type III radiation was mostly observed at twice the local plasma frequency and that the electrons which produced them travelled along spiral lines of force. Some trajectories were mapped out of the ecliptic; some left the Sun in one hemisphere and crossed the ecliptic afterwards (Fitzenreiter et al., 1977).

We shall limit ourselves to recent observations, mostly those made using the ISEE-3 radio receiver. Bougeret et al. (1984b) tracked noise storm sources for several days and were thus able to map the spiral magnetic field lines of force using a single spacecraft. From the spiral shape (Figure 9) they could also derive the solar wind velocity at several radial distances and in some cases the solar wind acceleration (Bougeret et al., 1983). This was done using only the ISEE-3 determined source azimuths (in the ecliptic), but will be extended to include the elevations.

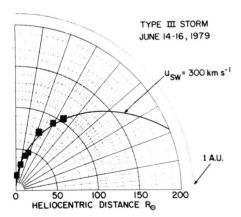

Figure 9. The spiral shape of the interplanetary line of force which guided the Type III storm electrons on June 14-16, 1979 (After Bougeret et al., 1984b).

Weber et al. (1977) observed a few isolated Type IIIs with Helios 1 and 2 and RAE B in lunar orbit. They analyzed differences in arrival times at frequencies larger than 500 kHz or so; they concluded that the positions determined from direction finding and differences in arrival times were essentially the same, if some corrections for refraction and scattering were made to the data.

Systematic observations of the same nature were made (Steinberg et al., 1984) which used the full direction (two angles) determined from ISEE-3 and the differences in arrival times measured between ISEE-3 and Voyager 1 and 2, more than 5 AU away. Large discrepancies were reported between the measured differences in arrival times and those which were predicted from the expected actual positions of the Type III sources based upon other types of informations (density models, active center position, arrival to the Earth of the Type III electrons). These differences, when fully understood, might bring some knowledge of the interplanetary structures which, most probably, produced them.


## 3. CONCLUSION

Radio astronomy has contributed a great deal to our description and understanding of the 3-D structure of the corona and heliosphere. The main reason is that radio waves are emitted in regions and structures which are not or hardly observable with visible light, EUV or X-rays.

In regions which are accessible to both radio and optical observations, radio astronomers do not yet identify all the radio counterparts of optical objects. Observations from radio spectrographs, multifrequency radio heliographs and polarimeters carried out over much longer time periods will be necessary to reach that goal.

The study and tracking of radio bursts should allow one to map arches which are not observable by any other methods because they are too high. Maps of open lines of force are obtained from the tracking of Type III bursts, up to the Earth orbit, something which could not be done by any other technique.

From space, there is a wealth of information still to be gained from the existing data, particularly from the ISEE-3 spacecraft. Many results have already been obtained and the measurement of elevations above or below the ecliptic should enable the true 3-D mapping of interplanetary field lines and interplanetary streamers. The tracking of isolated Type IIIs unveiled difficulties most probably due to propagation problems which are beginning to be understood. When they are fully understood, these very difficulties might be put into use to study the 3-D structures which produce them.

Due to these propagation problems, it is hard to accurately map the interplanetary field lines of force which guide Type III electrons. Another difficulty comes from the fact that Type III sources are now known to be large, 40° in longitudinal size, and broadened by scattering (Bougeret et al., 1985; Steinberg et al., 1985). Thus, we shall only be able to map an "average" line of force which guides the centroid of the radio source and we will lose the trace of all wiggles of the actual lines of force. In any case, the Earth environment is certainly not the best place to map spirals which are not much inclined out of the ecliptic. Radio instruments will be much more efficiently used from the ULYSSES space probe. Then we shall hopefully be able to directly plot the spiral shape of these lines of force.

We have seen that it is always much safer to use two spacecraft to get a 3-D map. It is a pity that the second ULYSSES probe has been cancelled by NASA, all the more so because the second spacecraft was to carry essential instruments, in particular, a coronagraph. A great opportunity has been missed by that decision, but we have many reasons to hope that much can still be achieved using our sole European spacecraft.

4. BIBLIOGRAPHY.

Alissandrakis C.E.,Lantos P. and Nicolaidis E.: 1985, Preprint.
Axisa F., Avignon Y., Martres M.J., Pick M. and Simon P.: 1971, Solar Phys. 19, 110-127
Baumback M.M., Kurth W.S. and Gurnett D.A.: 1976, Solar Phys., 48, 361-380
Bougeret J.L., Fainberg J. and Stone R.G.: 1983, Science, 222, 506-508
Bougeret J.L., Fainberg J. and Stone R.G.: 1984a, Astron. Astrophys.,136, 255-262
Bougeret J.L., Fainberg J. and Stone R.G.: 1984b, Astron. Astrophys., 141, 17-24
Bougeret J.L., King J.H. and Schwenn R.: 1984, Solar Phys. 90, 401-412
Bougeret J.L., Lin R.P., Fainberg J. and Stone R.G.: 1985, Preprint, to be submitted to Astron. Astrophys.
Brueckner G.E.: 1983, Solar Phys. 85, 243-265
Burnell J.: 1969, Nature, 224, 356-357
Dennison P.A. and Wiseman M.: 1968, Proc. Astron. Soc. Australia, 1,142-145
Dulk G.A. and Sheridan K.V.: 1974, Solar Phys., 36, 191-202
Dulk G.A., Sheridan K.V., Smerd S.F. and Withbroe G.L.: 1977, Solar Phys., 52, 349-367
Dulk G.A.: 1985, preprint, to appear in Ann. Reviews
Duncan R.A.: 1979, Solar Phys. 63, 389-398
Fainberg J. and Stone R.G.: 1970, Solar Phys. 15, 222-233 and 433-445 and 1971, Solar Phys. 17, 392-401
Fitzenreiter R.G., Fainberg J., Weber R.R., Alvarez H., Haddock F.T. and Potter W.H.: 1977, Solar Phys. 52, 477-484
Gapper G.R., Hewish A., Purvis A. and Duffet-Smith P.J.: Nature, 296, 633-636
Gurnett D.A., Baumback M.M. and Rosenbauer H.: 1978, J. Geophys. Res., 83, 616-622
Hewish A., Tappin S.J. and Gapper G.R.: 1985, Nature, 314, 137-140
Houminer Z. and Hewish A.: 1972, Planet. Sp. Sci., 20, 1703-1716
Jackson B.V., Sheridan K.V. and Dulk G.A.: 1979, Proc. Austral. Soc. Astron., 3, 387-389
Kai K., and Sheridan K.V.: 1974, Solar Phys. 35, 181-192
Kerdraon A., Pick M., Trottet G., Sawyer C., Illing R., Wagner W., and Kundu M.R.: 1982, Rep. Progr. Phys., 45, 1435-1541
Kundu M.R., Gergely T.E., Turner P.J., and Howard R.A.: 1983, Astrophys. J. 269, L67-71

Lantos-Jarry M.F.: 1970, Solar Phys. 15, 40–47
Lantos P. and Avignon Y.: 1975, Astron. Astrophys., 41, 137–142
Lantos P., Pick M. and Kundu M.R.: 1984, Astrophys. J.,283, L71–74
Leblanc Y.: 1970, Astron. Astrophys., 4, 315–330
Leblanc Y., Poquérusse M., and Aubier M.G.: 1983, Astron. Astrophys. 123, 307–315
Leblanc Y., and Hoyos M.: 1985, Astron. Astrophys. 143, 365–373
Mac Lean D.: 1970, Proc. Astron. Soc. Australia, 1, 315–316
Mercier C.: 1973, Solar Phys. 33, 177–186
Mercier C.: 1975, Solar Phys. 45, 169–179
Pick M., Trottet G. and Mac Queen R.M.: 1979, Solar Phys., 63, 369–377
Sheridan K.V., Labrum N.R., Payten W.J., Nelson G.J., and Hill E.R.: 1983, Solar Phys. 83, 167–177
Steinberg J.L., Dulk G.A., Hoang S., Lecacheux A. and Aubier M.: 1984, Astron. Astrophys. 140, 39–48
Steinberg J.L., Hoang S. and Dulk G.A.: 1985, in press in Astron. Astrophys.
Stewart R.T.: 1985, Solar Phys. 96, 381–395
Stewart R.T. and Vorpahl J.: 1977, Solar Phys. 55, 111–120
Stone R.G. and Tsurutani B.T.(eds): 1985, Collisionless shocks in the Heliosphere: A tutorial volume, Geophysical Monograph Series No 34, American Geophysical Union, Washington D.C., USA.
Tappin S.J., Hewish A. and Gapper G.R.: 1983, Planet. Sp. Sci., 31, 1171–1176
Tsurutani B.T. and Stone R.G. (eds): 1985, Collisionless shocks in the Heliosphere: Reviews of current Research, Geophysical Monograph Series No 35, American Geophysical Union, WashingtonD.C., USA
Trottet G., Pick M., House L., Illing R., Sawyer C. and Wagner W.: 1982, Astron. Astrophys., 111, 306–311
Vlasov V.I.: 1979, Astron. Zh., 56, 96–105, transl. in Sov. Astron., 23, 55–59
Watanabe T. and Kakinuma T.: 1972, Publ. Astron. Soc. Japan. 24, 459–467
Weber R.R., Fitzenreiter R.J., Novaco J.C. and Fainberg J.: 1977, Solar Phys., 54, 431–439
Wild J.P.: 1970, Proc. Astron. Soc. Australia, 1, 365–370

LATITUDE DISTRIBUTION OF INTERPLANETARY MAGNETIC FIELD LINES ROOTED IN
ACTIVE REGIONS.

G.A. Dulk[1], J.L. Steinberg[2], S. Hoang[2] and A. Lecacheux[3]
[1]Department of Astrophysical, Planetary and Atmospheric
Sciences,CB 391,University of Colorado,Boulder,CO 80309,USA
[2]DESPA, Observatoire de Paris, 92195, Meudon, France,
[3]DASOP, Observatoire de Paris, 92195, Meudon, France.

ABSTRACT. Electrons, accelerated in solar active regions to $\geqslant$ 10 keV,
escape along open magnetic field lines and generate plasma radiation
(Type III bursts) at frequencies ranging from $\approx$ 100 MHz near the Sun to
$\approx$30 kHz near the Earth. With the radio astronomy receiver on the ISEE-3
spacecraft we have measured the positions of 120 Type III bursts in the
range 0.03 to 1 AU (2 MHz to 30 kHz). Here we describe the latitude
distribution of the field lines traced out by the Type III electrons.
For the years 1980-1981 we find that about half of the field lines from
the active regions were confined within 15° to 20° of the ecliptic, and
about 90% were within 40° of the ecliptic. At the same time, the
latitudes of active regions on the Sun averaged about 15°. Many field
lines apparently bend over at large distances ($\sim$ 0.3 AU) and continue
outward at low ecliptic latitudes.

1. INTRODUCTION.

Type III radio bursts occur when electrons are accelerated to $\gtrsim$10 keV in
solar active regions and then travel outward through the corona and
solar wind at speeds of 0.1 to 0.5 c. At each successive height the
electrons excite plasma oscillations (Langmuir waves) at the local
plasma frequency $f_p$ ($\propto \sqrt{n_e}$) and some of the Langmuir wave energy is
converted to radio waves at frequencies $f=f_p$ and $f=2f_p$. Thus the
progress of the electrons to regions of successively lower density is
traced out by radio waves progressing to successively lower frequency.
    The fast electrons are forced to follow the magnetic field lines
rooted in the active region in which acceleration occurs. While
sometimes the field lines are "closed" and lead back to another location
on the Sun (giving rise to "inverted U bursts"), the field lines of
interest here are "open", spreading outward and permeating a limited
volume of the solar wind. Therefore, if we can observe the locations of
the radio bursts at progressively lower frequency we can trace out the
open field lines.
    The joint French-U.S. radio experiment on the ISEE-3 spacecraft has
recorded tens of thousands of Type III bursts during the years 1978-

R. G. Marsden (ed.), The Sun and the Heliosphere in Three Dimensions, 229–233.
© 1986 by D. Reidel Publishing Company.

1985. The instrument (Knoll et al., 1978) has the capability of
measuring several burst parameters in the frequency range 1980 to 30 kHz
(arising in the radial distance range ≈ 0.03 to ≈ 1 AU). Here we
describe the two-dimensional position measurements of the burst
centroids: azimuth east or west of the Sun in the ecliptic and elevation
north or south of the ecliptic. The data set utilized consists of 120
intense and isolated bursts which occurred during the interval January
1980 to August 1981, when ISEE-3 was located at the Lagrangian point
about $10^6$ km sunward of Earth.


2.  OBSERVATIONS.

Because we measure only two coordinates, we directly determine positions
on the plane of the sky and not the distance of the burst source along
the line of sight. However, we can estimate the radial distance from the
Sun from a knowledge of the average electron density profile of the
solar wind or certain empirical observations.

2.1. Type III burst trajectories.

Figure 1 shows the inner portion of the trajectories of Type III
electrons for each of the 120 bursts, where we utilized only the data in
the range 1980 to 160 kHz (≈ 0.03 to ≈ 0.4 AU). Several points on
Figure 1 are notable:

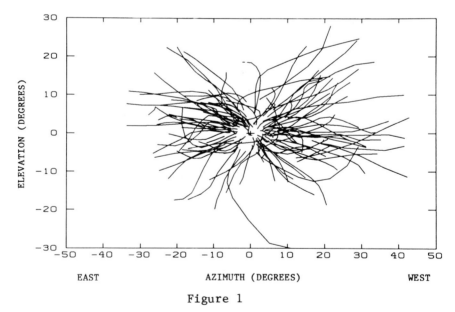

Figure 1

        1) Most of the trajectories lie fairly close to the ecliptic, with
only a very few being directed mainly to the north or south. Most lie
within 30° of the ecliptic.
        2) Many trajectories tend to bend over toward the horizontal, thus

becoming parallel to the ecliptic.

3) Several trajectories start off to the west and then cross over to the east, probably as a result of the Archimedean spiral form of the magnetic field.

4) The end points of the trajectories (corresponding to 160 kHz) lie at a variety of radial distances, and trajectories to the west tend to end at larger central distances than those to the east.

Figure 2 shows the trajectories for that subset of bursts that had azimuth and elevation measurements to a frequency of 60 kHz, where the sources lie at nearly 1 AU. On Figure 2 the most notable feature is that almost all trajectories lie near the ecliptic plane or bend over to become parallel to the ecliptic. The two exceptional bursts travelled nearly toward Earth but somewhat south of the ecliptic, and the elevation angle tends to large values when the source is close to the observer.

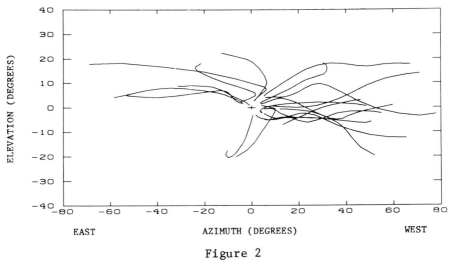

Figure 2

## 2.2. Ecliptic latitudes of field lines from active regions.

As we wish to proceed to determine the latitude distribution of the interplanetary field lines that determine the electron trajectories, we need to know the 3-dimensional location of the burst sources. Unfortunately this is generally not known on a case by case basis. We know only on a statistical basis that the average distance of burst sources at frequency f is considerably above the plasma level for that frequency. For example, for f = 233 kHz, the plasma level is located at R = 0.11 AU in the average solar wind according to the "Helios model" of density (Bougeret et al., 1984), but the observed 233 burst sources are located somewhere near R = 0.4 AU (Steinberg et al., 1984), apparently because of the propagation effects, mainly scattering (Steinberg et al., 1985).

We have proceeded to determine the latitude distribution of interplanetary field lines as follows:

1) At a given frequency f we <u>assume</u> that the burst sources are

located at radial distance $R_f$, where $R_f$ is a parameter to be varied.

2) For each burst we determine the source azimuth and elevation and their uncertainties from a least squares solution at frequency f and neighboring frequencies.

3) From the intersection of the direction of the source with the Sun-centered sphere of radius $R_f$, we determine the source latitude.

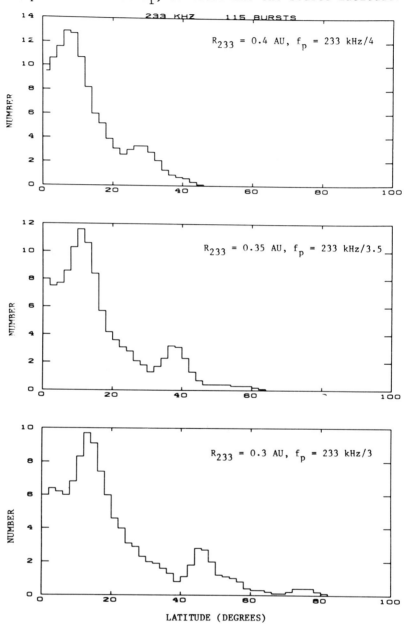

LATITUDE (DEGREES)

Figure 3

For the smaller assumed values of $R_f$ there are some bursts of azimuth and/or elevation so large that there is no solution for the source latitude. This is a warning that the assumed $R_f$ is too small. If the discrepancy is large the bursts are disregarded. For marginal cases the minimum possible $R_f$ is assumed; this does not seriously affect the derived latitude.

Figure 3 shows the ecliptic latitude distributions of 233 kHz sources which result from assumptions of $R_{233}$ = 0.3, 0.35, and 0.4 AU (corresponding to radial distances where $f_p$ = 233 kHz/3, 233 kHz/3.5 and 233 kHz/4).

For the larger assumed values of $R_{233}$ the latitude distribution is more confined. We can reject the results for $R_{233}$ = 0.3 AU on the basis that half of the bursts have azimuths and/or elevations too large to be consistent. There remain the distributions for $R_{233}$ = 0.35 and 0.4 AU (or possibly even larger). We know of no way to choose between these; however, the latitude distribution differs little from one to the other. Basically, the distribution peaks at a latitude of 10° to 15° and falls off rapidly near 15° or 20°. A few bursts were observed at latitudes higher than 35° to 45°, well beyond the latitude range of active regions, and indicate a non radial divergence of some field lines from some active regions.

3.  CONCLUSIONS.

During the years 1980-1981 the solar activity had just passed its peak and was starting to decline. The latitudes of active regions were, on average, about 15° (Solar Geophysical Data 1981). This is approximately the latitude of the peak in the interplanetary field line distribution as indicated by Figure 3, demonstrating that the field lines from active regions are often nearly radial to distances of $\approx$ 0.3 AU. Further out there is an indication that they frequently bend over to become approximately parallel to the ecliptic, as demonstrated in Figures 1 and 2.

4. REFERENCES.

Bougeret J.L., King J.H. and Schwenn R.: 1984, Solar Phys.,90, 401-412
Knoll R., Epstein G., Hoang S., Huntzinger G., Steinberg J.L., Fainberg J., Grena F., Mosier S.R. and Stone R.G.: 1978, IEEE Trans. Geosci. Electronics, GE-16, 199-204
Steinberg J.L., Dulk G.A., Hoang S., Lecacheux A. and Aubier M.G.: 1984, Astron. Astrophys., 140, 39-48
Steinberg J.L., Hoang S. and Dulk G.A.: 1985, Astron. Astrophys., in press.

HELIOSPHERIC STRUCTURE AND MULTISPACECRAFT OBSERVATIONS OF TYPE III
RADIO BURSTS.

J.L. Steinberg[1], S. Hoang[1] and A. Lecacheux[2]
[1]Département de Recherche Spatiale. [2] Département
d'Astronomie Solaire et Planétaire, Observatoire de Paris,
92195 Meudon Principal Cedex, France.

ABSTRACT  We have tried to obtain 3-D positions of Type III radio burst
sources from their direction measured from ISEE-3 and the difference in
arrival times at ISEE-3 and the Voyager probes, 5 to 8 AU away from the
Sun. When the difference in solar longitudes of the spacecraft does not
exceed 40° or so, we get positions in agreement with those obtained from
other data sets. When that stereo angle exceeds 80-120°, the difference
between the observed and expected positions can be as large or even
larger than 0.5 AU at 290 kHz. This is an indication that spacecraft at
very different locations do not see the same radio source centroid at
the same time. We analyze one such event which is seen behind the limb
on May 1, 1979; and suggest that most of the observed effects are due to
scattering on the large scale density structures of the heliosphere.

1.  INTRODUCTION.

It is generally agreed that Type III radio bursts are produced by
energetic electrons ejected from the Sun and guided along open magnetic
lines of force; all over their path, they generate high levels of plasma
waves at the local plasma frequency $f_p(kHz) = 9 \sqrt{N_e}$ where $N_e$ is the
plasma density in $cm^{-3}$. The plasma waves are scattered and partially
transformed into radio waves at $f_p$ and $2f_p$.
     By tracking the Type III radio source at several frequencies, one
can thus map a magnetic line of force as well as the electron density
along it. At frequencies below 5 MHz, the source region is in the
interplanetary medium and observations must be carried from space
because of the ionospheric cut-off. Several spacecraft are necessary if
one is to obtain positions in three dimensions; such observations were
made by Fitzenreiter et al. (1977), Gurnett et al., (1978) and Weber et
al., (1977). These last authors did not only use direction finding as
the others, but also analyzed the differences in arrival times among 3
spacecraft. They concluded that all position measurements were in good
agreement at frequencies higher than 1 MHz.
     Here, we report some observations made in 1979-1980 from ISEE-3,
which can measure the direction of a radio source (2 angles) at 24

235

*R. G. Marsden (ed.), The Sun and the Heliosphere in Three Dimensions, 235–240.*
© *1986 by D. Reidel Publishing Company.*

frequencies (Knoll et al., 1978; Fainberg et al., 1985), and the two
Voyager probes which were then 5 to 8 AU away from the Sun and have no
direction finding capability. The difference in solar longitudes between
ISEE-3 and the Voyager probes (stereo angle) varied from 0 to 120
degrees in absolute magnitude. Differences in arrival times ($\Delta t$) and
intensity ratios $R_S$ ($R_S = I_I/I_V$) were measured by comparing the burst
time profiles at the 2 or 3 spacecraft, either manually (using
transparent paper) or numerically on a computer. The frequency range was
500 to 100 kHz; 45 bursts were observed.

By analyzing a few observations we show in the present paper that,
these cannot be interpreted without taking into account the large scale
structure of the heliosphere, including such overdense regions as the
neutral sheet and interaction regions. More observations, their
statistical analysis and a discussion of their interpretation will be
published elsewhere.


## 2. OBSERVATIONS.

Because the Voyager probes are much farther away from the Sun than ISEE-
3 and because they carry shorter antennas (7m equivalent dipole length
against 45 m for ISEE-3), the receiving threshold of their radio
instrument is about 2000 times higher than that of ISEE-3. Our
observations are therefore limited to intense events.

When scanning both sets of data, it is found that most bursts
received on board Voyager with a signal/noise ratio of more than 10-15
dB have a counterpart on ISEE-3. This undoubtedly means that most
intense events can be recorded even when they take place behind the limb
from the observing site, as was found by Mac Dowall (1983). Those events
with a signal/noise ratio of more than ≈ 20 dB at ISEE-3 and more than
10-15 dB at Voyager were retained for analysis, a total number of 45.

The main results of this analysis are the following:
When the stereo angle is small (~ 20°) and events are not seen more
than 20° away from the Voyager-ISEE-3 baseline direction, the
differences in arrival times $\Delta t$ are approximately what they should be
(Fig.1), if one takes into account the known source direction and the
known range of radial distances of the Type III sources obtained from
the statistical study of Steinberg et al., 1984. The rms deviation
between the observed and predicted values of $\Delta t$ is about 30 seconds,
which is approximately the accuracy with which any one $\Delta t$ can be
measured. This is a proof that there are no timing errors involved in
the analysis.

However, for large stereo angles, 80 to 120 degrees, and assuming
that the spacecraft see the same centroid, the differences in arrival
times as well as the intensity ratios attain values which cannot be
reconciled with any plausible geometry. For a number of bursts, we have
independent estimates of source location based on observations of
flares, active centers, or 2-D radioheliograms at meter wavelengths. In
each case there is an anomalous delay in one of the paths from source to
spacecraft, and this anomalous delay path is always in the path to ISEE-

3. Here, we shall illustrate this result by analyzing one burst, that of 1979 May 1.

Figure 2 shows the geometry of the observations as projected onto the ecliptic (upper panel) and in the plane of the sky as seen by ISEE-3 (lower panel). Each hyperbola is the locus of possible source positions for which Δt has the value indicated. From the measured Δt, and the source azimuths measured by ISEE-3, the apparent source locations are derived; these are shown by triangles and error bars for a few frequencies. The derived 3-D positions are clearly behind the limb for

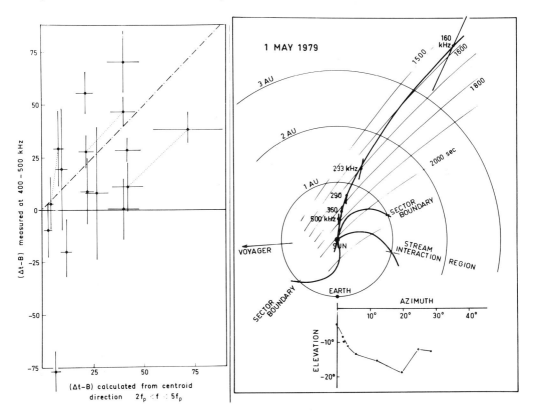

Figure 1: Plot of Δt−B where Δt is the difference in arrival time of a given burst profile at Voyager and ISEE-3 and B the baseline length counted in light-seconds. The vertical error bars correspond to the change in Δt which would result in a threefold increase of the σ of the fit of the burst time profiles; the horizontal ones to the uncertainty in source radial distance.

Figure 2: Geometry of the 1979 May 1 event as projected onto the ecliptic (upper panel) and on the plane of the sky as seen from ISEE-3.

both ISEE-3 and Voyager as already inferred by Mac Dowall (1984) for this same event. The lower panel shows that the sources were located almost directly south of the Sun at the higher frequencies ($\sim 300$ kHz) and progressively further west at the lower frequencies; this is incompatible with what we know of the interplanetary magnetic field structure where an event in front of the Sun always progresses to the east with decreasing frequency. Moreover, there was no associated meter wave event in the Culgoora data and no observed flare within 1 hour of the event. Thus all lines of evidence lead us to conclude that the event was south of the ecliptic and behind the Sun. A similar conclusion is drawn for a nearly identical burst which occurred about three hours earlier than the one described here.

However, when examined in details, the derived source positions are physically impossible. For instance, the radial distance of the 290 kHz radio source is found to be greater than 1 AU, thus implying an electron density of 260 to 1040 $cm^{-3}$ (assuming radiation at $2f_p$ and $1f_p$ respectively) at 1 AU very much higher than the densities observed along the Earth orbit. Moreover, the ratio $R_S$ of the flux densities at ISEE-3 and Voyager was measured to be about 2, whereas it should be about 20 based on the distances from the derived source positions to the spacecraft (assuming the source was behind the Sun; if the source were in front of the Sun, its distance to ISEE-3 would be at least a factor 2 smaller and this would worsen the situation). Clearly the flux density at ISEE-3 is smaller than expected given the derived position. (The position deduced from the flux density ratio places the source even farther away from ISEE-3 than that deduced from the time delay).

How then might the excessive time delay and intensity ratio be explained? In our view the most likely explanation involves scattering of the radio waves by small scale irregularities, and that the scattering is especially strong along the path to ISEE-3. Published Monte-Carlo calculations show that, on the average, rays tend to avoid regions of high scattering power where the angular deviation per unit path length is maximum, that is, regions of large density. Scattering in a spherically symmetric heliosphere does not seem, in general, to be capable of explaining how a source can be seen from a spacecraft when it is behind the Sun. In the present case, some rays could reach ISEE-3 because the source is considerably south of the Sun. However, scattering in discrete, overdense structures in the interplanetary medium provide a promising alternative. Indeed, the source of 1979 May 1 was in fact located behind a sector boundary (neutral sheet) and a stream-stream interaction region whose positions are shown in Figure 2. The sector boundary locations were taken from the magnetic field data in King's book (King, 1983) and confirmed by Stanford solar magnetic field maps (Solar Geophysical Data). The latter, together with calculations by Hoeksema et al.(1983) indicate that the neutral sheet reaches high solar latitudes. Neutral sheets are marked by localized increases in plasma density as described by Gosling et al. (1981) and Borrini et al. (1981). The two of interest here had densities at 1 AU of 30 $cm^{-3}$ on April 21 and 15 $cm^{-3}$ on May 4, 1979. Moreover, on April 25 a stream/stream interaction region passed the Earth with a density of 40 $cm^{-3}$. As seen on Figure 2, two of these scattering screens were located between the probable radio source and ISEE-3, but none between the source and Voyager. Thus we tentatively attribute the anomalous time delay to ISEE-3 and anomalous flux density ratio to the effects of scattering in the neutral sheet and/or stream-stream interaction region.


3.   CONCLUSION.

We have several other examples of observations made when the path from the source to ISEE-3 crossed overdense sheets. But the May 1, 1979 case is instructive because all spacecraft saw the Type III sources "behind the limb" and ISEE-3 saw it through scattering screens.

  It thus appears that spacecraft at very different longitudes do not observe the same Type III centroids. Scattering by the "average" heliosphere as well as by discrete structures must be considered a necessary ingredient of any model designed to explain the observations. The fact that the anomalous delay has so far always been found in the ISEE-3 leg might be explained as follows: a) ISEE-3 is closer than Voyager to the radio source, so that the average ray from the source to ISEE-3 travels closer to the Sun's equatorial plane where the density might be, on the average, larger than at higher solar latitudes; b) there is a selection effect in the choice of events: since the Voyager probes are less sensitive than ISEE-3, the selected events are likely to be in favourable positions to be seen from the Voyagers, i.e. roughly at the center of the Sun as seen from Voyager, while the source may be seen on or behind the limb by ISEE-3.

  Situations where different centroids are viewed by spacecraft in different directions are probably quite common. Bougeret et al. (1984) measured the radial distance of 25 long lasting Type III storm sources using their apparent rates of rotation near the time when the sources crossed the central meridian. Figure 10 of that paper shows clearly that the apparent altitude of a source becomes larger when the source longitude exceeds some value (a few days after meridian crossing); in the case of the 1979 May 27 to June 9 storm, the apparent altitude increase was 60% at a frequency of 290 kHz. The implication is that a spacecraft viewing a source on the limb would see it considerably higher than would another spacecraft viewing it near the Sun's center. Possibly related is the fact (Steinberg et al., 1984) that the angular diameter of Type III sources is larger at the limb than near the center of the Sun.

  The Voyager ISEE-3 observations at large stero angles show that interplanetary Type III radio sources can be seen at unexpected positions, positions which differ for the two spacecraft. A systematic case by case and statistical analysis should yield information on propagation through scattering screens and the properties and geometry of overdense heliospheric structures.

## 4. REFERENCES:

Borrini G., Gosling J.T., Bame S.J., Feldman W.C. and Wilcox J.M.: 1981, J. Geophys. Res., 86 , 4565-4573

Bougeret J.L., Fainberg J. and Stone R.G.: 1984, Astron. Astrophys., 141, 17-24

Fainberg J., Hoang S. and Manning R.: 1985, in press in Astron. Astrophys.

Fitzenreiter R.G., Fainberg J., Weber R.R., Alvarez H., Haddock F.T. and Potter W.H.: 1977, Solar Phys., 52, 477-484

Gosling J.T., Borrini G., Asbridge J.R., Bame S.J., Feldman W.C. and Hansen R.T.: 1981, J. Geophys. Res., 86, 5438-5448

Gurnett D.A., Baumback M.M. and Rosenbauer H.: 1978, J. Geophys. Res., 83, 616-622

Hoeksema J.T., Wilcox J.M. and Scherrer P.H.: 1983, J. Geophys. Res., 88, 9910-9918

King J., Interplanetary medium data handbook, Suppl. 2, 1978-1982, NSSDC, WDC-A and S, document 83-01

Knoll R., Epstein G., Hoang S., Huntzinger G., Steinberg J.L., Fainberg J., Grena F., Mosier S.R. and Stone R.G.: 1978, IEEE Trans. Geosc. Electronics, GE-16, 199-204

Mac Dowall R.J.: 1983, Directivity measurements of kilometric Type III radio bursts, M Sc. Thesis, University of Maryland, College Park, USA

Steinberg J.L., Dulk G.A., Hoang S., Lecacheux A. and Aubier M.: 1984, Astron. Astrophys., 140, 39-48

Weber R.R., Fitzenreiter R.J., Novaco J.C. and Fainberg J.: 1977, Solar Phys., 54, 431-439

THE RELATIONSHIP OF THE LARGE-SCALE SOLAR FIELD TO THE
INTERPLANETARY MAGNETIC FIELD: WHAT WILL ULYSSES FIND?

J. Todd Hoeksema
Center for Space Science and Astrophysics
Stanford University
Stanford, California 94305
U. S. A.

ABSTRACT. Photospheric magnetic fields observed at the Wilcox Solar
Observatory at Stanford together with a potential field model can be
used to calculate the configuration of the interplanetary magnetic field
(IMF). The IMF is organized into large regions of opposite polarity
separated by a neutral sheet (NS). Small quadripolar warps in a roughly
equatorial NS produce the four-sector structure commonly observed in the
IMF at Earth near solar minimum. Soon after minimum the latitudinal
extent of the NS increases substantially. Near maximum the structure is
more complex, occasionally including multiple neutral sheets. The IMF
structure simplifies and becomes very stable during the declining phase.
Large-scale structures persist for up to two years during the entire
cycle. From 1978 through 1983 the NS extended to at least $50^{\circ}$. Like
coronal holes, the computed coronal field shows evidence of decreased
differential rotation. Patterns in the inferred IMF polarity from solar
cycles 16 - 21 are very similar. This suggests that structures observed
by Ulysses during the coming cycle will be similar to those in the past.

1. Introduction

Ulysses' flight over the sun's polar regions provides a unique opportunity to explore a region of the heliosphere never before observed in
situ. Spacecraft have traveled over a wide range in heliocentric distances and all longitudes, but until now, all direct measurements of the
interplanetary medium have been carried out within a few degrees of the
ecliptic plane. Except for comets, the only probes of the solar wind
over the poles have been indirect, such as observations of the low
corona or analysis of interplanetary radio scintillation.
    It is possible to calculate the approximate magnetic structure of
the coronal and interplanetary magnetic field (IMF) using observations
of the photospheric field and a model for the behavior of the fields
above the sun's surface. The potential field model was first developed
independently by Schatten et al. (1969) and Altschuler & Newkirk (1969).
Two assumptions are made about the magnetic field in this model: 1) at
a certain height above the solar surface, called the source surface, all

241

*R. G. Marsden (ed.), The Sun and the Heliosphere in Three Dimensions, 241–254.*

the field lines are radial and open; and 2) between the source surface
and the photosphere the field is a potential field, i.e.  there are no
currents.  Observations of the photospheric field and these assumptions
are sufficient to allow a determination of the coronal field structure
by solving Laplace's equation.

To describe the structure of the IMF we make the further assumption that the magnetic structure on the source surface is carried radially outward by the solar wind.  See the paper in this volume by Fry &
Akasofu for examples of kinematic extensions of simple neutral sheet
configurations into the interplanetary medium.  See also Suess et al.
(this volume) for cautions about the effects of varying solar wind speed
on this simple first approximation.

While none of these three assumptions is strictly true, they are
adequate to reproduce the large scale features of the coronal field and
the IMF.  Later improvements of the model by modifying these assumptions
produce better detailed agreement of the fine scale field, but do not
greatly affect the large scale organization of the IMF (e.g.  Levine,
1982 and references therein.) In this paper I will present results of
this model for the present solar cycle and make a qualitative prediction
of the IMF structures Ulysses may encounter.

Measurements of the photospheric field began at the Wilcox Solar
Observatory in May 1976 and have continued to the present.  Daily magnetograms are combined to form synoptic charts of the line-of-sight

Photospheric Magnetic Field        0, ±100, 200, 500, 1000, 2000 MicroTesla

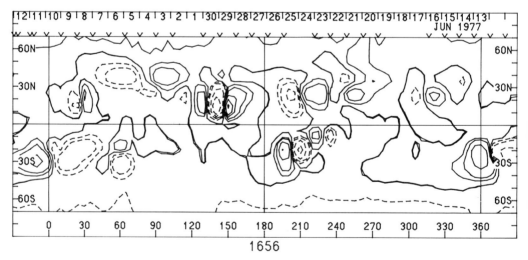

Figure 1:  The observed line-of-sight magnetic field at the  photosphere
for  Carrington Rotation 1656.  Magnetic contours are 0, ± 100, 200, 500
central meridian passage.  Inverted carets show dates of magnetograms.

magnetic field as shown in Figure 1.  In 1977 the field organization was
relatively simple.  There were small strong active regions and larger
weak-field regions which showed a predominance of one polarity.  Because
of the large 3' aperture of the Stanford instrument the spatial resolution in the polar regions is relatively poor.  The last observed scan
point is centered at about 70° latitude, therefore the area above 70° is
blank.
      Photospheric data for 360° of longitude are used in the potential
field model calculation.  The calculation is repeated each ten Carrington degrees and the results combined to produce the best estimate of the
coronal field at each location.  Correlations with the IMF polarity
measured near Earth have been used to find the best location for the
source surface:  2.5 solar radii.  The large aperture and the necessity
of measuring the line-of-sight component of a nearly radial field near
the limbs require a substantial correction to the polar data near solar
minimum.  See Hoeksema et al. (1982, 1983) for a more complete description of the model and the method.

2.  The Interplanetary Magnetic Field Configuration Through the Solar
Cycle

The results for the radial field at the source surface are assembled
into synoptic charts much like the photospheric charts.  Figure 2 shows

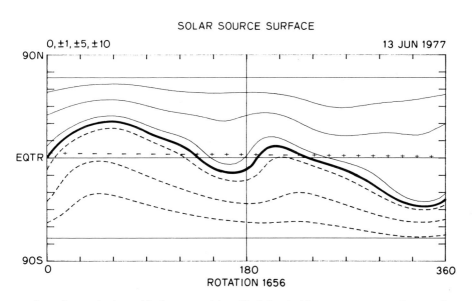

Figure 2:  Computed radial magnetic field at the source surface  for  CR
1656.  Magnetic  contours  are  0,  ±  1, 2, 5, ... µT; dashed contours
represent field directed toward the Sun.  The + and - symbols  show  the
daily IMF polarity measured near Earth.

the results for the same Carrington Rotation, CR 1656. The field at the source surface is much weaker and much simpler. The heavy line is the neutral line. This field pattern is drawn out by the solar wind into the interplanetary medium. Note the limited latitudinal extent of the neutral sheet. Even so, the neutral sheet already extends to higher latitude than at solar minimum in 1976 (Smith et al., 1978). This is typical of the field configuration just after solar minimum and corresponds to the early part of Ulysses' journey to Jupiter.

The IMF polarity measured by spacecraft near Earth is shown for reference. There is a very good agreement with the polarity predicted by the model. The kinds of structures observed are sensitive to the latitude of the observer. Above 15° two unequal length sectors would have been observed rather than the four actually observed at Earth. Above 30° the IMF would always have been the same polarity.

During the rising phase of the cycle the latitudinal extent of the neutral sheet increases rapidly to higher latitudes. Figure 3, showing Carrington Rotation 1665, less than a year later, indicates that the neutral sheet extends to nearly 60° in each hemisphere. The IMF would show four sectors everywhere between 60° N and 60° S. This configuration, typical of most of 1978, would not be very sensitive to the latitude of the observer until the observer approached 60°. One would expect Ulysses to see the same sector structure as the Earth during the corresponding part of the next solar cycle (2 to 3 years after minimum), some time after its encounter with Jupiter. Depending on when the new

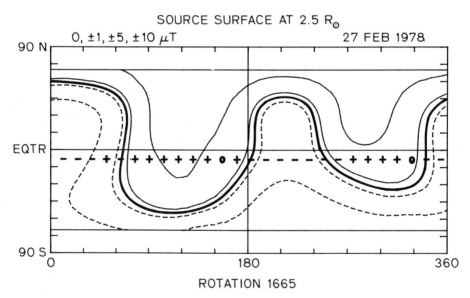

Figure 3: The radial field at the source surface for CR 1665 is typical of the rising phase of the solar cycle. Note the increase in latitudinal extent since solar minimum.

## HELIOSPHERIC CURRENT SHEET STRUCTURE: 1978-1980

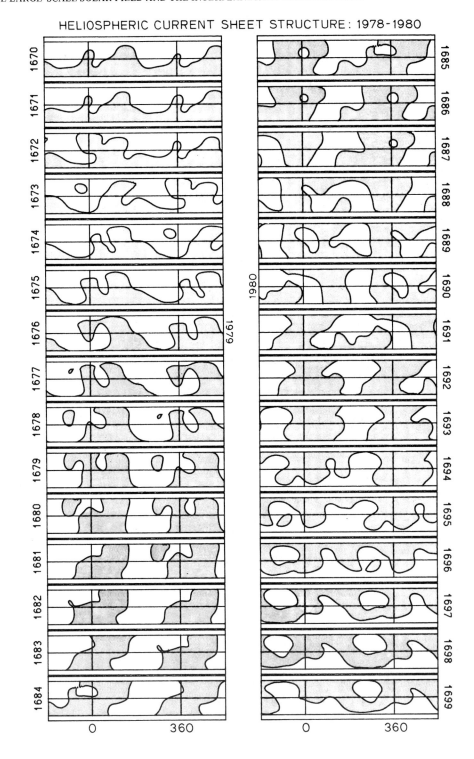

cycle actually begins, Ulysses may pass above the neutral sheet soon
after the Jupiter encounter and then be passed by the neutral sheet
sometime later as the sheet's latitudinal extent increases rapidly.

The times of greatest interest will be when Ulysses is at high
latitudes over the solar poles. This should occur in the year or so
preceding solar maximum. Figure 4 shows the computed location of the
neutral sheet for each rotation in the period around the last solar maximum: CR 1670 - CR 1699; mid 1978 - late 1980. The strong polar fields
are gradually fading and the neutral sheet extends above $70^{\circ}$ through
most of the period. The polar field reverses around the end of 1979,
but the evolution from CR 1670 to CR 1699 is smooth -- there is no sudden change in the polar field. The structure is more complex during
this interval than in the rising and declining phases of the cycle.
There are often multiple neutral sheets. In spite of the complexity,
many large features can be followed for long periods of time, just as
during the rest of the cycle.

The greatest changes in the location of the neutral line generally
occur far from the ecliptic. Throughout the interval the changes near
the equatorial plane are small. The Earth usually experiences only two
sectors and there are few sudden changes in the IMF sector structure
near the ecliptic. Note that the shape of the neutral line changes more
rapidly than the pattern of the large field regions because of its sensitivity to relatively small changes in regions where the field is weak.

The IMF structure often differs greatly at different latitudes
during this interval. Consider, for example, Carrington rotation 1679.
There the structure in the northern hemisphere is fairly complex compared with the southern hemisphere. At the highest latitudes there is
probably a single IMF polarity, but at slightly lower latitudes there
are two sectors, while near the equator there are four or more. There
are multiple neutral sheets which will be detectable over only a limited
range of latitudes. Another example of this is CR 1698. Such a
heliospheric structure will look much like the stream of water from a
rotating sprinkler.

During other intervals the structure is very similar over large
latitude ranges, e.g. CR 1684 or CR 1693. An observer at high latitude
would see the same structures observed in the ecliptic but at shifted
longitudes.

Ulysses will be at high latitudes during just this period of the

---

Figure 4 (Preceeding Page): The heliospheric neutral sheets for CR 1670
- CR 1699. Negative polarity regions are shaded. Each panel shows the
labeled rotation plus an additional half rotation on each side, thus
each box is two rotations wide. Vertical lines show rotation boundaries. Horizontal lines show $\pm 70^{\circ}$ and the Equator. The polar fields
seem to reverse near solar maximum around CR 1690 in late 1979.

next cycle when there will be a great deal of evolution in the IMF configuration over the polar regions. The spacecraft will be racing the
neutral sheet to the poles and may experience the change in polar field
direction in either the north or south or both. It will be interesting
to compare solar wind parameters at high and low latitudes just at this
time when the polar coronal holes are disappearing and the field direction changing. It will be important to remember how different the
structures may be in the northern and southern hemispheres.

For completeness, Figure 5 shows the structure typical of the declining phase of the cycle. The neutral sheet still extends over a large
range in latitudes, but the structures are much simpler and much more
stable. During almost all of 1982 the field looked much as pictured
here. There was a very stable four-sector structure extending to about
60 degrees during most of 1983. Perhaps such structures will be sampled during Ulysses' second orbit.

## 3. Similarity of Solar Cycles

To make predictions there must be some confidence that the future will
be like the past. Comparisons of the K-coronameter data and the potential field model show a good agreement near solar minimum (Wilcox & Hundhausen, 1983), but the coronameter data is more difficult to interpret
near solar maximum. The same applies to interplanetary scintillation
data (Rickett & Coles, 1983.) Furthermore, the time period over which

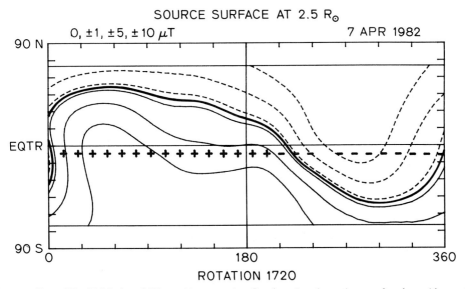

Figure 5:  CR 1720 typifies the neutral sheet structure during the declining  phase of the cycle with a large latitudinal extent and a simple,
slowly varying configuration.

INFERRED SOLAR MAGNETIC SECTOR STRUCTURE DURING FIVE SUNSPOT CYCLES

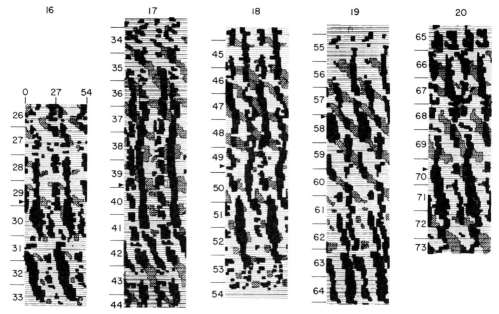

26.84 DAYS  CALENDAR SYSTEM  STARTING FEB 19, 1926

Figure 6:  A plot of the inferred solar magnetic structure  during  sunspot cycles 16 - 20.  Two successive rotations are displayed horizontal-
ly to aid in pattern recognition.  Sectors with field polarity  directed
toward  the  sun  are  shaded black if they are judged to be part of the
four-sector pattern and have dashed shading if  part  of  the  28.5  day
structure.  The evolution of the pattern is much the same from one cycle
to the next and was similar in Cycle 21.  (Svalgaard & Wilcox, 1975)

such data exist is limited to the past several years.  Potential field
model calculations of the source surface fields have been made starting
with data from 1959 (Newkirk et al., 1973.) and show similar patterns to
the present analysis, though comparison is somewhat difficult because of
the relatively large zero-level errors.
    Geomagnetic activity can be used to infer the polarity of the IMF.
Svalgaard & Wilcox (1975) have compiled the polarity data from solar
cycles 16 through 20 into Bartels charts, as shown in Figure 6.  In such
a chart a box represents each day.  Each box is shaded to represent the
assigned polarity.  Each row shows the polarity for 54 days.  Successive
rows begin at intervals of 26.84 days.  In this system, patterns recurring with a 26.84 day period produce vertical structures.  Patterns with
other recurrence times will have different slopes.  Svalgaard & Wilcox
found two predominant recurrence times at about 27 days and 28.5 days.

Negative polarity days judged to be part of the 27 day, four-sector
structure were shaded black and stand out as vertical bars. Negative
polarity days judged to be part of the 28.5 day, two-sector structure
have dashed shading. Similar structures are observed in each solar
cycle. Similar patterns can be observed in solar cycle 21. The structures have very long lifetimes and are very stable.

The potential field model can be used to predict the daily IMF
polarity. The predicted polarity is correct about 80% of the time. The
structures predicted are almost identical to those observed in the IMF
in the current cycle. This gives us confidence in the results of the
model. This also gives us some confidence that the next solar cycle
will be similar. Of course similarity near the ecliptic does not
guarantee that higher latitudes will be the same, so we should be ready
for some surprises.

We should also note that the model parameters have been adjusted
to produce the best agreement with ecliptic measurements of the IMF
polarity. Ulysses will provide the first real test of our 3-dimensional
prediction of the IMF structure. It will be extremely valuable to calibrate the model for higher latitudes. This will improve the predictive
capability of the model for all time periods.

4. Analysis of the Solar Cycle

One way to get a quantitative grasp of the evolution of the heliospheric
field through a solar cycle is to consider the variations of the multipole components of the field at the source surface. Figure 7 shows
the RMS contribution to the field at the source surface due to the
dipole, quadrupole, and octupole components. The higher order multipoles are much smaller in magnitude at the source surface because they
decrease much more rapidly with height above the photosphere.

Near solar minimum the dipole component dominates the field configuration. There are small contributions from the quadrupole and octupole terms which affect the position of the neutral line near the equator enough to produce the four-sector structure observed at Earth.

During the rising phase of the cycle the dipole component
decreases in magnitude, reaching a minimum about the time of solar maximum. After maximum there is a gradual increase in the dipole field.
There is a large temporary increase in the dipole component in 1982
corresponding to the very stable two-sector structure observed
throughout 1982 in the IMF. If the IMF could be described as a rotating
dipole as suggest by Saito et al. (1978), the magnitude of the dipole
should remain constant. This together with an analysis of the individual components of the dipole suggests that the polar dipole fades away
during the rising phase of the cycle and reestablishes itself after maximum. The equatorial dipole component evolves independently.

The other multipoles gradually increase in magnitude, reaching a
maximum value in the years following the maximum in solar activity. For
several years, from 1979 through 1981, the quadrupole component is of
comparable magnitude to the dipole. Even the octupole contributes substantially during this period. The simple description of the IMF as a
dipole is not accurate during these years. Even during the rest of the

# MORPHOLOGY OF POLAR FIELD REVERSAL

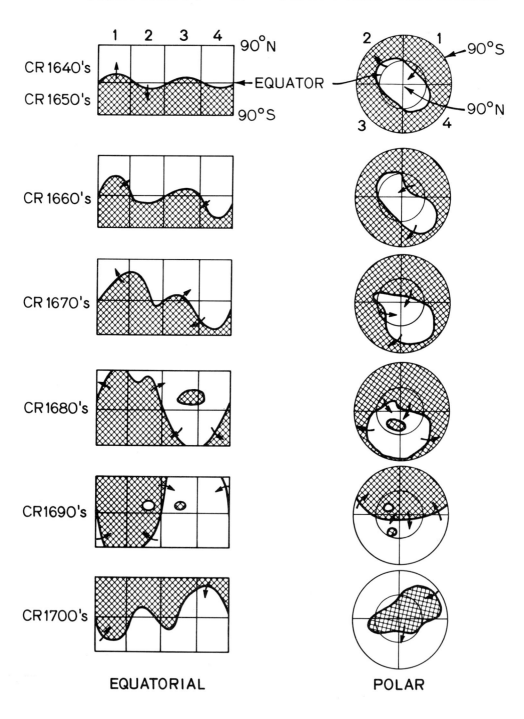

EQUATORIAL                              POLAR

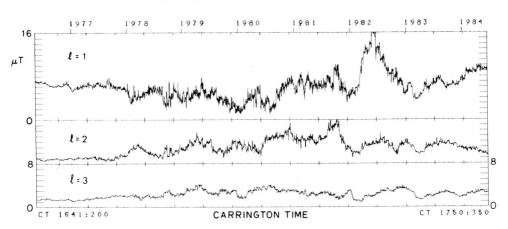

Figure 7: The RMS value of the field at the source surface due to the dipole (1=1), quadrupole (1=2), and octupole (1=3) multipoles from May 1976 through June 1984. All are plotted on the same scale.

cycle the higher order components are important in shaping the IMF structure observed near the ecliptic. See for example Figure 3.

All this suggests that the evolution of the IMF structure is very smooth in time. There is no sudden reorganization of the field configuration at the reversal of the polar fields. Rather, there is a slow ordered transition from one state to the other corresponding to the gradual reorganization of the large-scale photospheric magnetic field. It is difficult to conceptualize the reversal of the polar field, partly because of the format in which the fields are typically shown. To help illustrate this, Figure 8 shows the position of the neutral sheet in two projections. One is the standard projection corresponding to the figures shown previously. The other projection emphasizes the continuity

Figure 8 (Opposite Page): The panels at the left show the evolution of the field configuration before and after maximum in the standard projection. Each frame typifies the field structure for the indicated rotations. The panels at the right show the polar azimuthal equidistant projection of the same field configurations. This illustrates the smooth evolution of the neutral sheet in the northern hemisphere from minimum through maximum and the beginning of the declining phase of the cycle. The arrows suggest the movement of the neutral line with time.

of the structures over the north pole.  This is a cartoon abstracted
from the neutral lines computed for each rotation and extrapolated into
the polar regions.  The northern hemisphere lies within the inner circle.  The southern hemisphere is distorted and lies between the inner
and outer circles.  The south pole is the outer circle.

Near minimum, corresponding to CR 1640 - 1660, the neutral sheet
lies near the equator with two warps north and two warps south of the
equator.  As the cycle progresses the latitudinal extent of the sheet
gradually increases.  Near solar maximum, (CR 1680 - CR 1700) the neutral sheet reaches the polar region and gradually passes over it.  This
corresponds to the reversal of the polar field in the IMF which is seen
to be a natural consequence of the migration of the neutral sheet
through the polar regions.  During the declining phase the neutral sheet
again begins to approach the equator and looks like the configuration
seen earlier, during the rising phase of the cycle, but with the opposite polarity.

Ulysses will be near the poles about the time of polar reversal
and should reveal in more detail the actual configuration of the IMF in
the polar regions near reversal.

## 5.  Coronal Rotation

Examination of the neutral sheet computed for each rotation shows
reduced differential rotation, much like that observed in coronal holes.
To investigate this an autocorrelation has been performed on the source
surface field at each latitude.  A maximum in the autocorrelation occurs
about one rotation later.  Figure 9 shows the rotation rate determined
at each latitude with $\frac{1}{2}$ symbols.  The best fit to the data gives a rotation of $13.2 - 0.5 \sin^2 \phi$ degrees/day, where $\phi$ is latitude. Also plotted
are the rotation rates determined for coronal holes (Bohlin, 1977);
solar features such as prominences, coronameter enhancements, magnetic
field patterns, white light features, and 5303 enhancements (Bohlin,
1977); and the Newton & Nunn recurrence rate for long-lived sunspots.

This demonstrates that the coronal field does not generally participate in the differential rotation of the photospheric features.  The
rotation rate of the coronal field is almost the same as the rotation
rate of coronal holes.  This implies that the higher latitude fields
rotate at nearly the same rate as the low latitude fields.  The IMF at
high latitudes will have only a slightly slower recurrence rate than the
lower latitude fields.

## 6.  Conclusion:  What will Ulysses Find?

Ulysses presents the first opportunity to measure the solar wind at
latitudes significantly different from that of the Earth.  Assuming that
the next solar cycle will be qualitatively similar to the present cycle,
we can make a prediction of the kinds of interplanetary magnetic field
structures the spacecraft will encounter.  Comparison of the ecliptic
structures for the last several cycles gives us some confidence that
such predictions are plausible.

Launch will occur near solar minimum.  Simple low latitude neutral

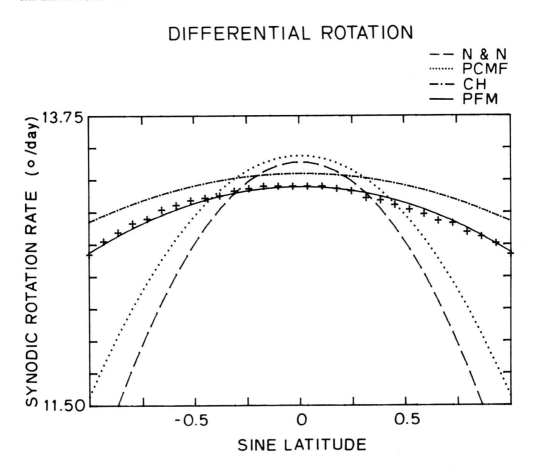

Figure 9:  The differential rotation curves for various solar  features.
The  long dashed line shows the Newton and Nunn curve for recurrent sunspots; the dotted and dash-dot lines show the synodic rotation rate  for
photospheric  features  and  coronal  holes,  respectively  (Bohlin, 1977);
and  the solid line shows the best fit rate of the coronal fields  determined  from  the  source  surface  field.  The + symbols show the actual
rates  determined from an autocorrelation analysis.  Each  curve  has  an
error of about 0.1 degree/day.

sheets characterize this time.  En route to Jupiter the spacecraft will
remain at low latitudes and will therefore experience much the same IMF
structure as the Earth.  There will likely be a four-sector structure
very near the ecliptic.  Even small differences in latitude between the
spacecraft and Earth could make a substantial difference in the polarity
structures observed.
    A couple of years after solar minimum the latitudinal extent of
the neutral sheet increases rapidly to at least 60°.  About this time

Ulysses will encounter Jupiter and begin its voyage out of the ecliptic.
Depending on the timing of the next cycle the spacecraft may emerge
above the neutral sheet for a time before being passed again as the
sheet moves to higher latitudes.

The neutral sheet reaches to the poles near maximum.  Ulysses will
pass over the poles just at this very interesting time.  There exists
the possibility of multiple neutral sheets and the spacecraft should
sample some of these regions not directly observed in the ecliptic during it journey.  The structure at high latitudes will sometimes be similar to that at Earth and sometimes quite different.  We should also
expect substantial differences between the fields in the northern and
southern hemispheres.  The importance of ecliptic measurements of solar
wind parameters for comparison with Ulysses data cannot be overstated.

Results presented here suggest that the polarity transition over
the poles is smooth and regular.  It will be interesting to see the
results of Ulysses measurements of the solar wind at this time.  The
coronal field and by extension the interplanetary field does not rotate
at the same differential rate as the photosphere.  Thus we should expect
the recurrence rate of features at high latitudes to be very similar to
that at lower latitudes.

This flight provides a new opportunity for us to test our ideas
about the heliosphere in a new region.  We will be able to learn about
the solar wind over the poles at maximum, of course.  We will also be
able to evaluate our methods of inferring the 3-dimensional structure of
the heliosphere and improve them for application to other time periods.

Acknowledgements: This work was supported in part by the Office of
Naval Research under Contract N00014-76-C-0207, by the National Aeronau-
tics and Space Administration under Grant NGR5-020-559, and by the
Atmospheric Sciences Section of the National Science Foundation under
Grant ATM77-20580.

# 7.  References

Altschuler, M.D. and G. Newkirk, Jr.  1969, Solar Phys., 9, 131.
Bohlin, J.D.  1977, in J.B. Zirker (ed.), Coronal Holes and High Speed
      Solar Wind Streams, Colo. Assoc. Univ. Press, Boulder, 27.
Fry, C.D. & S.-1. Akasofu 1986, Proc. 19th ESLAB Sym., D. Reidel.
Hoeksema, J.T., J.M. Wilcox, & P.H. Scherrer 1982, J.Geoph.R.,87,10331.
Hoeksema, J.T., J.M. Wilcox, & P.H. Scherrer  1983, J.Geoph.R.,88, 9910.
Levine, R.H.  1982, Solar Phys., 79, 203.
Newkirk, G.Jr., D.E. Trotter, M.D. Altschuler, and R. Howard  1973
      NCAR-TN/STR-85.
Rickett, B.J. and W.A. Coles  1983, in M. Neugebauer (ed.), Solar Wind
      Five, NASA Conf. Pub., 2280, 323.
Saito,T., T. Sakurai, and K. Yumoto  1978, Plant. Space Sci., 26, 431.
Schatten, K.H., J.M. Wilcox and N.F. Ness  1969, Solar Phys., 6, 442.
Smith, E.J., B.T. Tsurutani, & R.L. Rosenberg  1978, J.Geoph.R.,83, 717.
Suess, S.T., P.H. Scherrer, & J.T. Hoeksema 1986, Proc. 19th ESLAB
      Symp., D. Reidel.
Svalgaard, L. and J.M. Wilcox  1975, Solar Phys, 41, 461.
Wilcox, J.M. and A.J. Hundhausen  1983, J. Geophys. Res., 88, 8095.

THE LARGE-SCALE STRUCTURE OF THE HELIOSPHERIC MAGNETIC FIELD

A. Balogh
The Blackett Laboratory
Imperial College of Science and Technology
London SW7 2BZ
England

ABSTRACT. The heliospheric magnetic field, embedded in the solar
wind, originates in the highly non-uniform magnetic structures of the
solar corona. Its source function is not easily discernible, but a
brief description of the photospheric and coronal fields can give an
indication of its temporal and spatial boundary conditions. Direct
observations of the interplanetary magnetic field have only been made
in and near the solar equatorial plane. These have confirmed
Parker's original model as a good first approximation. The most
significant feature of the equatorial structure is the current sheet
separating the two polarity regions on the sun and extending deep into
interplanetary space. This feature has been successfully modelled and
remains the major achievement among efforts to relate solar and
coronal structures to interplanetary observations. On the other hand,
the heliolatitude dependence of the structure of the field remains
unknown. The paper concludes with a model list of questions which will
only be answered by the in situ observations planned during the forthcoming Ulysses mission.

1. INTRODUCTION

The tenuous magnetized plasma which constitutes the interplanetary
medium is the result of the instability and consequent expansion of
the solar corona. The original theoretical model of the supersonic
solar wind, dragging out with it magnetic field lines from the corona
was proposed by Parker (1958, 1963). Its basic features have been
confirmed by in situ observations in interplanetary space, and it has
remained the principal framework for most subsequent investigations
into the subject. Much theoretical work and the many observational
programmes undertaken over the past two and a half decades have
provided an increasingly refined insight into the structures and
physical processes of the solar wind and the heliospheric magnetic
field. However, many questions, some of them of basic importance,
remain to be answered.

*R. G. Marsden (ed.), The Sun and the Heliosphere in Three Dimensions, 255–266.*
© *1986 by D. Reidel Publishing Company.*

The principal source of difficulty is that the corona is highly
non-uniform in structure and variable in time, leading to a non-
uniform, time-dependent source function of the solar wind.  Inhomogeneities in the corona, in turn, generate a complex set of dynamical
effects in interplanetary space, thus preventing a simple interpolation
between interplanetary observations and coronal features.  Parker
assumed a spherically symmetric source function; the real source
function, at this stage of development of solar observations and of
theories of solar wind generation cannot be evaluated.  Hence there is
a need for repeated iterations between interplanetary and solar observations, in order to relate solar atmospheric features to different
solar wind regimes and magnetic field structures.

Knowledge of the large scale structure of the heliospheric
magnetic field is based on observations over the heliocentric distance
range 0.3 to about 25 AU, but only in and near the solar equatorial
plane; also on the mainly qualitative and occasionally only tentative
relationships which have been established between the sun and the
interplanetary medium.  Time dependences have been investigated on the
basis of observations made over approximately one complete 22-year
solar magnetic cycle.  However, all indications point to a strong
heliolatitude dependence of the parameters describing the interplanetary magnetic field and its structures.

As essentially no direct observations have been made out of the
ecliptic, the current view of large scale structures, therefore this
paper also, are heavily biased towards a two-dimensional approach.
This pre-Ulysses paper needs no excuses for this  restricted view.

After a brief review of solar and coronal magnetic fields, the
paper summarises the results obtained so far on spatial and temporal
dependences of the interplanetary magnetic field.  These are compared
to Parker's model which proves to be an excellent first order approximation.  The solar "equatorial" current sheet leading to the sector
structure of the interplanetary magnetic field is probably the most
intensively – and successfully – investigated feature relating coronal
structures to interplanetary observations, and is therefore discussed
in some detail.  Mention is made of recent results concerning the
radial dependence of the field magnitude and of the potential consequences if the less-than-expected magnetic flux in the solar
equatorial region is confirmed.

Finally the aims of the magnetic field investigation on Ulysses
are described briefly, in the context of the questions which need to
be answered before a truly three-dimensional view can be formed of the
heliospheric magnetic field.


## 2.   PHOTOSPHERIC AND CORONAL MAGNETIC FIELDS

The solar surface (photospheric) magnetic fields cannot be described
simply in terms of global parameters. (For reviews cf. Howard 1977,
and Zwaan, 1981).  Observations with increasing spatial resolution
indicate that at the level of the photosphere the magnetic field is
radial and concentrated into discrete elements of very high – greater

than 100 mT - strength.  Most of the fields are bipolar, i.e. opposite
polarity regions are closed by loops low in the solar atmosphere.  With
reduced resolution, the average field of larger scale regions is
measured, which is considered to be of greater relevance to extrapolating field patterns to the corona and into interplanetary space
(Levine 1979).  An alternative way of stating this is to say that the
lowest orders in a spherical harmonic description of the solar magnetic
field - the dipole and quadruple terms - are the most influential in
shaping the distant corona and the interplanetary field (Schultz 1973,
Bruno et al. 1982, Hoeksema 1984).  The relative magnitude of these
terms, and the presence of higher order terms are a strong function of
the solar cycle.

On the large scale, low- and mid-latitudes are dominated by
complex bipolar active regions during the years of high solar activity.
Outside active regions, large unipolar magnetic regions are observed,
where the average net flux density is relatively low (0.1 to 1 mT).
An important characteristic feature of these regions is their
longevity: their evolution can be followed over many solar rotations.
The two polar caps are also unipolar regions, of opposite polarity in
the two hemispheres.  The area they occupy is highly dependent on the
phase of the solar cycle.  At maximum activity, the polar caps essentially disappear, only to re-emerge with polarities of the opposite
sign.  Polarity reversal occurs a year or so after solar maximum, but
the two poles do not reverse their signs at the same time (Webb et al.
1984).  Unipolar magnetic regions, including the polar caps, are
apparently related to the slow break-up of active region fields,
followed by poleward drift and diffusion of the magnetic flux over the
solar surface (Howard and Labonte 1981).  Recent theoretical work
(De Vore et al. 1984) has shown that the observations are consistent
with the poleward drift restricted to low latitude active region
remnants.  The details of the relationship of large scale and polar
fields to sunspot and active region fields have not been completely
established, and the role played by giant cells in the solar convection
zone remains to be clarified (McIntosh and Wilson 1985).

The technique of measuring solar surface magnetic fields is not
applicable in the corona, as thermal broadening masks the Zeeman
splitting of spectral lines caused by the magnetic field.  Following
the field patterns through specific brightness features from the photosphere, through the transition region, and into the lower corona has
shown that the originally tightly bundled flux tubes fan out considerably with height.  As the magnetic field strength decreases less
rapidly with height than the plasma density, magnetic structures define
and delineate plasma structures in the corona.  Coronal densities and
temperatures are highly non-uniform and vary considerably over the
solar cycle.  This is clearly shown in the two eclipse photographs,
taken in white light, in Figure 1.

The difference between solar maximum and solar minimum conditions
is directly related to the solar cycle dependence of coronal magnetic
fields (cf. Sime 1985).  At the maximum of the activity cycle closed
structures dominate at (almost) all latitudes, populated by hot plasma,
giving a bright appearance to the corona around the solar limb.

Figure 1.  The appearance of the white light corona at (a) solar
maximum and (b) solar minimum.  (Photographs courtesy of High Altitude
Observatory).

     Streamers are radial structures extending from the top of closed
loop systems far out into the distant corona.  Their structure and
nature are not well understood, but they are likely to play some role
in interplanetary space.
     The evolution from maximum to minimum activity is accompanied
by the disappearance of the bright corona first over the solar poles,
then also at somewhat lower latitudes.  At minimum, streamers are
restricted to the region close to the solar equator.  The dark areas
in the eclipse photographs, the so-called coronal holes, are in fact
probably of most direct relevance to the interplanetary medium.  It
is now widely accepted that coronal holes, with their open magnetic
field structure, are the source of high speed solar wind streams, and
possibly that of the low speed wind as well (Hundhausen 1977).  It
therefore appears that the heliospheric magnetic field also has its
origin mainly in the regions of the photosphere underlying coronal
holes.
     Densities and temperatures in coronal holes are about an order of
magnitude lower than the surrounding regions.  (For a review cf.
Zirker 1981).  As observed during the Skylab mission (Bohlin 1977)
close to last solar minimum, coronal holes can cover a latitude range
from close to the solar equator to the polar regions.  Their cross-
sectional area increases with height much faster than would be expected
on the basis of radial expansion alone.  Based on Skylab measurements,
Munro and Jackson (1977) found that a particular coronal hole had a
cross-sectional area at two solar radii seven times greater than would
have been the case if the field lines delineating it had remained
radial.  Using potential field models, theoretical extrapolations of
coronal field lines also show, possibly to an even greater extent, this
same "nozzle" effect (cf Levine 1980).  Recent theoretical models of
coronal holes (Osherovich et al. 1985) and of solar wind generation

based on polar coronal holes (Whang 1983) have confirmed both the
nozzle effect and its importance in defining the source function of the
interplanetary medium.


## 3. THE HELIOSPHERIC MAGNETIC FIELD

### 3.1. The sector structure and its solar origin

Since the first systematic observations of the interplanetary magnetic
field in the vicinity of the Earth, its organisation into polarity
sectors has been recognized as a major feature of its large scale
structure (Wilcox and Ness 1965). The average magnetic field vectors
are, on the whole, aligned along the Archimedean spiral direction,
deduced simply from Parker's theory, using the actually measured solar
wind speed. (For further discussion, see below). At 1 AU, and for
average solar wind conditions, the (outward directed) spiral makes an
angle of about 135° with the sun-Earth direction. The radial component
of the field can point either towards or away from the sun. The sector
structure consists of regions lasting a significant fraction of a solar
rotation period in which the field, on average, points either away from
or towards the sun. A solar rotation period is normally divided into
two or four alternating polarity sectors, and the pattern evolves only
slowly from one rotation to the next. The sector pattern measured over
the descending phase of solar cycle 20 is shown in Figure 2 (Hedgecock
1975).

   High speed solar wind streams, showing closely related recurrence
characteristics, are normally included within a given polarity sector.
(For a review, cf. Smith 1979). Alternatively stated, magnetic field
lines within a sector can be traced back to the coronal hole associated
with the high speed stream contained in that sector (Burlaga et al.
1978).

   The sector structure observed in the ecliptic, that is within
7.25° on either side of the solar equatorial plane, is now considered
to be the result of the projection of the dominant polarities of the
two solar hemispheres into interplanetary space. In this picture
(Schulz 1973) the heliosphere is divided into two regions, roughly
along the solar equator: the radial component of the magnetic field
vectors in one region point away from the sun, in the other, towards
the sun. The two regions are separated in interplanetary space by a
vast current sheet. The sector structure arises from the fact that the
current sheet is in fact non-planar (it has been likened to a
ballerina's skirt) and as the sun rotates, an observer close to the
solar equatorial plane passes through the current sheet an even number
of times (two or four) in every solar rotation (Thomas and Smith 1981).

   More successful results were obtained when the heliolatitude of
Pioneer 11 reached 16° north, and observations showed the gradual disappearance of the sector structure at that latitude, during a period
close to solar minimum (Smith et al. 1978). Careful analysis of Helios
1 and 2 data during the same period in the inner solar system (Villante
et al. 1979) has also led to essentially the same picture of the shape

HELIOGRAPHIC LATITUDE SEPARATION
(EARTH-PIONEER 10)

IMF SECTOR STRUCTURE

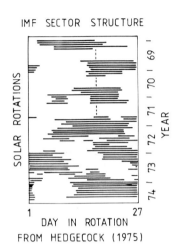

DAY IN ROTATION
FROM HEDGECOCK (1975)

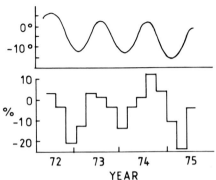

YEAR

DIFFERENCE IN PERCENTAGE OF
POSITIVE POLARITY
(EARTH-PIONEER 10)

FROM THOMAS AND SMITH (1980)

Figure 2. The sector structure
of the interplanetary magnetic
field in the vicinity of the
Earth observed during the
second half of solar cycle 20.

Figure 3. Measurement of the
relative percentage of toward and
away sectors at Pioneer 10 and
near the Earth compared to the
separation in heliolatitude.

of the current sheet and its role in separating opposite polarity
regions. Further analysis of Pioneer data and its comparison with
near-Earth measurements shown in Figure 3 (Thomas and Smith 1980)
provided further confirmation that the current sheet is indeed
relatively near the solar equator, at least during years of low solar
activity.

In order to establish the solar origin of the interplanetary
current sheet, two parallel approaches have been used. Both use the
observation that large scale solar and coronal polarity patterns appear
to be matched by the sector structure of the interplanetary magnetic
field near the Earth if the solar wind transit time, 4 to 5 days, is
taken into account.

One method is based on the observation (Hundhausen 1977) that
the boundaries of coronal holes appear to coincide with the contour
lines of maximum brightness on synoptic maps of coronal polarization.
This method has been used, among others by Bruno et al. (1982) and
Behannon et al. (1983) to relate interplanetary sectors to single
polarity regions on the sun.

The alternative method, used by Hoeksema et al. (1982) over the
early portion of solar cycle 21, and through the last solar maximum by
Hoeksema et al. (1983), is based on the measurement of the photospheric magnetic field and its extrapolation, using a potential field
model to a source surface at which the field is assumed to be radial.
First introduced by Schatten et al. (1969) and by Altschuler and

Newkirk (1969), the technique is based on the assumption that below some critical radius, variously set between 1.6 and 2.6 solar radii, the magnetic field dominates the plasma motions and therefore can be derived from a potential. Several improvements had to be applied to the early form of the method, the main one being the inclusion of polar fields, before satisfactory fits could be obtained.

The two methods yield similar contours near the sun. A comparative study by Wilcox and Hundhausen (1983) concluded that major discrepancies between them could arise only if the solar magnetic field changes significantly on a timescale of several days, as the potential field method is sensitive to structures close to the central meridian, whereas the maximum coronal brightness method is based on measurements over the solar limb.

As discussed by Levine (1980), matching of solar and coronal field structures to interplanetary structures has to remain an approximation as details of the coupling of coronal fields into the solar wind are largely unknown. Furthermore, the current sheet may well be deformed in interplanetary space by dynamical effects (Thomas and Smith 1981). However, even if only approximately related to coronal structures, the interplanetary current sheet is the best established major structural feature of the heliosphere.

## 3.2. Radial and temporal dependence in the ecliptic

Parker's model predicts a simple dependence of the field configuration in the solar equatorial plane. Measurements made in the ecliptic have confirmed that field lines follow an Archimedean spiral over a large range of heliocentric distances (Behannon 1978, Thomas and Smith 1980). The results of Thomas and Smith (1980), obtained from a substantial data set consisting of hourly averages measured by the Pioneer 10 and 11 spacecraft 1 AU to 8.5 AU, are shown in Figure 4. These results are shown in a coordinate system aligned with the local Parker spiral direction at all heliocentric distances. As is apparent in Figure 4, there is clearly no systematic deviation from the spiral structure. The analysis of the north-south component, shown also in Figure 4, confirms that the field, on average, lies in the plane of the spiral.

The radial dependence of the magnetic field strength has also been compared to the Parker model over the same radial interval. Results appear to be less consistent than in the case of the spiral structure. On the one hand, Burlaga et al. (1984) find that the field strength follows the functional dependence predicted by Parker. These results are shown in Figure 5. These authors concluded that the Parker model is a valid first order approximation to the measurements made by the two Voyager spacecraft, covering the time interval from 1977 to 1981.

On the other hand, Slavin et al. (1984) have concluded, on the basis of joint temporal and radial analysis of field data, that the field strength decreases more rapidly, mainly beyond 1 AU, than predicted by Parker. A more detailed analysis has been performed recently by Thomas et al. (1985). These authors use the radial dependence of the azimuthal component of the magnetic field as a

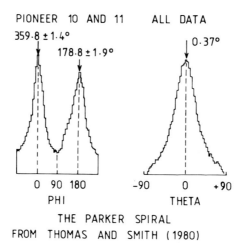

PIONEER 10 AND 11     ALL DATA

THE PARKER SPIRAL
FROM THOMAS AND SMITH (1980)

Figure 4.   Histogram of the azimuth and latitude angle of hourly
averages of the magnetic field vector in spiral coordinates.

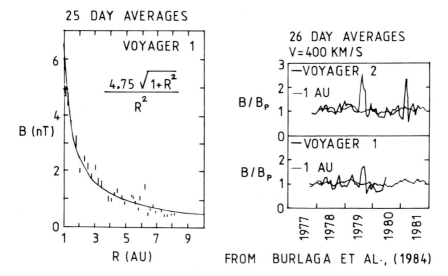

FROM   BURLAGA ET AL·, (1984)

Figure 5.   The amplitude of the average magnetic field vector as a
function of radial distance and time, compared to the Parker model.

measure of the field strength, as the radial component becomes very
small at larger heliocentric distances, and therefore more susceptible
to uncertainties due to wave activity.   Their results, comparing the
azimuthal component measured at the Earth with that measured at Pioneer
11, as a function of time, are shown in Figure 6.

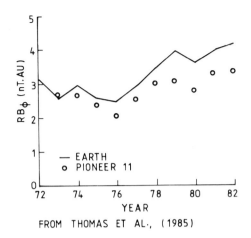

FROM THOMAS ET AL., (1985)

Figure 6.   Comparison of the normalized azimuthal component of the magnetic field vector measured at 1 AU and at increasing radial distances by Pioneer 11.

The apparent discrepancy between these results is not easy to resolve, except by more observations which will be carried out, as far as Jupiter, by the Ulysses and Galileo spacecraft in the next few years, essentially close to solar minimum.  It is not impossible that the missing flux effect, if confirmed, is more apparent at low solar activity.

A possible explanation of the implied meridional flux transport is the higher magnetic pressure at low heliolatitudes (Nerney and Suess, 1975, Suess et al. 1985).  If such meridional flux transport does indeed exist, off-ecliptic effects are likely to be observable by the Ulysses mission.

The variation of the interplanetary magnetic field strength as a function of the solar cycle has been examined by many authors (cf. King 1979, King 1981, Slavin and Smith 1983).  No simple solar cycle dependence has been found which could be related to magnetic changes on the sun.  While there was a significant decrease in field strength in 1975-76, coincident with the solar minimum, followed by a significant increase (as also apparent in Figure 6) during the following solar maximum years, the effect was not present in the data during the previous solar cycle.  Further work is needed to examine the relationship between solar and interplanetary fields to relate large temporal and spatial changes in the two data sets.

4.   HIGH LATITUDE STRUCTURE: THE MAIN QUESTIONS

Many of the problems of the large scale structure of magnetic fields in the heliosphere await the in situ observations at high helio-latitudes planned for the second half of this decade with the Ulysses mission.  The conspicuous success of Parker's model in describing the

large scale structure in the ecliptic has been noted in the previous
section.   Out of the ecliptic plane, the picture may prove more
complex.   In particular, solar cycle dependent phenomena may be more
apparent and more influential than is the case in the ecliptic.   The
fact that Ulysses is scheduled to sweep the range of heliolatitudes as
the solar cycle changes from minimum to maximum will inevitably lead
to the need to consider the solar cycle dependence of the observations
and any models which will be derived.

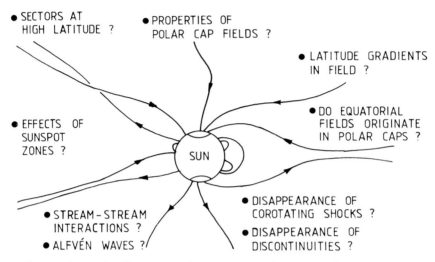

Figure 7.   Schematic illustration of questions concerning the high
latitude structure of the heliospheric magnetic field, to be answered
by the Ulysses mission.

        A summary of the questions raised in connection with the helio-
latitude dependence of field structures is illustrated in Figure 7.
(Balogh et al. 1983).   First, the identification of the solar roots of
magnetic field lines remains to be established.   If all or most of the
solar wind flow is generated in coronal holes, their disappearance at
low latitudes as solar activity increases should lead to a variation
of flux as a function of the cycle.   This effect has not been
established in the ecliptic.   However, a relatively fast (compared to
the solar cycle) latitude sweep over the first ten to twenty degrees
at a time when low latitude coronal holes are being filled may produce
a measureable effect.
        As sunspot zones develop, together with their coronal signatures,
their effect far from the sun may well be the generation of high
latitude current sheets.   While essentially impossible to predict with
any certainty, the existence of such current sheets could well lead to
large-scale, single-polarity regions at high latitudes, akin to the
sector structure in the ecliptic.
        Polar caps recede as solar activity increases.   The properties
of fields originating in the polar caps are likely to be very
different at high latitudes, as the geometric spiralling effect is

considerably lessened.  Both small scale phenomena (waves, discontinuities etc,) and large scale structures are likely to be affected
by the different geometry of the field at these latitudes.  Measurement
of the field strength above the poles will be the first direct evidence
concerning the polar fields of the sun.

Although such speculations about the off-ecliptic magnetic field
are far from exhaustive or secure, they concern the major questions
about the large scale structure of the heliospheric field.  The complex
interplay of coronal magnetic polarities is likely to be reflected in
structures at large distances from the sun.  Nevertheless, it is likely
that the Ulysses observations can still be organised in the framework
of Parker's original model.

# REFERENCES

Altschuler, M.D., and Newkirk, G., Jr.: 1969, Solar Phys. 9, 131.
Balogh, A., Hedgecock, P.C., Smith, E.J., and Tsurutani, B.T.: 1983,
    in The International Solar Polar Mission - Its Scientific Invest-
    igations, K.P. Wenzel, R.G. Marsden, and B. Battrick, editors,
    ESA SP-1050, 27.
Behannon, K.W.: 1978, Rev. Geophys. Space Phys. 16, 125
Behannon, K.W., Burlaga, L.F., and Hundhausen, A.J.: 1982, J. Geophys.
    Res. 88, 7837
Bohlin, J.D.: 1977, in Coronal Holes and High Speed Wind Streams,
    J. Zirker, editor, Boulder, Colorado Associated University Press,
    27
Bruno, R., Burlaga, L.F., and Hundhausen, A.J.: 1982, J. Geophys. Res.
    87, 10,339
Burlaga, L.F., Behannon, K.W., Hansen, S.F., Pneuman, G.W., and
    Feldman, W.C.: 1978, J. Geophys. Res. 83, 4177
Burlaga, L.F., Klein, L.W., Lepping, R.P., and Behannon, K.W.: 1984,
    J. Geophys. Res. 89, 10,659
DeVore, C.R., Sheeley, N.R., Jr., and Boris, J.P.: 1984, Solar Phys.
    92, 1
Hedgecock, P.C.: 1975, Solar Phys. 44, 205
Hoeksema, J.T.: 1984, Ph.D. Dissertation, Stanford University
Hoeksema, J.T., Wilcox, J.M., and Scherrer, P.H.: 1982, J. Geophys Res.
    87, 10, 331
Hoeksema, J.T., Wilcox, J.M., and Scherrer, P.H.: 1983, J. Geophys.Res.
    88, 9910
Howard, R.: 1977, Ann. Rev. Astron. Astrophys. 15, 153
Howard, R., and Labonte, B.J.: 1981, Solar Phys. 74, 131
Hundhausen, A.J.: 1977, in Coronal Holes and High Speed Wind Streams,
    J. Zirker, editor, Boulder, Colorado Associated University Press,
    298
King, J.H.: 1979, J. Geophys. Res. 84. 5983
King, J.H.: 1981, J. Geophys. Res. 86, 482
Levine, R.H.: 1979, Solar Phys. 62, 277
Levine, R.H.: 1980, in Solar and Interplanetary Dynamics, M. Dryer and
    E. Tandberg-Hanssen, editors, IAU Symposium No. 91, D. Reidel
    Publishing Co., Dordrecht, 1.

McIntosh, P.S., and Wilson, P.R.: 1985, Solar Phys. 97, 59
Munro, R., and Jackson, B.V.: 1977, Astrophys. J. 213, 874
Nerney, S.F., and Suess, S.T.: 1975, Astrophys. J. 200, 503
Osherovich, V.A, Gliner, E.B., Tzur, I., and Kuhn, M.L.: 1985, Solar
    Phys. 97, 251
Parker, E.N.: 1958, Astrophys. J. 128, 664
Parker, E.N.: 1963, Interplanetary Dynamical Processes, Interscience
    Publishers, New York
Rosenberg, R.L., and Coleman, P.J., Jr.: 1969, J. Geophys. Res. 74,
    5611
Rosenberg, R.L., Kivelson, M.G., and Hedgecock, P.C.: 1977, J. Geophys.
    Res. 82, 1273
Schatten, K.H., Wilcox, J.M., and Ness, N.F.: 1969, Solar Phys. 6, 442
Schultz, M.: 1973, Astrophys. Space Sci. 24, 371
Sime, D.G.: 1985, in Proc. ESA Workshop on Future Missions in Solar,
    Heliospheric and Space Plasma Physics, ESA SP-235, 23
Slavin, J.A., and Smith, E.J.: 1983, in Solar Wind 5, ed. by M.N.
    Neugebauer, NASA CP-2280, 323
Slavin, J.A., Smith, E.J., and Thomas, B.T.: 1984, Geophys. Res. Lett.
    11, 279
Smith, E.J.: 1979, Rev. Geophys. Space Phys. 17, 610
Smith, E.J., Tsurutani, B.T, and Rosenberg, R.L.: 1978, J. Geophys.
    Res. 83, 717
Suess, S.T., Thomas, B.T, and Nerney, S.F.: 1985, to be published in
    J. Geophys. Res.
Thomas, B.T., and Smith, E.J.: 1980, J. Geophys. Res. 85, 6861
Thomas, B.T., and Smith, E.J.: 1981, J. Geophys. Res. 86, 11, 105
Thomas, B.T., Slavin, J.A., and Smith, E.J.: 1985, to be published in
    J. Geophys. Res.
Villante, U., Bruno, R., Mariani, F., Burlaga, L.F., and Ness, N.F.:
    1979, J. Geophys. Res. 84, 6641
Webb, D.F., Davis, J.M., and McIntosh, P.S.: 1984, Solar Phys. 92, 109
Whang, Y.C.: 1983, Solar Phys. 88, 343
Wilcox, J.M., and Ness, N.F.: 1965, J. Geophys. Res. 70, 5793
Zirker, J.B.: 1981, in The Sun as a Star, S. Jordan, editor, NASA
    SP-450, 135
Zwaan, C.: 1981, in The Sun as a Star, S. Jordan, editor, NASA SP-450,
    163

# THE HELIOSPHERIC CURRENT SHEET: 3-DIMENSIONAL STRUCTURE AND SOLAR CYCLE CHANGES

Edward J. Smith, J. A. Slavin
Jet Propulsion Laboratory
California Institute of Technology
Pasadena, California 91109

B. T. Thomas
University of Bristol
Computer Science Department
Bristol DS81TW, UK

ABSTRACT. The reversal in polarity of the interplanetary current sheet/sector structure is investigated during the recent solar maximum. Multipoint observations by ISEE-3 and Pioneer 11 show that a simple two sector or occasional four sector structure persisted throughout the maximum and out to distances of 10 AU. The polarity reversal occurred between March 1979 and October, 1980 without any indication of an abrupt transition that might permit a more precise timing. The reversal coincided approximately with the reversal in the sun's polar cap fields. The current sheet appeared to be highly inclined during the ascending and descending phases of the solar cycle but was apparently too complex to describe as a simple inclined current sheet during solar maximum.

## 1. INTRODUCTION

One of the largest scale, three dimensional structures in the heliosphere is the current sheet which is responsible for the sector structure (Smith et al., 1978). Since the early 1970's, it has been known that the sector structure reverses polarity at, or following, sunspot maximum (Wilcox and Scherrer, 1972). The field above the current sheet is outward for 11 years, or one half of a solar magnetic cycle, and inward for the succeeding 11 years. The recent solar maximum of 1979 has provided another opportunity to further study phenomena associated with the polarity reversal.

There are three principal questions of scientific interest. When did the reversal occur? How did it occur, i.e., what was the topology of the current sheet during the reversal? What was the relation of the change in polarity to changes in the solar magnetic field, such as the polar cap fields?

We have been seeking answers to these questions using data acquired near 1 AU, principally by ISEE-3, as well as further out in the heliosphere, obtained by Pioneer 11. The multispacecraft observations have been used to distinguish temporal from spatial variations

267

R. G. Marsden (ed.), The Sun and the Heliosphere in Three Dimensions, 267–274.
© 1986 by D. Reidel Publishing Company.

and to investigate the coherence and evolution of the heliospheric current sheet over distance scales of up to 10 AU.

COHERENCE OF THE SECTOR STRUCTURE WITH TIME AND DISTANCE

To begin, the sector structure at 1 AU is shown in the left half of Figure 1 between 1973 and 1982. The figure is based on ISEE-3 observations, supplemented by earlier measurements obtained by IMP (King, 1983). This interval includes sunspot maximum which occurred in December 1979 and contains the reversals in both the sun's polar caps and in the polarity of the interplanetary field. Half-day polarities are shown in a conventional format consisting of pluses (outward from the sun) and minuses. The polarities are organized by rows, each row corresponding to the 27 days of a Bartels rotation as indicated in the second column. The two left-hand columns are the year ("74" identifies the beginning of 1974) and the bartels rotation number (e.g., 1974). The two right hand columns are the year and heliographic latitude of Pioneer.

Only two well defined sectors were present throughout most of the maximum phase (1979-1980). Four sectors can be seen for an interval of about 6 months in 1980. The sector structure continued to be simple and well defined throughout solar maximum. There is no tendency for a filamentary structure to develop nor is the sector structure significantly disrupted by the many transient events (flares, coronal transients, etc.) which characterize solar maximum.

The right half of Figure 1 shows that the simple sector structure observed near the ecliptic persisted out to large heliocentric distances. Pioneer 11 was launched in April 1973 and the polarity is shown each half day from then until the end of 1982. In order to emphasize the gross structure, the polarities at both 1 AU and at the location of Pioneer have been smoothed using running averages of 11 successive intervals. This smoothing has suppressed occasional data gaps and multiple current sheet crossings. No allowance has been made for either the corotation delay of up to 13.5 days, associated with the longitudinal separation of Pioneer and earth, or the delay of up to one month associated with the radial separation.

A reasonably close correspondence between the two sets of polarities exists throughout the entire interval showing that the sector structure is coherent over distances that exceed 10 AU by the end of the interval. (A notable difference in polarity can be seen in 1975-76 when Pioneer 11 ascended to a heliographic latitude of 16°, was above the current sheet and observed only a single positive polarity (Smith et al., 1978).) The interval of interest here is represented by the lower half of the figure which includes solar maximum. The sector structure at Pioneer 11 is very similar to that observed at 1 AU. Neither the evolution of the solar wind with distance nor the occurrence of transient solar wind disturbances (such as flare ejecta) produced a major disruption of the sector structure at large distances.

# INTERPLANETARY SECTOR STRUCTURE

DAY OF BARTELS ROTATION

Figure 1.   Sector structure from 1973–1982 at two heliospheric locations.

THE REVERSAL IN POLARITY

Figure 1 does not show when the polarity reversal occurred. It is necessary to analyze the observations as a function of latitude in order to determine when the field above the current sheet changed from outward to inward. A technique which has worked successfully in the past (Rosenberg and Coleman, 1969; Wilcox and Scherrer, 1972) has been to compare the heliographic latitude of observing spacecraft with the fractional positive polarity per solar rotation, defined by $P(+) = N(+)/N_T$ where $N(+)$ is the number of days with positive polarity and $N_T$ is the total number of observing days per rotation (nominally 27).

Plots of $P(+)$ and the heliographic latitude of ISEE-3, $\delta$, show sinusoidal variations in both (Figure 2). The fractional polarity increases at north latitude in late 1978 and early 1979 and decreases at north latitudes in late 1980 and early 1981. The change from an "in-phase" to "out-of-phase" occurs by means of a "long" cycle which extends from mid-1979 to mid-1980. The exact time of reversal cannot be specified more precisely than having occurred between mid-1979 and mid-1980.

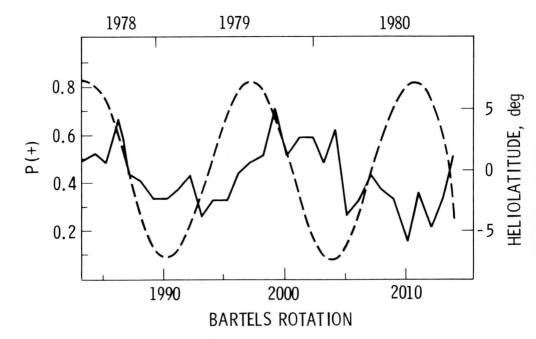

Figure 2. Fractional positive polarity and heliographic latitude at ISEE-3. The sinusoidally varying latitude is the dashed curve. Unsmoothed values of $P(+)$ for each Bartels rotation are connected by solid lines.

In principle, the use of multi-spacecraft observations might be expected to discriminate against time variations and reveal more precisely when the polarity reversal took place. Figure 3 shows the difference in P(+) at Pioneer 11 and ISEE-3 plotted as a function of time during solar maximum. Also shown for comparison is the difference in the heliographic latitudes of the two spacecraft.

Figure 3 shows that whenever $\Delta\delta = 0$, i.e., the two spacecraft are at the same latitude (not necessarily on the solar equator), the polarity difference, $\Delta$ P(+), is also zero. This correspondence shows explicitly that the current sheet is coherent over distances of ~ 10 AU; when the spacecraft are at the same latitude, they observe the same fractional polarity. The differences in P(+) at other times are then presumably spatial differences which accurately indicate the sense of the field above or below the current sheet.

The comparison shows the in-phase and out-of-phase correlations before mid-1979 and after mid-1980, respectively. However, no simple

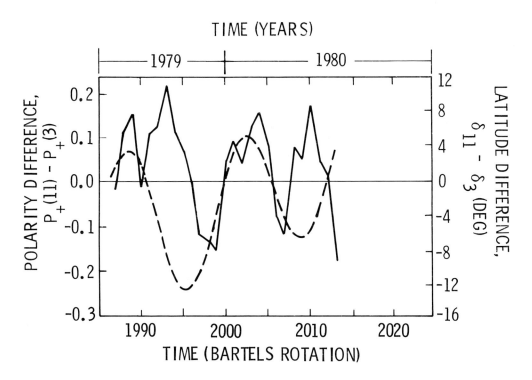

Figure 3. Differences in fractional polarities and latitudes at two heliospheric locations. The difference in fractional polarities at Pioneer 11 and ISEE-3, designated $P_+(11) - P_+(3)$, is shown as the solid line. The difference in the helio-latitudes of Pioneer($\delta$11) and ISEE($\delta$3) is shown as the dashed line.

transition from an in-phase to an out-of-phase relation can be discerned. Both phases are evident in the intervening interval.

In an effort to make further progress, mathematical modelling has been resorted to. An accurate model might be able to reveal precisely when the reversal occurred and could lead to a characterization of the three dimensional structure of the current sheet in regions other than those near the solar equator being sampled by spacecraft.

The simplest conceivable model involves a gradually increasing inclination of the current sheet until it eventually passes over the poles and returns to lower latitudes (Saito, 1975; Saito et al., 1978). This model, which is equivalent to a rotating magnetic dipole, has also been considered in attempts to account for cosmic ray modulation by invoking drifts in the large scale gradients associated with the Heliospheric magnetic field and the current sheet (Jokipii and Thomas, 1981). In terms of solar magnetic fields, a highly inclined magnetic dipole could correspond to asymetrically placed magnetic poles (possibly associated with coronal holes) located at low, rather than polar, latitudes.

This model is easily formulated mathematically and P(+) can be derived for a given $\delta$, once the changing inclination of the current sheet is specified as a function of time. It has been assumed that the current sheet reaches 90° inclination in January 1979, i.e., midway through the interval 1976-1982, and that the inclination changes at a constant rate of 24°/yr. The change in fractional polarity associated with such a model is shown in Figure 4 along with P(+) observed by ISEE-3. Smoothed values of P(+) based on 5 successive

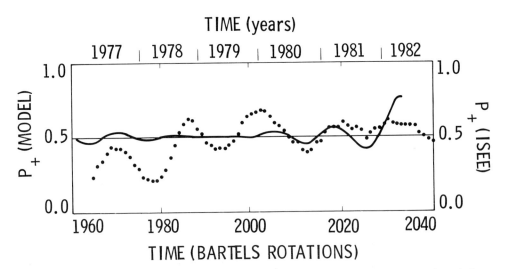

Figure 4. Observed fractional positive polarity compared with a model. The smoothed ISEE-3 measurements are shown as dots. The values derived from the model lie along the solid curve.

solar rotations are shown in order to suppress short period fluctuations. Clearly, this model (or one like it that involves a different
time dependence) cannot account for the "long cycle" that is a significant feature of the observations. For high inclinations, the
dependence of P(+) on δ is greatly reduced and, in the limit of
90° inclination, P(+) = 0.5, independent of latitude. This qualitative feature leads to a long interval of nearly constant P(+) which
is distinctly unlike the observations.

The model can also be tested by computing current sheet contours as
a function of time and comparing them with the source surface contours based on the photospheric magnetic field that have been published recently by Hoeksema et al (1983). Such a comparison has
been carried out and there are serious discrepancies between the
predictions of the model and the contours inferred from the magnetograph observations. One of several serious departures is the occurrence of Stanford contours which slope in the same sense in both
solar hemispheres, indicative of a significant twisting (torsion) of
the current sheet. A more important discrepancy is the presence of
multiple current sheets. Finally, some of the Stanford contours are
suggestive of a tilted quadrupole rather than tilted dipole. Thus
the simple rotating dipole model fails on the basis of this comparison as well.

DISCUSSION

In light of the preceeding analysis we can now attempt to answer the
three basic questions posed in the introduction. As to when the
change in polarity occurred, it can only be stated with confidence
that it took place sometime during the one and one half year interval
between March 1979 – October 1980. The "long cycle" in P(+) at 1 AU
(figure 2) and the inconstant relation between Δ P+ and Δδ (figure
3) fail to exhibit any well defined event that might be interpreted
as indicating an abrupt transition from one polarity to the other.
It seems that the entire interval is characterized by an ambiguous or
irregular topology.

Our inability to provide a precise answer to the first question is
closely related to the second question of how the change took place.
A simple model consisting of a wavy tilted current sheet seems to
account adequately for both the observed P(+) and for the source
surface contours prior to April 1979 and following September 1980.
There is reasonable support for a highly inclined current sheet
during the ascending and descending phases. However, during sunspot
maximum, it does not appear possible to characterize the current
sheet using this model. Instead, multiple current sheets may be
present and the global configurations of the solar field may correspond to a quadrupole (or higher order multipoles) much more than to a
dipole during intervals of several solar rotations. If a solar
dipole is still present, it is presumably overwhelmed by the larger
more complex field configurations.

Regarding the third question, the complex three dimensional topology of the current sheet during solar maximum is consistent with the rapidly evolving magnetic fields in the solar photosphere and corona. It is undoubtedly significant that the reversal in the sector structure once again coincided approximately (i.e., to within the achievable precision) with the reversal in the polar cap fields which occurred between early and late 1980 (Howard and Labonte, 1981; Hoeksema et al., 1983). Both reversals occurred one to three years earlier than in the previous four solar cycles when a significant delay was observed between sunspot maximum and the reversals. This shift in both the interplanetary and polar cap reversals to an earlier phase of solar activity lends strong support to this correlation.

## ACKNOWLEDGEMENT

Mary Requarth assisted in the data analysis, in particular modelling of the rotating current sheet. The research reported here was carried out by the Jet Propulsion Laboratory, California Institute of Technology, under contract with the National Aeronautics and Space Administration.

## REFERENCES

Hoeksema, J. T., J. M. Wilcox, and Scherrer, P. H.: 1983, J. Geophys. Res, 88, 9910.

Howard, R., and Labonte, B.: 1981, Solar Phys., 74, 131.

Jokipii, J. R., and Thomas, B.: 1981, AP. J., 243, 1115.

King, J. H.: 1983, Interplanetary Medium Data Book, Goddard Space Flight Center Report.

Rosenberg, R. L., and Coleman, P. J., Jr.: 1969, J. Geophys. Res., 74, 5611.

Saito, T.: 1975, Sci. Rept. Tohoku Univ., Ser. 5, 23, 37.

Saito, T., Sakurai, T., and Yumoto, K.: 1978, Planet. Space Sci., 26, 413.

Smith, E. J., B. T. Tsurutani, and Rosenberg, R. L.: 1978, J. Geophys. Res., 83, 717.

Wilcox, J. M., and Scherrer, P. H.: 1972, J. Geophys. Res., 77, 5385.

SOLAR WIND SPEED AZIMUTHAL VARIATION ALONG THE
HELIOSPHERIC CURRENT SHEET

S. T. Suess
Space Science Lab, ES52
NASA - Marshall Space Flight Center
Huntsville, Alabama 35812, U.S.A.

P. H. Scherrer and J. T. Hoeksema
Center for Space Science and Astrophysics
Via Crespi, ERL 328
Stanford University
Stanford, California 94305

ABSTRACT. We report on analysis of the speeds measured by Voyager 1 and 2 while skimming along a horizontal (east-west) portion of the current sheet over several days in 1977. The results demonstrate that in this case speed variations exist and are large enough to significantly deform the sheet within a few AU or less if the current sheet were anything but perfectly horizontal. The spatial scale of the speed variation ranges from the smallest measureable scale using 1 hour averaged data up to tens of degrees in longitude. A deformation example is given under the assumption that the observed velocity variation exists on a current sheet that is initially perpendicular to the heliographic equator.

1.   INTRODUCTION

The shape of the heliospheric current sheet (HCS) near the Sun is estimated from source surface models of coronal structure (Hoeksema, Wilcox and Scherrer, 1983) and from coronagraph observations of density (Behannon, Burlaga and Hundhausen, 1983). In the highly idealized case of uniform, radial solar wind flow, the pattern of the HCS is easily computed and is exactly imaged by radial projection throughout the solar system - with the image at one radial distance only being displaced in longitude relative to the image at a different distance because of the rotation of the Sun. However, the solar wind is observed to be nonuniform - the velocity varies from point to point along the HCS, causing a distortion of progressively greater amplitude with increasing distance from the Sun. Significant and observable distortion is produced by gradients in velocity of less than 5 km/s/deg. Examples demonstrating the principle, the magnitude, and the character of the effect are given by Suess and Hildner (1985) (hereafter SH). In their conclusion, SH list several important problems that have yet to be addressed. One is analysis of solar wind velocity data from a spacecraft skimming along a

*R. G. Marsden (ed.), The Sun and the Heliosphere in Three Dimensions, 275–280.*
© *1986 by D. Reidel Publishing Company.*

horizontal current sheet in order to document the HCS velocity gradient in a specific case.

It is difficult to measure velocities on the HCS because spacecraft trajectories rarely lie on that surface. Previous studies have been statistical or depended on different spacecraft intersecting the same current sheet. Rhodes and Smith (1975) conducted a multi-spacecraft study of the solar wind velocity at sector boundaries and concluded that an average gradient of approximately 10 km/s/deg existed for a sample of 18 sector boundaries. Their analysis suffered, however, from a large difference in radius between the spacecraft observing at different latitudes on the HCS. Statistical studies cannot directly determine a gradient, but show the velocity dispersion on the HCS - a useful number because it gives an idea of the magnitude of the velocity difference that might be encountered on different parts of the HCS. Newkirk and Fisk (1985) conducted this type of analysis in connection with another problem, finding that the solar wind velocity on the HCS averages about 325 km/s and has a dispersion of approximately 50 km/s.

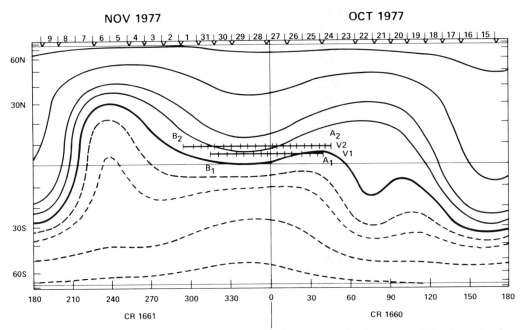

Figure 1. Magnetic field contours at +/- 1, 2, 5, 10, and 20 microtesla on the 2.5 solar radii source surface of a coronal model using Wilcox Solar Observatory data. Coordinates are Carrington Longitude (Carrington Rotations (CR) 1660 and 1661) versus sine of latitude. V1 and V2 are the paths of Voyager 1 and 2 - shifted to the source surface radius using observed solar wind velocities and the constant velocity approximation. $A_1$, $A_2$, $B_1$, and $B_2$ are the beginning and end times for the data sets - days 304.0 to 312.6 for V1 and days 308.5 to 314.25 for V2.

In this paper, we analyze solar wind velocities for a period when the HCS lay approximately at the latitude of the observing spacecraft over a heliographic longitude of 70 degrees (or about 5 days). This means that the spacecraft was actually on, or very near, the HCS for that entire time. This period in 1977 was first identified by Behannon, Burlaga, and Hundhausen (1983), who noted that Voyager 1 and 2 (V1 and V2) observed a state of mixed IMF polarity throughout much of the interval and that the maximum brightness contour in the corona indicated a current sheet which was approximately in an east-west direction and close to the equatorial plane.

Examining the magnetic field at the source surface of a coronal model for this period, based on Wilcox Solar Observatory data, we find support for the current sheet being nearly horizontal. The source surface data are shown in Figure 1, along with the tracks of V1 and V2 after being shifted to the proper Carrington longitudes using the solar wind velocities at the two spacecraft and the "constant velocity approximation." Figure 1 suggests caution in totally accepting the statement that Voyager 1 and 2 are in the HCS for this period. Instead, it is more accurate to say that they were in the "close vicinity" of the HCS. At this time, V1 and V2 were separated by about .02 AU in radius and 2 degrees in longitude (V2 being at the larger radius and longitude).

## 2.   DATA ANALYSIS

The data are hourly average, solar wind velocities at Voyager 1 and 2 and are shown in Figure 2. The velocities at V1 and V2 are noticeably different, possibly due to the presence of a transient or, more likely, to the two spacecraft being separated in latitude by 2 degrees. While the velocity signature at V2 is more like that expected in the HCS - between 300 and 350 km/s with a 50 km/s dispersion (Newkirk and Fisk, 1985), we will analyze both sets of data equally.

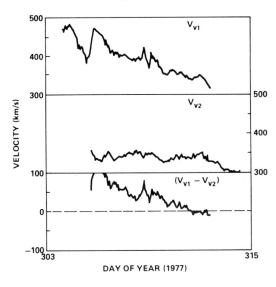

Figure 2. Solar wind flow speeds at Voyager 1 and 2 (V1 and V2) are plotted in the top two panels for the intervals given in Figure 1. The bottom panel shows the difference in velocities between V1 and V2, taking into account any delays between the two spacecraft.

The parameter determining the deformation of the HCS is the velocity gradient along the current sheet. We compute this at each hourly value for which the data are given in Figure 2. Because of the usual, small-scale velocity fluctuations, the resulting gradients are large and often change sign. These computed velocity gradients are shown at the top of Figure 3. The small scale gradients are, however, of little interest for macroscopic deformation of the HCS because their dynamic interaction time is a few hours - they do not coherently retain their character over anything approaching an AU. We remove the small-scale gradients by taking running 6-, 12-, and 24-hr averages - shown in .the bottom three panels of Figure 3.

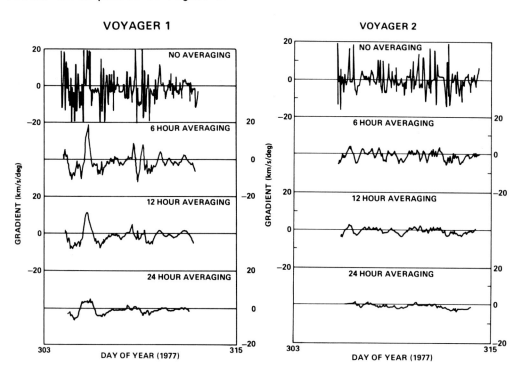

Figure 3. Computed velocity gradients at Voyager 1 and 2, using the data shown in Figure 2. The top panel shows the simple gradient; the following three panels show the data after smoothing, with running 6-, 12-, and 24-hr averages, respectively.

We focus on the 12-hr averaging gradients. This average is sufficiently long that the dynamic interaction time is several days. Thus, we can consider the gradients to be quasi-steady. The average magnitude is over 5 km/s/deg for V1 and is over 2 km/s/deg for V2. An expanded-scale plot of the 12-hr average gradient for V2 is given in Figure 4.

SH have shown that if gradients of the magnitude shown in Figure 4 were imposed on anything other than a perfectly horizontal current sheet, then the sheet would be strongly deformed within a few AU. We will now demonstrate this, using the data shown in Figure 2, by making the ergodic assumption that the character of this data is typical of conditions on the HCS at any random time.

Imposing the velocities observed at V1 and V2 on a current sheet

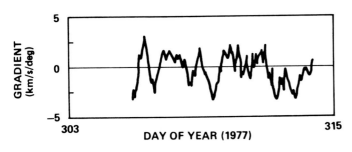

Figure 4. 12-hr averaged, solar wind velocity gradients for V2, taken from Figure 3, plotted on an expanded scale to show that the average value over the total interval is between 2 and 3 km/s/deg.

that is initially vertical at the Sun, we compute the sheet's subsequent deformation. In Figure 5, we show this deformation for varying distances from the Sun.

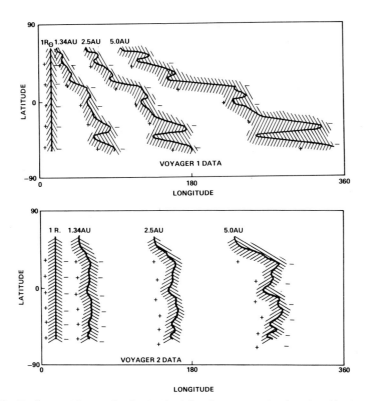

Figure 5. Deformation of hypothetical current sheets that are initially vertical at the Sun and that have imposed on them the velocity variations shown in Figure 2. Results are given at 1.34 AU (the location of the Voyagers), 2.5 AU, and 5 AU.

Figure 5 demonstrates substantial deformation of the hypothetical HCS within a few AU of the Sun. But, even though the deformation is

large enough to cause order one modifications to the local normal to the
current sheet, the timing of sector boundary passages at the earth is
changed by less than a day. Thus, published reports that source surface
models or the maximum brightness contour in the corona can be used to
predict sector passages at the earth to within a +/- 1 day uncertainty
do not invalidate our findings. In this analysis our only assumption was
that the velocity variations observed on a horizontal HCS during 1977
are not unlike those occurring on any HCS.

We finally note that the character of the ruffles illustrated in
Figure 5 is essentially similar to that suggested to exist on the HCS by
Behannon, Neubauer, and Barnstorf (1981) from their own data analysis.
Those sorts of variations could be manifestations of speed variations
like those reported here (Behannon, priv. comm.), although it has also
been suggested that they are surface waves (Hollweg, 1982).

## 3.   SUMMARY AND CONCLUSIONS.

We have analyzed the speed variations as measured by Voyager 1 and 2
while skimming along a horizontal (east-west) portion of the HCS over a
period of several days in 1977. We find that, for either V1 or V2, the
magnitude of the large scale velocity gradient along the current sheet
is at least 2 km/s/deg. This gradient, if it existed on anything other
than a perfectly horizontal current sheet, has already been shown by
Suess and Hildner (1985) to be sufficient to strongly deform the HCS
within a few AU of the Sun. Assuming, then, that the velocity variations
seen at V1 and V2 are typical of current sheet conditions in general, we
illustrate the character and magnitude of the subsequent deformation on
a current sheet that is initially vertical at the Sun.

ACKNOWLEDGMENTS.

We thank L. Burlaga and K. Behannon for their generous help and we thank
the MIT solar wind group for data from the Voyager plasma science
experiments - H. S. Bridge, PI.

This research was supported by NASA's Office of Solar and
Heliospheric Physics and Office of Space Plasma Physics, by the Office
of Naval Research, by the Atmospheric Sciences Section NSF, and by the
Max C. Fleischmann Foundation.

REFERENCES.

Behannon, K. W., Burlaga. L. F., and Hundhausen, A. J.  1983, J.
    Geophys. Res., 88, 7837.
Behannon, K. W., Neubauer, F. M., and Barnstorf, H.  1981, J. Geophys.
    Res., 86, 3273.
Hoeksema, J. T., Wilcox, J. M., and Scherrer, P. H.  1983, J. Geophys.
    Res., 88, 9910.
Hollweg, J. V.  1982, J. Geophys. Res., 87, 8065.
Newkirk, G., Jr., and L. A. Fisk  1985, J. Geophys. Res., 90, 3391.
Rhodes, E. J., and E. J. Smith  1975, J. Geophys. Res., 80, 917.
Suess, S. T., and Hildner, H.  1985, J. Geophys. Res., in press.

THREE-DIMENSIONAL STRUCTURE OF THE HELIOSPHERE AS INFERRED FROM
OBSERVATIONS WITH A JAPANESE HALLEY SPACECRAFT

T. Saito, K. Yumoto      K. Hirao, I. Aoyama      E.J. Smith
Onagawa Mag. Obs.        Space Sci. Lab.          J.P.L.
Geophysical Institute    Tokai University         Oak Grove Drive
Tohoku University        Hiratsuka                Pasadena
Sendai   980             Kanagawa   259-12        CA  91103
Japan                    Japan                    U.S.A.

ABSTRACT.  A sinusoidal neutral line with a pair of giant regions appeared on the sun about one year before the launch of "Sakigake", the
first of two Japanese Comet Halley spacecraft. The Sakigake
magnetometer data during the early part of the mission (February -
March, 1985) are well interpreted by an eastward shift of the tilting
neutral sheet. The shift is further explained by an effect of a new
giant region appearing at ~10° heliolatitude and ~50° Carrington
longitude in August 1984. The toward polarity ratio of IMF observed by
Sakigake changed from ~22% in February to ~62% in early June and then
increased rapidly up to 98%. This ratio is interpreted as a decrease of
the tilt angle of the sheet down to only ~4°. It is the first
spacecraft observation of "the disappearing sector structure" with such
small tilt angle.

1.  INTRODUCTION

     According to the two-hemisphere model (Saito, 1975 and 1985,
Saito, et al., 1978), the heliosphere evolves during the course of the
solar cycle as shown in Fig. 1. In sunspot minimum years, a nearly
horizontal warped neutral sheet separates the heliosphere into two
hemispheres with toward and away polarities (Fig. 1Ca). Associated with
the reversal of the heliopolar-cap magnetic polarities, the neutral
sheet undergoes a rotational reversal (b→c→d→e) in sunspot maximum
years. The sheet tends to have a fairly large tilt angle with respect
to the ecliptic plane in sunspot declining years (f). In the case of
the last solar cycle, for example, the tilt angle oscillated starting
from 4° (in November, 1972), via 26° (April, 1973), 9° (October, 1973),
and 46° (May, 1974), to 22° (November, 1974). Finally the sheet returns
to the nearly horizontal state (f→g) in sunspot minimum years. From the
viewpoint of the heliomagnetospheric structure, one solar cycle is
divided into four cyclic phases; aligned, reversing, pseudo-aligned,
excursion, and aligned again (Fig. 1B). When the quadrupole component
of the photospheric magnetic field survives to the solar source
surface, the main neutral sheet is accompanied a by a subcone sheet

R. G. Marsden (ed.), The Sun and the Heliosphere in Three Dimensions, 281–286.
© 1986 by D. Reidel Publishing Company.

(Fig. 1Cd), which appears only within a few solar rotations in the
reversing phase. The surviving quadrupole component gives rise to a
saddle-type sheet, especially in the pseudo-aligned phase (Fig. 1Ce).
Along with the neutral sheet evolution, the macroscopic structure of
the coronal holes and the high-speed solar wind source evolve as
exhibited in Fig. 1D. Detailed observational and theoretical evidence
to support this solar cycle evolution is given in Saito (1985).

     The present paper is an initial report of IMF observations with
Sakigake, the first of two Japanese Comet Halley spacecraft. Evidence
supporting the two-hemisphere model will be presented using the Sakigake
magnetometer data.

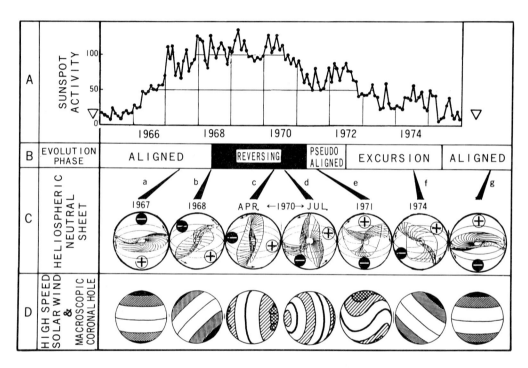

Fig. 1. Solar cycle evolution of the three-dimensional heliosphere.

## 2. SAKIGAKE MAGNETOMETER

     Sakigake was launched on 8 January 1985, and the 2-m boom antenna
was deployed on 19 February. The magnetometer, a triaxial ring-core
sensor, is installed at the tip of the boom and started observation on
the same day. Resolution and the sampling rate are 0.03nT and 1/4
seconds, respectively. Because of the restriction of the orbit,
Sakigake will not encounter directly with Comet Halley; rather, the
purpose of the mission is to study the solar wind structure. The data
providing evidence to support the two-hemisphere model are presented in
the following section.

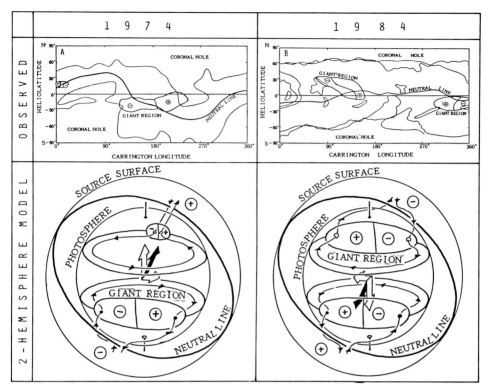

Fig. 2. Coronal holes, giant regions, and magnetic neutral lines observed in 1974 and 1984, and interpretation by the two-hemisphere model.

## 3. EASTWARD SHIFT OF THE NEUTRAL SHEET BY A NEW GIANT REGION

Based on the two-hemisphere model, the heliosphere at the time of the Sakigake launch should have been in the excursion phase with two giant regions, each of which consists of a pair of ± UM regions, on the sun. Indeed, at the beginning of 1984 the situation resembled that in 1974, as predicted (Fig. 2), and remained so up to the time of the Sakigake launch.

The observed IMF polarities are plotted in Fig. 3D against the Carrington longitude from which the observed solar wind started. When a solar flare takes place on the sun, the flare plasma rushes out in the heliosphere disturbing the basic sector polarities. The polarity of the $B_X$-component of IMF tends to be disturbed more than that of the $B_Y$, though the disturbance is generally expressed by a local belly of the basic warped neutral sheet. Taking into consideration of this disturbance effect, the basic sector boundary is drawn with the thick broken line in the figure. Figure 3B shows an example of magnetic field on the photosphere (thin lines) and the assumed neutral line on the source surface (thick line). The expected sector polarities are expressed by the thin (away) and the thick (toward) curves with the

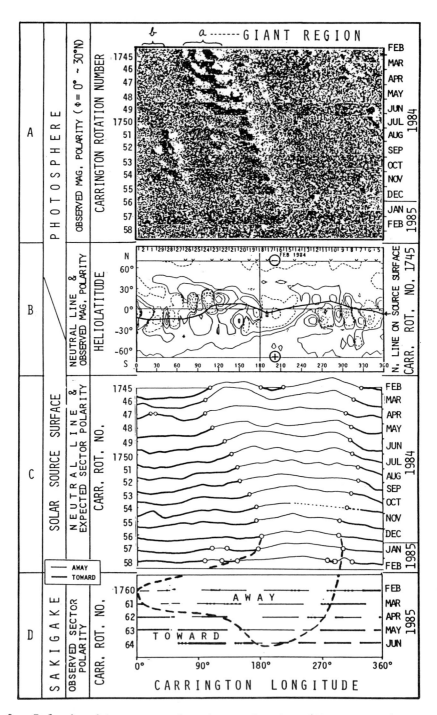

Fig. 3.   Relationships among the observed solar (A, B, and C) and interplanetary (D) phenomena.

circles of the expected sector boundary crossings in Fig. 3C. In this
way all the expected crossings are displayed till the latest available
rotation from Solar-Geophysical Data.

The expected boundary at 300 ~ 330° longitude (C) corresponds well
with the observed boundary (D) and a giant region in the southern low
latitudes (see Fig. 2), while the other boundary made a gradual
westward shift (~110°/1745 to ~180°/1756) and a dramatic eastward shift
(~180°/1756 to ~120°/1758, see Fig. 3C), and is connected to the observed boundary (D). It is clear from the chronological map for the
northern low-latitude SMF in Fig. 3A that the westward shift (C)
corresponds well with that of the giant region (a) in Fig. 3A, while
the dramatic eastward shift can be interpreted well by the effect of
the appearance of the new giant region (b) and of the disappearance of
the region (a). This interpretation appears to be reasonable, since a
similar situation occurred in the last solar cycle.

4. DISAPPEARANCE OF THE SECTOR STRUCTURE

Figure 4 indicates the toward polarity ratio per solar rotation
obtained by Sakigake. The change tends to agree with the curve of the
expected ratio for a constant tilt angle of the neutral sheet. The
last point of about 100% means that the tilt angle became as small as
~4°. It is the first time that a spacecraft has encountered such a
small tilt angle of the neutral sheet. Interpretation in terms of the
two-hemisphere model must be appropriate, because many geophysical
phenomena are losing the 27-day recurrence tendency, as was the case
in 1954.

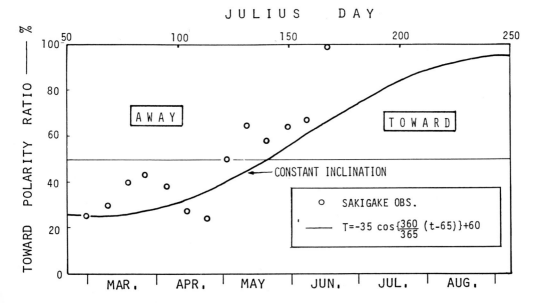

Fig. 4.  Toward polarity ratio observed by Sakigake.

## 5. CONCLUSION

There appear to be two possibilities concerning the present heliospheric evolution: the excursion phase continues, giving rise to one or two more excursions, or the excursion phase ends by entering into the aligned phase.

ACKNOWLEDGEMENTS. The authors wish to express sincere thanks to many PI's, Co I's, scientists, and technicians who have cooporated earnestly on the Sakigake mission.

REFERENCES

Saito, T.: 1975, Sci. Rept. Tohoku Univ., Ser. 5, 23, 37.
Saito, T., Sakurai, T., and Yumoto, K.: 1978, Planet. Space Sci., 26, 413.
Saito. T.: 1985, Space Sci. Rev. to be submitted.

# THREE-DIMENSIONAL STRUCTURE OF THE HELIOSPHERIC CURRENT SHEET

Craig D. Fry, and S.-I. Akasofu
Geophysical Institute
University of Alaska
Fairbanks, Alaska 99775-0800
U.S.A.

ABSTRACT. Three-dimensional structure of the heliospheric current sheet, HCS, out to 5 AU is constructed for actual observed source field configurations.

## 1. INTRODUCTION

The kinematic method developed by Hakamada and Akasofu (1982), the HA method, can be used to model the first order variations in solar wind speed and interplanetary magnetic field. We studied solar wind speed as a function of source surface magnetic field configuration, and phase of the solar cycle (Fry and Akasofu, 1985). This variation has been parameterized and incorporated into the HA code. The Stanford source surface field analysis of Hoeksema et al. (1982, 1983) was used. By following those solar wind streams which originated near the magnetic neutral line on the source surface, the heliospheric current sheet, HCS, can be mapped out into interplanetary space. This is done for several Carrington rotations corresponding to solar minimum, the rising phase, and shortly after solar maximum. Since the kinematic method may be used to simulate solar wind conditions at any point in the heliosphere and at any phase of the solar cycle, we are able to offer predictions for the Ulysses observations.

## 2. THE KINEMATIC METHOD

In the kinematic method, the solar wind flow is characterized on an inner boundary, the solar wind source surface, by a velocity distribution which can be varied in space and time. The velocity distribution on the source surface depends upon the source surface magnetic field configuration. Specifically, the source wind velocity is directed radially outward and is a function of the magnetic latitude of the source region. The speed is a minimum at the magnetic neutral line, increasing to maxima at the poles. Consider the idealized case of a dipolar source field. Magnetic latitude, $\lambda_m$, for this dipole case is found from

*R. G. Marsden (ed.), The Sun and the Heliosphere in Three Dimensions, 287–293.*
© *1986 by D. Reidel Publishing Company.*

$$\sin \lambda_m = B/B_{max} \qquad (1)$$

where B is the source field strength at the solar wind source, and $B_{max}$ is the value at the magnetic pole of the same sign as B. The source magnetic neutral line is inclined with respect to the heliographic equator at some angle, $\chi$. Thus, as the sun rotates, the magnetic dipole "wobbles" and the neutral line moves up and down with respect to a fixed observer out in space. This causes successively fast and slow streams to be emitted along a radial line. The kinematic method adjusts for this stream-stream interaction and allows for the formation of shocks. The radial distance a solar wind fluid parcel attains, R, is then a function of its emission time, $\tau$, and the simulation time, t. A velocity profile is constructed along any radial line by taking the difference between R-$\tau$ relations at successive time steps. The IMF magnitude and direction are determined by using the frozen field conditions, along with magnetic flux and mass flux conservation.

3. LATITUDINAL DEPENDENCE OF SOLAR WIND SPEED

Accurate modeling of actual solar wind conditions requires a realistic specification of the boundary conditions. We used the Stanford "potential analysis" magnetic field information provided by Hoeksema et al. (1982, 1983) to specify the source surface field. Various researchers have investigated the variation of solar wind speed as a function of angular distance from the source surface neutral line (cf. Zhao and Hundhausen, 1981, 1983; Hakamada and Akasofu, 1981). For a dipolar field, angular distance from the neutral line is equivalent to the magnetic latitude discussed above. The solar wind speed observed near the earth was used to map the flow back to the source region near the sun. Scatter diagrams were made of solar wind speed as a function of magnetic latitude. Then, best fit curves were drawn through the scatter of points. These resulting curves showed minima near the neutral line (magnetic equator) and maxima far from the neutral line.

As a first-guess of the velocity function to use in the kinematic code, we fit the lower envelope of the scatter diagrams to the relationship

$$V = V_0 + V_1 (1-\cos^n \lambda_m). \qquad (2)$$

It can be seen that at solar minimum, solar wind speed increases away from the neutral line more rapidly than at the maximum of the solar activity cycle (Fry and Akasofu, 1985). This agrees with the recent results of Newkirk and Fisk (1985). Near solar minimum, when the velocity profile increases most rapidly away from the neutral line, our velocity function, Equation 2, yields values of n equal to 16-32. Near solar maximum, when there is a broad minimum around the neutral line, n decreases to 2-4.

A first guess v-vs-$\lambda_m$ is inserted into the kinematic code and the simulated solar wind speed and IMF are compared with the King data

set. Through an iterative process, small adjustments to $V_0$ and $V_1$ are
made until a best fit is obtained. If care is taken in locating the
earth, and in specifying the velocity function and source magnetic
field, the quiet-time corotating structure can be simulated.

## 4. THE HELIOSPHERIC CURRENT SHEET

By keeping track of all solar wind streams originating near the
neutral line, we use the frozen field condition to extend the neutral
line out into interplanetary space and map the realistic, quiet-time
HCS. For example, Carrington rotation 1647, near solar minimum, was
characterized by a four-sector IMF pattern. The upper part in Figure
1 shows the Stanford potential field analysis for this rotation. In
the lower part of Figure 1, the corresponding HCS determined by the
kinematic method is displayed out to 5 AU. The observer's view angle
is from longitude = 30° and latitude = 20°. The gap in the HCS
corresponds to the zero longitude in this figure, representing
Carrington longitude = 0° at the start of the Carrington rotation.
Even small excursions of the neutral line above and below the equatorial plane, cause an outwardly propagating wavy structure in the
current sheet.

As the neutral line extends to higher latitudes, the current
sheet becomes increasingly distorted. The upper part of Figure 2
shows the source surface field strength pattern for Carrington rotation 1663, during the rising phase of the 11-year solar activity
cycle. The corresponding HCS is shown in the lower part of Figure 2
and Figure 3 out to 2 AU and 5 AU, respectively. The location of the
observer is at longitude = 300° and latitude = 20°. Compression of
the high latitude extensions (or ridges and troughs) of the HCS occur
due to the solar wind stream-stream interaction. This distortion of
the HCS was not predicted by earlier conceptual sketches (Svalgaard
and Wilcox, 1976; Smith et al., 1978; Kaburaki and Yoshii, 1979;
Thomas and Smith, 1981; Jokipii and Thomas, 1981).

Carrington rotation 1719, occurring shortly after solar maximum,
is displayed in the upper part of Figure 4, and the resulting HCS in
the lower part of Figure 4 and Figure 5. A two-sector IMF structure
dominated during this period, with latitudinal excursions of the neutral line extending to ±45°. Even in this simple source field configuration, the stream-stream interaction begins to distort the HCS at
several AU.

## 5. CONCLUSION

Our method is powerful in constructing the heliospheric current
sheet by taking into account the stream-stream interaction.

## 6. ACKNOWLEDGEMENTS

This work was supported in part by U.S. Air Force Contract F19628-81-
K-0024 and by a grant from the National Aeronautics and Space Administration NSG 7447.

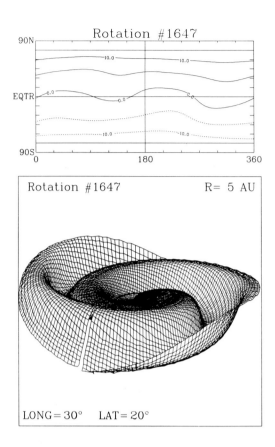

Figure 1. Upper part: Contour map of source surface magnetic field
strength for Carrington rotation 1647, based upon the Stanford potential analysis data from Hoeksema et al. (1982).  Contours are shown
for 0, +5 and +10 micro–Tesla.  Lower part: Heliospheric current
sheet obtained by extending the Stanford (potential analysis) neutral
line out into interplanetary space to a distance of 5 AU using the
kinematic method.  Observer's view angle is from longitude = 30°,
latitude = 20°.  The 'zero' longitude (marked by the gap in the current
sheet) corresponds to Carrington longitude = 0° at the beginning of
Carrington rotation 1647.

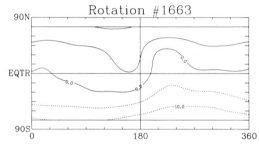

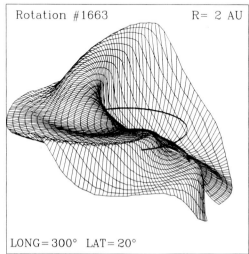

Figure 2. Upper part: Stanford source surface field strength map for Carrington rotation 1663. Lower part: Heliospheric current sheet for Carrington rotation 1663, out to a distance of 2 AU. Observer's view angle is from longitude = 300° and latitude = 20°.

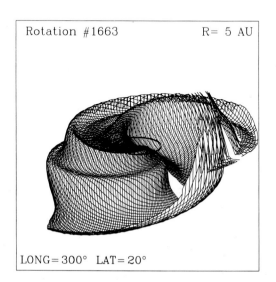

Figure 3. Heliospheric current sheet for Carrington rotation 1663, out to a distance of 5 AU. Observer's view angle is from longitude = 300° and latitude = 20°.

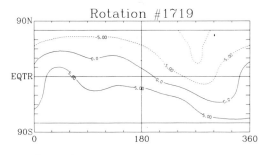

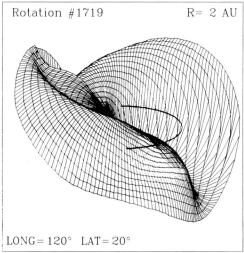

Figure 4. Upper part: Stanford source surface field strength map for Carrington rotation 1719. Lower part: Heliospheric current sheet for Carrington rotation 1719, out to a distance of 2 AU. Observer's view angle is from longitude = 120° and latitude = 20°.

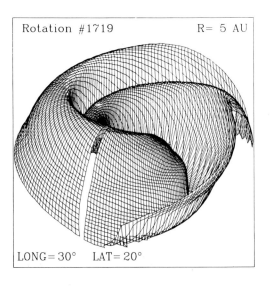

Figure 5. Heliospheric current sheet for Carrington rotation 1719, out to a distance of 5 AU. Observer's view angle is from longitude = 30° and latitude = 20°.

7.  REFERENCES

Fry, C. D., and Akasofu, S.-I., 1985, Planet. Space Sci., 33, 925.
Hakamada, K., and Akasofu, S.-I., 1982, Space Sci. Rev., 31, 3.
Hoeksema, J. T., Wilcox, J. M., and Scherrer, P. H., 1982, J. Geophys.
    Res., 87, 10331.
Hoeksema, J. T., Wilcox, J. M., and Scherrer, P. H., 1983, J. Geophys.
    Res., 88, 9910.
Jokipii, J. R., and Thomas, B. T., 1981, Astrophys. J., 243, 1115.
Kaburaki, O., and Yoshii, Y., 1979, Solar Phys., 64, 187.
Newkirk, G. R., Jr., and Fisk, L. A., 1985, J. Geophys. Res., 90, 3391.
Smith, E. J., Tsurutani, B. T., and Rosenburg, R. L., 1978, J. Geophys.
    Res., 83, 717.
Svalgaard, L., and Wilcox, J. M., 1976, Nature, 262, 766.
Thomas, B. T., and Smith, E. J., 1981, J. Geophys. Res., 86, 105.
Zhao, X.-P., and Hundhausen, A. J., 1981, J. Geophys. Res., 86, 5423.
Zhao, X.-P., and Hundhausen, A. J., 1983, J. Geophys. Res., 88, 451.

SECTION VI:  ENERGETIC PARTICLES

# SEPARATION AND ANALYSIS OF TEMPORAL AND SPATIAL VARIATIONS IN THE 10 APRIL 1969 SOLAR FLARE PARTICLE EVENT

R. Reinhard
Space Science Dept. of ESA
ESTEC
2201 AG Noordwijk
The Netherlands

E. C. Roelof and R. E. Gold
Johns Hopkins University
Applied Physics Laboratory
Laurel, MD 20707
USA

**ABSTRACT:**

We analysed the 10 April 1969 energetic particle event with five spacecraft separated over 180° in azimuth. The particle fluxes are mapped back to the coronal injection longitudes which are estimated from the instantaneous solar wind velocities and connected at the same time of observation to give the longitudinal particle gradients. Cutting these gradients at representative longitudes gives the temporal variations at fixed coronal longitudes which turn out to be power-law decays. This shows that the exponential rise and decay observed by spacecraft is due to the spatial gradient sampled by the spacecraft as the connection longitude changes due to the Sun's rotation.

## 1. INTRODUCTION

In terms of energetic charged particle propagation, a spacecraft in interplanetary space is connected to the Sun by the spiral magnetic field lines. Due to the Sun's rotation, the connection longitude continuously shifts eastward in the Sun's reference frame and the observed temporal flux variation at a single spacecraft is a mixture of a spatial azimuthal and a true temporal variation. Only with several spacecraft well distributed in azimuth is it possible to disentangle the spatial and temporal variation. The last opportunity for multispacecraft observations of that kind was in April 1969 when a major flare - the second largest of the previous solar cycle - was observed by five spacecraft distributed over 180° of solar azimuth (Table 1).

In order to make use of multispacecraft observations, the experiments on the various spacecraft have to be intercalibrated, correcting for the different energy channels and electron sensitivities. This was done in a very careful manner by selecting time periods in late 1968 and March 1969 when Pioneer 9 and Explorer 34 were <5° separated in azimuth. We find that the differences in the electron sensitivity

297

**Table 1**

| spacecraft | azimuthal separation from flare | heliocentric distance ($10^8$ Km) | energetic particle detectors |
|---|---|---|---|
| Explorer 34 (IMP 4) | 114° | 1.5 | > 10 Mev |
| Pioneer 6 | -36° | 1.45 | 7.4-44 MeV |
| Pioneer 7 | 20° | 1.68 | 7.2-47.4 MeV |
| Pioneer 8 | 89° | 1.57 | 7.4-38.5 MeV |
| Pioneer 9 | 146° | 1.13 | 7.4-38.5 MeV |

and the energy thresholds partially cancel each other so that only a correction factor of ~20% should be applied, which is negligible compared to the much larger spatial and temporal variations we find in this event and, therefore, we decided not to apply any correction factors.

The most likely location of the flare site is N22 E114 with 11 UT on 10 April 1969 for the time of particle acceleration. The time-intensity profiles of the 10 April particle event observed by the five spacecraft (Fig. 1) rise and decay exponentially and show several kinks. The event has a number of unusual features:

1)  the fluxes rise rapidly at Pioneer 6 and Pioneer 7, while the rise is delayed by a day at the other three spacecraft;
2)  at Pioneer 7 two peaks are observed with the second rise starting on 11 April;
3)  after 12 April the fluxes at Pioneer 8, Explorer 34 and Pioneer 9 are 1 to 3 orders of magnitude higher than at Pioneer 6 and Pioneer 7, which are better connected to the flare site;
4)  the intensity difference persists until very late in the event in contrast to conventional views of diffusive equilibrium (the whole solar system being uniformly filled with particles late in the event)

Note that the last two features could only be detected by multi-spacecraft observations.

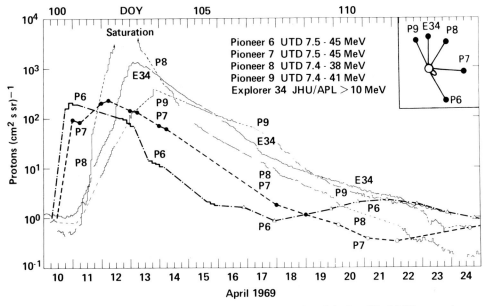

Figure 1.   Energetic particle fluxes for the 10 April 1969 event as
observed by Pioneer 6 - 9 and Explorer 34.

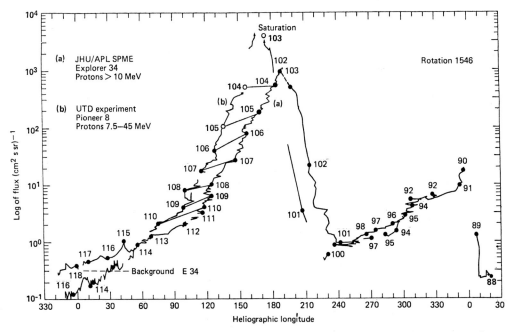

Figure 2.   Energetic particle fluxes mapped back to the high coronal
injection longitudes and connected at the same time (DOY)
of observation.

Fig. 3. The variation of longitudinal particle gradients with time. 12:18 means 12 April, 18 UT.

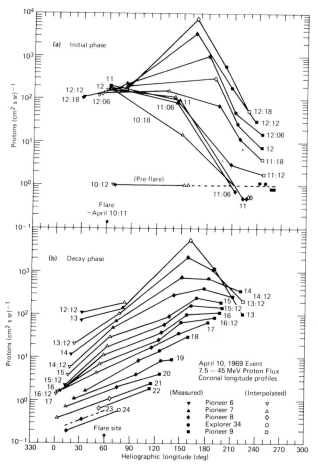

## 2. SEPARATION OF SPATIAL AND TEMPORAL VARIATIONS

To separate the spatial from the temporal variations we map all fluxes back to the corona using the instantaneous solar wind velocity measured on board the various spacecraft (mapped-back fluxes from Explorer 34 and Pioneer 8 are shown in Fig. 2 as an example). We make the approximation of non-accelerating radial solar wind transport with frozen-in magnetic field lines (Nolte and Roelof, 1973) and that the particle intensity gradients along interplanetary field lines are relatively small (Roelof, 1986). Connecting the fluxes at the same time of observation gives the longitudinal particle gradients, as shown in Figure 2 from DOY 104 to 110. This is done in the same way for the fluxes from all five spacecraft, and connecting all the fluxes at the same time of observation (we left out the mapped back profiles for clarity) gives the longitudinal particle gradients shown in Figure 3. All the unusual features of Figure 1 have disappeared and we observe a very clear pattern throughout the event and observed by all spacecraft: the fluxes rise

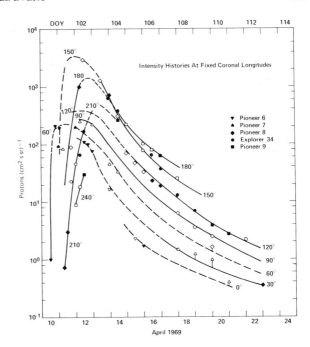

Fig. 4. Temporal variations of the particle fluxes at fixed coronal longitudes. The flare site is at 60° heliographic longitude.

and decay exponentially with an e-folding length of 30° persisting up to 14 days after the flare. However, the main peak is not observed at the flare site but at ∼150° heliographic longitude, 90° further to the east. Figure 3 shows the spatial variation at any time during the event. It also shows the temporal variation at any longitude. To show this more clearly we plotted the intensities as a function of time at 30° intervals from 0° to 240° (Fig. 4). For some of the longitudes we only have a few data points but for the others we see a non-exponential decay, which is close to a power-law. This is in striking contrast with the exponential decays observed at the spacecraft (Fig. 1). Obviously the exponential decays observed by spacecraft are dominated by the exponential spatial gradient. If this observation is typical we can draw the following general conclusion: if the duration of the event is short, the temporal behaviour dominates and we observe a power-law decay; if the event is long-lasting (mostly eastern hemisphere events), the spatial gradient dominates and we observe an exponential decay.

## 3. DISCUSSION

We interpret the power law decays shown in Fig. 4 as the signature of interplanetary propagation. Consequently, up to the time of the onset of the power law decay the temporal variations are dominated by coronal injection, which is found to be long-lasting (of the order of

Fig.5. "Solar Activity Chart" for rotations
1544-47. Open circles denote N flares,
solid circles B flares; the size of the
circle corresponds to the size of the Hα
flaring region, a square indicating particle acceleration.

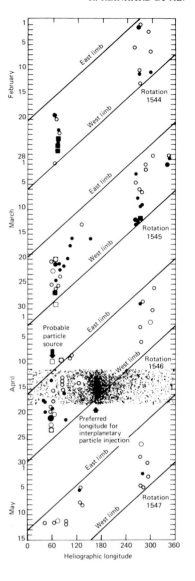

days). This is consistent with the anisotropies observed by Pioneer 8 and 9 (not
shown): streaming from the Sun persists at
least until 13 March. Figure 3 also shows
that the highest fluxes escape ~1.5 days
after the flare from a region around 150°.
The solar activity chart (Fig. 5) shows the
region of preferential release as the
heavily speckled area. There is no flare
activity in that region for four rotations
indicating that the region is quiet and
stable with weak magnetic fields. It is
clear that the particles were not accelerated at the site of preferential release
but were accelerated in the highly active
region at 90° and then propagated to 150°.

Figure 6 gives the normalised intensity
histories of 0.5-1.1 MeV electrons, 19-80
MeV and 6-19 MeV protons, and 1.0 - 1.6
MeV/nuc alphas observed by different detectors on Explorer 34. The electron and
proton data were taken from Van Hollebeke
et al. (1974), the α -particle data were
obtained from Lanzerotti (private communication). The high energy protons are saturated during 13 April, the electrons are
saturated from mid-12 April until mid-15
April. If normalised during the decay,
the rises display a velocity dispersion:
the electrons rise first, then the high
energy protons, then the medium energy
protons and, finally, the low energy
alphas. Since the rises are dominated by coronal processes these
latter must be velocity-dependent. The decays are independent of
energy, rigidity and charge which is interpreted as the signature of
a large-scale spatial structure in interplanetary space controlling
the particle propagation. The particles are confined by an expanding
volume with the converging magnetic field lines close to the Sun as
the one boundary and the interplanetary shock wave with the disturbed
magnetic fields behind it as the other boundary. The decay is
controlled by the expanding shock volume; at a given time we find
equal fluxes even on the same field line at different heliocentric
distances, and different fluxes on different field lines even at the

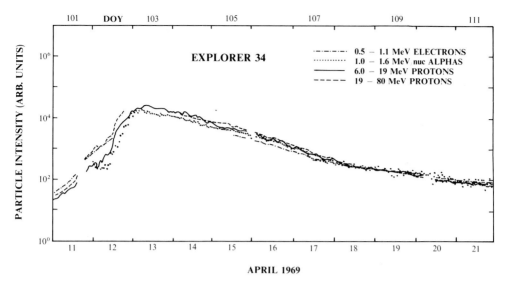

Figure 6.   Normalised intensities of electrons, high and medium
            energy protons and low energy alphas observed by Explorer
            34 during the 10 April 1969 event.

same heliocentric distance.   The whole population is slowly decaying
with time.

Although we cannot completely exclude the possibility of some
interplanetary acceleration in association with the shock we believe
that the effect is only minor for the following reasons:

1)   The shock passed Pioneer 8 on 12 April at 11 UT.   Particle
     streaming from the sun at Pioneer 8 (not shown) persisted until
     at least 13 March.
2)   Shock acceleration is most effective at low energies ($\lesssim 1$ MeV)
     (Wenzel et al. 1981), therefore the effect on the 10 MeV protons
     should be small.   Also the low-and the high-energy protons show
     the same profile.
3)   Efficient shock acceleration is not observed for electrons;
     however, the electrons show the same profile as the protons.
4)   The shock is weaker at Explorer 34 and Pioneer 9 than at Pioneer
     6 and 7, which are closer to the flare site, yet the fluxes are
     higher at the spacecraft which are further away (the shock is
     hardly noticeable at Explorer 34 and Pioneer 9).

We conclude that there is at most a small contribution by
interplanetary acceleration to the observed fluxes.   Subtracting a
small contribution would not change the profiles significantly, and,
therefore, would not change the results based on their interpretation.

## 4.   CONCLUSIONS

We draw the following general conclusions for the interpretation of energetic charged particle data to be obtained by the Ulysses spacecraft:

1) Conclusions drawn from the analysis of profiles of long-lasting events observed by a single spacecraft can be very misleading; the usefulness of deriving parameters using models which do not include coronal propagation is limited.

2) Coronal azimuthal transport exists and it is quite possible that coronal latitudinal transport exists as well so that energetic particles can be expected at high latitudes.

3) Regions of preferential release of particles from the corona into the interplanetary medium exist, and they may not be at or close to the flare site.

4) The decay is controlled by interplanetary structure, most likely by the expanding shock wave and the disturbed region behind it.

5) Over the solar poles the effect of solar rotation disappears; the spacecraft is connected to the same longitude, even if the solar wind velocity varies, and we expect to observe the pure temporal variation as in Fig.4.

## REFERENCES

Nolte, J.T. and Roelof, E.C.: 1973, Solar Phys., 33, 241.
Roelof, E.C.: 1986, this volume.
Van Hollebeke, M.A., Wang, J.R. and McDonald, F.B.: 1974, GSFC
      preprint X-661-74-27.
Wenzel, K.-P., Reinhard, R. and Sanderson, T.R.: 1981, Adv. Space
      Res., Vol.1, 105.

ACCELERATION OF ENERGETIC PARTICLES AT SOLAR WIND SHOCKS

Martin A. Lee
Space Science Center
University of New Hampshire
Durham, NH  03824
USA

ABSTRACT.  A review of the acceleration of energetic ions at
interplanetary shocks is presented with an emphasis on the theory of
diffusive shock acceleration and interplanetary traveling shocks.  The
basic theory is discussed briefly including wave excitation.  Ten
predictions of the theory as outlined by Kennel et al. (1985) are
presented and found to compare favorably with the observations of the
11, 12 November 1978 event.  Some problems are presented which should
be addressed by future theoretical/experimental work.  A simple
illustrative application of diffusive acceleration theory is made to
the acceleration of ions at shocks in the distant heliosphere and
compared qualitatively with the intensity profiles observed by Pioneer
10.  Finally some brief thoughts on shock acceleration at high solar
latitudes are presented.

1.  INTRODUCTION

The acceleration of energetic particles at shock waves in the solar
wind has been an exciting and fruitful area of research over the last
ten years.  Observations by many satellites over a range of
heliocentric radial distances, $r$, in the ecliptic from $\sim$ 0.4 AU to >35
AU have revealed a close correspondence between energetic particle
enhancements and shock waves.
    Pioneers 10 and 11 observed increasing fluxes of $\sim$ 1 MeV/nucleon
ions with increasing $r$ which recurred with the period of solar rotation
and peaked at the forward and reverse shocks bounding corotating interaction regions (McDonald et al., 1976; Barnes and Simpson, 1976;
Tsurutani et al., 1982).  A shock origin accounts for many of the
observed features (Palmer and Gosling, 1978; Fisk and Lee, 1980).
    A variety of energetic ion populations in the energy range 10-150
keV/nucleon have been observed and measured intensively upstream of
Earth's bow shock by the three ISEE satellites (e.g. Paschmann et al.,
1979, 1981; Ipavich et al., 1979, 1981; Bonifazi and Moreno, 1981;
Wibberenz et al., 1985).  The "reflected", "intermediate", and "diffuse" ion distributions observed can be understood as a consequence of

R. G. Marsden (ed.), The Sun and the Heliosphere in Three Dimensions, 305–318.
© 1986 by D. Reidel Publishing Company.

time-dependent shock acceleration initiated on a given interplanetary
magnetic field line when it first connects to the bow shock (e.g., Bame
et al., 1980;  Eichler, 1981; Ellison, 1981; Lee, 1982; Lee and
Skadron, 1985).  Similar ion populations were observed upstream of
Jupiter's bow shock by Voyagers 1 and 2 (Zwickl et al., 1981; Baker et
al., 1984) and have been interpreted as shock accelerated distributions
(Smith and Lee, 1985).

Energetic storm particle (ESP) events at energies $\sim$ 1 MeV/nucleon
have been observed in association with interplanetary traveling shocks
near 1 AU since the early 1960's (e.g. Bryant et al., 1962).  Recent
detailed measurements by ISEE 3 at energies from $\sim$ 10 keV to $\sim$ 1 MeV
have provided strong support for their shock origin (e.g. Scholer et
al., 1983; Van Nes et al., 1984; Tsurutani and Lin, 1985; Lee, 1983;
Kennel et al., 1985).  The same shocks, early in their evolution as
coronal shocks, may be the origin of the energetic solar flare particles preceding the shock into interplanetary space (Lee and Ryan, 1985;
Ellison and Ramaty, 1985).

A more speculative, but nevertheless plausible, association exists
between the cosmic ray anomalous component and the termination shock of
the solar wind (Pesses et al. 1981).

In this paper we first present a brief review of the process of
shock acceleration and the basic theory.  Following Kennel et al.
(1985) we outline the success of the theory in predicting the behavior
of the energetic particle distribution and the ultra-low-frequency
(ULF) fluctuation spectrum observed in association with the large
interplanetary traveling shock of 11, 12 November 1978.  We then
outline some theoretical problems which remain unanswered.  As an
illustrative example of the theory of diffusive shock acceleration we
present a time-dependent solution which qualitatively accounts for the
MeV-proton enhancements associated with traveling shocks observed in
the distant heliosphere (Pyle et al., 1984).  Finally we speculate
briefly on the shock-accelerated energetic particle distributions which
are likely to be observed by Ulysses.


## 2.  BASIC IDEAS AND THEORY

The fundamental process of shock acceleration at a fast shock is
simple.  Consider an infinite planar shock (between homogeneous
upstream and downstream states) in the deHoffman-Teller frame in which
the shock is stationary and $\underline{E}$ (= $- \underline{V} \times \underline{B}/c$) vanishes.  Here $\underline{E}$ and $\underline{B}$ are
the average electric and magnetic fields, and $\underline{V}$ is the average plasma
velocity.  In the absence of pitch-angle-scattering irregularities an
energetic particle encounters the shock once and is either transmitted
or reflected ("mirrored" from the increased downstream magnetic field).
In the deHoffman-Teller frame the particle energy is conserved; in any
other frame the particle gains or loses energy.

In the presence of scattering irregularities energetic particles
are scattered back and forth across the shock and are coupled to the
fluids (more precisely, the irregularities) on either side.  They are
thus coupled to the compression of the shock and accelerated by a

"first-order Fermi mechanism" as they scatter between the converging
irregularities upstream and downstream of the shock.  If the energetic
particle speeds, v, are much greater than the fluid speeds, $|V|$, and if
scattering is efficient so that the energetic particle distribution is
nearly isotropic, then the transport equation derived to describe the
solar modulation of galactic cosmic rays (Parker, 1965; Gleeson and
Axford, 1967) also describes diffusive shock acceleration:

$$\frac{\partial f}{\partial t} + (\underline{V} + \underline{V}_d)\cdot\nabla f - \nabla\cdot\underline{\underline{K}}\cdot\nabla f - \frac{1}{3}\nabla\cdot\underline{V}\, p\, \frac{\partial f}{\partial p} = Q. \tag{1}$$

In equation (1) $f(p,\underline{r},t)$ is the energetic particle omnidirectional
distribution function as a function of momentum p, spatial coordinate $\underline{r}$
and time t, $\underline{\underline{K}}$ is the symmetric spatial diffusion tensor, $\underline{V}_d$ is the
drift velocity (Jokipii et al., 1977), and Q represents particle
sources and sinks.  Equation (1) is a statement of overall particle
conservation as particles convect, drift or diffuse through a given
volume of space and gain or lose energy.  Particle energy changes are
described by the last term on the left side of equation (1):  If
$\nabla\cdot\underline{V} > 0$, as in the diverging solar wind, particles lose energy.  If
$\nabla\cdot\underline{V} < 0$, as at a shock, particles gain energy.  At a shock of
negligible thickness $\nabla\cdot\underline{V} \propto \delta(x)$, where x is distance from the shock
surface parallel to the shock normal.  In this case the shock
contribution to this term can be incorporated into boundary conditions
(continuity of f and particle differential flux) connecting solutions
of equation (1) on either side of the shock (e.g. Drury, 1983).
         If scattering is inefficient over the time scales of interest,
particles may only encounter the shock one or two times and exhibit
anisotropic distributions.  This occurs, for example, for small field
line connection times upstream of Earth's bow shock and for the
reacceleration of solar flare particles with very large scattering mean
free paths at interplanetary traveling shocks.  In these cases equation
(1) does not apply.  Particle distributions must be calculated by
explicitly following particle trajectories (Decker, 1981; Pesses et
al., 1982; Decker and Vlahos, 1985).  In view of the absence of a
kinetic theory for the acceleration in these cases, we will not
consider them further.  However, "reflected" and "intermediate" ion
distributions upstream of Earth's bow shock, and "shock-spike" events
at quasi-perpendicular interplanetary shocks near 1 AU, can apparently
only be understood in this context.
         At first estimate diffusive shock acceleration might appear to be
unimportant in interplanetary space where scattering mean free paths,
$\lambda$, over a large range of rigidities are inferred from studies of solar
flare particle propagation to be large, $\lambda > 0.01$ AU (e.g. Fisk, 1979).
The time scale for diffusive shock acceleration is approximately
$\tau \sim 3K_1/[V_1(V_1-V_2)]$, where subscripts "1" and "2" refer to upstream and
downstream, respectively (Forman and Drury, 1983).  With $V_1 \sim V_{SW}$,
where $V_{SW}$ is the solar wind speed, $\tau \gtrsim (v/V_{SW})(\lambda/V_{SW})$.  Even with
$\lambda = 0.01$ AU, we find $\tau$ on the order of several hours, far greater, for
example, than the typical connection time of several minutes upstream
of Earth's bow shock.  However, the actual scattering mean free paths

upstream of interplanetary shocks are substantially reduced in
comparison with this estimate due to the enhanced ULF turbulence
observed upstream of interplanetary shocks (Hoppe et al., 1981;
Tsurutani et al., 1983; Viñas et al., 1984). The resulting
acceleration time scales are well within time scales of interest.

Indeed these ULF wave enhancements are excited by the accelerated
protons themselves (Tademaru, 1969; Barnes, 1970; Lee, 1982): The
energetic particle gradient upstream of the shock ($x < 0$) is determined
by a balance between convection and diffusion: $f \propto \exp [-V_1 \int_x^0 dx'/$
$(K_1(x'))]$. The normal component of the differential diffusive flux in
the upstream wave frame is then $S_x = - K_1 \partial f/\partial x = - V_1 f < 0$. Thus the
energetic particles stream upstream relative to the upstream wave
frame. Transverse waves which can interact with the particles tend to
scatter them toward isotropy in the wave frame. Viewed in the upstream
fluid frame, in which the energy of bulk motion of the massive fluid is
negligible during the interaction, the particles lose energy if in the
wave frame they stream in the direction of wave propagation. That
energy is transferred to the waves. The momentum is transferred to the
bulk motion of the fluid. Thus, the accelerated protons excite
upstream propagating waves. It can be shown from these considerations
of energy and momentum conservation and from characteristic particle
and wave lifetimes upstream of the shock that, just upstream of the
shock, the ratio of the energetic proton energy density, $U_p$,
to the enhanced magnetic field fluctuation energy density,
$<|\delta \underline{B}|^2>/8\pi$, satisfies (Lee, 1984)

$$U_p [<|\delta \underline{B}|^2>/8\pi]^{-1} \propto V_1/V_A \qquad (2)$$

where $V_A$ is the Alfvén speed.

The transverse waves which interact most strongly with energetic
particles are those which are cyclotron resonant with them:
$\omega - kv_\shortparallel + \Omega = 0$, where $\Omega$ is the particle gyrofrequency, $\omega$ and $k$ the
wave frequency and wavenumber, and $v_\shortparallel$ is the particle velocity
component along the magnetic field. This condition is satisfied if a
particle and circularly polarized wave have the same helical sense and
the particle guiding center moves through one wavelength in one
gyroperiod so that the particle is secularly influenced by the wave
perturbation fields. As is apparent in Figure 1, which shows an
energetic proton and wave in resonance, this condition has important
consequences for the polarization of the excited wave: Since the
unstable waves, which attempt to propagate upstream away from the
shock, generally dominate the wave spectrum, right polarized waves in
the fluid frame are excited by protons leaving the shock (depicted in
Figure 1) and left polarized waves are excited by protons returning to
the shock. The same polarizations result for both orientations of the
ambient magnetic field. However, the wave polarizations in the shock
frame (spacecraft frame in the case of Earth's bow shock) are reversed
since the wave frequencies are dominated by the Doppler shift.

Since the energetic particles satisfy $v \gg \omega/k$, the cyclotron
resonance condition yields $\omega \ll \Omega$ and the unstable waves are Alfvén and
magnetosonic waves.

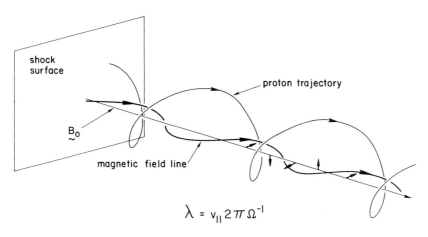

Figure 1. An upstream propagating proton trajectory and a resonant circularly polarized wave (depicted by its helical field line). Also shown are the average and perturbation magnetic field vectors. If the wave attempts to propagate upstream it is right-polarized in the fluid frame and left-polarized in the shock frame.

The proton-excited waves may be described by a wave kinetic equation (Kadomtsev, 1965)

$$\frac{\partial I_i}{\partial t} + V_{-g,i} \cdot \nabla I_i - \nabla \omega_i(k) \cdot \frac{\partial I_i}{\partial k} = 2\,\gamma_i(k)\,I_i \qquad (3)$$

where $I_i(k, r, t)$ is the magnetic field spectral distribution associated with mode i: $\Sigma_i \int d^3k\, I_i(k, r, t) = <|\delta B|^2>$. Here $V_{-g,i}$ and $\omega_i(k)$ are the mode group velocity and frequency and $\gamma_i(k)$ is the growth rate due to proton streaming.

Since the proton spatial diffusion tensor $\underline{\underline{K}}$ depends on $I_i(k,r,t)$ and $\gamma_i(k)$ depends on $-\underline{\underline{K}} \cdot \nabla f(r,p,t)$ (the proton diffusive streaming in the fluid frame), equations (1) and (3) describe the coupled wave-proton system: the waves are excited by the protons and in turn confine them near the shock for efficient acceleration. The only free parameters in a proton-acceleration wave-excitation theory based on equations (1) and (3) are the shock parameters (strength and field orientation), which may be determined by accompanying plasma and field measurements, and the particle "injection" rate Q. Thus a theory based on equations (1) and (3) has considerable predictive power.

Shocks will accelerate any particles which can traverse the shock in both directions. The most copious supply of "seed" particles in interplanetary space is clearly the solar wind. However, for weak and/or low-β and/or quasi-perpendicular shocks few shock-heated solar wind ions are able to return upstream to become legitimate seed ions for further acceleration. The condition that a few percent of the downstream solar wind ions can return upstream may be written (Kennel, 1984; Edmiston et al., 1982)

$$2 \, C_2 \cos \theta_2 \geq V_2 \tag{4}$$

where $C_2$ and $V_2$ are the downstream sound speed and fluid velocity component parallel to the shock normal, and $\theta_2$ is the downstream angle between the magnetic field and shock normal. Condition (4) defines a second critical Mach number, M**, such that for M > M** condition (4) is satisfied, where M is the fast magnetosonic Mach number.

Condition (4) is generally not satisfied by interplanetary traveling shocks at 1 AU which are often weak or quasi-perpendicular. Such subcritical shocks, however, will reaccelerate solar flare particles which happen to be present. Since these solar flare particles have large scattering mean free paths, their reacceleration at 1 AU may not be a diffusive process. Many subcritical shocks have no associated particle enhancement (Van Nes et al., 1984). Strong quasi-parallel traveling shocks accelerate solar wind ions efficiently. Two examples are the shocks of 27 August 1978 and 11, 12 November 1978 (Gosling et al., 1981; Scholer et al., 1983; Kennel et al., 1984a, 1984b, 1985).

The quasi-parallel bow shock accelerates solar wind ions into the "diffuse" population (Ipavich et al., 1984) but will also accelerate any other energetic ion such as an ESP ion associated with an approaching interplanetary traveling shock (Scholer and Ipavich, 1983).

3. THE 11, 12 NOVEMBER 1978 EVENT

The predictions of the theory of Lee (1983) based on equations (1) and (3) have recently been compared in detail with the ULF wave and energetic ion spectra observed upstream of the 11, 12 November 1978 interplanetary traveling shock by Kennel et al. (1985). This paper followed two previous papers (Kennel et al., 1984a,b) which presented composite ISEE-1 and -3 observations and inferred shock parameters for this event.

Kennel et al. (1985) list 10 separate predictions of the theory and compare each with the observations. In all but two cases the agreement is very good. Here we summarize the 10 predictions and the observations:

1. The theory predicts $f \propto p^{-\beta}$ at or just downstream of the shock for $p_0 < p < p_1$, where $p_0$ is a characteristic injection momentum and $p_1$ is a characteristic upper momentum based on shock lifetime. The observed proton spectrum is a power law in the energy range 35-200 keV.

2. The index $\beta$ is predicted to satisfy $\beta^{-1} = \frac{1}{3}(1-V_2/V_1)$ where $V_1$ and $V_2$ are the normal components of the upstream and downstream velocities of those frames of reference in which the ions are scattered in pitch angle. Taking these quantities to include the upstream-directed propagation of the ULF waves relative to the solar wind, Kennel et al. (1985) find $\beta \cong 4.2$, in excellent agreement with the observed index (Scholer et al., 1983).

3. The scalelength for the decay of the upstream protons just upstream

of the shock is predicted to satisfy $L(p) \propto p^{\beta-3}$, in agreement with
the observed energy dependence of the upstream "ramp".

4. The magnitude of $L(p)$ is predicted and is proportional to $[f(p)]^{-1}$
   at lower energies. At 17 keV both theory and observations
   (inferred via the derived shock speed) yield $L \cong 1.6 \times 10^5$ km.

5. The upstream anisotropy in the solar wind frame is predicted to be
   independent of distance upstream and to be proportional to $v^{-1}$; the
   downstream anisotropy should vanish. Anisotropy observations are
   in essential agreement with the predictions including the magnitude
   of the upstream anisotropy (Scholer et al., 1983; Lee, 1984).

6. The upstream wave phase and group velocities are predicted to be
   directed upstream, consistent with Point 2. The predicted wave
   growth rates are largest for wavevectors parallel to the ambient
   field (Gary, 1985), in accord with the observed clustering
   of the directions of minimum variance within $20^\circ$ of the ambient
   field direction. With nearly isotropic proton distributions the
   average wave polarization should be weak as observed (Kennel et
   al., 1985).

7. The scalelength of upstream wave decay, $L'(k)$, is predicted to
   satisfy $L'(k) \cong L(p)$ where $k$ and $p$ satisfy the cyclotron resonance
   condition. Since a range of proton energies is cyclotron resonant
   with a given $k$, however, a qualitative correspondence is predicted
   and is observed. For the 5 April 1979 event Sanderson et al.
   (1985) find $L'(k) \cong L(p)$ upstream of the shock if higher energy
   protons ($\sim$200 keV) with larger pitch angles dominate the wave
   excitation.

8. The predicted ratio of energetic particle to wave fluctuation
   energy densities, approximately given by equation (2), is in accord
   with the observations.

9. The predicted wave power spectrum, $P(\nu)$, at higher frequencies
   where the waves are predominantly ion-excited, satisfies
   $P(\nu) \propto \nu^{\beta-6}$. Here $\nu$ is the wave frequency in Herz. With $\beta = 4.2$
   the predicted wave power spectrum satisfies $P(\nu) \propto \nu^{-1.8}$. However,
   in the relevant range of frequencies ($10^{-2}$ Hz $< \nu < 10^{-1}$ Hz) the
   observed spectrum upstream of the shock is flat or even peaked
   near $\cong 5 \times 10^{-2}$ Hz.

10. The predicted proton-excited power drops precipitously to
    interplanetary levels at frequencies above that resonant with a
    proton propagating parallel to the upstream magnetic field with the
    minimum energy necessary to propagate upstream. In this event that
    frequency is about $10^{-1}$ Hz (Kennel et al., 1985). However, the
    observed wave power enhancement just upstream of the shock at
    frequencies above $10^{-1}$ Hz is substantial.

The quantitative agreement between the theoretically predicted
wave and ion spectra and the observed spectra for the 11, 12 November
1978 event is impressive. Only the detailed form of the wave spectrum
appears to be at odds with quasilinear theory. However, it should be
emphasized that the theory of Lee (1982, 1983) only approximately
includes the contribution of particles with larger pitch angles to the
wave growth rate. A more careful numerical evaluation of the wave

growth rate could improve the agreement between the observed and
predicted wave spectra, particularly at higher frequencies
($\nu \gtrsim 5 \times 10^{-2}$ Hz) where larger pitch angle particles play an important
role. However, the discrepancy could also arise from effects of
particle trapping or a nonlinear cascade of wave power to higher
frequencies, both of which are neglected in quasilinear theory.

## 4. SOME REMAINING QUESTIONS

In spite of the success of the coupled quasilinear theory for particle
diffusive shock acceleration and hydromagnetic wave excitation at
supercritical quasi-parallel traveling shocks near 1 AU, several
interesting unsolved problems remain. The nonlinear evolution of the
excited waves is one, which also happens to be crucial to understanding
cosmic ray acceleration at interstellar shocks (Völk and McKenzie,
1981).
 "Injection" rates cannot be addressed in the context of the
diffusive transport theory. Nevertheless, their determination is
required for a self-consistent description of shock structure in which
particle acceleration and ULF wave excitation provide a channel for
dissipation of the ordered upstream flow. The upstream foreshock
described by the theory presented defines one dissipative scalelength
(of many) of a quasi-parallel supercritical collisionless shock. The
dependence of injection rates on $\theta_1$ needs to be investigated. For
example, it is unclear whether the relatively weak quasi-perpendicular
shocks bounding corotating interaction regions can accelerate ions out
of the solar wind. However, there is no other clearly available seed
population for the corotating ion events.
 The time-dependent coupled evolution of protons above $\sim$ 200 keV
and their resonant waves of lower frequency at interplanetary traveling
shocks must be understood to predict the proton spectrum at higher
energies.
 The diffusive shock acceleration of electrons is a problem of
great interest. Although diffusive distributions of accelerated
electrons are not observed at interplanetary traveling shocks near 1 AU
or Earth's bow shock, they are observed as solar flare particles and
may be accelerated at coronal shocks (Ellison and Ramaty, 1985).
Electrons with energies less than $\sim$ 1-10 MeV resonate with high
frequency whistler waves, whose phase speeds and growth rates in the
presence of a streaming electron distribution are quite different from
their hydromagnetic proton-excited counterparts, and should exhibit new
interesting characteristics.
 Acceleration of "diffuse" ions upstream of Earth's bow shock is
well accounted for by the coupled quasilinear theory (Lee, 1982), as
are the upstream diffuse-ion-associated hydromagnetic waves.
Nevertheless, following Eichler (1981) the theory invokes perpendicular
diffusion to magnetic field lines disconnected from the shock front in
order to reproduce the observed exponential spectrum in energy per
charge. Since inclusion of the perpendicular diffusive losses could
not be accomplished with rigor in the theory but the observed spectrum
is very close to exponential, it is possible that other loss mechanisms

play a role.  Jokipii (1982) has suggested that particle drift to
disconnected field lines could account for the observed spectrum but
has not performed detailed calculations.

## 5.  ESP EVENTS AT LARGE HELIOCENTRIC DISTANCES

Another neglected problem is the evolution of ESP events to large
heliocentric distances.  Although shock acceleration time scales are
longer, the tendency for the shock to become nearly perpendicular
should drastically reduce the injection rate of shocked solar wind
plasma and leave only previously accelerated particles or solar flare
particles for further acceleration.

Traveling shocks and ESP events in the distant heliosphere can be
impressive.  A recent paper by Pyle et al. (1984) has documented 3
large events observed by Pioneer 10 between 24 and 28 AU which are
shown in Figure 2.  During proton events A and C strong shocks provided
additional proton (and apparently helium and electron) acceleration, in
contrast with event B, which is composed of solar flare particles only.
The events apparently represent a coalescence of flare particles and
shocks from many individual flares during 3 separate periods of solar
activity (actually 4, if the small event in the 11-20 MeV proton
channel during April, 1982, is included).  Interestingly the 4 events
reveal clearly the 154-day periodicity of solar activity found by
Rieger et al. (1984), which was extracted rather delicately from the
flare record at 1 AU.

The sharply-peaked shock-accelerated particles do not exhibit the
upstream ramp and constant downstream intensity characteristic of
1-dimensional steady-state calculations.  In order to illustrate the
theory of diffusive shock acceleration based on equation (1) and to
show that it provides qualitatively for the profiles shown in Figure 2,
we now present an illustrative example.  We consider time-dependent
shock acceleration at an infinite planar shock at which protons are
injected with momentum $p_o$ at a rate proportional to $S(t) e^{-\mu t}$, where
$S(t)$ is the standard step function.  Equation (1) may then be rewritten
as

$$\frac{\partial f}{\partial t} + V_i \frac{\partial f}{\partial x} - \frac{\partial}{\partial x} (K_i \frac{\partial f}{\partial x}) + \frac{1}{3} (V_1 - V_2) \delta(x) p \frac{\partial f}{\partial p}$$

$$= qS(t) e^{-\mu t} \delta(x) \delta(p - p_o) \qquad (5)$$

where $V_1$ $(V_2)$ and $K_1$ $(K_2)$ are the upstream (downstream) normal
component of the fluid velocity relative to the shock and the upstream
(downstream) $(x,x)$-component of the diffusion tensor.  If we take for
simplicity $V_i$ and $K_i$ to be independent of x and t for $x \neq 0$ and $K_i$
independent of p, and $V_1^2 K_2 = V_2^2 K_1$, then simple solutions to equation
(5) can be obtained analytically.  If we furthermore assume that
$\mu = V_1^2 / (4K_1)$ then we obtain for $t > 0$ and $x < 0$ (upstream)

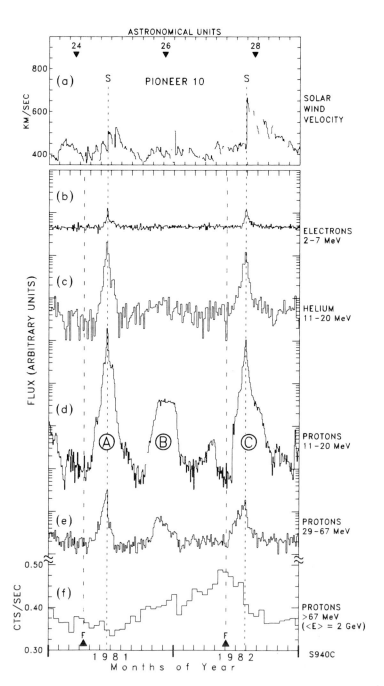

Figure 2. Pioneer 10 solar wind velocity and energetic particle profiles between 24 and 28 AU (from Pyle et al. 1984). Solid triangles, F, indicate periods of major solar activity; S indicates times of strong shock passage.

$$f = \frac{3}{V_1 - V_2} \frac{q}{p_o} \left(\frac{p}{p_o}\right)^{-3/2} \exp\left(\frac{V_1}{2K_1} x - \mu t\right) \operatorname{erfc}\left[(K_1/t)^{\frac{1}{2}} V_1^{-1}\right].$$

$$\cdot\left(\frac{3}{2}\frac{V_1+V_2}{V_1-V_2} \ln\frac{p}{p_o} + \frac{V_1|x|}{2K_1}\right)] \tag{6}$$

and for $t > 0$ and $x > 0$ (downstream)

$$f = \frac{3}{V_1 - V_2} \frac{q}{p_o} \left(\frac{p}{p_o}\right)^{-3/2} \exp\left(\frac{V_2}{2K_2} x - \mu t\right) \operatorname{erfc}\left[(K_1/t)^{\frac{1}{2}} V_1^{-1}\right].$$

$$\cdot(\frac{3}{2}\frac{V_1+V_2}{V_1-V_2} \ln\frac{p}{p_o} + \frac{V_2 x}{2K_2})] \tag{7}$$

where $\operatorname{erfc}(z)$ is the standard complementary error function.

Although solutions (6) and (7) neglect adiabatic deceleration and the spherical geometry of the solar wind, they should portray the basic characteristics of ESP events in the outer heliosphere where these effects are small. With $x = -R + V_s t$ they predict the time profile observed by a spacecraft a distance $R$ from the Sun due to a shock leaving the Sun at $t = 0$. Figure 3 shows that time profile in arbitrary units for $\ln (p/p_o) = 2.5$ and a reasonable set of parameters: $V_1 = 200$ km/s, $V_2 = 100$ km/s, $V_s = 600$ km/s, $K_1 = 10^{20}$ cm$^2$/s, $R = 20$ AU and $\mu = 10^{-6}$/s. The condition on $K_2$ yields $K_2 = 0.25$ $K_1$, a reasonable decrease in the diffusion coefficient across the shock. The characteristic width of the intensity peak ($\sim 10$ days) is similar to that observed and depicted in Figure 2. If p is the momentum of a 10 MeV proton, then $K_1 = 10^{20}$ cm$^2$/s corresponds to a radial scattering mean free path of $\sim 0.01$ AU. This value is smaller than characteristic values at 1 AU of $0.1 - 0.2$ AU (Fisk, 1979) and may account for a proton-excited wave enhancement.

6.  ESP EVENTS AT HIGH LATITUDES

Although flares and solar activity are confined to lower latitudes we expect shock waves to propagate to higher latitudes (Burlaga, 1986). The shocks may be weaker at high latitudes. But since the magnetic field should be approximately radial over the solar poles, if the shocks have radial shock normals, then they should be quasi-parallel, a configuration which favors their being supercritical in the sense of equation (4). Thus we expect large shocks to produce ESP events over the solar poles. Indeed the radial field yields a rapidly decreasing Alfvén speed with increasing heliocentric distance which increases the shock Mach number, and could tend to increase acceleration efficiency with radial distance over the poles. Unfortunately Ulysses is limited to a heliocentric distance of $\sim 1$ AU over the poles.

Actually ESP events may be simpler and easier to study in detail at high latitudes. Due to the regular magnetic fields in polar coronal holes it is probably difficult for solar flare particles to propagate

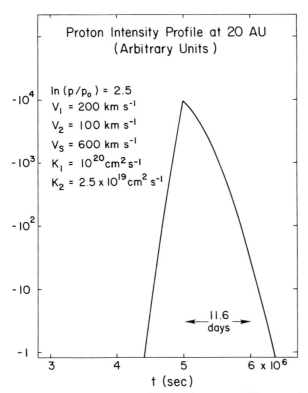

Figure 3. A predicted proton intensity profile at 20 AU based on equations (6) and (7) for the parameters shown.

to high latitudes. Thus ESP events should be accelerated out of the solar wind and not be contaminated by solar flare particles as they are at 1 AU. Ulysses can expect clean ESP events amenable to direct comparison with a theory based on solar wind seed particles.

Acknowledgements. The author wishes to thank K.-P. Wenzel and R.G. Marsden for the efficient and very successful organization of the ESLAB Symposium in Les Diablerets and publication of the proceedings. He also wishes to thank G.E. Morfill, M. Scholer and colleagues for the kind hospitality extended to him during his stay at the Max-Planck-Institut für Physik und Astrophysik, Institut für Extraterrestrische Physik, during which this work was initiated. This work was supported, in part, by NSF Grant ATM 8311241 and NASA Solar-Terrestrial Theory Program Grant NAGW-76.

References.

Baker, D.N., Zwickl, R.D., Krimigis, S.M., Carbary, J.F. and Acuña, M.H.: 1984, J. Geophys. Res. 89, 3775.
Bame, S.J., Asbridge, J.R., Feldman, W.C., Gosling, J.T., Paschmann, G. and Sckopke, N.: 1980, J. Geophys. Res. 85, 2981.

Barnes, A.: 1970, Cosmic Electrodyn. 1, 90.
Barnes, C.W. and Simpson, J.A.: 1976, Astrophys. J. 210, L91.
Bonifazi, C. and Moreno, G.: 1981, J. Geophys.Res. 86, 4397.
Bryant, D.A., Cline, T.L., Desai, U.D. and McDonald, F.B.: 1962, J.
   Geophys. Res. 67, 4983.
Burlaga, L.F.: 1986, this volume.
Decker, R.B.: 1981, J. Geophys. Res. 86, 4537.
Decker, R.B. and Vlahos, L: 1985, J. Geophys. Res. 90, 47.
Drury, L. O'C.: 1983, Rep. Progr. Phys. 46, 973.
Edmiston, J.P., Kennel, C.F. and Eichler, D.: 1982, Geophys. Res.
   Lett. 9, 531.
Eichler, D.: 1981, Astrophys. J. 244, 711.
Ellison, D.C.: 1981, Geophys. Res. Lett. 8, 991.
Ellison, D.C. and Ramaty, R.: 1985, 'Shock Acceleration of Electrons
   and Ions in Solar Flares', Astrophys. J., in press.
Fisk, L.A.: 1979, in 'Solar System Plasma Physics: A Twentieth Anni-
   versary Overview', edited by C.F. Kennel, L.J. Lanzerotti, and E.N.
   Parker, North-Holland, Amsterdam.
Fisk, L.A. and Lee, M.A.: 1980, Astrophys. J. 237, 620.
Forman, M.A. and Drury, L. O'C.: 1983, Proc. 19th Int. Conf. Cosmic
   Rays 2, 267.
Gary, S.P.: 1985, Astrophys. J. 288, 342.
Gleeson, L.J. and Axford, W.I.: 1967, Astrophys. J. 149, L115.
Gosling, J.T., Asbridge, J.R., Bame, S.J., Feldman, W.C., Zwickl, R.D.,
   Paschmann, G., Sckopke, N. and Hynds, R.J.: 1981, J. Geophys. Res.
   86, 547.
Hoppe, M.M., Russell, C.T., Frank, L.A., Eastman, T.E. and Greenstadt,
   E.W.: 1981, J. Geophys. Res. 86, 4471.
Ipavich, F.M., Gloeckler, G., Fan, C.Y., Fisk, L.A., Hovestadt, D.,
   Klecker, B., O'Gallagher, J.J. and Scholer, M.: 1979, Space Sci.
   Rev. 23, 93.
Ipavich, F.M., Galvin, A.B., Gloeckler, G., Scholer, M. and Hovestadt,
   D.: 1981, J. Geophys. Res. 86, 4337.
Ipavich, F.M., Gosling, J.T. and Scholer, M.: 1984, J. Geophys. Res.
   89, 1501.
Jokipii, J.R.: 1982, Astrophys. J. 255, 716.
Jokipii, J.R., Levy, E.H. and Hubbard, W.B.: 1977, Astrophys. J. 213,
   861.
Kadomtsev, B.B.: 1965, Plasma Turbulence, Academic Press, New York.
Kennel, C.F.: 1985, in 'Collisionless Shock Waves in the Heliosphere',
   Proceedings of AGU Chapman Conf., Napa Valley, 1984.
Kennel, C.F., Scarf, F.L., Coroniti, F.V., Russell, C.T., Wenzel,
   K.-P., Sanderson, T.R., Van Nes, P., Feldman, W.C., Parks, G.K.,
   Smith, E.J., Tsurutani, B.T., Mozer, F.S., Temerin, M., Anderson,
   R.R., Scudder, J.D. and Scholer, M.: 1984a, J. Geophys. Res. 89,
   5419.
Kennel, C.F., Edmiston, J.P., Scarf, F.L., Coroniti, F.V., Russell,
   C.T., Smith, E.J., Tsurutani, B.T., Scudder. J.D., Feldman, W.C.,
   Anderson, R.R., Mozer, F.S. and Temerin, M.: 1984b, J. Geophys. Res.
   89, 5436.
Kennel, C.F., Coroniti, F.V., Scarf, F.L., Livesey, W.A., Russell,

C.T., Smith, E.J., Wenzel, K.-P. and Scholer, M: 1985, 'A Test of the Quasi-linear Theory of Proton Fermi Acceleration by Collisionless Shocks', J. Geophys. Res., submitted for publication.

Lee, M.A.: 1982, J. Geophys. Res. 87, 5063.

Lee, M.A.: 1983, J. Geophys. Res. 88, 6109.

Lee, M.A.: 1984, Adv. Space Res. 4, 2-3, 295.

Lee, M.A. and Skadron, G.: 1985, J. Geophys. Res. 90, 39.

Lee, M.A. and Ryan, J.M.: 1985, 'Time-dependent Coronal Shock Acceleration of Energetic Solar Flare Particles', Astrophys.J., submitted for publication.

McDonald, F.B., Teegarden, B.J., Trainor, J.H., von Rosenvinge, T.T. and Webber, W.R.: 1976, Astrophys. J. 203, L149.

Palmer, I.D. and Gosling, J.T.: 1978, J. Geophys. Res. 83, 2037.

Parker, E.N.: 1965, Planet. Space Sci. 13, 9.

Paschmann, G., Sckopke, N., Bame, S.J., Asbridge, J.R., Gosling, J.T., Russell, C.T. and Greenstadt, E.W.: 1979, Geophys. Res. Lett. 6, 209.

Paschmann, G., Sckopke, N., Papamastorakis, I., Asbridge, J.R., Bame, S.J. and Gosling, J.T.: 1981, J. Geophys. Res. 86, 4355.

Pesses, M.E., Jokipii, J.R. and Eichler, D.: 1981, Astrophys. J. 246, L85.

Pesses, M.E., Decker, R.B. and Armstrong, T.P.: 1982, Space Sci. Rev. 32, 185.

Pyle, K.R., Simpson, J.A., Barnes, A. and Mihalov, J.D.: 1984, Astrophys.J. 282, L107.

Rieger, E., Share, G.H., Forrest, D.J., Kanbach, G., Reppin, C. and Chupp, E.L.: 1984, Nature 312, 623.

Sanderson, T.R., Reinhard, R., Van Nes, P., Wenzel, K.-P., Smith, E.J. and Tsurutani, B.T.: 1985, J. Geophys.Res. 90, 3973.

Scholer, M. and Ipavich, F.M.: 1983, J. Geophys. Res. 88, 5715.

Scholer, M., Ipavich, F.M., Gloeckler, G. and Hovestadt, D.: 1983, J. Geophys. Res. 88, 1977.

Smith, C.W. and Lee, M.A.: 1985, 'Coupled Hydromagnetic Wave Excitation and Ion Acceleration Upstream of the Jovian Bow Shock', J. Geophys. Res., in press.

Tademaru, E.: 1969, Astrophys. J. 158, 959.

Tsurutani, B.T. and Lin, R.P.: 1985, J. Geophys. Res. 90, 1.

Tsurutani, B.T., Smith, E.J., Pyle, K.R. and Simpson, J.A.: 1982, J. Geophys. Res. 87, 7389.

Tsurutani, B.T., Smith, E.J. and Jones, D.E.: 1983, J. Geophys. Res. 88, 5645.

Van Nes, P., Reinhard, R., Sanderson, T.R., Wenzel, K.-P. and Zwickl, R.D.: 1984, J. Geophys. Res. 89, 2122.

Viñas, A.F., Goldstein, M.L. and Acuň, M.H.: 1984, J. Geophys. Res. 89, 3762.

Völk, H.J. and McKenzie, J.F.: 1981, Proc. 17th Int. Conf. Cosmic Rays 9, 246.

Wibberenz, G., Zöllich, F., Fischer, H.M. and Keppler, E.: 1985, J. Geophys. Res. 90, 283.

Zwickl, R.D., Krimigis, S.M., Carbary, J.F., Keath, E.P., Armstrong, T.P., Hamilton, D.C. and Gloeckler, G.: 1981, J. Geophys. Res. 86, 8125.

# SHOCK ACCELERATION OF NUCLEONS AT $\geq 16°$ SOLAR LATITUDE ASSOCIATED WITH INTERPLANETARY COROTATING INTERACTION REGIONS

J.A. Simpson[*]
Enrico Fermi Institute and Dept. of Physics, Univ. of Chicago
and
E.J. Smith and Bruce Tsurutani
Jet Propulsion Laboratory, California Institute of Technology

ABSTRACT: The analysis of magnetic field and nuclei measurements in the ~1 to ~10 MeV/nucleon range for protons and helium at ~16°N solar latitude near solar cycle minimum (1975-76) shows both: a) that nucleons are accelerated locally in corotating shocks, and, b) that the associated recurring corotating interaction regions (CIRs) at ~16° latitude have approximately the same temporal and spatial parameters found for CIRs in the equatorial plane at radial distances < 5 AU. The evidence indicates that both the CIR structures and nucleon acceleration by shocks extend to intermediate solar latitudes and are likely to be observed during the Ulysses Mission out of the ecliptic plane.

## 1. INTRODUCTION

The Pioneer 10 and Pioneer 11 spacecraft missions near the equatorial plane have led to the identification of corotating interaction regions bounded by forward and reverse shocks beyond 2 AU. These corotating regions which recur for many solar rotations near the time of the solar cycle minimum (1973-76) have their origin in high speed solar wind streams (from coronal holes) overtaking the ambient solar wind in the heliosphere (Smith and Wolfe, 1976; Hundhausen, 1977). Fluxes of protons and helium nuclei in the energy range ~ 1-10 MeV/nucleon were discovered in association with these recurring CIRs and increased in intensity with radial distance out to ~ 4-5 AU (McDonald, et al., 1976; Barnes and Simpson, 1976; Pesses, Van Allen and Goertz, 1978) and with decreasing intensity beyond this distance (e.g., Van Hollebecke et al., 1978; Christon and Simpson, 1979). The peak intensities of these accelerated nucleons were found to be correlated with the locations of the forward and reverse shocks (Barnes and Simpson, 1976; Pesses et al., 1978) and later proven to be acceleration at the shocks when the angle of the interplanetary magnetic field direction relative to the shock normal was $\gtrsim 85°$; the conditions required for nucleon acceleration in a quasi-perpendicular shock (Tsurutani et al., 1982).

[*]John Simon Guggenheim Fellow, 1985

*R. G. Marsden (ed.), The Sun and the Heliosphere in Three Dimensions, 319–324.*
© *1986 by D. Reidel Publishing Company.*

It is the purpose of this paper to investigate the properties of these CIRs and shock particle acceleration as a function of solar latitude with a view toward predicting the observations which may be expected from the Ulysses Mission that will cover a wide range of latitudes inside 5 AU. The only opportunity known to us to investigate the latitudinal dependence of CIRs and nucleon acceleration by their associated shocks, with the constraint that the radial range be less than 5 AU, arose when the transfer trajectory of Pioneer 11 between Jupiter and Saturn encounters brought the spacecraft to ~16° north solar latitude at ~3.8-4 AU during the presence of two long-term CIR series near solar cycle minimum. Smith, Tsurutani and Rosenberg (1978) have shown that interplanetary magnetic field sector structure (e.g., the crossing of the wavy current sheet) essentially disapppeared during this period with the outward directed north polar (+) magnetic field being observed on Pioneer 11 greater than 90% of the time.


2.    OBSERVATIONS AND ANALYSIS

2.1   CIRs and Proton Fluxes

The proton measurements (shown in Figure 1) in the energy range 0.5-1.8 MeV were obtained on Pioneer 10 and Pioneer 11 from the Low-Energy Telescope (LET) of the University of Chicago instrument described by Simpson et al. (1974). The magnetic field observations in Figure 1 were derived from the Jet Propulsion Laboratory vector helium magnetometer described by Smith et al. (1974).
       The delays in the arrival of a CIR at Pioneer 10 near the equatorial plane after passing Pioneer 11 are shown in Figure 1 by the slant lines between Figures 1B and 1A for solar wind velocities of 400 km/sec (dashed) and 500 km/sec (solid). The resulting correlation leads to what we shall define as the principal series of approximately 26 day recurring proton intensity increases (symbol $\Delta$) which are a factor of 10 to 100 above background. In Figure 1 we also show that this principal series of recurring enhancements ($\Delta$) is associated with the most intense recurring series of magnetic field enhancements (CIRs) extending to 16° latitude, which confirms the conclusion reached by Hamilton and Simpson (1979) from recurrent Jovian electron modulation that this is a recur- rent CIR. Forward (F) and reverse (R) shocks, which have been identified by jumps in $|B|$, are shown by the arrows in Figure 1C. They are identi- cal to the shocks associated with CIRs near the equatorial plane in approximately the same radial range. The lack of continuous data coverage prevented the positive identification of additional shocks which appear to be present in the data in Figure 1C so that the absence of such a designation does not mean no other shocks were present. Furthermore, it is clear from Figures 1B and 1C that the CIR associated with the $\Delta$ series contained enhanced magnetic fields of the same polarity as that of the interplanetary magnetic field both ahead of and behind the CIR (namely, the North Polar Field [+]). The principal CIR series, therefore, has no sector boundary (or field reversal region) within or near the CIR. The origin of the high-speed solar wind stream

which led to the formation of this CIR is associated with a long-lived coronal hole of positive magnetic field polarity extending to low latitude in the north solar hemisphere (cf., Hundhausen, 1977).

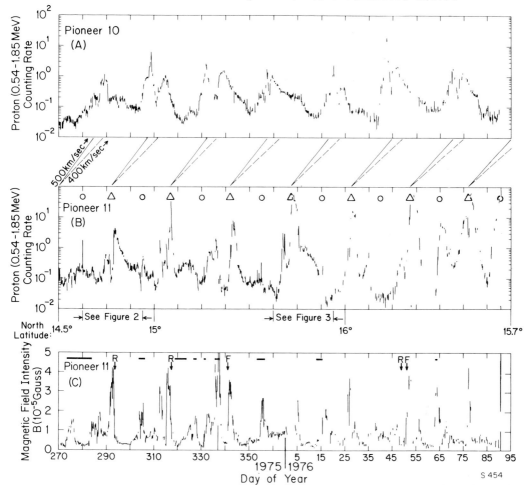

**FIG. 1:** **A)** Proton intensity averaged over 1 hour intervals in the heliocentric radial range 8.2-9.7 AU; **B)** Proton intensity in the radial range 3.7-3.9 AU. The principal series of accelerated protons is indicated by Δ. The minor series of enhanced proton fluxes is indicated by 0; **C)** Interplanetary magnetic field intensity B. Time intervals within which the magnetic field is negative are shown as horizontal bars at the top of the Figure. Forward (F) and reverse shocks (R) are indicated by ↓. The heliographic latitude is indicated at the top of this panel.

We also find a second series of recurring proton intensity increases (with approximately a factor 2-5 intensity increase above background) identified by symbol (0) which is associated with recurring enhanced magnetic field regions of <u>opposite</u> (−) polarity embedded in the

surrounding positive polarity interplanetary magnetic field. We call
this the <u>minor</u> recurring series. Figures 1B and 1C reveal that the
small percentage of negative magnetic field polarity reported by Smith
et al. (1978) is associated with this recurring minor series. The origin
of the related, high-speed solar wind stream is a long-lived, negative
field coronal hole in the southern (-) solar hemisphere (Sheeley and
Harvey, 1978).

## 2.2 Evidence for Acceleration at High Latitudes

We now investigate the question of the source region of nucleon acceleration for the high latitude principal series (Δ) in Figures 1A and
1B which, as we have shown, is associated with a CIR that clearly is
free from the signatures of an equatorial, warped neutral current sheet
surrounding the Sun. A basic question is: Are the physical characteristics of the recurring CIR at high latitudes the same as those observed
in the equatorial zone; <u>or</u>, are we observing in Figure 1B the fluxes of
nucleons arriving by propagation from distant acceleration regions –
such as in the equatorial zone neutral sheet beyond the radial range of
Pioneer 11? To decide between these alternatives, we have examined the
physical characteristics of the principal series at $> 14°$ and compared
them with observations in the equatorial zone. Figures 2 and 3 illustrate two intervals taken from Figure 1 which are compared with
events of comparable magnitude occurring in the equatorial zone ($<6°$) in
the heliocentric radial range 3.75-4.5 AU.

All of the
characteristics
which have been
identified by Barnes
and Simpson (1976)
and Tsurutani et al.
(1982) from the
equatorial zone
analysis are present
in the events at 15°
N, including: 1)
the double peaks
close to the edges
of the leading and
trailing shocks of
the CIR (in 60% of
all cases); 2) the
proton intensity
decrease inside the
CIR where the field
is most compressed;
3) the declining
flux extending into
the interplanetary
rarefaction regions
behind the CIR;

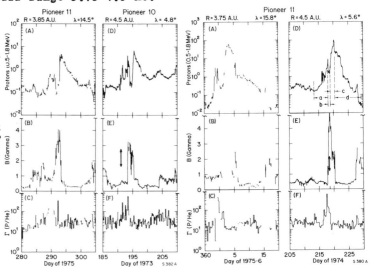

**FIG. 2:** $\left(\frac{P}{He}\right)$ is the ratio of proton to helium
intensity in the same energy interval per
nucleon. λ is heliographic latitude.
Magnetic sector boundary crossings are shown
as $\updownarrow$.
**FIG. 3:** The time intervals for Table 1 are
shown as a,b,c, and d in (D) above.

4) the large proton/helium ratio of intensities (> 10 fold increase in
equal kinetic energy per nucleon) at the leading edge of the CIR; and
5) (not shown) a significant increase in the slopes of the energy
spectra near the forward peak for many of the CIR-shock events.

   To compare further the time-intensity characteristics at > $15°$ and
< $6°$ we have averaged the times (in days) for the four time intervals
a,b,c,d,, as defined in Figure 3D. The results are summarized in Table 1
clearly showing that there are no discernible differences in the general
time dependent or spatial characteristics of the proton intensity increases in these two latitude zones. The reader should recall that these
rates of intensity rise and decay outside the CIR translate into a
spatial distribution of proton fluxes on interplanetary field lines
which enter or connect with the CIR at a greater radial distance than
the position of the spacecraft. For those periods where the anisotropy
can be measured, we find that there is a net Sun-ward flow along the
magnetic field away from the leading and trailing edges of the CIR. No
evidence has been found of an 'upward' flow from the equatorial zone.
Finally, we note from Figures 1A and 1B that the proton peak intensity
is significantly larger at 3.7-3.9 AU (Pioneer 11) than at 8-10 AU
(Pioneer 10) as reported earlier for the radial dependence of proton
intensity by Van Hollebeke et al. (1978).

### TABLE 1

|                        | 12 CIRs in Equatorial Plane | | 9 CIRs At $\sim 15°$ Latitude | |
|------------------------|------------|------------|------------|------------|
|                        | rise       | fall       | rise       | fall       |
| Single Peaks           | $3\pm0.9$  | $7.7\pm2.5$ | $3.5\pm1.5$ | $7.3\pm1.8$ |
|                        | rise       | fall       | rise       | fall       |
| All Peaks              | $3.2\pm0.8$ | $7.9\pm0.2$ | $3.4\pm1.3$ | $7.7\pm2.5$ |
|                        | a          | c          | a          | c          |
| Double Peaks           | $3.3\pm0.8$ | $2.0\pm0.9$ | $3.2\pm1.2$ | $2.8\pm0.9$ |
| (Fig.3D)               | b          | d          | b          | d          |
|                        | $1.0\pm0.5$ | $7.7\pm1.7$ | $1.5\pm0.3$ | $8.1\pm3.5$ |

## 3. DISCUSSION AND CONCLUSIONS

We confirm the evidence already published that recurring CIRs and associated nucleon intensity increases extend to beyond $16°$ N solar
latitude (e.g., Van Hollebecke et al., 1978; Smith et al., 1978;
Christon and Simpson, 1979; Hamilton and Simpson, 1979) and, from our
analysis in this paper, conclude furthermore:
1) The temporal and spatial distribution of magnetic field, shock and
   accelerated particle characteristics of CIRs in the equatorial plane
   extend in latitude to beyond $16°$ N solar latitude, essentially
   unchanged;
2) Nucleons are accelerated to energies of order ~1-10 MeV/n locally at
   ~$16°$ latitude by the shocks associated with the CIRs, presumably by

the same quasi-perpendicular shock mechanism which occurs in the
equatorial plane (Tsurutani et al., 1982);

3) The neutral current sheet is not a significant source of CIR-
associated nucleons in the MeV energy range.

Based on the above evidence it is clear that near solar minimum
conditions the acceleration of nuclei in corotating, CIR-generated
shocks is likely to be an important phenomenon at much higher latitudes
than reported here for the Pioneer 11 measurements. Instruments on the
Ulysses Mission may be expected to encounter these recurrent phenomena
as the spacecraft rises in latitude inside 5 AU after its Jovian
encounter near mid-1987, especially if this is a period near the end of
the solar activity cycle. There is, indeed, independent but indirect
evidence from interplanetary scintillation studies that CIRs extend to
relatively high latitudes ($\gtrsim 40°N$) (e.g, Tappin, Hewish and Gapper,
1984).

We thank Robert Hogan for his effort to increase the telemetry
coverage for Pioneer 10 and 11 and appreciate the assistance of J.M.
Cherneff, Steven Christon, Elaine Dobinson, and Peter Kruley in preparing the data and its analysis. The research at the University of
Chicago was supported in part by the NASA/Ames Contract NAS 2-65551 and
NASA Grant NGL 14-001-006. Portions of the work described in this paper
were carried out at the Jet Propulsion Laboratory, California Institute
of Technology under contract with NASA.

## 4. REFERENCES

Barnes, C.W. and Simpson, J.A.: 1976, Ap.J. (Letters), 210, L91.

Christon, S. and Simpson, J.A.: 1979, Ap.J. (Letters), 227, L49.

Hamilton, D.C. and Simpson, J.A.: 1979, Ap.J. (Letters), 228, L123.

Hundhausen, A.J. in J.B. Zirker (ed.), Coronal Holes and High Speed Wind
        Streams , Colorado Assoc. Univ. Press, Boulder, CO., p. 225, and
        references therein.

McDonald, F.B., Teegarden, B.J., Trainor, J.H., Von Rosenvinge, T.T. and
        Webber, W.R.: 1976, Ap.J. (Letters), 203, L149.

Pesses, M.E., Van Allen, J.A., Goertz, C.K.: 1978, J. Geophys. Res., 83,
        553.

Sheeley, N.R., Jr. and Harvey, J.W.: 1978, Solar Physics, 59, 159, and
        references therein.

Simpson, J.A., Hamilton, D.C., McKibben, R.B., Mogro-Campero, A., Pyle,
        K.R. and Tuzzolino, A.J.: 1974, J. Geophys. Res., 79, 3522.

Smith, E.J., Davis, L., Jr., Jones, D.E., Coleman, P.J., Jr., Colburn,
        D.S., Dyal, P., Sonett, C.P. and Frandsen, A.M.A.: 1974, J.
        Geophys. Res., 79, 3501.

Smith, E.J., Tsurutani, B.T. and Rosenberg, R.L.: 1978, J. Geophys.
        Res., 83, 717.

Smith, E.J. and Wolfe, J.H.: 1976, Geophys. Res. (Letters), 3, 137.

Tappin, S.J., Hewish, A. and Gapper, G.R.: 1984, Planet. Sp. Sci., 32,
        1273.

Tsurutani, B.T., Smith, E.J., Pyle, K.R. and Simpson, J.A.: 1982, J.
        Geophys. Res., 87, 7389.

Van Hollebeke. M.A.I., McDonald, F.B., Trainor, J.H. and Von
        Rosenvinge, T.T.: 1978, J. Geophys. Res., 83, 4623.

# LATITUDE DEPENDENCE OF CO-ROTATING SHOCK ACCELERATION IN THE OUTER HELIOSPHERE

R. E. Gold
The Johns Hopkins University
Applied Physics Laboratory
Laurel, MD 20707, USA

L. J. Lanzerotti and C. G. Maclennan
AT&T Bell Laboratories
Murray Hill, NJ 07974, USA

S. M. Krimigis
The John Hopkins University
Applied Physics Laboratory
Laurel, MD 20707, USA

ABSTRACT. Energetic particle observations ($E_i \geq 30$ keV) have been made in the outer heliosphere ($\geq 12$ A.U.) by the LECP instruments on the Voyager 1 and Voyager 2 spacecraft. These observations show a definite latitude dependence of the number and intensities of particle enhancements produced by co-rotating interplanetary regions during a lengthy interval (approximately 200 days) in 1983-84 when no energetic solar particle events were observed. The particle enhancements are significantly fewer in number and tend to be less intense at higher (20°-23°) heliolatitudes. The energy spectral shapes of the accelerated particles at the two spacecraft are similar, indicating that the acceleration process is the same at the two latitudes, but less intense at the higher latitudes.

## 1. INTRODUCTION

Particles can be accelerated in various regions of the heliosphere, for example planetary bow shocks and magnetospheres, interplanetary traveling shocks, co-rotating interaction regions. The acceleration processes and regions have been studied extensively near Earth (1 A.U.) [e.g., review by Gloeckler, 1984]. The progression of the two Voyager spacecrafts, with essentially identical instrumentation, deep into the outer heliosphere, beyond the orbit of Saturn, provides the opportunity to study interplanetary shock acceleration processes in the outer heliosphere as well as to search for heliolatitude dependences of the acceleration processes. This report presents the results of some initial analyses of Voyager observations made during an interval when the Voyager 1 (V1) spacecraft was rising out of the ecliptic plane to higher heliolatitudes. The data are from the Low Energy Charged Particle (LECP) instrument flown on both of the spacecraft [Krimigis et al., 1967]. Ion and proton data from selected sub-systems of the instrument are used. Preliminary discussions of these results were presented in Gold et al. [1985a].

The radial and latitudinal locations of the two spacecraft, in the heliosphere, projected into the ecliptic-Z plane, are shown in Figure 1. At the beginning of 1984 the two spacecrafts were

*R. G. Marsden (ed.), The Sun and the Heliosphere in Three Dimensions, 325–329.*
© *1986 by D. Reidel Publishing Company.*

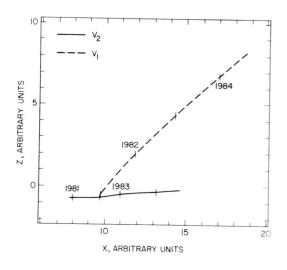

Figure 1. Locations of the two Voyager spacecraft from 1981 through mid-1984 projected into the ecliptic (X)–Z plane. The vertical lines indicate the first day of the indicated years.

separated vertically by ~7 A.U. (~22°) and in radial distance by ~4 A.U. The heliolongitude separation at this time was ~20°.

## 2.  OBSERVATIONS

Shown in Figure 2 are the daily average fluxes of low energy ions and higher energy protons for the interval 1981 to mid-1984. Increases in the proton fluxes provide information on the times of occurrence of solar energetic particle events. The lower energy ion fluxes, when their enhancements are unaccompanied by corresponding increases in the higher energy protons, provide information on the times of occurrence of co-rotating interplanetary regions which can accelerate particles. From

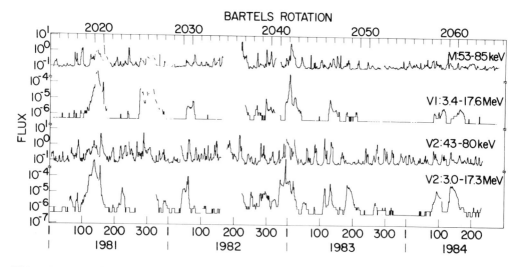

Figure 2.  Plot of daily average fluxes in ion and proton channels from the LECP experiment on Voyager 1 and Voyager 2 spacecraft, 1981 to mid-1984.

the Figure, it can be seen that enhanced co-rotating particle fluxes occur sporadically at each spacecraft location.

While there are several interesting intervals in the data that warrant study (see Gold et al., 1985b, for discussion of some other aspects of these data), we concentrate here specifically on the ~ 200 days of data between day 243, 1983, and day 75, 1984. During this interval of time essentially no enhancements indicative of transient solar particle events were seen in the 3.4-17.6 MeV proton fluxes on Voyager 1. Voyager 2 (V2), approximately in the ecliptic plane, observed some proton activity at about day 270. In general, however, increases in the fluxes of the higher energy protons during this interval were considerably suppressed compared to those occurring before and after.

In contrast to the quiescent behavior of the higher energy protons, the low energy ion fluxes have increased intensities sporadically throughout the interval of interest. The intensity enhancements, produced by co-rotating interplanetary interaction (shock) regions, are significantly less numerous and less intense at Voyager 1 than at Voyager 2 (Figure 2).

The statistical distinction between the ion and proton fluxes for time intervals before and after the interval of special interest (i.e., day 243, 1983 to day 75, 1984) is illustrated by the box plots of Figure 3. The intervals denoted as (a), (b), and (c) correspond to 200 days preceding the special interval, the interval of interest, and 150 days following. Each box represents 50% of the distribution of particle fluxes during each time interval, with the median indicated by the line across the middle or lower edge of the box. The vertical "error bars" indicate the extent of the distributions of fluxes for the upper 25% and lower 25% of the fluxes. A background, consisting of linear interpolations between the lowest flux rates measured in each respective Bartels rotation, has been subtracted from the data before compiling the statistics.

The statistical box-plot distributions show clearly that the low energy ion fluxes have essentially the same distributions for each time interval at each spacecraft. However, the distributions of the higher energy protons are vastly different, quantitatively, in interval (b) than in the other two time intervals. At Voyager 1, these proton fluxes in interval (b) were all basically near background, as is also qualitatively evident visually in Figure 2.

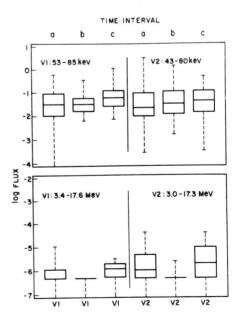

Figure 3. Statistical distributions (represented as box plots) of fluxes in ion and proton channels on the two Voyager spacecraft for three time intervals. The interval (b) correspond to day 243, 1983, to day 75, 1984. Intervals (a) and (c) correspond to a 200 day period preceding and a 150 day period following interval (b).

Plotted in Figure 4 are daily average ion spectra from the Voyager 1 LECP instrument on February 1, 1984 and from the Voyager 2 instrument on January 4, 1984. The two spectra are each taken from a different co-rotating region and are representative of ion spectra during co-rotating events during the analysis interval [Gold et al., 1985b]. The spectral shapes at the two locations for the two different interaction regions are very similar. This comparison holds for the other co-rotating regions during the interval as well [Gold et al., 1985b]. The similarity in the spectra together with the statistical result showing decreased amplitudes at higher latitudes indicates that the co-rotating acceleration process is the same in and out of the ecliptic, but less effective at higher latitudes.

Lanzerotti et al. [1985] suggested that the relatively undisturbed (by transient events) outer heliosphere which existed during this approximately 200 day interval facilitated the detection of enhanced levels of 3 kHz plasma waves reported by Kurth et al. [1984]. These authors attributed the waves to plasma processes produced at the heliosphere boundary by the interaction of the solar wind with the interstellar medium. Plotted in Figure 5 are the Voyager 1 plasma wave data from Kurth et al., together with the low energy LECP data from both Voyagers. The low energy VLF plasma waves were reported to be first observed on day 243, 1983, and to have disappeared in the spring of 1984 [EOS, 1985]. On about day 75, 1984, the outer heliosphere was again disrupted by transient events, as measured by the LECP instrument.

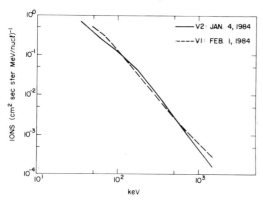

Figure 4. Daily average ion spectra from Voyager 1 on February 1, 1984 and from Voyager 2 on January 4, 1984.

## 3. DISCUSSION

During the approximately 200 days of quiescence in the outer heliosphere, when the interplanetary environment (as observed by both Voyager spacecraft) was essentially undisturbed by intermittent solar energetic particle events and the accompanying transient interplanetary disturbances, the particle enhancements associated with co-rotating interaction regions were more frequent and more intense in the ecliptic plane than at latitudes $\geq 20°$ above the plane. This conclusion was also reached by the Voyager cosmic ray instrument team studying $> 0.5$ MeV protons [Christon and Stone, 1985]. The acceleration process, however, appears to be the same at both locations since the spectra of the co-rotating ions were similar as measured on both Voyagers.

For the $\sim 200$ days, the medium may have resembled the prediction of Burlaga [1983], who speculated that beyond $\sim 10$ A.U. the large scale dynamics of the interplanetary medium could be dominated by pressure waves. That is, the originating coronal signatures would be lost so that the mapping back to the sun of the co-rotating regions, as attempted by Christon and Stone [1985], would not be possible. The solar wind would therefore be more homogeneous on a large scale [see also Smith et al., 1985, and Whang and Burlaga 1985].

Because of the fact that the Ulysses spacecraft will always be within $\sim 5$ A.U. of the sun, it is unlikely that the probe will encounter an interplanetary situation such as that proposed by Burlaga [1983] and encountered by the Voyager spacecraft. Nevertheless, the results presented

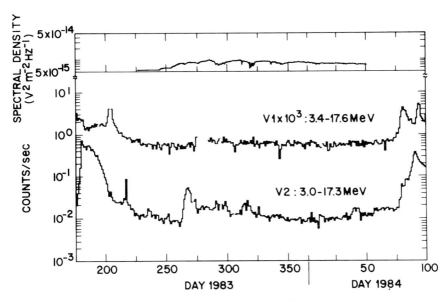

Figure 5. Plasma data (top trace) from Voyager 1 [Kurth et al., 1984] plotted together with Voyager 1 and Voyager 2 low energy ion data around the interval when the enhanced plasma waves were observed.

herein showing a latitude dependence of particle enhancements associated with co-rotating interaction regions (which occur throughout the heliosphere) are probably representative of interplanetary features that will be evident in low energy ion data acquired by the Ulysses spacecraft.

## 4. ACKNOWLEDGEMENTS

The work at The Johns Hopkins University Applied Physics Laboratory was supported by NASA under Contract N00024-83-C-5301.

## 5. REFERENCES

Burlaga, L. F.: 1983, *J. Geophys. Res., 88,* 6085-6094.
Criston, S. E., and Stone, E. C.: 1985, *Geophys. Res. Lett., 12,* 109-112.
Gloeckler, G.: 1984, *Adv. Space Res., 4,* 127-137.
Gold, R. E., Lanzerotti, L. J., Maclennan, C. G., and Krimigis, S. M.: 1985a, *Proc. 19th Internat. Cosmic Ray Conf., 4,* 186.
Gold, R. E., Lanzerotti, L. J., Maclennan, C. G., and Krimigis, S. M.: 1985b, in preparation.
Krimigis, S. M., Armstrong. T. P., Axford, W. I., Bostrom, C. O., Fan, C. Y., Gloeckler, G., and Lanzerotti, L. J.: 1977, *Space Sci. Rev., 21,* 329-354.
Kurth, W. S., Gurnett, D., and Scarf, F. L.: 1984, *Nature, 312,* 27-31.
Lanzerotti, L. J., Maclennan, C. G., and Gold, R. E.: 1985, *Nature, 316,* 243-244.
Smith, Z. K., Dyer, M., and Steinolfson, R. S.: 1985, *J. Geophys. Res., 90,* 217-220.
Whang, Y. C., and Burlaga, L. F.: 1985, *J. Geophys. Res., 90,* 221-232.

# THREE-DIMENSIONAL GRADIENTS OF SOLAR PARTICLES INSIDE 5 AU

Edmond C. Roelof
Johns Hopkins University/Applied Physics Laboratory
Johns Hopkins Road
Laurel, Maryland  20707  U.S.A.

ABSTRACT.    The  spatial  gradients  (field-aligned,  longitudinal  and
latitudinal) in solar flare energetic particle events are described in
terms  of  their  occurrence  in  the  three  phases  of  the  event  (rise,
maximum and decay).   The information that  can  be  derived  from  these
phases is discussed in the context of the Ulysses mission.

## 1.  INTRODUCTION

The question of the spatial distribution of solar flare energetic particle events must be posed within the context of the temporal evolution
of each individual event.   This is particularly true of the region of
the heliosphere between the radii of Earth and Jupiter wherein the
Ulysses spacecraft will spend its lifetime.   No other companion spacecraft will accompany Ulysses out of the ecliptic plane, and the in-
ecliptic coverage will be restricted mainly to near-earth spacecraft.
Therefore our success in inferring the acceleration and injection processes of the particles at the sun and their subsequent response to the
structure of the inner and outer interplanetary medium will hinge upon
our recognition of the signatures of these processes in the energetic
particle events themselves.
    The events are best thought of in terms of three phases:  the rise,
the maximum, and the decay.   In the rise phase, the flux anisotropy is
usually strong and field-aligned (although not without exception), and
hence the time history is dominated by the coronal injection process.
In the maximum phase, the anisotropy usually decreases, either because
of the arrival of particles back-scattered from interplanetary magnetic
field structures beyond the spacecraft (as is the case at near-relativistic energies for electrons and ions), or because of the arrival at
the spacecraft of the flare plasma ensemble (FPE) composed usually of
the flare-associated shock and the flare plasma ejecta that produced
it.   The lower the energy of the particles, the greater the effect of
the FPE on the particle population.   Finally, in the decay phase, the
anisotropy is usually much diminished and the particles are, to a first
approximation, being convected out of the inner heliosphere by the
combination of the $\underline{E}$ x $\underline{B}$ drift (field line motion) and residual (but
much reduced) field-aligned streaming.   Although there are vestiges of
the  coronal  transport  still  detectable  in  this  final  stage,  its
evolution is dominated by the IMF structure beyond the spacecraft.

331

*R. G. Marsden (ed.), The Sun and the Heliosphere in Three Dimensions, 331–340.*
*© 1986 by D. Reidel Publishing Company.*

The three phases of solar particle events were identified by
McCracken and Rao and their coworkers (McCracken and Rao, 1970) on the
basis of the anisotropy signatures, but our understanding of the
importance of the FPE has increased since their seminal work some 15
years ago. The major effect of the passage of the FPE is to reduce the
field-aligned gradient of the energetic particle population behind the
FPE all during the decay phase.

For Ulysses, the significance of the small field-aligned gradient
(along with the concommitant circumstance of negligible cross-field
diffusion), is that the intensity measured at Ulysses during the decay
phase is approximately the same anywhere on that field line inside the
spacecraft, and hence the Ulysses intensity may be compared directly
with that measured (in the decay phase) by a spacecraft near Earth,
giving us a useful two-point measurement. Likewise, but for a completely different reason, intensities can be compared at Ulysses and
Earth in the rise phase whenever the anisotropies are strong at both
spacecraft, because both spacecraft are quite directly sampling the
coronal injection (although at different field line coronal foot
points). We now turn to several examples to illustrate the foregoing
concepts.

## 2. RISE PHASE

As a first example, we examine in Figure 1, from Roelof (1982), an
isolated solar flare particle event in 5 MeV protons measured by the Low
Energy Charged Particle (LECP) telescope on the Voyager spacecraft which
was at 1.7 AU during late November, 1977. On November 22, 1977 (Day
326), a 2B flare erupted in McMath plage region 15031 at N23, W40. The
Hα intensity increased at 0946 UT attaining its maximum at 1006 UT. A
detailed multi-spacecraft analysis of this FPE event using five spacecraft (Helios 1/2, IMP 8, and Voyager 1/2) has been published by Burlaga
et al. (1980). The shock front was demonstrably non-spherical, arriving
at Voyager 1 at 2226 UT, November 27 (Day 331), yielding an average
radial speed of 369 km s$^{-1}$ (compared with a local computed shock speed
of 302 km s$^{-1}$). The interplanetary medium at IMP 8 was quiet, and
Voyager 1 lay close to the Earth-Sun line. Consequently, the flare
particle event is a good candidate for being magnetically well-connected
through a relatively undisturbed interplanetary medium.

As expected, Figure 1 reveals a very fast, strongly anisotropic
rise phase, with the strong anisotropy persisting (rather unusually for
5 MeV protons) right into the maximum. The format of the figure is
somewhat novel. The anisotropies have been transformed into the local $\underline{E}$
x $\underline{B}$ frame using the measured plasma velocity and field direction. Consequently the transformed anisotropy should be approximately field-
aligned. The LECP scans 8 sectors in a plane that at this time was not
much inclined to the ecliptic. The normalized anisotropies (2-hour
averages) are presented in the left panel. The sector averages are
grey-shaded in 16 tones linearly from the minimum (non-zero) sector to
the maximum. Isolated squares mark the field direction (projected on
the scan plane). The arrow at the top of the panel marks the projected
look-direction to the sun. The right-hand panel presents the minimum

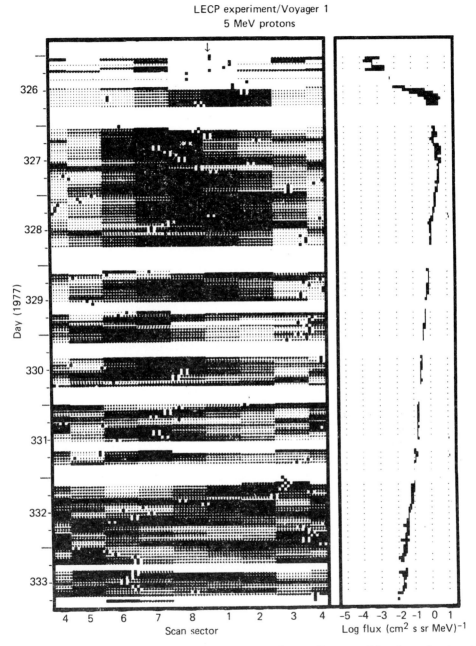

LECP experiment/Voyager 1
5 MeV protons

Fig. 1. Left: Grey shade representation of normalized anisotropies (transformed into the E x B frame) of 5 MeV protons measured in eight sectors by Voyager 1 at 1.7 AU. Right: fluxes of 5 MeV protons (2h averages) in the sectors with minimum and maximum rates; bar height gives anisotropy ratio. From Roelof (1962).

and maximum sector flux, connected by a solid bar. Since the flux scale is logarithmic, the "front-to-back" (or, more properly "max-to-min") anisotropy can be read directly off the figure. It is nearly 10:1 at flux maximum some 10 hours after flare Hα maximum.

Let us first establish that the rise phase of this event can not be described by diffusion. A strong anisotropy at flux maximum is inconsistent with the conditions under which a diffusive transport equation is valid. In the usual nomenclature, the particle density U is proportional (in simple homogeneous three-dimensional diffusion) to the function $t^{-3/2}$ $\exp(-r^2/4 \kappa t)$. This means that the anisotropy $\xi = 3S/Uv$, where $S = -\kappa \, \partial U/\partial r$, is just given by $\xi = 3r/2vt$. Now the time of intensity maximum is $t_m = r^2/6\kappa$, so that the anisotropy at that time is $\xi_m = 9\kappa/rv$, or writing $\kappa = \lambda v/3$, we have the simple relationship

$$\xi_m = 3 \, \lambda/r \tag{1}$$

The path length travelled by any particle between injection at $t = 0$ and event maximum $t = t_m$ is $S_m = vt_m = r^2/2\lambda$. Thus we have a direct relation between the path length (expressed as a ratio to the radial distance r) and the anisotropy at maximum

$$S_m = 3 \, r/2 \, \xi_m \tag{2}$$

The inconsistency of a strong anisotropy at maximum is now apparent. If $\xi_m > 1$, then (1) implies that the observer is less than three mean-free-paths from the source. Moreover, (2) implies that the path length itself is less than 3/2 the distance from the source, or the particle has travelled less than 4 or 5 mean-free-paths by the time of event maximum! These arguments can be generalized to a case of anisotropies, non-homogeneous diffusion. Parker (1963) gave an analytic Green's function for diffusion in a flux tube whose area varied as $r^\alpha$ with $\kappa$ varying as $r^{-\beta}$. The equivalent formulas replace the 3 in (1) by $(\alpha+1)$ and the 2 in (2) by $(2-\beta)$.

Diffusion equations are valid only after many mean-free-paths have been travelled, especially if the observer is close to the source. For this particularly "clean" event, propagation inside 1.7 AU must have been relatively "scatter-free". The persistence of the strong anisotropy to event maximum (well beyond the first arrival time of about 3 hours for 5 MeV protons) implies extended particle acceleration/injection on a time scale $\sim 10$ hours from either the corona or the outward-moving shock. Since the shock did not arrive at Voyager 1 until late on Day 331 (note the change in the anisotropy at this time), the reduction in the anisotropy must have been due to a combination of the cessation of injection and back-scatter from beyond Voyager 1.

This class of event could possibly be observed on Ulysses, either enroute to Jupiter or as it nears perihelion. The main information contained in the rise phase is the injection history at the inner foot of the interplanetary field line. The intensity gradient outward along the field line must be strongly negative at this time (nearly like $r^{-2}$) due to the focussing effect of the diverging magnetic field. What of the three-dimensional gradients after the maximum?

## 3.  DECAY PHASE

In the November 22, 1977 event, we fortunately had good "stereo" coverage from the JHU/APL CPME instrument on IMP 8 near Earth. When the 5
MeV proton fluxes from both spacecraft are "mapped-back" to their high
coronal connection longitudes using techniques my colleagues and I
introduced a dozen years ago (Roelof and Krimigis, 1973) the result is
Figure 2. As expected, the maximum flux on Day 327 is higher at 1 AU
than at 1.7 AU. However, post-maximum (Day 330 and afterwards), the
flux is higher at Voyager 1. If this were interpreted as a radial
gradient, it would imply one that is positive, steady and very large
(about 260%/AU). However, if the three-dimensional gradient were purely
azimuthal, it would yield an exponential dependence on connection
longitude of the form $j \propto \exp(\phi/\phi_o)$ with $\phi_o \simeq 27°$.

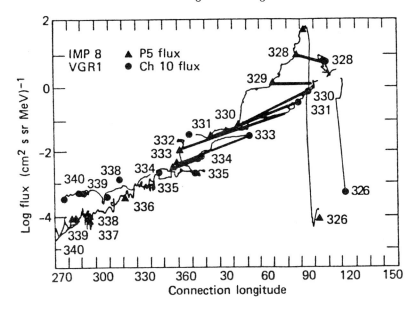

Fig. 2.  Fluxes of 5 MeV protons measured at Voyager 1 (1.7 AU) and IMP
8 "mapped-back" to their high corona connection longitudes using solar
wind velocities measured at the spacecraft. From Roelof (1982).

There are two other multi-spacecraft events in the literature, this
time using the Pioneer 10 and 11 spacecraft, along with IMP, that suggest that the decay phase gradient is mainly azimuthal (or, for Ulysses,
possibly meridional), and that the field-aligned gradient is small. The
events of November 3, 1973 and November 5, 1974, shown in Figures 3 and
4, were analyzed by Hamilton (1977) in terms of a diffusion model. However, Roelof and Krimigis (1977) pointed out the strong effects of the
flare plasma ensemble (FPE) on the particle history defined by the
measured solar wind stream-stream interactions. I would like to return
here to those two events because they both offer the unusual opportunity
to deduce the field-aligned gradient in the decay phase.

Figures 3 and 4, extracted and redrawn from Hamilton (1977), show in the upper panel 11–20 MeV proton fluxes over 15 days from spacecraft at two different helioradii. The lower panel gives the coronal connection longitude, computed from the solar wind velocity measured on the same spacecraft in the constant radial velocity approximation (Nolte and Roelof, 1973a, 1973b). These authors argued that the connection longitude "labelled" the field lines, so two spacecraft with the same connection longitudes were essentially in the same flux tube. In the November 3, 1973 event (Figure 3), the two spacecraft have very nearly the same connection longitude (which was nearly that of the flare on the day of the flare), except for Days 308–312 during which time the FPE was in transit between IMP 8 at 1 AU and Pioneer 11 at 2.67 AU. Once the FPE neared Pioneer 11, the fluxes, which had been consistently higher at IMP

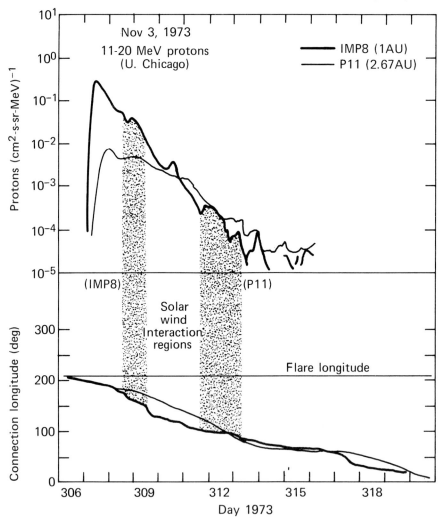

Fig. 3.   Two-spacecraft measurement of 11–20 MeV proton fluxes essentially along the same field line. Extracted from Hamilton (1977).

8, became comparable, within counting statistics, and the fluxes continued their decay together, exhibiting no measurable field-aligned
gradient.

In the second event, November 5, 1974 (Figure 4), the two spacecraft (Pioneer 10 at 6.08 AU, Pioneer 11 at 4.85 AU) began the event
well separated from each other in connection longitude. After Day 312,
both spacecraft were in the long-lived rarefaction of a recurrent solar
wind stream (i.e., a "dwell" in connection longitude) that is typical of
the region ~ 5 AU during the decline of the solar cycle. Their
connection longitudes differed by $\Delta\phi$ ~ 10°-20° for Days 313-318 until
Pioneer 11 entered the next recurrent stream (just as Pioneer had done
on Day 310). Pioneer 11 saw a smooth rise and maximum transition, but
Pioneer 10 was not well-connected to the flare particle emission region

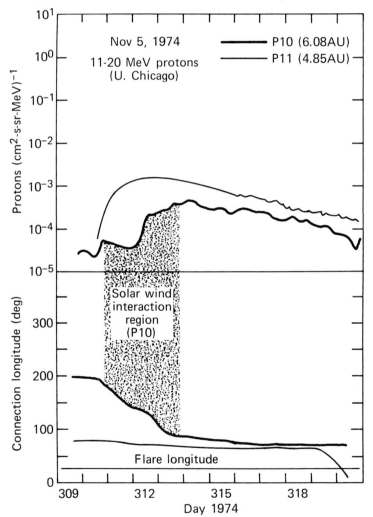

Fig. 4. Two-spacecraft measurements of 11-20 MeV proton fluxes at
nearly the same connection longitudes. Extracted from Hamilton (1977).

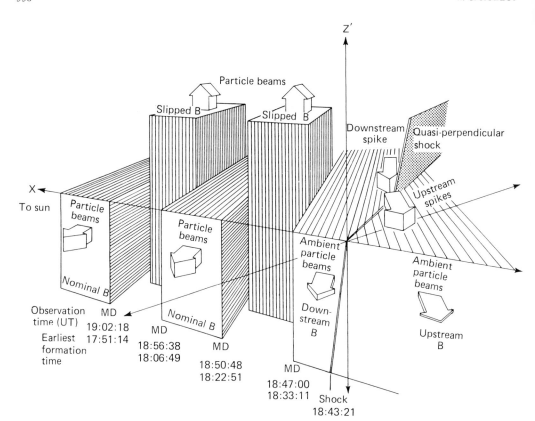

Fig. 5.  Stylized visualization of the shock/multiple-magnetic discontinuity ensemble that controls the energetic particles.  The +X axis
points toward the sun, and the Y′ axis lies in the plane of the magnetic
field upstream of the shock.  Observation times for each element are
indicated, but in order to demonstrate that each must be a dynamic,
evolving structure, the times of earliest possible formation by the
shock passage are listed below the corresponding observation time.  All
structures are populated with ambient particle fluxes (35–1600 keV) that
are strongly anisotropic.  From van Nes et al. (1985).

until  Day  313.    Only  when  their  connection  longitudes  were  close
together  were  their  decay  rates  the  same.    If  their  flux  differences
were  attributed  to  a  radial  gradient,  it  would  be  sustained,  large  (–
58%/AU)  and  negative.    However,  if  we  interpret  it  as  an  azimuthal
dependence  of  the  form  $\exp(\phi/\phi_o)$,  it  is  in  the  correct  sense  (Pioneer
11  is  connected  closer  to  the  flare  site  than  Pioneer  10)  and  $\Delta\phi \simeq 10°$–
$20°$  implies  $\phi_o \simeq 15°$–$30°$,  a  range  of  values  commonly  found  in  azimuthal
gradients;  for  instance,  we  found  $\phi_o = 27°$  from  Figure  2.

4.   MAXIMUM PHASE

The  maximum  phase  of  the  event  is  perhaps  the  most  sensitive  of  all  to
the  state  of  interplanetary  medium  in  the  vicinity  of  the  spacecraft.

Only in a high energy event (near-relativistic particles) that propagate into a relatively undisturbed medium will have a simple, smooth transition in intensity and anisotropy from the rise into the decay. As one goes to lower energies (e.g., ions ~ 1 MeV/nucleon and electrons ~ 10 keV), the protons are strongly affected by the large-scale FPE structure and the electrons by wave-particle interactions. This phase is often further complicated by an inseparably concomitant effect, namely, the rapid "sweep" of the connection longitude eastward across the corona as the high velocity flare-heated plasma arrives behind the FPE. This can either increase or decrease the number of particles injected on the field lines before and after the FPE. In addition, the shock accelerates ions and the FPE confines both flare and interplanetary-accelerated ions behind it. The further out in helioradius the spacecraft, the more dramatic these effects can be.

Even though Roelof and Krimigis (1977) showed, in event after event at low energies, the FPE actually shaped the maximum phase history, the FPE has seldom been examined in detail for its effects on particles. The FPE-associated shocks have been studied, but the FPE itself is much more extensive (taking more than a day to pass over spacecraft like Ulysses well beyond the Earth). Recently van Nes et al. (1985) studied the FPE in a major flare-associated event observed at ISEE-3 (June 6, 1979). Using the Imperial College/ESTEC measurements of 35-1600 keV ions, they demonstrated that these particles passed nearly adiabatically through the shock (which was within a few degrees of being an ideal perpendicular shock), but were accumulated, stored and channeled by the complex, evolving FPE structure behind it. A schematic representation of the particle and magnetic field behavior is given in Figure 5. These authors aver that such structures are not uncommon, and that they must dominate the evolution of flare particle events at low energies. They must therefore also affect the higher energies as well to a significant degree.

## 5. SUMMARY

For the Ulysses mission, the implications of the foregoing discussion of three-dimensional gradients in solar flare particle events are:

Rise Phase. While the strong anisotropy lasts, fluxes are controlled by the duration and longitude/latitude dependence of coronal injection as well as subsequent possible interplanetary acceleration and storage by the flare plasma ensemble (FPE). The field-aligned gradients are strong and negative; transport across interplanetary field lines is restricted mainly to $\underline{E} \times \underline{B}$ convection. The information contained is mainly about coronal and near-sun propagation.

Maximum Phase. At high energies, the anisotropic phase ends (although it can sometimes extend well into the decay) due to the cessation of coronal injection and the back-scatter of particles from beyond the spacecraft. At lower ion energies, the maximum phase occurs as the FPE envelopes the spacecraft, due to its dual role of interplanetary acceleration and (perhaps more importantly) storage. The information contained in this case is mainly local – on the scale of the FPE.

Decay Phase. Once the FPE has passed over the spacecraft, the field-aligned gradients are small and transverse transport is by $\underline{E} \times \underline{B}$ convection. The information contained in the intensities can therefore be "mapped" along field lines inward to the corona (or the next-arriving FPE) and outward to the last FPE that passed over the spacecraft.

In conclusion, this discussion (which has made no pretense of being anything other than one worker's view of the phenomena), has attempted to identify the kind of information on solar flare particle events that can be extracted from the Ulysses mission. The nature of that information depends upon the phase of the event itself. The discussion should have made obvious the historical fact that our knowledge increased exponentially when we had the opportunity to compare measurements among different spacecraft. Let us hope for good base-line measurements as Ulysses sails out of the ecliptic plane.

## 6. ACKNOWLEDGEMENTS

This work was supported by the Air Force Geophysics Laboratory under program ZF10, Task I of Contract N000-24-85-C-5301. This paper would not have appeared without the patience and understanding of the Editor, Dr. R. G. Marsden.

## 7. REFERENCES

Burlaga, L., Lepping, R., Weber, R., Armstrong, T., Goodrich, C., Sullivan, J., Gurnett, D., Kellog, P., Keppler, E., Mariana, F., Neubauer, F., Rosenbauer, H. and Schwenn, R.: 1980, J. Geophys. Res. 85, 2227.

Hamilton, D.C. 1977, J. Geophys. Res. 82, 2157.

McCracken, K.G. and Rao, U.R.: 1970, Space Sci. Rev. 11, 155.

Nolte, J.T. and Roelof, E.C.: 1973a, Solar Phys. 33, 241.

Nolte, J.T. and Roelof, E.C.: 1973b, Solar Phys. 33, 483.

Parker, E.N.: 1963, Interscience Publishers, John Wiley and Sons (New York).

Roelof, E.C.: 1982, Air Force Geophysics Laboratory Technical Report AFGL-TR-82-0342.

Roelof, E.C.: 1973, J. Geophys. Res. 78, 5375, 1973.

Roelof, E.C. and Krimigis, S.M.: 1977, Study of Travelling Interplanetary Phenomena/1977, ed. M. A. Shea, D. F. Smart and S. T. Wu, D. Reidel (Dordrecht), 343.

van Nes, P., Roelof, E.C., Reinhard, R., Sanderson, T.R. and Wenzel, K.-P.: 1985, J. Geophys. Res., 90, 3981.

# A SPATIALLY CONFINED, LONG-LIVED STREAM OF SOLAR PARTICLES

K. A. Anderson*
and W. M. Dougherty
Space Sciences Laboratory
University of California
Berkeley, CA 94720
*also Physics Department

ABSTRACT. Studies of low energy electrons and ions of solar origin lead us to conclude:
(1) That on one occasion in 1979 interplanetary field lines connected Earth at 2°S heliographic latitude to a solar active region situated at 16°N latitude. The width of this field line bundle was 2.5 $\times 10^6$ km at 1 AU. A linear extrapolation gives a 12,000 km width at the Sun.
(2) That electrons and ions were continuously injected onto this bundle of field lines but not outside it for at least 6 hours and probably more than 20 hours. We hypothesize that flare accelerated particles are trapped in a magnetic structure high in the corona and escape over periods of time on the order of one day.

## 1. INTRODUCTION

In this paper we discuss certain characteristic changes in solar-interplanetary particle intensity found in space at a distance of 1 AU from the Sun. On approximately 40 occasions in 1978 and 1979 we found well-defined ("square wave") changes in the intensity of low energy electrons or ions and sometimes both (see Figure 1). These intensity changes lasted from 1 to 6 hours. The intensity changes could be either increases or decreases. The duration of a few hours corresponds to a corotating spatial structure of about 3 $\times 10^6$ km in width in close agreement with the widths of "filaments" reported by *Bartley et al.* (1966) and *McCracken and Ness* (1966). However, those authors reported that only the direction of the particle flow changed due to changes in the field direction. They reported that the particle intensity did not change as a "filament" was crossed. It is therefore not clear at

341

*R. G. Marsden (ed.), The Sun and the Heliosphere in Three Dimensions, 341–348.*

this point that the particle effects reported here are related to those reported by *Bartley et al.* (1966). One important difference between the two experiments is the energy range of the particles measured: we report ion measurements in the range 40 to several hundred keV, while *Bartley et al.* measured ions having energies of tens of MeV. *Domingo et al.* (1976) reported several examples of count rate changes in solar particle fluxes associated with direction changes in the interplanetary magnetic field. These changes lasted from 1 to 2 hours and had the square wave appearance of the events we discuss here. Their measurements were made on electrons of about 1 MeV energy and protons in the range 1 to 36 MeV.

Although the electrons we measure are of quite low energy (2 to 40 keV), such particles have very high speeds (9% to 35% the speed of light), but small gyroradii. In a 5 nT magnetic field at 60° pitch angle the gyroradii are in the range 15 to 70 km. The proton gyroradii are on the order of several thousand kilometers.

## 2. OBSERVATIONS

Figure 1 shows a two-day period in which three well-defined particle intensity changes having a square wave appearance occur. In the first of these the low energy electron fluxes increase while the ion flux decreases. In the second electrons increase but ions show no change, and in the third case (1100 to 1300 UT on 3 June), the electron intensity decreases and the ion flux also appears to decrease. It is possible that some of the many variations seen at other times on these two days represent the same phenomenon but are not so well defined as the three examples just described.

Well-defined correlations exist between some of the particle changes and solar wind and IMF parameters. For example, in the case of the particle intensity variation from 0240 to 0525 UT on 2 June, the magnetic field decreases and the solar wind density increases. However, the correlations differ greatly from event to event and in such cases definite signatures in the solar wind and IMF cannot be identified. Because of this, we do not use the term filament to describe the magnetic field lines on which these intensity changes occur. We prefer to say the particle effects occur in particle propagation channels with no implication that a particular magnetic field structure is involved. However, it is difficult to believe that the channeling was not established by some magnetic field structure at some time and place.

We next discuss an example which allows some conclusions about the origin and nature of one particular particle propagation channel. On May 20 and 21, 1979, several small solar flares occurred in McMath plage region 16014, located at about N16 W66. One of the flares in this region accelerated electrons from ≤2 keV to above 40 keV and ions ≤40 keV to above 270 keV energy (see Figure 2). The electron intensity *versus* time

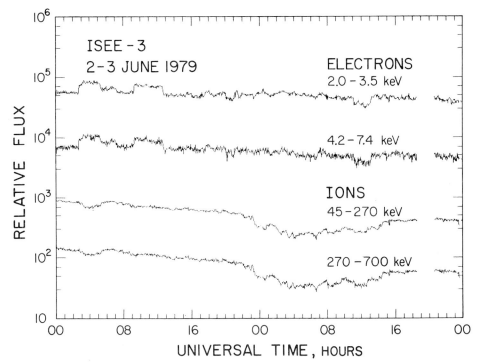

Figure 1. Changes in the intensity of solar-interplanetary low energy electrons and ions having a "square wave" appearance were a fairly common occurrence over 15 months of observations in 1978 and 1979. In the two-day period shown at least three such events occur. In some cases the intensity changes are associated with solar wind or interplanetary magnetic field changes. For some there is no clear signature.

profile is characteristic of an impulsive injection of particles at the Sun into a medium whose mean free path for pitch angle scattering is on the order of 1 AU. The slow decay phase seen in Figure 2 is consistent with a brief injection phase followed by scattering in the interplanetary medium (*Lin,* 1974). This view is supported by the measured pitch angle distributions: Initially, for some tens of minutes, the electrons are highly anisotropic, then become nearly isotropic.

The most remarkable feature during these two days is the sudden increase in the electron and ion intensities at about 2000 UT on 20 May, followed about six hours later by a sudden return to the earlier slow decay. The increase is simultaneous at all particle energies. This is also the case for the decrease. During this 6 hour period the pitch angle distributions became highly anisotropic showing strong flow of electrons away from the Sun (Figure 3). Evidently the interplanetary field lines have connected to a source region in the solar atmosphere able to continually inject electrons and ions into the interplanetary medium for at least 6 hours. We suppose that this injection also occurred before and after this six hour interval. From the observed slow decay of particle intensity during this time, the source could

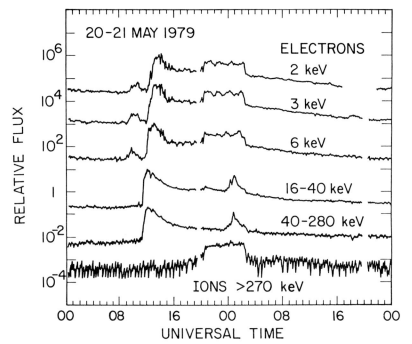

Figure 2. An importance 1N flare began at 1107 UT on 20 May 1979 and injected electrons from 2 to 100 keV and low energy ions into interplanetary space. The spacecraft entered a particle propagation channel at 2030 UT. This channel was 2.5 ×10⁶ km in width at 1 AU.

have supplied particles for intervals as long as 20 hours or more.

From the solar wind speed and direction of the interplanetary magnetic field we find the width of the particle propagation channel to be $2.5 \times 10^6$ km. Linear extrapolation of this width to the Solar surface gives a size of 12,000 km. The magnitude and topology of the magnetic field near the Sun no doubt will affect the actual size, but we appear to be dealing with a dimension which is larger than the size of a small flare but smaller than the plage region. Both at the Sun and in interplanetary space this structure is relatively small. One consequence of this is that one or more such structures may be associated with active regions so that they are a common feature at 1 AU, but are not often seen because of their small size.

While the spacecraft is located in the particle propagation channel three impulsive, flare-like injections of particles occur (Figure 3). They are identified by their impulsive appearance and their velocity dispersion, consistent with about a 1.3 AU travel distance. One of these electron events can be firmly identified with an optical flare in region 16014. No optical association could be found for the other two particle injections. This is not surprising since we have found from a study of many impulsive low energy electron events that those having very low intensity often have no reported

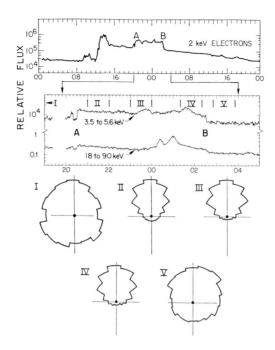

Figure 3. About nine hours after the solar flare that began at 1107 UT, highly anisotropic fluxes of electrons appeared for several hours (A to B). Before and after this time the electron fluxes were nearly isotropic. The vertical lines through the pitch angle distributions at the bottom of the figure lie along the interplanetary magnetic field. The direction of the Sun lies somewhere in the upper half of these diagrams. The pitch angle distributions are indexed to the middle panel by means of roman numerals.

H$\alpha$ association (*Potter et al.*, 1980). However, it is unusual that three low energy solar electrons events occur in a six hour interval. This leads us to believe that during this interval the spacecraft is situated on field lines connected to the site of flaring on the Sun. If this is the case the field lines guiding the electrons and ions away from the Sun from 2030 UT on 20 May to 0230 on 21 May have their origin in or near McMath plage region 16014, located at N16 W66 at this time. Since Earth is at a heliographic latitude of 2°S at this time the field lines would have been deflected 18° southward. This result is consistent with the suggestions of *Schulz* (1973) and by *Svalgaard et al.* (1975) and *Svalgaard and Wilcox* (1976), that the current sheet is related to the equator of a solar magnetic dipole. In such a view field lines drawn out by the solar wind near the equator could trace back to latitudes well removed from the equator in the manner shown in Figure 7 of *Smith et al.* (1978).

The fact that energetic particles stream away from the Sun for periods of time on the order of days has been known for some time. (See, for example, *Simnett*, 1971 and *Anderson et al.*, 1982.) There are two general views on the mechanisms behind long-lived emission of solar particles. The first of these is storage of flare accelerated particles in the corona and their subsequent escape. The other hypothesis is continuous acceleration. Present observational evidence is not sufficient to resolve this issue, and there are conceptual difficulties with each hypothesis. Radio wave observations give the best evidence in favor of coronal storage, although the configurations of the magnetic fields have not been made clear in this way. The difficulty with coronal storage is the rapid rate of energy loss of the fast particles to background electrons in the coronal plasma. Only at very high coronal altitudes is the rate of energy loss low enough to permit storage over periods of days (*Krimigis and Verzariu, 1971*). The problem with the continuous acceleration hypothesis is that no physical mechanism for it has been identified, and the best understood mechanisms involve shock waves and are therefore impulsive in character. Impulsive acceleration is noisy in the sense of copious X-ray and radio wave emission whereas these emissions are largely absent during periods of long-lived streaming. The present observations provide some additional information on the problem of long-lived emission of solar particles. In the first place only flares of small size are involved here whereas in the past the process has been generally associated with large flares. In the case at hand the largest flare was importance 1N and it accelerated only low intensities of ions in addition to the rather large fluxes of electrons. However, this flare occurred in an active region above which type III bursts frequently appeared, indicating that beams of fast electrons were present in the corona over a period of 1 or 2 days.

Secondly, the streaming of solar particles on 20-21 May 1979 is restricted to a spatial region whose dimension at the Sun is on the order of $10^4$ km. The near-perfect confinement of the streaming particles must be associated with discontinuities or major changes in the topology of magnetic fields in the solar atmosphere. However no magnetic feature could be identified with the particle channel on this occasion. In particular, nothing suggesting a neutral sheet appeared.

The third point we would make using data from the 20-21 May 1979 interval concerns the energy spectrum of the electrons. We noted above that the intensity of the 2 to 10 keV electrons changed very little over a 6 hour period. If the small change in intensity is interpreted as due to energy loss of these electrons to background electrons we can set a lower limit to the altitude at which the electrons must reside if indeed they are trapped in magnetic structures. A 10 to 20% energy loss in 6 hours at the lowest energies requires that the electron density not exceed $10^5$ cm$^{-3}$. This corresponds to a heliocentric distance of 4 solar radii for the quiet Sun and 19 solar radii if the coronal region is a streamer with enhanced density (*Fainberg and Stone, 1971*).

We have also compared the energy spectrum of the electrons in the impulsive injection from the 1N flare that began at 1107 UT on 20 May with the spectrum of electrons streaming in the propagation channel. The results are given in Table I. We have fitted the spectral data to a power law for 30 one-hour data samples and obtained the power law exponent for each sample. Ten samples are taken during times preceding entry into the particle channel, 10 samples in the channel and 10 samples following exit from the channel. The average value for each set of ten samples and the standard deviations are given in the Table. We find that there is no difference in the low energy electron spectra between particles in the propagation channel and those on field lines outside the channel. There can be little doubt that the electrons outside the channel were accelerated by the 1N flare. We take the similarity of the energy spectra to be significant but not conclusive evidence for storage and subsequent escape over a period of many hours of electrons accelerated over a brief period of time by the importance 1N flare which began at 1107 UT on 20 May 1979.

## ACKNOWLEDGEMENTS

This work is the result of research supported by NASA grant NAG5-376.

Table I.

| Time<br>20-21 May 1979 | Number of Intervals | Spectral Index, $\gamma$ |
|---|---|---|
| 14:00-19:00 | 10 | $3.77 \pm 0.08$ |
| 21:00-02:00 | 10 | $3.39 \pm 0.29$ |
| 03:00-08:00 | 10 | $3.49 \pm 0.07$ |

Comparison of the energy spectra of 2 to 10 keV electrons inside the particle propagation channel (2100 to 0200 UT) with spectra calculated before and after entry into the channel. The electrons outside the channel are certain to have been accelerated by an importance 1N flare, although at this time they are no longer streaming from the Sun. The close similarity of the spectral indices suggests that the particles streaming away from the Sun in the propagation channel were accelerated by the same flare but then were trapped and released over a period of many hours.

# REFERENCES

Anderson, K. A., R. P. Lin and D. W. Potter, 1982, *Space Science Rev.* **32**, 169.

Bartley, W. C., R. P. Bukata, K. G. McCracken and U. R. Rao, 1966, *J. Geophys. Res.* **71**, 3297.

Domingo, V., D. E. Page and K.-P. Wenzel, 1976, *J. Geophys. Res.* **81**, 43.

Fainberg, J. and R. G. Stone, 1971, *Solar Phys.* **17**, 392.

Krimigis, S. M. and P. Verzariu, 1971, *J. Geophys. Res.* **76**, 792.

Lin, R. P., 1974, *Space Sci. Rev.* **16**, 189.

McCracken, K. G. and N. F. Ness, 1966, *J. Geophys. Res.* **71**, 3315.

Potter, D. W., R. P. Lin and K. A. Anderson, 1980, *Astrophys. J.* **236**, L97.

Schulz, M., 1973, *Astrophys. Space Sci.* **24**, 371.

Simnett, G. M., *Solar Phys.* **20**, 448, 1971.

Smith, E. J., B. T. Tsurutani and R. L. Rosenberg, 1978, *J. Geophys. Res.* **83**, 717.

Svalgaard, L. and J. M. Wilcox, 1976, *Nature* **262**, 766.

Svalgaard, L., J. M. Wilcox, P. H. Scherrer and R. Howard, 1975, *Solar Phys.* **45**, 83.

SUPER-EVENTS IN THE INNER SOLAR SYSTEM AND THEIR RELATION TO THE
SOLAR CYCLE

R. Müller-Mellin, K. Röhrs, and G. Wibberenz
Institut für Reine und Angewandte Kernphysik, Universität Kiel
Olshausenstr. 40 - 60
2300 Kiel 1
Federal Republic of Germany

ABSTRACT. We define a super-event in the energetic particle population
by the following characteristics:
a)  Its onset and decay is much longer than for usual flare-associated
    solar particle events, the intensity may remain above background
    for a period of the order of 40 days.
b)  The intensity variation with heliolongitude is small.
c)  The events are observed in nucleons and electrons simultaneously.
The super-events are rare, we have found only three cases between 1978
and 1981. It is interesting that they occur at time periods where the
long-term variation of galactic cosmic rays shows relative intensity
minima. As one hypothesis for the origin and storage mechanism for
these events we suggest shock waves which accelerate particles also at
heliolongitudes far away from the flare site, and act as a barrier for
galactic cosmic rays as well as for particles generated in the inner
solar system.

1. INTRODUCTION

Increases in the intensity of low energy cosmic rays which could not be
associated with individual solar flares are a well-known phenomenon
since several solar cycles. Several types of such increases with different characteristics could meanwhile be identified and clearly related
to different causes. The quiet time increases of electrons found by
Cline et al. (1964) could clearly be identified as electrons from the
Jovian magnetosphere (Chenette et al. 1974; Teegarden et al. 1974).
These electron  increases were not accompanied by enhancements in the
intensity of nuclei. In contrast the corotating events described in
detail by McDonald and Desai (1971) were later found to be of interplanetary origin (McDonald et al. 1975) and are meanwhile clearly identified with particle acceleration at corotating shocks. These corotating
events have a typical duration of 5 - 7 days, they are confined to a
sector in the interplanetary magnetic field, the spectrum is relatively
weak and can be described by an exponential in velocity (Barnes and
Simpson 1976; Gloeckler et al. 1979). For fixed energy per nucleon the

*R. G. Marsden (ed.), The Sun and the Heliosphere in Three Dimensions, 349–354.*
© *1986 by D. Reidel Publishing Company.*

proton to alpha ratio is typically 50, the events are not accompanied
by electrons.

In contrast to these intensity increases of non-solar origin, Fan
et al. (1968) postulated long-lived activity centers on the sun as the
origin of long-lived events. They find that a) the heliographic longitudinal range over which enhanced proton fluxes in the MeV range are
continuously observed in interplanetary space may be as great as $\sim 180^{\circ}$;
b) this flux exhibits definite onsets and cutoffs; c) the interplanetary regions within which these low energy protons are confined also mo-
dulate the galactic cosmic radiation; d) superimposed on this flux level
are occasional large and discrete flare-produced intensity increases extending in energy to more than 50 MeV.

The various types of increases can be classified according to their
temporal structures, energetic and chemical composition, radial gradients and association with other phenomena within the solar system. In
this paper we shall describe the discovery of a new phenomenon which cannot be understood in terms of the explanations given above. We have
tentatively named these increases as "super-events" because of their
long duration and high intensity. We describe the appearance of these
events in the inner solar system based on measurements of HELIOS-1 and
-2. Correlation with measurements at and beyond 1 AU will be discussed
in a subsequent paper.

## 2. OBSERVATIONS

HELIOS-1 and -2 are spinning deep space probes with elliptical heliocentric orbits between 1 and 0.3 AU. Their position in the three periods when the super-events were observed is shown in Figure 1 using
a coordinate system with the earth-sun-line fixed.

Solar activity in cycle 21 had started in September 1977 and by the time of
April 1978 flaring activity had become so
frequent that it was no longer possible
to unambiguously associate flux increases
with specific regions of the sun, e.g.
there were 58 active regions reported for
Carrington rotation 1667 (April 9 to May
6, 1978). In Figure 2 we present time-
intensity profiles of electrons above
0.3 MeV and protons and alpha-particles
between 4 and 13 MeV/N along with their
ratio. The plot contains hourly averages
of counting rates measured on board
HELIOS-1 during a time period of 50 days
including the period from April 20 to May
10, 1978, when the intensity of protons
stayed above a threshold of $10^3$ particles/
$m^2$ s sr MeV indicated by the hatched area.
Although this threshold is chosen somewhat arbitrarily, it is already more than

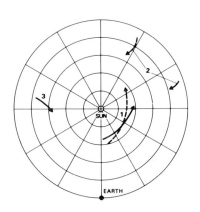

Fig. 1. Trajectories of HE-
LIOS-1 (solid line) and HELI-
OS-2 (dashed line) during super-events 1,2, and 3 in a
rotating coordinate system
with a fixed earth-sun-line.

two orders of magnitude above the
quiet time background in this
channel. During this prolonged
enhancement period HELIOS-1 had
scanned more than $180^{\circ}$ in heliolongitude after correction for
its own not negligible angular
velocity.

The event also shows up at
somewhat higher energies as
shown in Figure 3, where we have
compared simultaneous flux measurements of 13 to 27 MeV protons
taken on board HELIOS-1 and -2
and IMP-8. HELIOS-1 moved from $9^{\circ}$
west of the earth-sun-line to $94^{\circ}$
west just behind the limb. HELIOS-
2 was ahead of HELIOS-1 by $10^{\circ}$ to
$30^{\circ}$ and IMP-8 stayed at $0^{\circ}$. There
is good agreement in the gross
features of the time profiles.
Differences in some details can
be explained by considering the
magnetic connections of the three
spacecraft to the high corona. In
the beginning of the event HELIOS
-2 and IMP-8 were magnetically
lined up, so both spacecraft see
some flare indication on April 20,
but not HELIOS-1. On April 30 HE-
LIOS-1 and IMP-8 were magnetically lined up, so they measure the
same decay profile whereas HELIOS
-2 does not match.

Note that the absolute fluxes
are comparable at 0.3 AU and at 1
AU, so there is no significant
gradient inside 1 AU. Spectra of
the form $dJ/dE = AE^{-\gamma}$ have been
eye-fitted to the proton fluxes
in the energy range 4 to 50 MeV.
The dependence on time of the
spectral exponent $\gamma$ shows a gradual softening from a value of
2.0 on April 20 to 2.5 on April
28 before a 3b flare produced a
harder spectrum.

The frequency of occurrence
for such type of events is low.
Between December 74 and December
81 we have found only 3 examples.

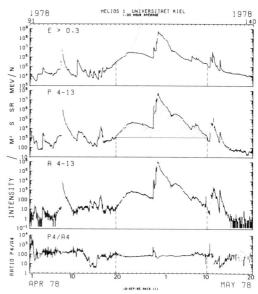

Fig. 2. Time-intensity profiles of
electrons > 0.3 MeV, and protons and
helium nuclei and their ratio in the
energy range 4 - 13 MeV/N measured
on board HELIOS-1 at the time of
super-event 1.

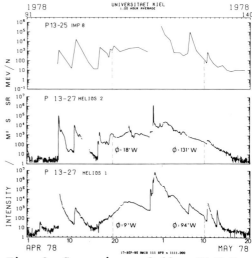

Fig. 3. Comparison of 13 - 27 MeV
proton fluxes on board HELIOS-1,
HELIOS-2, and IMP-8. Also indicated
are the heliolongitudes of HELIOS-1
and -2 at the beginning and end of
super-event 1.

Figure 4 presents a super-event
from September 8 to October 5,
1979. HELIOS-1 stayed close to
108° west and HELIOS-2 did not
move very much either (153° west,
see Figure 1). At the time of the
super-event from April 26 to May
20, 1981 only HELIOS-1 data were
available as shown in Figure 5.

Based on our observations
we define a super-event in the
energetic particle population by
the following characteristics:
1. Its onset and decay is much
   longer than usual for flare-
   associated solar particle
   events. Exponential rise and
   decay times have time constants in excess of 30 hours.
2. The intensity may remain a-
   bove background for a period
   of 30 to 40 days, i.e. more
   than one solar rotation.
3. The events are observed in
   nuclei and electrons simultaneously.
4. The proton to alpha ratio of
   about 100 remains constant
   throughout the super-event.
   For quiet times this ratio is
   about 10, for corotating events
   about 50.

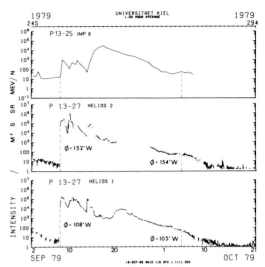

Fig. 4. Comparison of 13 - 27 MeV
proton fluxes on board HELIOS-1, HE-
LIOS-2, and IMP-8. Also indicated are
the heliolongitudes of HELIOS-1 and
-2 at the beginning and the end of
super-event 2.

3.    DISCUSSION

We want to show that the super-
events are a class of energetic
particle phenomena which cannot
be understood in terms of existing acceleration or propagation
schemes. After subtraction of solar flare accelerated particles
characteristically different at
the three locations (see Figure 3)
we are left with an underlying
population with no indication
for corotating effects. We rather
find a common temporal structure
at all locations during event 1.
A similar feature is also found

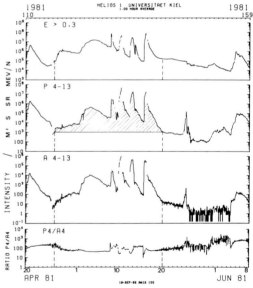

Fig. 5. Time-intensity profiles of
electrons >0.3 MeV, and protons and
helium nuclei and their ratio in the
energy range 4 - 13 MeV/N measured on
board HELIOS-1 at the time of super-
event 3.

in the long-lasting simultaneous decay on all three s/c during event 2
(see Figure 4). Fan et al. (1968) had suggested a magnetic field configuration rooted around the center of flaring activity and spread out
over a range of ~180° in longitude. The small variation with longitude
which we find precludes an interpretation where the particle source is
connected with a specific activity center on the sun. In principle, a
larger number of activity centers distributed around the sun and producing many energetic particle events could simulate a slowly varying
source. In this case the solar origin should produce a strong radial
variation. A careful examination of radial gradients has not yet been
performed. However, during event 1 HELIOS-1 and -2 are close to 0.3 AU,
and for a solar source we would expect about an order of magnitude smaller intensities near 1 AU. This is not observed (see Figure 3).

We suggest an interplanetary origin of the super-events related to the
effects of many systems of transients travelling outwards, piling up to
nearly spherical shells as one shock overtakes the other. Shock waves
can accelerate particles and are found to propagate around the sun over
more than 90° in longitude (Chao and Lepping, 1974). During the 50 day
period from day 91 to 140, 1978, containing the first super-event in Figure 2 many interplanetary phenomena are observed which make this period
unique. In TABLE 1 we give the number of shock waves as observed by the
plasma experiment on board HELIOS-1 (R. Schwenn, private communication)
as well as the number of type II and IV radio bursts taken from Solar-
Geophysical Data. For comparison the respective numbers for periods of
equal length adjacent to this period are given.

TABLE 1.

| Interplanetary Phenomena | Day of Year in 1978 | | |
|---|---|---|---|
| | 41 – 90 | 91 – 140 | 141 – 190 |
| Shock Waves | 3 | 18 | 2 |
| Type II Radio Bursts | 16 | 49 | 25 |
| Type IV Radio Bursts | 6 | 33 | 10 |

Further studies are required to
show whether the periods of the
two other HELIOS super-events
can be characterized by similarly large number of shocks. We
obtain an indirect proof for our
suggestion by studying the long-
term variation of galactic cosmic rays. Figure 6 shows 5-month
running means of the Kiel neutron monitor countrate. The fil-
tering smoothes out individual
Forbush decreases. We see a series of relative minima, sometimes considered as parts of
"mini solar cycles". The occurrence of the HELIOS super-events
close to the relative minima as

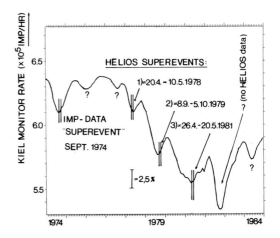

Fig. 6. Five-month running means
of Kiel neutron monitor data.

indicated by the pairs of vertical bars in Figure 6 is very striking.
This pattern was very suggestive to look for super-events during the
other periods of relative minima (see the question marks in Figure 6)
when HELIOS data were not available. Preliminary information from the
University of Chicago experiment on IMP-8 (courtesy R.B. McKibben) indicates indeed the presence of long-lasting intense particle enhancements
near the pronounced minima in 1974 and 1982.

The correlation between systems of large interplanetary shocks and
galactic cosmic ray modulation is clearly established (see e.g. Lookwood
and Webber, 1984). We take the existence of minima in the galactic cosmic rays as indication for effective barriers set up outside the orbit
of earth. The nearly simultaneous occurrence of the super-events supports
our hypothesis of a shock origin for this population. This does not mean
that we see the local effects of interplanetary shocks as in energetic
storm particle events. In this case we would expect large temporal variations and drastic differences from one spacecraft to the other. In our
conception the collision and superposition of a variety of flare initiated shocks (see Table 1) outside the orbit of earth leads to the efficient acceleration, connected with little longitudinal and radial variation of the energetic particle intensity. Based on our present information we cannot distinguish whether the gradual decay of the super-
events is caused by a diminished acceleration efficiency, a gradual reduction of the barrier effect, or both.

## ACKNOWLEDGEMENTS

We wish to express our thanks to H. Kunow, G. Green, and B. Iwers for
their contributions to the successful and long-lasting operation of the
University of Kiel cosmic ray experiment on HELIOS and for useful and
stimulating discussions. This work was supported by the Bundesministerium für Forschung und Technologie, contract 01 QC 086.

## REFERENCES

Barnes, C.W. and Simpson, J.A., 1976, Ap. J., 210, L91 - L96.
Chao, I.K. and Lepping, R.P., 1974, J. Geophys. Res., 79, 1799.
Chenette, D.L., Conlon, T.F., and Simpson, J.A., 1974,
    J. Geophys. Res., 79, 3551 - 3558.
Cline, T.L., Ludwig, G.H., and McDonald, F.B., 1964
    Phys. Rev. Letters, 13, 786.
Fan, C.Y., Pick, M., Pyle, R., Simpson, J.A., and Smith, D.R., 1968,
    J. Geophys. Res., 73, 1555 - 1582.
Gloeckler, G., Hovestadt, D., and Fisk, L.A., 1979,
    Ap. J., 230, L191 - L195.
Lookwood, J.A. and Webber, W.R., 1984, J. Geophys. Res., 89, 17.
McDonald, F.B. and Desai, U.D., 1971, J. Geophys. Res., 76, 808 - 827.
McDonald, F.B., Teegarden, B.J., Trainor, J.H., von Rosenvinge, T.T.,
    and Webber, W.R., 1975, Ap. J., 203, L149 - L154.
Teegarden, B.J., McDonald, F.B., Trainor, J.H., Webber, W.R., and
    Roelof, E.C., 1974, J. Geophys. Res., 79, 3615 - 3622.

# THE MAXIMUM ENTROPY PRINCIPLE IN COSMIC RAY TRANSPORT THEORY

P. Hick, G. Stevens, J. van Rooijen
Laboratory for Space Research Utrecht
Beneluxlaan 21
3527 HS   Utrecht
The Netherlands

ABSTRACT. We describe a procedure to obtain an approximate solution to the cosmic ray transport equation, which, contrary to the familiar diffusion approximation, is valid also for large anisotropies. Using some moments of the distribution function an approximation is constructed, in accordance with Jaynes' principle of maximum entropy. We apply the procedure to the case of the one-dimensional transport equation and compare the resulting maximum entropy approximation and the diffusion approximation with the numerical solution. We find, that there is a qualitative agreement between the m.e.-approximation and the numerical solution, particularly also close to the particle source where the diffusion approximation breaks down.

## 1. INTRODUCTION

In general solutions to a cosmic ray transport equation are not known in closed form. A numerical approach is always possible, of course, but is complicated and elaborate. A procedure which generates a reasonable approximation to the distribution function would be a welcome alternative. An often used approximation is the so-called diffusion approximation, which assumes that the first order anisotropy $f_1$ is small and proportional to the gradient of the isotropic component, $f_1 \propto \text{grad} \, f_0$. This means, however, that the diffusion approximation breaks down close to the particle source, where the anisotropies may be large. We present an approximation which is also valid for large anisotropies.

   This approximation is based on the recognition that the lower order moments of the distribution function satisfy relatively simple differential equations. Consider, for example, the one-dimensional transport equation

$$\frac{\partial f}{\partial t} + \mu v \frac{\partial f}{\partial x} = \frac{1}{2} \nu \frac{\partial}{\partial \mu} \left(1 - \mu^2\right) \frac{\partial f}{\partial \mu} \quad , \quad f(x \to \pm\infty, \mu, t) = 0 \qquad (1)$$

where $f(x, \mu, t)$ is the distribution function, $v$ the particle velocity, $\mu$ the pitch angle cosine, $\nu$ the scattering frequency. The scattering time is defined as $\tau = \nu^{-1}$, the mean free path as $\lambda = v\tau$. The moments of

R. G. Marsden (ed.), The Sun and the Heliosphere in Three Dimensions, 355–358.

$f(x,\mu,t)$, e.g. $<x^2>$, are defined as

$$<x^2> = \int_{-\infty}^{\infty} dx \int_{-1}^{+1} d\mu\ x^2\ f(x,\mu,t) \tag{2}$$

Appropriate integrations of equation (1) over phase space give[1]:

$$\frac{d}{dt}<x^2> = 2v<\mu x> \quad,\quad \frac{d}{dt}<\mu x> = v<\mu^2>-\nu<\mu x> \quad,\quad \frac{d}{dt}<1> = 0 \tag{3}$$

For given initial conditions we can use this system of equations for the moments $<1>$, $<x^2>$ and $<\mu x>$ to construct an approximation to the distribution function, based on Jaynes' principle.


## 2. JAYNES' PRINCIPLE OF MAXIMUM ENTROPY

Jaynes' principle formulates a procedure to find a representation for a distribution function (more general, a density function) when the available information by itself (e.g. some of its moments) does not uniquely determine this function. The principle demands that the representation meets two requirements:
1.   It should be consistent with the information i.e. it should reproduce the correct moments.
2.   It should be as non-committal as possible in any other aspect.
The mathematical formulation of this statement says, that the representation, satisfying both requirements, maximizes the entropy

$$S = -\int_{-\infty}^{\infty} dx \int_{-1}^{+1} d\mu\ f\ \log\ f, \tag{4}$$

subject to the constraints implied by the available moments (Jaynes, 1957a,b; Hobson,1971).
Applying this to the 1-dimensional transport equation (1), using the moments $<1>$, $<x^2>$ and $<\mu x>$, the m.e.-approximation takes the form (using Lagrange multipliers):

$$f_{me}(x,\mu,t) = \exp\{\rho_0(t)+\rho_1(t)\mu x-\rho_2(t)x^2\} \tag{5}$$

This expression for $f_{me}$ is substituted in (3), which can be solved for the Lagrange multipliers $\rho_0,\rho_1$ and $\rho_2$ for given initial conditions.


## 3. THE ONE-DIMENSIONAL TRANSPORT EQUATION: A CASE STUDY

The numerical solution to eq. (1) was calculated using an Alternating Direction Implicit Method (see e.g. Johnson and Riess, 1982). We present results for the initial condition

---

[1] It is possible to close the system of equations (3) with respect to the moments by introducing the moment $<\mu^2>$: $d/dt<\mu^2> = \nu(<1>-3<\mu^2>)$. In that case it is possible, for given initial conditions, to find explicit expressions for the moments as functions of time. Note, however, that our method does not require a closed system.

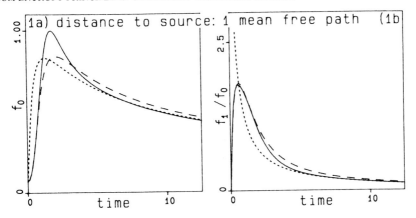

Fig. 1. Comparison of maximum entropy solution (dashed line) and diffusion approximation (dotted line) with numerical solution (full line) at a distance of 1 mean free path from the particle source. a) isotropic component, b) first order anisotropy. Time is in units of the scattering time τ.

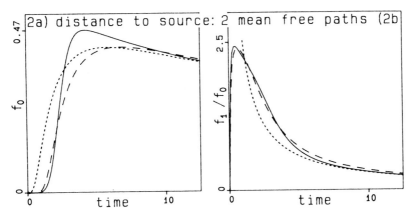

Fig. 2. Idem, for 2 mean free paths from the source.

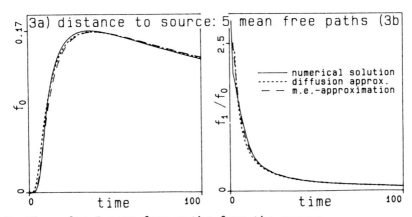

Fig. 3. Idem, for 5 mean free paths from the source.

$$f(x,\mu,t=0) = \frac{2}{\lambda\sqrt{\pi}} \exp\{-[\frac{2x}{\lambda}]^2\}$$ (6)

i.e. at t=0 a particle distribution, which is isotropic in $\mu$ and normal in x with a width of 0.5 mean free paths, is injected. In Figures 1 through 3 the m.e.- and diffusion approximations are compared with the numerical solution. Figures 1a, 2a, 3a show the isotropic components $f_0$, and Figures 1b, 2b, 3b the first order anisotropies $f_1/f_0$, for various distances to the source.

$$f_0 = \frac{1}{2} \int_{-1}^{+1} f \, d\mu \quad , \quad f_1 = \frac{3}{2} \int_{-1}^{+1} \mu f \, d\mu \quad ,$$ (7)

We conclude that there is qualitative agreement between m.e.-approximation and numerical solution at all times and for all distances to the source. As already mentioned, the diffusion approximaton is not valid for large anisotropies, which occur for small distances at small times after the injection of particles (Figures 1b, 2b for $t<5\tau$). In this range the m.e.-approximaton is a significant improvement to the diffusion approximation, especially in its description of the first order anisotropy. For large distances ($>5\lambda$) and/or times ($>5\tau$) the m.e.- and diffusion approximations are almost identical to the numerical solution, so that the m.e.-approximation can be regarded as including the diffusion approximation.


4. CONCLUSION

It seems that the m.e.-approximation is capable of handling the large anisotropies as observed e.g. close to particle sources. This feature makes the m.e.-approach a potentially valuable tool for studying cases where the particle transport close to the source is important, for instance in the case of moving sources sweeping past the observer (such as interplanetary shocks).

REFERENCES

Hobson, A.: 1971, Concepts of Statistical Mechanics, Gordon and Breach.
Jaynes, E.T.: 1957a,b 'Information theory and Statistical Mechanics, Paper I and II', Phys. Rev., **106**, 620; **108**, 171.
Johnson, L.W. and Riess, R.D.:1982, Numerical Analysis, 2nd edition, Addison-Wesley.

# SECTION VII:  COSMIC RAYS

MODULATION OF GALACTIC COSMIC RAYS IN THE HELIOSPHERE

R. B. McKibben
Enrico Fermi Institute
933 E. 56th St.
Chicago, Ill., 60637
U.S.A.

ABSTRACT: Observations of the intensity of galactic cosmic rays and anomalous components near earth and in the outer heliosphere are providing important tests for theories of solar modulation. The most recent observations show effects that seem to require that models for modulation include gradient and curvature drifts as well as the conventional processes of convection, adiabatic deceleration, and diffusion. New observations are required to define the interplay of these processes in the three-dimensional heliosphere.

1. INTRODUCTION

It has been only about 30 years since Forbush (1954) discovered an anticorrelation between the level of solar activity and the intensity of the relativistic cosmic radiation at the surface of the Earth. Since that time, the variation of the cosmic ray intensity has been followed through almost 3 solar cycles (Figure 1), and a large body of observations and theoretical ideas concerning the modulation of the cosmic ray intensity by solar activity has been amassed.

In the years following Forbush's discovery, although there was at first some uncertainty as to whether the modulation was caused by the effects of solar activity on the geomagnetic field or by more general interplanetary effects, it soon became clear that the observed intensity variations represented real variations in the interplanetary cosmic ray intensity. In 1956, Morrison (1956) suggested that the modulation was caused by plasma ejected from the Sun, and Meyer, Parker, and Simpson (1956) used observations of particles from the February, 1956 solar flare to infer the existence of a scattering medium for charged particles outside the orbit of Earth.

A theoretical basis for modulation came with the prediction (Parker, 1958), and subsequent observation (e.g. Gringauz et al., 1963; Bonetti et al., 1963; Snyder and Neugebauer, 1964) of the solar wind. In 1965, Parker (1965) described the effect of the steady state solar wind on the cosmic ray density, U, using a Fokker-Planck equation,

R. G. Marsden (ed.), The Sun and the Heliosphere in Three Dimensions, 361–374
© 1986 by D. Reidel Publishing Company.

$$\frac{\partial U}{\partial t} + \frac{\partial}{\partial x_i}\left(Uv_i\right) + \frac{\partial}{\partial T}\left(U\frac{\partial T}{\partial t}\right) - \frac{\partial}{\partial x_i}\left(\kappa_{ij}\frac{\partial U}{\partial x_j}\right) = 0 \qquad (1)$$

where the terms represent, in order, the time rate of change of the
density, the convection of cosmic rays by a solar wind with velocity v,
the change in energy, T, produced by adiabatic deceleration in the
expanding solar wind, and the diffusion of particles through the
irregular interplanetary magnetic field. As later pointed out by
Jokipii and Parker (1970), the antisymmetric terms of the diffusion
tensor describe the charged particle's guiding center drift in response
to the large-scale gradient and curvature of the interplanetary field.

Today, equation 1 is generally believed to provide a locally
complete and correct description of the physical processes affecting the
cosmic radiation in the solar wind. The problem is that, while equation
1 is locally valid, the modulated intensity at a given point depends
upon values of $\kappa$ , v , and T integrated over trajectories of particles
in the solar wind from the boundary of the modulation region to the
point of observation. For most of the history of the study of
modulation, we have had observations of the cosmic ray intensity and of
the interplanetary medium only near the earth. As a result, analysis of
modulation has for the most part relied on simplified models, which
normally neglected drifts and assumed spherical symmetry, quasi-steady
conditions (i.e. $\partial U/\partial t = 0$), and somewhat ad hoc forms for the variation
of $\kappa$ with particle magnetic rigidity and position in the heliosphere.
Nevertheless, such models, often called conventional models, have been
successful in describing the modulation of cosmic rays at 1 AU (e.g.
Fulks, 1975; Garcia-Munoz et al., 1985). Recent reviews of modulation
theory and observations have been given by Jones (1983a,b) and Quenby
(1984). This paper will focus on the most recent observations of cosmic
ray intensity variations, their relation to current theories of
modulation, and their implications for the Ulysses mission.

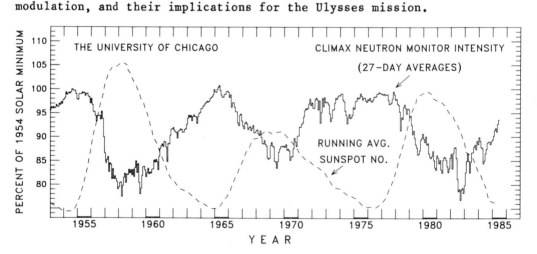

Figure 1: Relativistic (mean proton energy = 10 GeV) cosmic ray
intensity compared to the running average sunspot number.

## 2. MODEL PREDICTIONS

The most controversial question in modulation theory today concerns the importance of drifts in the modulation process. Since drifts are an intrinsically 3-dimensional process, it is appropriate to give them a rather central role in this discussion. In an extensive series of papers, Jokipii and his coworkers (e.g. Jokipii and Kota, (1985) and references therein), and, more recently, Potgieter and Moraal (1983, 1985) and Potgieter (1985a,b) have explored the effects of drifts on modulation of cosmic rays.

In drift-dominated models, the paths of cosmic rays in the heliosphere depend primarily on the global magnetic structure of the heliosphere, and change dramatically when the polarity of the solar magnetic field, and, hence, of the interplanetary magnetic field changes every 11 years. When the north pole of the sun is magnetically positive, as it was from 1970-1980, positive particles drift from the poles towards the equatorial current sheet, and then drift rapidly outwards along the current sheet. Negative particles, on the other hand, drift inwards along the current sheet and drift poleward in latitude. When the sun's north pole is magnetically negative, these

TABLE 1

Status of Predictions from Drift-Dominated Modulation Models

| Observable | Prediction | Confirmation | Selected Refs. Theory | Expt. |
|---|---|---|---|---|
| Radial Gradients | Increase on change from qA<0 to qA>0 | No | 1,2 | 8,11 |
| Latitude Gradients | Intensity minima always above or below current sheet Gradient larger for qA<0 | Uncertain | 3 | 9 |
| Anomalous Components | Intensity lower near ecliptic for qA<0 | Yes | 4 | 8,12 |
| Nucleon/electron Ratio | Depends on sign of qA | Yes | 5 | 10 |
| Solar Cycle Time-intensity Profiles | Intensity maxima wide for qA>0, narrow for qA<0 | Proton: yes Electron: no | 6,7 | 10 |

References: 1. Potgeiter (1985b) 2. Kota and Jokipii (1983) 3. Jokipii and Kota (1985) 4. Pesses et al. (1981) 5. Potgeiter and Moraal (1985) 6. Jokipii and Thomas (1981) 7. Kota and Jokipii (1983) 8. McKibben et al. (1985) 9. Fisk and Newkirk (1985) 10. Garcia-Munoz et al. (1986) 11. Webber and Lockwood (1985) 12. Webber et al. (1985)

motions are reversed.  Therefore, significant changes in cosmic ray
modulation should be expected when the sun's magnetic polarity changes.
      Several such predicted changes, testable with currently available
observations, are listed in Table 1, where the notation qA>0 is used to
indicate drifts from the pole to the equator.  For positive particles
(q>0), this corresponds to positive magnetic polarity at the north pole.
The primary assumptions underlying these predictions are that the Parker
spiral model provides a good description of the average magnetic field
throughout the heliosphere and that the interplanetary magnetic polarity
is determined by the dipole component of the sun's magnetic field.
These assumptions have been verified in all regions so far explored
(e.g. Smith et al., 1978; Thomas and Smith, 1981), and their extension
to unexplored regions seems on fairly sure ground.
      For conventional modulation models, which neglect drifts, such
clear predictions are not possible.  For conventional models, modulation
depends upon the local interaction of cosmic rays with irregularities in
the interplanetary field all along their paths in the heliosphere.
These paths are best described as random walks, and are without the
large scale regularity characteristic of drift-dominated paths.  For the
conventional models, the model parameters are insensitive to the field
polarity and there is no clear-cut event such as the reversal of the
field polarity.  Furthermore, there is no solid theoretical or
observational basis for extrapolating to unexplored regions what is
known about the model parameters from spacecraft observations.  As a
result, analysis of modulation using conventional models tends to be
more explanatory than predictive.  Such predictions as are possible
arise when, having found a set of parameters that describes the
modulated intensity for one species, the behavior of other species with
different rigidity, charge, and charge to mass ratio may be predicted.
Tests of the models thus generally consist of a) determinations of
whether it is possible within the constraints of the model to find a set
of parameters that will describe the behavior of at least one species,
and b) tests of the mutual consistency of the behavior of various
particle species using these parameters.


3. OBSERVATIONS

3.1 Time-Intensity Profiles

The most readily available observations relevant to modulation are
records of the cosmic ray intensity vs. time.  The observations in
Figure 1 clearly show the connection between solar activity and the
cosmic ray intensity, but do not yield much information concerning the
physical mechanisms by which the intensity changes are produced.
      In recent years, observations from spacecraft well-removed from the
earth in the heliosphere have provided powerful insights into the
mechanisms of modulation.  There are now four active spacecraft
returning measurements of the cosmic ray intensity from distant regions
of the heliosphere.  Figure 2 shows the trajectories of Pioneer 10 and
11 since their respective launch dates in 1972 and 1973, and of Voyager

1 and 2 since their flybys of Saturn. By mid-1985, Pioneer 10 had reached a radial distance of 35 AU from the sun near the ecliptic in the direction of the hypothetical heliomagnetic tail. In the opposite direction, toward the solar apex, Pioneer 11 had reached a distance of 18 AU at a heliographic latitude of 15°. At the same time, Voyager 1 had climbed to a latitude of nearly 30°, while Voyager 2 remained near the ecliptic.

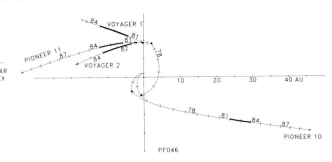

Figure 2. Trajectories, projected onto the ecliptic, of active spacecraft in the outer heliosphere.

The oldest of these spacecraft, Pioneer 10, has been returning data from beyond 5 AU for more than a complete solar cycle. Figure 3A contains 27-day averages of the integral intensity of galactic cosmic rays from the University of Chicago experiment on Pioneer 10, together with similar measurements from University of Chicago instruments on IMP spacecraft in earth orbit (McKibben et al., 1985a,b). Other integral flux measurements have been reported by McDonald et al. (1981, 1985), Fillius and Axford (1985), Fillius et al. (1985), Van Allen and Randall (1985), Webber and Lockwood (1985), and Venkatesan et al. (1985).

As of 1985, Pioneer 10, at 35 AU, remained deep in the modulation

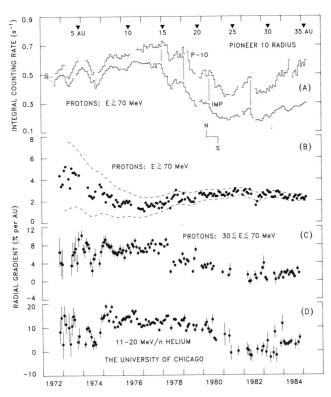

Figure 3. A) 27-day averages of quiet-selected integral cosmic ray intensity (E>70 MeV) at Pioneer 10 and at 1 AU. N,S indicate solar magnetic field reversal. B) 27-day average radial gradients for the integral flux. Dashed lines show probable systematic uncertainty. C,D) Same as B for low energy protons and helium.

region, as shown by the good correlation between intensity variations at
1 AU and at Pioneer 10. Furthermore, most of the modulation was
occurring outside of 35 AU, since the intensity at Pioneer 10 was
comparable to that observed at 1 AU at solar minimum.

As first pointed out by McDonald et al. (1981), in the period of
increasing modulation (1978-1980) the intensity decrease took place in a
series of steps separated by plateaus of nearly constant intensity. The
steps propagated outwards at near the solar wind speed, and in Figure
3A, the propagation delays are high-lighted by lines drawn at the times
of major decreases at 1 AU. The data suggest that discrete structures
in the solar wind are responsible for the modulation. Burlaga et al.
(1984) have shown that steps are associated with periods of frequent
transient flows and shocks in the solar wind, while during the plateau
periods, quasi-steady flows dominate the solar wind structure.

Burlaga et al. (1985) have recently investigated the association
between modulation and interplanetary structures in much more detail
using observations of the cosmic ray intensity, magnetic field strength,
and solar wind velocity from Voyager 1 and 2 in 1982-83. At this time
Voyager 2 was near 10 AU in the ecliptic, and Voyager 1 was near 15 AU
at latitudes ranging from $14^{\circ}$
to $20^{\circ}$. As shown by the
sample of their observations
reproduced as Figure 4, they
found nearly a one to one
correlation between intensity
decreases and regions of
enhanced magnetic field
strength. They suggested the
name Merged Interaction
Regions (MIRs) for these
regions to reflect their
origin as the coalescence at
large radii of propagating
shocks, stream interfaces, and
other disturbances in the
solar wind. Using a simple
model correlating the rate of
intensity decrease with the
strength of the magnetic field
in an MIR, and assuming a
constant recovery rate between
MIRs, they obtained
satisfactory fits to the
intensity profiles at both
Voyager 1 and 2.

These observations are
consistent with an hypothesis
which attributes the 11-year
modulation to the piling up of
decreases at solar maximum,
when successive disturbances

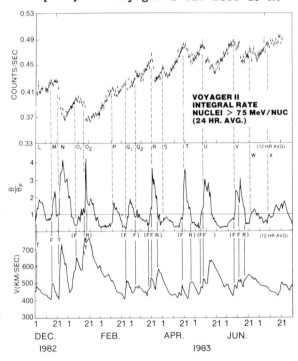

Figure 4. Top: Voyager 2 integral
intensity. Middle: Ratio of observed
(B) to Parker model (BP) magnetic
field. Bottom: Solar wind velocity.
From Burlaga et al. (1986).
(Copyright, American Geophysical
Union).

pass so frequently that full recovery from one is not possible before
another arrives (Lockwood, 1960). Satisfactory empirically based fits
to an entire 11-year cycle have been obtained using this idea in various
forms (e.g. Bowe and Hatton, 1982; Nagashima and Morishita, 1980.) Some
modification of conventional models is required by this hypothesis,
since the assumption of near-steady-state ( $\partial U/\partial t = 0$) is incompatible
with modulation produced by propagating structures in the solar wind,
which may require 6 months or more to reach the modulation boundary.
Perko and Fisk (1983) and Forman and Jones (1985) have recently begun to
investigate the effects of including time-dependence in the modulation,
but much work remains to be done.

While allowing that propagating disturbances are significant,
Jokipii and Thomas (1981) and Kota and Jokipii(1983) have suggested that
the underlying profile of the 11 year intensity is determined mainly by
the changing drift paths of particles in the heliosphere. Simply
stated, their hypothesis is that for periods when cosmic rays drift from
the poles toward the ecliptic (i.e. qA>0) they are little affected by
the structure of the current sheet (i.e. its inclination, waviness,
regularity, etc.). During such periods the cosmic ray flux in the
ecliptic should thus be relatively constant. For periods when particles
enter along the equatorial current sheet (i.e. qA<0), they are greatly
affected by the characteristics of the sheet. For solar minima with
qA<0, therefore, conditions of minimum modulation should be observed for
only a brief period very near minimum solar activity when the current
sheet is very simple and flat. Figure 1 shows that this accurately
describes the intensity profiles for relativistic protons during the
last two solar cycles. More recently, Smith and Thomas (1986) have
found that the cosmic ray intensity correlates with the inclination of
the interplanetary current sheet in the manner predicted by Kota and
Jokipii (1983).


## 3.2 Radial Gradients

### 3.2.1 Relativistic Nuclei: After measurements of the cosmic ray
intensity and its variation with time, the most basic measurement that
one can make relevant to modulation is of the variation of intensity
with position in the heliosphere. Figure 3B contains the radial
gradients, $g_r$, obtained from the counting rates shown in Figure 3A using
the formula $g_r = 100[\ln(P-10/IMP)]/\Delta R$. The most striking feature is
the relative constancy of the gradient as a function of time, in spite
of large changes in the level of modulation. Since 1974, the radial
gradient measured between Pioneer 10 and 1 AU has not varied outside the
range of 1 to 3 %/AU. In particular, contrary to the predictions of
drift-dominated models (cf. Table 1), there was no increase in the
gradient at the time of the solar magnetic polarity reversal in 1980.

From 1972 to 1974, the gradient was larger than at later times, as
shown in Figure 3B, and as confirmed using independent measurements by
Webber and Lockwood (1985). In more recent data, other observers see a
less pronounced minimum in the gradient in 1976-77 (Webber and Lockwood,
1985) and a stronger decrease in 1983-85 (Webber and Lockwood, 1985;

Fillius et al., 1985; Venkatesan et al., 1985). These disagreements may arise from systematic uncertainties in the various measurements of the integral cosmic ray intensity. From published reports, we have found that ratios between integral counting rates measured at the same position with essentially identical thresholds may show gradual long-term variations of up to 20%. The dashed lines in Figure 3B indicate the range of uncertainty in the gradient corresponding to an assumed ±10% uncertainty on the 1 AU baseline flux, and suggest that such variations may be sufficient to explain the apparent discrepancies.

3.2.2 Low Energy Galactic Protons: At lower energies, the radial gradient has been more variable in time, and is also found to depend on particle species and energy. Figure 3C shows the gradients obtained from 27 day averages of the intensities of 30-70 MeV protons measured at Pioneer 10 and 1 AU. No marked change is observed in the radial gradient upon reversal of the solar magnetic field. If there is any change associated with the polarity reversal, it is a decrease rather than the increase predicted by drift-dominated models.

A most interesting feature of these data is the correlation between the strength of modulation and the size of the radial gradient. Using the integral intensities shown in Figure 3A as a measure of the strength of modulation, almost every time the intensity decreased in response to increasing modulation, the gradient decreased as well. Especially clear examples are in 1974, during the so-called "mini-cycle" of modulation, and in 1978, upon the renewal of solar activity following solar minimum. At solar maximum (1981-82), the radial gradient was nearly 0%/AU, so that almost the same flux of 30-70 MeV protons was measured at 1 AU and near 30 AU. However, similar observations reported by McDonald et al. (1985) do not show show such a low value for the gradient in 1981-82. The cause of this discrepancy is under investigation.

3.2.3 Anomalous Components: Gradients for low energy helium in an energy range which includes the anomalous helium component are shown in Figure 3D. The anomalous components are generally believed to be interstellar neutral atoms which have been photo-ionized to a singly charged state and accelerated in the solar wind (Fisk et al., 1974) or at the solar wind termination shock (Fisk, 1982; Pesses et al., 1981; Potgieter et al., 1985). Anomalous helium was first observed in 1972, and anomalous components of nitrogen, oxygen, and neon have also been identified. The anomalous helium was continuously present at 1 AU from 1972 until 1979 (Garcia-Munoz et al., 1983) when it disappeared as the result of the increasing modulation. In the outer heliosphere, the component did not disappear until 1980 (McKibben et al., 1985; McDonald et al., 1985). Anomalous oxygen, however, remains present in the outer heliosphere in the most recent data available (Webber et al., 1985).

The anomalous components are potentially very useful probes of the modulation process. Their spectrum, which falls steeply towards high energies, renders them extremely sensitive to adiabatic deceleration (Von Rosenvinge and Paizis, 1981), and, if they are singly charged, their unique charge to mass ratios provide powerful tests of the dependence of modulation on the magnetic rigidity and velocity of charged particles.

The observations in Figure 3D show that, until the anomalous component disappeared at Pioneer 10, the radial gradient for low energy helium was much larger than that for galactic protons, possibly as a result of the greater sensitivity of the anomalous component to adiabatic deceleration. As for the protons, the size of the gradient varied inversely with the strength of the modulation. In contrast to the protons, however, while the low energy helium gradient did not show a sharp decrease at the onset of renewed solar activity in 1978, a significant decrease was observed in 1980 when the component disappeared at Pioneer 10. For slightly higher energy helium (30-70 MeV/n), the gradient also decreased, but remained finite at about 4 %/AU (McKibben et al., 1985a,b; McDonald et al., 1985). For the anomalous oxygen, although the intensity declined sharply in 1980, the component remained present, and the radial gradient remained larger than expected for galactic cosmic rays (Webber et al., 1985).

It is of particular interest that the final disappearance of the anomalous helium occurred at the same time as the reversal of the solar magnetic field. Furthermore, even though in recent data intensities of galactic cosmic rays at Pioneer 10 have recovered to levels where the anomalous helium had been observed before 1980, the anomalous helium has not returned at Pioneer 10 or at Earth (McKibben et al., 1985b). Pesses et al (1981) have suggested that the anomalous components are accelerated at the high latitude solar wind termination shock and reach the ecliptic as a result of drift motions. Thus, for qA>0, the anomalous component intensity in the ecliptic would be higher than for qA<0, as observed. However, Potgieter et al. (1985) have argued that this model cannot explain the large positive radial gradients observed for the anomalous components. On the other hand, they show that acceleration near the equator, while producing the correct radial gradients, would predict a higher intensity in the ecliptic for qA<0 than for qA>0, contrary to observations. Thus, although the observations suggest that the anomalous components are sensitive to the solar magnetic polarity, the situation remains rather muddled.

## 3.3 Latitude Variations

Little information is available concerning persistent gradients in latitude. Several approaches have been used to measure such gradients, and Jones (1983a,b) has reviewed earlier results. The most direct approach uses observations from spacecraft whose trajectories reach latitudes significantly above or below the heliographic equator. Alternatively, one may suppose that the plane of symmetry for modulation is the interplanetary equatorial current sheet, which can be highly inclined to the heliographic equator. In this case, if the inclination of the current sheet can be determined, observations obtained in the ecliptic plane can be used to determine latitude gradients with respect to the current sheet rather than the heliographic equator.

For the measurement of gradients in heliographic latitude, relevant observations are available only from Pioneer 11, which briefly reached a maximum latitude of $16^O$ near R=4 AU in 1976, and which has returned to latitudes above $10^O$ at radii > 12 AU since 1982, and from Voyager 1, which has climbed to latitudes near $30^O$ at radii > 15 AU following its flyby of Saturn. For relativistic cosmic rays, no significant latitude gradients have been reported from either Pioneer 11 (McKibben et al., 1979; Van Allen, 1980) or Voyager 1 (Decker et al., 1984), where in both cases, the sensitivity of the measurements was to gradients of the order of 1%/degree or larger. At lower energies, McKibben et al. (1979) reported marginally significant evidence for a positive (intensity increasing poleward) latitude gradient of about 1%/degree for 30-70 MeV galactic cosmic ray protons from Pioneer 11, and clearer evidence for a positive gradient of about 2 to 3 %/degree for the low energy (11-20, 30-70 MeV/n) anomalous helium component. More recently, Webber et al. (1985) have reported evidence for a positive latitude gradient of about 3%/degree for the low energy (7.1-10.6 MeV/n) anomalous oxygen component from Voyager 1 data.

Measurement of gradients in heliomagnetic latitude is more difficult, since some means must be found to establish the orientation of the interplanetary equatorial current sheet. Recently Newkirk and Fisk (1985) have analysed neutron monitor observations of the relativistic cosmic ray intensity for the period 1964-1982 using K coronameter observations of the polar coronal hole boundaries to determine the location and inclination of the current sheet. A sample of their observations is shown in Figure 5. Although there is a large scatter in the data, the analysis suggests the existence of very small but statistically significant symmetrical gradients of intensity with respect to the current sheet. Typical values are of the order of -0.06%/degree, with maximum intensity at the current sheet. From a compilation of solar wind observations, they also found a clear increase in the average solar wind velocity away from the current sheet. Newkirk et al. (1985) have extended the analysis to lower energies using spacecraft, and for energies greater than 100 MeV, they find gradients of about -0.2%/degree. Searching for time variations, they find no significant changes in either the magnitude or rigidity dependence of the gradients between 1972 and 1984. In rough consistency with these measurements, Christon et al. (1985), using observations from Voyager 1 and 2 in 1982-83, have reported a gradient of -0.15 $\pm$ 0.05%/degree with respect to the current sheet.

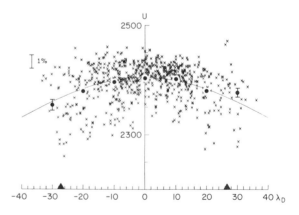

Figure 5. Daily average 5 GeV cosmic ray flux from Mt. Washington neutron monitor vs. angular distance from the equatorial current sheet. From Newkirk and Fisk (1985). (Copyright, American Geophysical Union).

As discussed by Newkirk and Fisk (1985), it is difficult to compare
the observations directly with model predictions, since the observations
average over a large range of current sheet inclinations.  Jokipii and
Kota (1985) have shown that, if the actual conditions of the measurement
are simulated, drift-dominated models can produce the sort of latitude
profiles shown in Figure 5 for both signs of the interplanetary magnetic
polarity.  However Newkirk and Fisk (1985) point out that for drift-
dominated models, a large change in the size of the latitude gradient
should have been expected on reversal of the solar magnetic polarity.
Although few observations were available after the reversal, no change
was apparent.  On the other hand, they also find it difficult to fit the
observations using conventional models which neglect drifts.


## 3.4 CHARGE DEPENDENT EFFECTS

A unique feature of drift-dominated models of modulation is the
discrimination between positively and negatively charged particles.  The
processes of convection, adiabatic deceleration, and diffusion do not
depend on the sign of the charge, but only on magnetic rigidity and
velocity.  Gradient and curvature drift velocities, however, have
opposite directions for positive and negative particles.  Thus, a change
in the relative modulation of nucleons and electrons coincident with the
change in solar magnetic polarity would be a clear indication of the
significance of drifts in modulation.

Recently, Garcia-Munoz et al. (1985) have searched for such an
effect by comparing the modulated intensities of helium nuclei and of
electrons at nearly equal magnetic rigidities over the period 1965-1985,
which includes two reversals of the solar magnetic field.  They find
(cf. Figure 6) that a change of about a factor of 2 in the He/e ratio
occurred at both reversals of the magnetic polarity, and that the He/e
ratio was highest during the period 1972-80.  Although this is also the
period when the anomalous helium was present, at energies > 70 MeV/n the
contribution of the anomalous component to the flux is negligible.  As
proof, even after the
anomalous component had
disappeared at 1 AU in 1979,
the He/e ratio remained high.
It thus appears that the He/e
ratio is sensitive to the
magnetic polarity of the
heliosphere, as predicted by
models which include drifts.

However, another
prediction, made by Kota and
Jokipii (1983), is not
fulfilled.  Since for
electrons qA<0 in 1972-80, as
in section 3.1 the intensity
profile of electrons should
have shown a sharp maximum

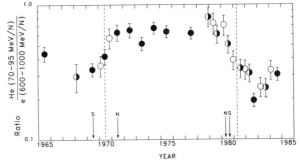

Figure 6.  He/e ratio vs. time.  N,S
indicate reversals of solar polar
magnetic fields.  From Garcia-Munoz et
al. (1986).  (Copyright, American
Geophysical Union).

near minimum solar activity rather than the broad maximum displayed by
the helium. Garcia-Munoz et al. (1985) show that, on the contrary, when
properly normalized and scaled, the intensity profile for electrons in
the period 1972-1980 was nearly identical to that of helium. It has yet
to be determined whether drift theories can accomodate this result.


## 4. IMPLICATIONS FOR ULYSSES

Two points summarize the observations and arguments above.
    1. Both drifts and conventional modulation processes seem to be
required to account for the observed modulation of cosmic rays, although
no model is successful in describing all observations. The importance
of drifts relative to the conventional processes of diffusion,
convection, and adiabatic deceleration is still unclear, and may vary as
a function both of time and of particle species and energy. Since
drifts are inherently three-dimensional, inclusion of drifts in
modulation models requires development of a three-dimensional model of
the heliosphere.
    2. There is at present no model which gives a reliable predictive
guide as to what the modulated intensity of cosmic rays at high
latitudes will be. While current models, both conventional and drift-
dominated, can reasonably account for temporal variations observed at a
single point, no model has yet satisfactorily accounted for the observed
behavior of the radial and latitude gradients. The only prediction on
which conventional and drift-dominated models seem to agree is that the
cosmic ray intensity should be lower at mid-latitudes than near the
equator or near the poles.
    If launched as planned in May 1986, Ulysses will be at heliographic
latitudes >45$^{o}$ during most of 1989-91, which should be a time of
increasing or near-maximum modulation. The magnetic polarity of the sun
should still have the same sign as at present, so that positively
charged particles will be drifting poleward. However, given that the
last reversal was observed in 1980, it is possible that the reversal of
the magnetic poles may occur while Ulysses is at high latitude.
    The most important role for Ulysses for the study of modulation
will be to gather data. At present there is almost a complete lack of
information concerning conditions at high latitudes for modulation of
cosmic rays. Lacking such information, it is not surprising that
modulation models are less than fully successful. Particularly
important will be measurements of:
    1. the latitudinal range of the shocks and other disturbances which
affect modulation near the ecliptic.
    2. the nucleon/electron ratio as a function of latitude, as a
critical diagnostic of the role of drifts in modulation.
    3. anomalous components as a function of latitude, to determine
their region of acceleration and their manner of propagation in the
heliosphere.
    4. solar particles and jovian electrons, as particles produced
primarily at low latitudes, to determine the extent and rapidity of
latitudinal propagation in the heliosphere.

It should be obvious that none of these measurements can be
properly interpreted without accompanying similar measurements in the
ecliptic plane to separate true latitude effects from the bewildering
variety of temporal effects that may be expected.  Unfortunately, at the
present time only one spacecraft, IMP 8, is returning relevant
measurements from interplanetary space, and no others are firmly planned
to be in place at the time of the Ulysses mission.  The Ulysses mission
will be exciting, and it offers  our best hope yet of coming to
understand the physics of the heliosphere.  However, if the mission is
to be fully successful, all possible efforts must be made to ensure that
comparable measurements from near the ecliptic are available.

5. ACKNOWLEDGMENTS:  This work was supported in part by NASA/Ames
Contract NAS 2-11126.

# 6. REFERENCES

Bonetti, A., Bridge, H.S., Lazarus, A.J., Rossi, B., and Scherb, F.:
    1963, Space Research 3, W. Priester, ed. (Wiley, New York), 540.
Bowe, G.A. and Hatton, C.J.: 1982, Solar Phys., 80, 351.
Burlaga, L.F., McDonald, F.B., Ness, N.F., Schwenn, R., Lazarus, A.J.,
    and Mariani, F.: 1984, J. Geophys. Res., 89, 6579.
Burlaga, L.F., McDonald, F.B., Goldstein, M.L., and Lazarus, A.J.: 1986,
    J. Geophys. Res., 90, in press.
Christon, S.P., Cummings, A.C., Stone, E.C., Behannon, K.W., and
    Burlaga, L.F.: 1985, Proc. 19th Int'l. Cosmic Ray Conf., La Jolla,
    CA., 4, 445.
Decker, R.B., Krimigis, S.M., and Venkatesan, D.: 1984, Ap. J., 278,
    L119.
Fillius, W., and Axford, I.: 1985, J. Geophys. Res., 90, 517.
Fillius, W., Axford, I., and Wood, D.: 1985, Proc. 19th Int'l. Cosmic
    Ray Conf., La Jolla, CA., 5, 189.
Fisk, L.A.: 1982, Trans. Am. Geophys. Un., 65, 1054.
Fisk, L.A., Koslovsky, B., and Ramaty, R.: 1974, Ap. J., 190, L39.
Forbush, S.E.: 1954, J. Geophys. Res., 59, 525.
Forman, M.A., and Jones, F.C.: 1985, Proc. 19th Int'l. Cosmic Ray Conf.,
    La Jolla, CA., 4, 400.
Fulks, G.J.: 1975, J. Geophys. Res., 80, 1701.
Garcia-Munoz, M., Pyle, K.R., and Simpson, J.A.: 1983, Ap. J., 274, L93.
---------: 1985, Proc. 19th Int'l. Cosmic Ray Conf., La Jolla, CA., 4,
    409.
Garcia-Munoz, M., Meyer, P., Pyle, K.R., Simpson, J.A., and Evenson, P.:
    1986, J. Geophys. Res., in press.
Gringauz, K.I., Bezrukikh, V.V., Balandina, S.M., Ozerov, V.D., and
    Rybchinsky, R.E.: 1963, Space Research 3, W. Priester, ed., (Wiley,
    New York), 602.
Jokipii, J.R., and Kota, J.: 1985, Proc. 19th Int'l. Cosmic Ray Conf.,
    La Jolla, CA., 4, 449.
Jokipii, J.R., and Parker, E.N.: 1970, Ap. J., 160, 735.

Jokipii, J.R., and Thomas, B.T.: 1981, Ap. J., **243**, 1115.
Jones, F.C.: 1983a, Proc. 18th Int'l Cosmic Ray Conf., Bangalore, India,
     **12**, 389.
---------: 1983b, Rev. Geophys. Sp. Phys., **21**, 318.
Kota, J., and Jokipii, J.R.: 1983, Ap. J., **265**, 573.
Lockwood, J.A.: 1960, J. Geophys. Res., **65**, 19.
McDonald, F.B., Lal, N., Trainor, J.H., Van Hollebeke, M.A.I., and
     Webber, W.R.: 1981, Ap. J., **249**, L71.
McDonald, F.B., Von Rosenvinge, T.T., Lal, N., Schuster, P., Trainor,
     J.H., and Van Hollebeke, M.A.I.: 1985, Proc. 19th Int'l Cosmic Ray
     Conf., La Jolla, CA., **5**, 193.
McKibben, R.B., Pyle, K.R., and Simpson, J.A.: 1979, Ap. J., **227**, L147.
---------: 1985a, Ap. J., **289**, L35.
---------: 1985b, Proc. 19th Int'l Cosmic Ray Conf., La Jolla, CA., **5**,
     198.
Meyer, P., Parker, E.N., and Simpson, J.A.: 1956, Phys. Rev., **104**, 768.
Morrison, P.: 1956, Phys. Rev., **101**, 1397.
Nagashima, K., and Morishita, I.: 1980, Planet. Sp. Sci., **28**, 177.
Newkirk, G.,Jr., and Fisk, L.A.: 1985, J. Geophys. Res., **90**, 3391.
Newkirk, G.,Jr., Lockwood, J.A., Garcia-Munoz, M., and Simpson, J.A.:
     1985, Proc. 19th Int'l Cosmic Ray Conf., La Jolla, CA., **4**, 469.
Parker, E.N.: 1958, Phys. Rev., **109**, 1874.
---------: 1965, Plan. Space Sci., **13**, 9.
Perko, J.S., and Fisk, L.A.: 1983, J. Geophys. Res., **88**, 9033.
Pesses, M.E., Jokipii, J.R., and Eichler, D.: 1981, Ap. J., **246**, L85.
Potgieter, M.S.: 1985a, Proc. 19th Int'l Cosmic Ray Conf., La Jolla,
     CA., **4**, 425.
---------: 1985b, Proc. 19th Int'l Cosmic Ray Conf., La Jolla, CA., **4**,
     429.
Potgieter, M.S., Fisk, L.A., and Lee, M.A.: 1985, Proc. 19th Int'l
     Cosmic Ray Conf., La Jolla, CA., **5**, 180.
Potgieter, M.S., and Moraal, H.: 1983, Proc. 18th Int'l Cosmic Ray
     Conf., Bangalore, India, **10**, 5.
---------:1985, Ap. J., **294**, 425
Quenby, J.J.: 1984, Sp. Sci. Rev., **37**, 201.
Smith, E.J., and Thomas, B.T.: 1986, J. Geophys. Res., **91**, in press.
Smith, E.J., Tsurutani, B.T., and Rosenberg, R.L.: 1978, J. Geophys.
     Res., **83**, 717.
Snyder, C.W., and Neugebauer, M.: 1964, Space Research 4, P. Muller,
     ed., (Wiley, New York), 89.
Thomas, B.T., and Smith, E.J.: 1981, J. Geophys. Res., **86**, 11105.
Van Allen, J.A.: 1980, Ap. J., **238**, 763.
Van Allen, J.A., and Randall, B.A.: 1985, J. Geophys. Res., **90**, 1399.
Venkatesan, D., Decker, R.B., and Krimigis, S.M.: 1985, Proc. 19th Int'l
     Cosmic Ray Conf., La Jolla, CA., **5**, 202.
Von Rosenvinge, T.T., and Paizis, C.: 1981, Proc. 17th Int'l Cosmic Ray
     Conf., Paris, **10**, 69.
Webber, W.R., Cummings, A.C., and Stone, E.C.: 1985, Proc. 19th Int'l
     Cosmic Ray Conf., La Jolla, CA, **5**, 172.
Webber, W.R., and Lockwood, J.A.: 1985, Proc. 19th Int'l Cosmic Ray
     Conf., La Jolla, CA., **5**, 185.

# EFFECTS OF THREE-DIMENSIONAL HELIOSPHERIC STRUCTURES ON COSMIC-RAY MODULATION

J. R. Jokipii
University of Arizona
Tucson, Arizona 85721
U. S. A.

ABSTRACT.  The theory of cosmic-ray transport in the heliosphere contains four distinct physical processes - diffusion, convection, adiabatic cooling, and drifts.  The last of these has only recently been evaluated.  Extrapolation of present understanding of the regions near the heliospheric equator to high heliographic latitudes leads to the conclusion that particle drift in the large-scale magnetic field plays an important role in cosmic-ray modulation.  The large-scale, three-dimensional structure of the interplanetary magnetic field is therefore very important in understanding cosmic rays.  Several key observed modulation effects are summarized, each of which is a natural consequence of drift, but which requires special assumptions if drift plays no role.  It is concluded that particle drifts play an important and possibly dominant role in transport in the heliosphere.

## 1.   Introduction.

Illustrated in figure 1 is the time dependence of the counting rate of the Climax neutron monitor, which measures protons with kinetic energies of a few GeV.  The major, eleven-year cyclic variation, in which the intensity is anticorrelated with the sunspot number is quite clearly evident. This modulation effect is caused by the solar wind with its imbedded magnetic field inhibiting the access of cosmic rays into the inner solar system.  The effect is larger at lower energies and becomes quite small at high energies ( > $10^{12}$ ev).

The solar wind blows radially outward from the sun with a speed of several times the sound speed, slowing down only when the ram pressure decreases to the back pressure of the interstellar medium, at some 50 to 200 A. U.  This occurs at a shock transition in which the velocity decreases by about a factor of four (corresponding to a strong shock).  Beyond this standing, spherical shock wave it can be shown that the flow is approximately incompressible, and the gas gets swept around the heliosphere.

*R. G. Marsden (ed.), The Sun and the Heliosphere in Three Dimensions, 375–387.*
© *1986 by D. Reidel Publishing Company.*

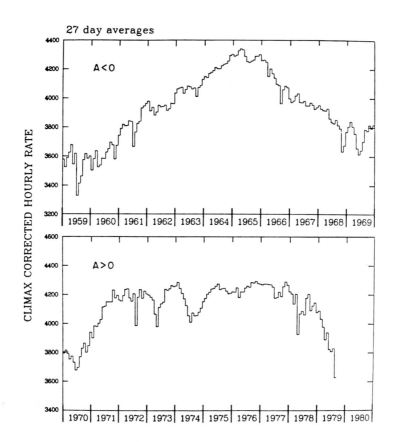

**Figure 1.** The counting rate of the Climax neutron monitor for two sunspot cycles (J. A. Simpson, private communication).

Within the heliosphere, the large-scale structure of the magnetic field has been clarified considerably by observations carried out on the Pioneer 10 spacecraft (see, e.g., Thomas and Smith, 1981), and the inferred relationship to observed coronal structure. During the years around each solar sunspot minimum, the field is generally organized into two hemispheres separated by a thin current sheet across which the field reverses direction. In each hemisphere the field is a Parker Archimedean spiral, with the sense of the field being outward in one hemisphere and inward in the other. At solar sunspot minimum, the current sheet is nearly equatorial. The field direction alternates with each 11-year sunspot cycle, so that during the 1975 sunspot minimum, the northern field was directed outward from the sun, but in 1965 the northern field pointed inward. Observations indicate that the inclination increases as a function of time away from sunspot minimum. The structure for the years near sunspot maximum is not simple. In the rest of the discussion, it will be assumed that the overall magnetic structure is given by this model, recognizing that it may not be a particularly good representation in the few years around maximum sunspot activity.


## 2.  The Model.

Available numerical simulations cannot handle the full complexity of the outer heliosphere.  In addition, the configuration beyond the solar-wind termination shock is uncertain.  For these reasons, a simpler model has been used in numerical simulations.

In this model, there is an outer, spherical boundary where the cosmic-ray distribution function takes on a specified "interstellar" value, which is taken to be a power law in total energy with an index corresponding to that observed at high energies.  Within this boundary the magnetic field configuration is as discussed above, and the cosmic rays random walk (diffuse) through the field, are convected with the solar wind, are adiabatically cooled, and drift through the large-scale magnetic field and current sheet.  Very recent computations, in which the shock in included and the interstellar spectrum is imposed at some boundary beyond the shock suggest that the neglect of the shock is a reasonable approximation for the bulk of the cosmic rays (Merenyi and Jokipii, 1985).

The motion of fast charged particles in the solar wind is determined by the (generally irregular) ambient magnetic field. Turbulence at the scale of the gyroradius scatters the particles in pitch angle and causes the distribution to relax toward isotropy.  If this scattering occurs quickly compared with respect to the macroscopic time variations, and if the scattering mean free path is small, the behavior of a distribution of particles can be described in the diffusion approximation (Parker, 1965, Axford, 1965; for reviews see Jokipii, 1971, Volk, 1975, or Fisk, 1978).  The theory then results in a diffusion-convection equation with energy change, which may be written as follows, with the various physical effects noted next to the corresponding terms:

$$\frac{\partial f}{\partial t} = \frac{\partial}{\partial x_i} K_{ij} \frac{\partial f}{\partial x_j} \qquad \text{(diffusion)}$$

$$- V_i \frac{\partial f}{\partial x_i} \qquad \text{(convection)} \qquad (1)$$

$$+ \text{div}(\underline{V}) \frac{\partial f}{\partial \ln(p)} \qquad \text{(adiabatic energy change)}$$

$$- V_{d,i} \frac{\partial f}{\partial x_i} \qquad \text{(drift)}$$

$$+ Q(\underline{r}, p, t) \qquad \text{(possible local source)}$$

where f($\underline{r}$, p, t) is the omnidirectional particle distribution as a
function of position vector $\underline{r}$, momentum magnitude p, and time t.  V is
the background convection velocity, $K_{ij}$ is the diffusion tensor which
may be expressed in terms of the spectrum of magnetic irregularities,
and Q is the local source, if any.  The drift velocity $V_d$ corresponds
to the gradient and curvature drifts in the large-scale magnetic field
B and for the isotropic distribution is $\underline{V}_d$ = (pc/3q) curl($\underline{B}/B^2$).

## 3. Numerical Solutions for Solar Modulation of Galactic Cosmic Rays.

Equation (1) may be applied to the modulation of the galactic cosmic
rays by the sun.  Here the solar system is regarded as being immersed
in a constant, isotropic bath of galactic cosmic rays.  In this case
the source Q = 0 and modulation may be regarded as a balance between
the inward random walk or gradient and curvature drifts caused by the
large-scale structure of the interplanetary magnetic field, and the
adiabatic cooling due to the radial expansion of the solar wind.
    The transport parameters used in our model configuration are as
follows:

**Wind Velocity V**:  Observed to average about 400 km/sec, independent
of distance from the sun.

**Diffusion Coefficient $K_{ij}$**: Not very well known.  Quasilinear theory
applied to the observed power spectrum of magnetic irregularities
gives K $\propto$ $K_0$ $p^{1/2}$ ß, where p is particle momentum and ß is the
ratio of particle speed w to the speed of light c (Jokipii, 1971).
This is taken to scale as the inverse of the magnetic field to give
a constant ratio of mean free path to gyroradius.

**Drift Velocity $V_d$**: Readily evaluated for Parker spiral field.  For
particles of charge Q one gets (Jokipii, Levy and Hubbard, 1977):

$$\underline{V}_d = \frac{2\,p\,c\,w\,r}{3\,Q\,A\,(1 + \Gamma^2)}\left\{ - \frac{r}{\tan(\theta)}\,\hat{e}_r + (2 + \Gamma^2)\,\hat{e}_\theta \right\}$$

where $\Gamma = \frac{r\,\Omega\,\sin(\theta)}{V}$ , and $\underline{B} = \frac{A}{r^2}[\,\hat{e}_r - \Gamma\,\hat{e}_\theta\,]$

Here A is a constant, $\Omega$ is the rotation angular velocity of the sun.
There is a delta-function at the current, where $\underline{V}_d$ changes sign.  If
the field is out in the northern hemisphere (A>0), the particles drift
in from the poles and out along the current sheet, and if the field is
in (A<0), the particles come in along the current sheet.
    We may examine the order of magnitude of the various terms in
equation (1) for roughly 1 GeV protons by setting the gradient of the
cosmic-ray intensity equal to the observed 5%/A.U., the wind velocity
equal to 400 km/sec, the drift velocity $\sim 10^8$ cm/sec, and the diffusion
coefficient $10^{22}$ cm /sec, we get in order of magnitude:

| Diffusion | | $3 \times 10^{-7}$ sec$^{-1}$ |
|---|---|---|
| Convection | | $3 \times 10^{-7}$ sec$^{-1}$ |
| Energy loss | | $1 \times 10^{-6}$ sec$^{-1}$ |
| Drift | | $1 \times 10^{-6}$ sec$^{-1}$ |

It is clear, then, that the various terms in equation (1) all contribute significantly to the equilibrium modulation, but that for the standard Parker Archimedean spiral magnetic field, the drift and energy-change terms seem to be somewhat larger. This suggests that drift effects should be important in solar modulation, and the numerical solutions discussed below support this.

If we are interested in the eleven-year sunspot cycle, we may estimate the time derivative as 1/11 years ($1 \times 10^{-8}$ sec$^{-1}$), which suggests that if the modulation is caused by the large scale structure, we may neglect the time derivative, and work with a quasi-time-independent model.

We have developed a series of numerical codes for solving equation (1). The most sophisticated of these solves the full, three-dimensional model with a heliospheric current sheet of essentially arbitrary shape, and with an arbitrary magnetic field. Hence a large variety of problems can be studied using this code. The only significant limitation to this model is that the solar wind velocity is assumed to be constant, independent of radius and heliographic latitude. The basic solutions were published and discussed in Kota and Jokipii (1982a, b).

Figure 2 illustrates the general nature of the solutions which have been obtained for a wide range of parameters. The calculated energy spectra are in excellent agreement with the observed spectrum. We have found that the general nature of the solutions is relatively insensitive to the values of the parameters, within relatively broad ranges.

A remarkable feature is an unexpectedly strong dependence of the solutions on the large-scale magnetic field. This is clearly seen in the contour plots shown in figure 2, where the contours of equal intensity are clearly influenced significantly by the sign of the magnetic field and the position of the current sheet. In addition, the solutions for qA>0 (northern hemisphere field out) differ from those obtained for the opposite sign of the magnetic field. This gives rise to a number of predicted 22-year solar magnetic cycle effects. The dependence on the sign of the magnetic field is another unique aspect of the drift term, and should be utilized in carrying out observational tests.

The solutions also show that, even for relatively large values of the diffusion coefficient, the drifts are quite effective in determining the paths through which particles enter the 'inner solar system. If the northern hemisphere magnetic field is outward (as in the 1975 solar minimum), positive particles are found to enter predominantly via the poles of the heliosphere and exit along the interplanetary current sheet, whereas if the northern field is negative, the particles enter along the current sheet, from the equatorial regions of the boundary, and exit via the poles.

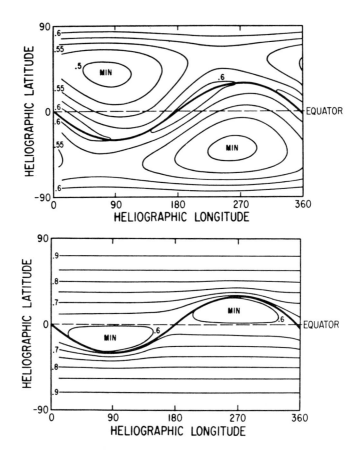

**Figure 2.** Contour plots of the intensity of 1.6 GeV protons as a function of latitude and longitude at a heliocentric radius of 1 A.U., for A<0 (top) and A>0 (bottom) (Kota and Jokipii, 1982a)

4.   Comparison of Model Calculations with Observations.

Consider next recent observations which focus on the effects of magnetic structure on the modulated cosmic-ray intensity.   These are of two general kinds - the organization of the intensity relative to the current sheet and 22-year, solar magnetic cycle effects.   Both effects are a consequence of the drift terms and would be absent without them.

4.1.   Current Sheet Effects.

As was shown in figure 2, the magnetic field organizes the cosmic-ray intensity relative to the heliospheric current sheet.   The observational analyses carried out by Newkirk and Lockwood (1981), Newkirk and Fisk (1985a) and Newkirk, Asbridge and Lockwood (1985)

attacked this problem statistically, using a large data set, in order
to minimize transient, time-dependent effects.   They have studied the
statistical dependence of the intensity of the Mt. Washington neutron
monitor (which responds to protons of about 5.3 GeV energy) on the
distance from the current sheet, for data obtained in the years
1973-1978, around the last solar sunspot minimum when the magnetic
field in the northern solar hemisphere was directed outward from the
sun.   They determined the position of the current sheet from coronal
white light data. Their figure (9) shows a scatter plot of the daily
intensity vs. the heliomagnetic latitude at the point of observation
(defined as the angular distance in degrees along a meridian to the
current sheet).   The data show considerable scatter, which is to be
expected since effects of transient disturbances, etc. may be expected
to disrupt the pattern.   Nontheless, there is a clear trend in the
data, and the solid line gives the best fit of the function

$$I = a_0 + a_1 \sin^2(\lambda_{mg}) \qquad (2)$$

to the data, where $\lambda_{mg}$ is heliomagnetic latitude.   They find that $a_0$
=2404 and $a_1$ = -112 $\pm$ 10.   In a subsequent paper, Newkirk, Asbridge and
Lockwood examined the energy dependence of the ratio $a_1/a_0$ by analyzing
data from a variety of other sources.   These data show that the effect
depends on energy approximately as $T^{-0.8}$.

In order to compare these data with the predictions of the
model, it is important to simulate the methods used to analyse the data
as accurately as possible, because otherwise other effects may obscure
the one desired. The following procedure was used in obtaining the
model predictions:

We note that the inclination of the current sheet varied
considerably over the time period spanned by the data set, so it would
not be appropriate to use only one simulation, with one current sheet
inclination.   Second, the orbit of the Earth carries the observer on a
trajectory which oscillates sinusoidally seven degrees above and below
the heliographic equator.   Since the maximum value of $\lambda$ mg in the data
set shown in figure (9) of Newkirk and Fisk is approximately 55°, a
maximum current sheet inclination of 45° would give a maximum $\lambda_{mg}$ of
52°.   The work of Jokipii and Thomas (1981) suggests a minimum
inclination of the order of 15° Hence, to simulate the varying
inclination of the current sheet, three different inclinations, 15, 30,
and 45 degrees were computed for each parameter set.   To simulate the
motion of the Earth in its orbit, the computed intensity was determined
at 60 different points, which were spaced in heliographic latitude and
longitude just as the Earth is in its orbit.   For each such point, in
addition to the computed intensity, j, the heliomagnetic latitude $\lambda_{mg}$ as
defined by Newkirk and Fisk was calculated.   The procedure was then
averaged over 20 different, equally spaced phases of the Earth's orbit
relative to the longitude of maximum inclination.

The resulting set of points $(j, \lambda_{mg})$, summed over three
inclinations and along the Earth's orbit is illustrated in figure (3a)
for the parameters shown in figure 3.   Note that there is considerable
scatter in the points, reflecting the fact that the computed intensity

depends on other parameters as well as $\lambda_{mg}$.  Nontheless, the
calculated values show a trend toward decreasing intensity as $\lambda_{mg}$
increases, similar to that found in the data by Newkirk and Fisk.
Figures (3 b, c, d)  show the scatter plot obtained for the individual
inclinations.  The general depression of the intensity as the
inclination of the current sheet increases, noted previously (Jokipii
and Thomas, 1981, and Kota and Jokipii, 1982), is evident.

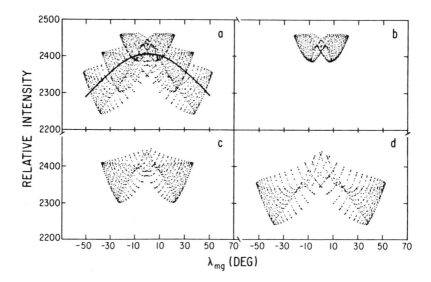

**Figure 3.**  Scatter plots of model calulations of intensity vs.$\lambda_{mg}$,
using an ensemble of current-sheet inclinations and following
the Earth in its orbit (Jokipii and Kota, 1985).

     To compare the computations quantitatively with the data, a
least squares fit of equation (2) to the synthesized scatter plot was
made to obtain the "theoretical value" of the coefficients $a_0$ and $a_1$
for each given set of parameters.  This procedure was repeated for each
set of parameters (diffusion coefficient normalization, energy, etc).
Since the normalization of the intensity at the boundary is arbitrary,
the absolute values are irrelevant, so the ratio $a_1/a_0$ is the most
relevant quantity.  This ratio was computed from the model calculations
for a range of diffusion-coefficient normalizations for particles of
energy 2.36 GeV. There is a broad range of diffusion coefficients for
which the computed value of $a_1/a_0$ is quite close to the value for this
energy given by Newkirk, Asbridge and Lockwood (1985).  We have carried
out this same calculation for outer boundary radii of 10 and 30 A. U.,
and find that the results are not very sensitive to this parameter.
Finally, we have determined the energy dependence of the ratio $a_1/a_0$
and find that although the calculated value does indeed decrease with
increasing particle energy, and agrees well with the data at a few GeV
energy, the functional dependence is not the simple power law seen in

the data.  We expect that changing the energy dependence of the
diffusion tensor could improve agreement here, but have not been able
to verify this.  In general, the extent of the agreement of this
simulation with the data must be regarded as a success of the model.

A second observation related to the organization of the intensity
by the interplanetary current sheet was published by Smith and Thomas
(1985).  They compared the inclination of the current sheet (or,
equivalently, its maximum excursion from the solar equator) as
determined from coronal measurements, with the Deep River neutron
monitor counting rate.  A simple plot of the two quantities vs. time
shows convincingly that periods of time during which the current sheet
had a large inclination (or excursion) relative to the heliographic
equator were associated with low counting rates, as predicted in the
theory.  Moreover, the theoretical calculations show that the effect of
the current-sheet inclination should be greater during sunspot cycles
in which the northern-hemisphere magnetic field is inward.  This is
clearly shown in the larger variation found for the intensity for a
given variation in the current sheet inclination.  This latter aspect
of this data seems to be a particularly strong element in support of
the theory since it seems to indicate a dependence on the sense of the
interplanetary magnetic field.

4.2. 22-year Solar Magnetic Cycle Effects.

As noted above, the theory predicts a number of effects related to
the change of sign of the solar magnetic field as one goes from one
sunspot cycle to the next.  A recently observed example of this was
discussed in the preceding paragraph.

The first effect of this kind was noted by Jokipii and Thomas
(1981) and confirmed in a more realistic calculation by Kota and
Jokipii (1982a).  They constructed a model 22-year solar magnetic cycle
by varying only sign of the northern magnetic field (which alternates
from one sunspot cycle to the next) and the tilt of the interplanetary
current sheet.  The years around solar maximum were not modelled
because of the expectation that the magnetic field geometry is much
more complicated then and thus cannot be represented by the model.  The
tilt of the current sheet and the sign of the field were the only
parameters allowed to change.  The time dependence of the computed
intensity of  GeV protons at the orbit of Earth and the observed
intensity over the last two sunspot cycles is striking.  The general
nature of the effect may be understood as a consequence of the fact
that during the 1975 solar minimum, the access of positive particles
was over the heliospheric poles, and hence relatively insensitive to
the tilt or warp of the current sheet.  On the other hand, during the
1965 solar sunspot minimum, the access of particles was along the
current sheet, and hence sensitive to its warp.

The time variation of electrons is more difficult to measure
than protons.  Available observations presented by Garcia-Munoz,
Evenson and Simpson (1985) indicate that while there is a clear
difference between the positively and negatively charged particles,
the electrons do **not** show the opposite behavior to the protons that

might be expected on the basis of the above discussion.  Evenson and
Meyer (1984) also reported that the electrons are recovering
significantly more rapidly than the protons as we go into the next
sunspot minimum, as expected. A full discussion of the expected
electron modulation has not yet been carried out.

Another significant observed effect which finds a ready and
natural explanation in terms of this model and which supports the
importance of drift effects is a careful study of the correlation of
the geomagnetic aa index with the counting rate of the Mt. Washington
neutron monitor by Shea and Smart (1980, 1985).  In the most-recent
analysis they found that the correlation was quite significant and
negative (-.86) during the years around the 1965 solar minimum, but
that it was much smaller and insignificant ( <.16) during the 1975
solar minimum.  As suggested by Shea and Smart and later confirmed in
model calculations by Jokipii (1981), this observation is a natural
consequence of the drift effects, if one takes the aa index to be a
measure of disturbances in the solar wind at the Earth, but not at
high heliographic latitudes. For if the particles come into the inner
solar system along the current sheet (as they did in the 1965 solar
sunspot minimum), then one expects disturbances in the ecliptic plane
to have a significant effect on the cosmic rays. In fact, disturbances
should impede cosmic-ray access and lead to a negative correlation, as
is observed.  On the other hand, if the access is from the poles, one
expects little effect of ecliptic perturbations, again as is observed.

## 4.3. The anomalous component and Modulation.

Pesses, Jokipii and Eichler (1981) pointed out that many features of
the anomalous component could be explained in terms of the acceleration
of freshly-ionized interstellar particles at the termination shock of
the solar wind, although a detailed model was not worked out.
Jokipii(1985) has presented results from detailed two-dimensional
(radius and latitude) models in which the full transport equation was
solved.

The acceleration of charged particles at the termination shock
depends on the structure of the wind and magnetic field beyond the
shock. Since the main emphasis here is on fast charged particle
transport, a crude representation of the velocity and magnetic field
beyond the shock was used.  The flow velocity was assumed to drop
suddenly by a the shock ratio r, (expected to be nearly 4.0) at the
shock, and then to continue to flow radially, varying as the inverse
square of the radius (required by the incompressibility approximation)
beyond the shock. This then determines the magnetic-field structure.
The transport of the charged particles is computed in this
configuration out to some outer absorbing boundary beyond the shock
which represents crudely the loss of particles to the interstellar
medium.  note that the discontinuity in the flow velocity at the shock
produces a discontinuous magnetic field as well. This implies that the
adiabatic energy change term in equation(1) which is proportional to
div(V) will have a delta function singularity, as will the drift
velocity.

As pointed out by Pesses etal (1981) the termination shock is essentially a perpendicular shock everywhere but near the poles, where it is quasiparallel.  The injection of low-energy particles is preferred there.  Although it would be desirable to start the particles just above the thermal proton energies, and follow them to several tens of MeV, this is impractical because of numerical constraints.  Instead, particles were followed for 2 or 3 decades in energy below about 100 Mev per nucleon, to cover the energies observed.  The particles chosen are singly-ionized helium and oxygen nuclei, following the initial suggestion of Fisk, etal (1974) that the particles are freshly-ionized interstellar neutrals.

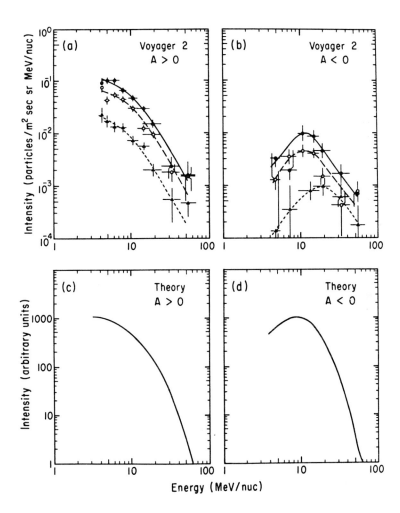

**Figure 4.**   Comparison of theoretical predictions and observations for the energy spectrum of anomalous oxygen for different signs of the interplanetary magnetic field.

   The energy spectra calculated near the heliospheric equatorial
plane, at 1 A. U. show a peak in intensity at an energy in the range
1-10 MeV, and it is found that the energy of the peak for A<0 is
approximately a factor of 3-5 higher than for A>0.  Figure ( 4 ) shows
this shift of the energy spectrum for singly-charged oxygen nuclei
injected at 0. 5 MeV/nucleon at polar angles θ < 5°, for the two signs
of the interplanetary magnetic field, compared with Voyager
observations of Cummings, et al.  The agreement between observation and
theory is clearly striking.  This persists over the entire range of
parameters explored.  The physical cause of the effect is apparently
the close association between drift and energy gain at the termination
shock.  The sense of drift at the shock is such that, for qA<0, the
particles drift to larger values of the polar angle   as they gain
energy.  Since they are injected at the poles, they therefore can gain
considerable energy as they move toward the equator where they have
easy access to the inner solar system via drift along the current
sheet.  On the other hand, for qA >0, the particles start out at the
poles, and hence do not have as much opportunity to gain energy before
propagating to the inner solar system.  In addition, this feature of
the model can perhaps explain observed features of the anomalous
component. This shift of the spectral peak may explain why anomalous
helium was not seen during the 1965 sunspot minimum, since the shift
may bring the helium into an energy region where it is hidden by the
normal galactic cosmic rays.
   Thus, there are a number of apparent 22-year solar magnetic
cycle effects which tend to support the importance of the drift motions
of the cosmic rays. However, the definite determination of such effects
must await analysis of more sunspot cycles.


5. Summary and Conclusions.

This paper has summarized the basic physics of the transport of
galactic cosmic rays in the heliosphere and explored its consequences
for solar modulation of galactic cosmic rays.  Consistent application
of the physics, together with a straightforward extrapolation of
observations of the interplanetary magnetic field and solar wind to the
polar regions of the heliosphere leads to the conclusion that particle
drifts must play a dominant role in transport.  Two main kinds of
effects have been studied - the organization of the intensity by the
interplanetary current sheet and 22-year solar-magnetic-cycle effects.

   Currently available observations show broad agreement with both
kinds of predictions of the theory, although the electron time
variation and the radial gradients remain as problem areas.
Particularly encouraging are several observations which are quite
naturally explained by the model, but which require special assumptions
if drifts are neglected.
   At present the chief uncertainty in the model calculations is the
unknown magnetic-field and solar-wind structure in the polar regions of

the heliosphere.   Observations of the magnetic field and solar wind carried out on the Ulysses spacecraft at high heliographic latitudes should make it possible to determine the accuracy of the extrapolations.   In addition, measurements of galactic cosmic rays and the anomalous component will provide checks of the theory by determining the latitude variation of the cosmic-ray intensity.

## Acknowledgements

This work was supported, in part, by the National Aeronautics and Space Administration under Grant NSG-7101 and by the National Science Foundation under Grant ATM-220-18.

## REFERENCES

Axford, W. I.: 1965, _Planet. Space Sci._, _13_, 115.
Cummings, A. C., Stone, E. C. and Webber, W. R.: 1985, to be published in _J. Geophys. Res._
Evenson, P. and Meyer, P.: 1985, Private communication.
Fisk, L. A.: 1978, in _Space Plasma Physics_, ed by Kennel, Lanzerotti and Parker.
Fisk, L. A., Kozlovsky, B., and Ramaty, R.: 1974, _Astrophys. J._, _190_, L35.
Garcia-Munoz, M., Evenson, P. and Simpson J. A.: 1985, private communication.
Gleeson, L. J. and Axford, W. I.: 1967, _Astrophys. J._, _149_, L115.
Jokipii, J. R.: 1971, _Rev. Geophys. Sp. Phys._, _9_, 27.
Jokipii, J. R.: 1981, _Geophys. Res. Lett._, _8_, 837.
Jokipii, J. R. and Thomas, B.: 1981, _Astrophys. J._, _243_, 1115.
Jokipii, J. R. and Kota, J.: 1985, _J. Geophys. Res._, submitted.
Jokipii, J. R., Levy, E. H. and Hubbard, W. B.: 1977, _Astrophys. J._, _213_, 861.
Kota, J. and Jokipii, J. R.: 1982a, _Astrophys. J._, _265_, 573.
--- : 1982b, _Geophys. Res. Lett._, _9_, 656.
Merenyi, E. and Jokipii, J. R.: 1985, in preparation.
Newkirk, G., Asbridge, John, and Lockwood, John A.: 1985, _J. Geophys. Res._, to be published.
Newkirk, G. and Lockwood, J. A.: 1981, _Geophys. Res. Lett._, _8_, 619.
Newkirk, G., and Fisk, L. A.: 1984, _J. Geophys. Res._, in press.
Parker, E. N.: 1965, _Planet. Space Sci._, _13_, 9.
Pesses, M., Jokipii, J. R., and Eichler, D.: 1981, _Astrophys. J._, _246_, L85.
Shea, M. A. and Smart, D. F.: 1981, _Advances in Space Res._, Vol. _1_, #3, 147.
--- : 1985, _Proceedings 19th Int. Cosmic Ray Conf._, La Jolla.
Smith, E. J. and Thomas: 1985, B., _J. Geophys. Res._, in press.
Volk, H. J.: 1976, _Rev. Geophys, and Sp. Phys._, _13_, 547.

MEASUREMENT OF RADIAL AND LATITUDINAL GRADIENTS OF COSMIC RAY INTENSITY
DURING THE DECREASING PHASE OF SUNSPOT CYCLE 21

D. Venkatesan[1,2], R. B. Decker[1] and S. M. Krimigis[1]

[1]Applied Physics Laboratory
The Johns Hopkins University
Laurel, Maryland 20707 USA

[2]The University of Calgary
Calgary, Alberta, T2N 1N4, Canada

ABSTRACT.    This  study  over  a  4-year  period  (1981-84)  during  the
decreasing  phase  of  sunspot  cycle  21  utilizes  data  from  spacecraft
Voyagers 1 and 2 and the earth-orbiting satellite IMP-8 to estimate the
radial  and  latitudinal  cosmic  ray  intensity  gradients.    The  detector
threshold  on  the  Voyagers  is  $\geqslant$  70  MeV/nuc  and  that  on  IMP-8  $\geqslant$  35
MeV/nuc.  Note that the contribution to the intensities between 35 and
70 MeV being only $\leqslant$ 0.1 % of the intensity integral > 70 MeV, is not
significant; thus the detector thresholds are considered to be identical
for  the  present  study.    The  position  of  Voyager  1  (2)  at  the  beginning
of  1981  is  given  by  r  $\approx$  10  AU  (8AU),  $\lambda$  $\approx$  -5°  (-5°),  and  that  at  the  end
of  1984  is  given  by  r  $\approx$  22  AU  (16  AU),  $\lambda$  =  +25°  (0°)  where  r  is  the
radial distance and $\lambda$ is the heliolatitude.  The heliolongitude separation between the two spacecraft changed from ~ 4° to ~ 26° during the
interval.    A  comparison  of  26-day  means  of  the  cosmic  ray  intensity
registered at the two Voyagers indicates that the data is consistent on
the  average  with  a  long-term  zero  latitudinal  gradient.    The  principal
result obtained from a comparison of Voyager data with IMP-8 (at 1 AU)
is  that  the  radial  gradient  over  this  period  decreased  at  the  rate
of ~ 0.4% per AU per year, reaching a value of ~ 2.0%/AU between 1 and
16 AU and a value of ~ 0.6%/AU between 16 and 22 AU.  If this decrease
continues  through  1987,  both  Voyagers  would  be  located  outside  the
principal cosmic ray modulation region at solar minimum at $\lesssim$ 22 AU.

I.    INTRODUCTION

Measurements  of  galactic  cosmic  ray  intensity  gradients  provide  vital
input to the formulation of theories connected with cosmic ray transport
in the heliosphere.  Most interplanetary observations in the past have
been  restricted  to  the  ecliptic  plane;  so  the  emphasis  has  been  on  the
radial gradient.  A unique opportunity for determining the heliolatitude
gradient  over  a  long  term  interval  became  possible  in  early  1981,  when
Voyager  1  took  off  towards  high  latitudes;  the  latitude  separation

389

R. G. Marsden (ed.), The Sun and the Heliosphere in Three Dimensions, 389–394.

between the two Voyagers reached a value of ~ 25° by the end of 1984. We use the terminology 'differential gradient' for inter-Voyager determinations and 'integral gradient' for those derived between IMP-8 and either of the two Voyagers.

Determinations of the integral radial gradient by several groups (for a summary, see Venkatesan et al., 1985) using data from Pioneers 10 and 11, Voyagers 1 and 2, and IMP-8 revealed relatively small values of 2-4% per AU. Decker et al. (1984) have estimated for the first time the heliolatitudinal gradient of galactic cosmic ray intensity during early-1981 through mid-1982. This period coincided with the initial phase of cosmic ray recovery following the minimum reached ~ late 1980. They have shown that the heliolatitudinal gradient to be 0%/deg over the 16° heliolatitudinal separation in the region 8-13 AU during that period of study, on the reasonable assumption that the radial gradient during 1981-82 cosmic ray intensity recovery period remained at the same ~ 2-4%/AU level as that determined from multi-spacecraft measurements during the 1977-80 period of cosmic ray intensity decline. We investigate here the cosmic ray radial and latitudinal gradients during the four year interval 1981-84.

## 2.    EXPERIMENT

A heavily shielded solid-state detector on the Voyager 1 and 2 Low Energy Charged Particle (LECP) experiment (Krimigis et al., 1977) provides a cosmic-ray channel responding to protons integral above 70 MeV. For this study, 26-day means of 1-hour count rates are computed to average over one solar rotation at the (essentially longitude-stationary) Voyagers. Total counts during a 26 day period are typically $\geqslant 2 \times 10^4$, yielding a statistical error of $\leqslant 0.7\%$. Contributions of high-energy particles from large solar flares to both Voyagers as well as Saturn encounter data of Voyager 2 during August 1981 have been removed. For details of data analysis, refer to Venkatesan et al. (1985).

## 3.    OBSERVATIONS

### 3.1.  Differential Gradient

At the top of Figure 1 we have shown a plot of the 26 day means of the cosmic ray intensity registered by both Voyagers. The Voyager 1 data has been shifted to the position of Voyager 2 using an average value of 500 km/sec for the propagation (solar wind) speed. The convection of cosmic ray features and the appropriateness of such a solar wind speed has been discussed in detail by Venkatesan et al. (1984, 1985). The close correspondence in the two intensity-time profiles is seen. In the middle of Figure 1, we have plotted the values of the radial gradient $G_r$ given by $G_r = [(\ln (R_1/R_2))/\Delta r] \times 100\%$, where $R_1$ and $R_2$ are the counting rates of Voyagers 1 and 2, and $\Delta r$ is the radial distance separation. These are point-by-point computations based on the assumption that the difference in intensity is entirely due to a radial gradient. The points are fitted with the line of least squares; note that $G_r$ decreased from ~ 2.1%/AU to ~ 0.6%/AU over the 1981-1984 interval giving a value for $\dot{G}_r$ of -0.38 ($\pm$ 0.09)%/AU/year. At the bottom of Figure 1, the values of the heliolatitudinal gradient $G_\psi$ are plotted, determined from the relation $G_\psi = [(\ln(R_1/R_2))/\Delta\psi] \times 100\%$ where $\Delta\psi$ is the latitudinal difference between the two spacecraft. Again, these are point-by-point computations based on the assumption that the observed difference in intensity is entirely due to a latitudinal gradient. The points are fitted with the line of least squares; note that $G_\psi$ also decreased from ~ 0.42%/deg to ~ 0.13%/deg over the four year period giving a value for $\dot{G}_\psi$ of -0.06 ($\pm$ 0.02)%/deg/year.

## 3.2 Integral Gradient

Figure 2 at the top shows the intensity time profiles for IMP-8 and the two Voyagers. The latter data were shifted to 1 AU to compensate for convection of cosmic ray features at the speed of 500 km/sec. Note the similarity among the three profiles.

The middle and lower part of Figure 2 show the integral radial gradient determinations from the two pairs, Voyager 1-IMP-8 and Voyager 2-IMP-8. Again these are point-by-point computations. The values of $G_r$ decreased from ~ 3.5%/AU to ~ 1.5%/AU and from ~ 3.8%/AU to ~ 2.0%/AU, respectively; the decreases in the radial gradient are given by $\dot{G}_r = -0.48$ ($\pm$ 0.5)%/AU/year and $\dot{G}_r = -0.43$ ($\pm$ 0.07)%/AU/year respectively.

## 4. DISCUSSION AND CONCLUSIONS

The following points can be made.

### 4.1

The value of 500 km/sec for the optimum solar wind speed to correct the data for convection of cosmic ray features to Voyager 1 and 2 in the heliospheric region ~ 8 AU to ~ 22 AU during 1981-84 is appropriate since the two intensity-time profiles at the top of Figure 1 agree fairly well.

### 4.2

The similarity in the long-term trend of $G_r$ and $G_\psi$ in Figure 1 is indicative of the fact that at least up to a radial distance of ~ 22 AU and heliolatitude of 25°, the heliosphere responds to cosmic ray modulation rather similarly. This is not surprising since it is clear from an earlier study over the period 1981-mid-1982 (Decker et al., 1984) that no significant long-term latitudinal gradient exists. Furthermore, as seen in Figure 2, the integral radial gradient obtained from the two pairs IMP-8-Voyager 1 and IMP-8-Voyager 2 are comparable, even though Voyager 1 had reached a heliolatitude of 25° by the end of 1984. It is quite obvious, therefore, that on a long-term basis, there is no

significant latitude gradient.    Note that the radial gradient $G_r$
of ~ 2.0%/AU is evident from other multispacecraft studies as well.
These results are not inconsistent with the observations of short-lived
(10-30 days) latitudinal gradients by Roelof et al. (1983).

4.3

A decrease in the cosmic ray intensity radial gradient as a function of
increasing radial distance and later epoch is clearly seen.    Note for
example, at the end of 1984, Voyagers 1 and 2 at radial distances of 16
and 22 AU respectively.    Comparing the middle part of Figure 1 with the
bottom portion of Figure 2, we find that the radial gradient $G_r$ in the
region 1-16 AU (IMP-8-Voyager 2 pair) is ~ 2%/AU and that in the region
16-22 AU (Voyager 2-Voyager 1 pair) is ~ 0.6%/AU, a distinctly lower
value.    This clearly demonstrates that a comparison of the integral
radial gradient over larger distances and at the same time over a
significant fraction of the solar cycle averages out and masks the
changes in value of the radial gradient.    A value of say 2.05%/AU for
the radial gradient over a complete solar cycle and out to a heliocentric distance of over 32 AU (Van Allen and Randall, 1985) could very
well consist of the mean of varying values of the radial gradient $G_r$
over different regions at different times including even a 0.0%/AU
gradient in the outermost region at the time of the approaching solar
minimum.    It is appropriate to mention that McKibben et al. (1985a) have
also observed a similar decrease in the radial gradient for low energy
cosmic rays (10 < E < 70 MeV) but not in the relativisitic cosmic ray
nuclei during 1981-83 (McKibben et al., 1985b).
    An independent check of measurements of the gradient at different
radial distances can be obtained from the cosmic ray transport equations
which enable the evaluation of the cosmic ray radial gradient from the
measurement of north-south anisotropy.    This technique has been
elaborated on by Yasue (1980).    Bieber and Pomerantz (1985), from a
study of cosmic ray north-south anisotropy over the 22-year interval
1961-1983, using neutron monitor data from Thule, Greenland and McMurdo,
Antarctica, obtain an average cosmic ray radial gradient of 1.7%/AU at a
rigidity of 10 GV; furthermore, they suggest that the radial gradient
(as observed at 1 AU) actually oscillates between values of 0.8%/AU and
2.6%/AU.    The decrease in the cosmic ray radial gradient at ~ solar
minima (1963-64, and 1974) is clearly seen from their study.

4.4

If the trend observed in $G_r$ over 1981-84 presists at the present rate,
the possibility of reaching a value of 0.0%/AU in the future (towards
solar minimum, ~ 1987) cannot be ruled out, at which time Voyager 1 and
2 would be respectively at distances of 29.1 AU and 21.7 AU.    $G_r$ =
0.0%/AU means that the intensity of cosmic rays is the same at both
places, i.e., all modulation at solar minimum may be taking place
inside ~ 22 AU.    Note that this is the first time that differential
measurements of the gradient are made in the outer heliosphere at solar

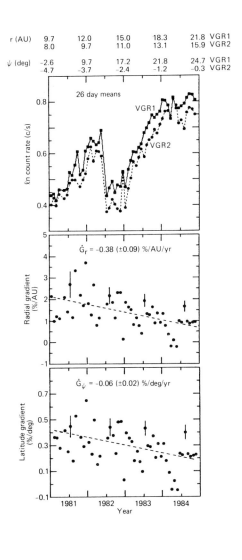

Figure 1

Intensity-time profiles of cosmic ray intensity at Voyagers 1 and 2; Voyager 1 data has been shifted to position of Voyager 2 using a solar wind speed of 500 km/sec. The middle and bottom portions provide the radial and heliolatitudinal gradients between the Voyagers.

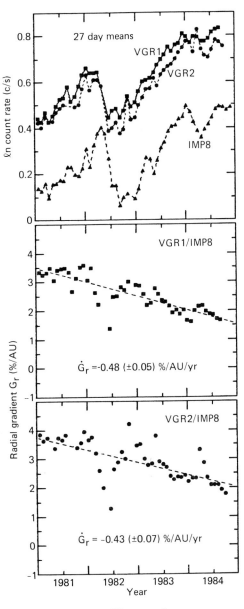

Figure 2

Intensity-time profiles of cosmic ray intensity at the Voyagers duly shifted to 1 AU using a solar wind speed of 500 km/sec and IMP 8; the radial gradients are between Voyager 1 and IMP 8 and Voyager 2 and IMP 8.

minimum.  As is evident from Figure 2, it is not possible to reach such a conclusion on the basis of comparison of intensities from a spacecraft at earth and a single spacecraft in the outer heliosphere.  Alternatively, it is possible that the low value of the radial gradient will persist through solar minimum, i.e., the Voyagers would still be inside the modulation region.

In summary, it should be pointed out that this is the first time radial gradient measurement is being reported during the recovery period of cosmic ray intensity.  Furthermore, it seems reasonable to conclude that the region of cosmic ray modulation shrinks or expands, possibly in step with the minimum and maximum of the sunspot cycle.

## 5.    ACKNOWLEDGEMENTS

This work was supported in part by NASA under Task I of Contract N000024-78-C-5384 and by NSF under Grant ATM-8305537.

## 6.    REFERENCES

Bieber, J. W. and Pomerantz, M. A.:    1985, Preprint, SH 4.2-6, 19th International Cosmic Ray Conference, Aug. 11-23.

Decker, R. B., Krimigis, S. M. and Venkatesan, D.:    1984, Ap. J., 278, L22.

Krimigis, S. M., Armstrong, T. P., Axford, W. I., Bostrom, C. O., Fan, C. Y., Gloeckler, G. and Lanzerotti, L. J.:    1977, Space Sci. Rev., 21, 329.

McDonald, F. B., Lal, N., Trainer, J. H., Van Hollebeke, M. A. I. and Webber, W. R.:    1981, Ap. J., 249, L71.

McKibben, R. B., Pyle, K. R. and Simpson, J. A.:    1985a, Ap. J., 289, L35.

McKibben, R. B., Pyle, K. R. and Simpson, J. A.:    1985b, 19th ESLAB Symposium, Les Diablerets, Switzerland, 4-6 June 1985, Abstract 2, Session F.

Roelof, E. C., Decker, R. B. and Krimigis, S. M.:    1983, J. Geophys. Res., 88, 9889.

Van Allen, J. A. and Randall, B. A.:    1985, J. Geophys. Res., 90, 1399.

Venkatesan, D., Decker, R. B. and Krimigis, S. M.:    1984, J. Geophys. Res., 89, 3735.

Venkatesan, D., Decker, R. B., Krimigis, S. M. and Van Allen, J. A.:    1985, J. Geophys. Res., 90, 2905.

Yasue, S.:    1980, J. Geomag. Geoelectr., 32, 617.

# NORTH/SOUTH ASYMMETRY IN SOLAR ACTIVITY AND ITS EFFECTS ON THE HIGH ENERGY COSMIC RAY DIURNAL VARIATION

M. A. Shea and D. F. Smart
Air Force Geophysics Laboratory
Hanscom AFB, Bedford, Massachusetts 01731, U.S.A.

D. B. Swinson
Department of Physics and Astronomy
The University of New Mexico
800 Yale N. E., Albuquerque, New Mexico 87131, U.S.A.

J. E. Humble
Department of Physics, University of Tasmania
Hobart, Tasmania 7001, Australia

ABSTRACT. Using a data base extending from 1955 through 1980 we find that the northern hemisphere of the sun was decidedly more active than the southern hemisphere for the period 1959 through 1970. From 1971 through 1980 there was no systematic asymmetry. Extending the time period through 1984, but using a different data base, we find an apparent change around October 1982, after which there are decidedly more flares occurring in the southern hemisphere of the sun than in the northern hemisphere. Examination of the diurnal variation amplitude present in the extremely high energy (approximately 100 GeV) cosmic radiation measurements indicates a consistent north/south asymmetry until about 1971 which we suggest is the result of the asymmetric solar activity.

## 1. SOLAR ACTIVITY ASYMMETRY

In studies of north-south asymmetry in solar activity, various researchers have used different phenomena as the measure of this asymmetry. See, for example, Dodson and Hedeman (1972), Waldmeier (1971), Bell (1961), Harvey and Bell (1968), and Roy (1977). We have assembled a data base extending through 1984 containing a number of different parameters including the comprehensive flare indices for "major" flares from 1955 through 1979 (Dodson and Hedeman, 1971; 1975; 1981) and additional unpublished comprehensive flare indices for 1980 (Dodson and Hedeman, private communication).

We determined the percentage of flares that occurred in the northern hemisphere each year by dividing the number of events in the northern hemisphere by the total number of flares that occurred in both

*R. G. Marsden (ed.), The Sun and the Heliosphere in Three Dimensions, 395–400.*
© *1986 by D. Reidel Publishing Company.*

hemispheres. For overlapping flares, the plage number was examined. If different plages were involved, the hemisphere of the location of each plage was counted separately. We included events for which no flare location was recorded; although optical measurements may not be available, the electromagnetic emissions of the flare are often recorded thereby indicating that a solar event has occurred. We have included these as "events with no known location" which generate the error bars illustrated in our Figure. These errors were evaluated in the following way. If we assume that all the unknown location flares were in the northern hemisphere, then we determine the absolute top limit of flares in the northern hemisphere by summing the number of known northern hemisphere flares with the unknown location flares and dividing this sum by the total of all flares on the sun for that year (i.e. northern, southern and unknown location flares). The absolute bottom limit was determined by assuming that all the unknown location flares were in the southern hemisphere. Thus the number of northern hemisphere flares was divided by the total number of all flares on the sun. Although it is unlikely that the unassigned flare locations would all be on one hemisphere or the other, this method does determine absolute limits of the maximum possible error. (Note that the location is known for all major flares that occurred in 1965; thus no error bars are shown.)

The results illustrated in Figure 1 show that more "major" flares occurred on the northern hemisphere of the sun than on the southern hemisphere from 1959 through 1970. For this 12-year period 77% of the flares occurred in the northern hemisphere; the maximum range of uncertainty is 68-80%. The results for the years 1955 through 1958 and 1971 through 1980 are mixed with no clearly defined preference for solar hemisphere. The shaded boxes in Figure 1 illustrates these results.

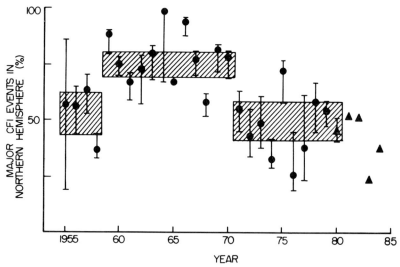

Figure 1. The percentage of "major" flares, as defined by the comprehensive flare index, that occurred in the northern hemisphere of the sun, 1955 - 1980. The triangles are preliminary data using solar flares from the six-station technique. See text for explanation of the errors.

TABLE 1

Comparison of Two Techniques for Determining the
Distribution of Solar Flares by Hemisphere

| Year | Percentage in the Northern Solar Hemisphere | |
| --- | --- | --- |
| | Grouped Flares | Six-Station Technique |
| 1979 | 58.0 | Not available |
| 1980 | 45.3 | 46.3 |
| 1981 | 51.6 | 51.9 |
| 1982 | Not available | 51.3 |
| 1983 | Not available | 23.4 |
| 1984 | Not available | 37.0 |

Comprehensive flare indices are not available for events after
1980. However, in a separate study, J. McKinnon of World Data Center A
(private communication) has been compiling statistics on the distribution of solar flares by hemisphere. Using a method based on the grouped
flare line that includes all solar optical observations, he has obtained
the results shown in the middle column of Table 1. To extend the data
base to more recent years, McKinnon developed an alternate technique
based on the solar flares reported by a set of six stations (Athens,
Holloman, Learmonth, Manila, Palehua and Ramey) located such that the
sun could be continuously monitored, local weather conditions and equip-
ment problems permitting. Using this method, and eliminating duplicate
reports, he has determined the statistics on the north/south asymmetry
as shown in the last column of Table 1. The similarity in results between these two methods for the two years of overlap gives a fair confidence in using the "Six Station Technique" to extrapolate the asymmetry
to more recent years. These points are shown as triangles in Figure 1.
    Combining the data from Figure 1 with the data in Table 1, it
appears that there was no dominant hemisphere for solar flares from 1971
through 1982; however, since 1982 there have been more solar flares in
the southern hemisphere of the sun than in the northern hemisphere.


2.  COSMIC RADIATION EFFECTS

    In the absence of spacecraft that can make extended observations of
cosmic ray intensity above and below the ecliptic plane, the only way to
observe cosmic ray gradients perpendicular to the ecliptic plane for
long periods of time is from ground-based cosmic ray detectors.
    There are two methods to detect a perpendicular cosmic ray density
gradient using cosmic ray detectors on the earth. One method is to use
data from cosmic ray neutron monitors which effectively sample the cosmic ray intensity at energies of a few GeV. At these energies the gyroradius of cosmic rays at 1 AU is only a few percent of an astronomical
unit, and so the cosmic ray detectors are effectively "viewing" in the
ecliptic plane. Since the earth has an excursion of ±7.25° in helio-

latitude during the year it is possible to look at variations in cosmic ray intensity in this small range of helio-latitudes. Using this method, Antonucci et al. (1978) found both annual and semi-annual waves in cosmic ray intensity which are dependent on the polarity configuration of the interplanetary magnetic field. Their results can be interpreted as indicating a north-south asymmetrical gradient pointing northward before the reversal of 1957-58 and after the reversal of 1969-71, and pointing southward between these two reversals.

A second method of observing a north-south asymmetrical gradient involves using higher energy cosmic rays so that detectors at various latitudes are truly looking away from the ecliptic plane. By using underground muon detectors whose mean energy of response is about 100 GeV, it is possible to sample directions which are truly above and below the ecliptic plane because the gyro-radius of these particles in the interplanetary medium is of the order of an astronomical unit. Employing these high energy measurements makes it possible to determine the contribution of the drift term $\overline{B} \times \overline{\nabla N}_p$ to the solar cosmic ray diurnal variation. ($\overline{B}$ is the interplanetary magnetic field vector, and $\overline{\nabla N}_p$ is the asymmetrical north-south cosmic ray density gradient.) This method has been described by Swinson (1970) and Hashim and Bercovitch (1972).

In a study of data from underground cosmic ray telescopes, Swinson et al. (1985) determined the yearly average solar diurnal variation for days when the interplanetary magnetic field was toward (T) and away (A) from the sun. The results for two stations are shown in Figure 2. The amplitudes determined for days when the magnetic field was away from the sun are joined by solid lines, and the amplitudes determined for days

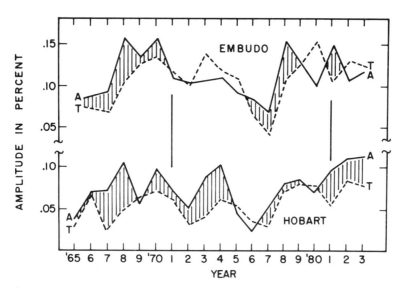

Figure 2. Yearly average amplitudes of the diurnal variations on away polarity days (solid line) and toward polarity days (dashed line) for underground muon telescopes at Embudo and Hobart, for 1965-1983. The shaded areas occur when the amplitude for the away polarity exceeds that for the toward polarity.

when the magnetic field was toward the sun are joined by dashed lines.
Shading occurs when the "away" amplitudes are greater than the "toward"
amplitudes, indicating a southward perpendicular gradient.  Prior to
1971 there is a definite predominance of "away" amplitudes over "toward"
amplitudes, which is consistent with a north-south asymmetrical gradient
pointing southward.  After 1971 the Embudo telescope, located in the
northern hemisphere, shows some slight but not statistically significant
dominance of "toward" amplitudes whereas the Hobart telescope, in the
southern hemisphere, shows little change from the pre-reversal situation.

      In a series of theoretical studies, Jokipii and his co-workers
(Kota and Jokipii, 1983 and references therein) have suggested the presence of a symmetrical north-south perpendicular gradient which changes
sign with the polarity reversal of the sun.  This symmetrical gradient
increased toward the poles after the 1969-1971 reversal and decreased
away from the solar equator for the opposite magnetic configuration.

      The high energy cosmic ray results could be explained if we assume
that the north-south symmetric gradient is a function of solar polar
reversal, but the north-south asymmetric gradient is a function of solar
activity and is possibly independent of the solar magnetic field polarity.  When the solar activity was high in the northern hemisphere, such
as between 1959-1970, there was both a strong southward pointing asymmetric cosmic ray gradient and an equatorward pointing symmetric cosmic
ray gradient as illustrated in Figure 3a.  For this time period the two
cosmic ray telescopes had increased diurnal variation amplitudes with the
interplanetary magnetic field away from the sun.  However, in 1971-1980
when there was no apparent north-south asymmetry in solar activity, the
asymmetric cosmic ray gradient essentially disappeared, leaving a poleward pointing symmetric cosmic ray gradient as illustrated in Figure 3b.

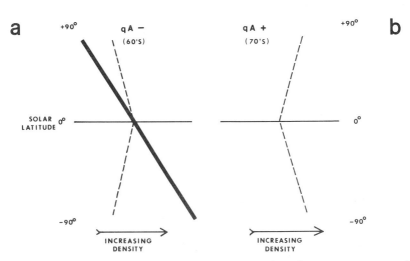

Figure 3.  Schematic representation of perpendicular cosmic ray density
gradients.  North-south symmetrical (dashed lines) and north-south asymmetrical (heavy solid line) gradients are shown for the two interplanetary magnetic configurations qA- and qA+.

Thus the telescope in the southern hemisphere still observed increased
diurnal variation amplitude with the interplanetary magnetic field away
from the sun, but the telescope in the northern hemisphere had the opposite measurement (i.e., increased diurnal variation amplitude with the
interplanetary magnetic field toward the sun). Furthermore, we suggest
that for the years studied, the asymmetric cosmic ray gradient present
from 1959-1971 as the result of strong solar activity in the northern
hemisphere of the sun was stronger than the symmetric gradient.

## 3.   CONCLUSIONS

From these results we conclude that from 1959 through 1970 there
was a strong predominance of solar activity from the northern hemisphere
of the sun, which disappeared from 1971 through 1982. Since 1982 there
appears to be a predominance of flares in the southern hemisphere; however, only time will tell if this is a trend that will last for several
years. It appears that strong solar activity concentrated in one hemisphere of the sun may be the source of an asymmetric gradient in the
high energy (median energy ~ 100 GeV) cosmic radiation as measured by
underground cosmic ray telescopes. Studies of the amplitude of the
diurnal variation of high energy cosmic radiation also suggest the presence of a symmetric density gradient which is weaker than the asymmetric
gradient established by asymmetric solar activity.

ACKNOWLEDGMENTS. The authors acknowledge the assistance of A. Bathurst
and D. Gaudette. DBS acknowledges support by the Atmospheric Sciences
Section, National Science Foundation, under grant ATM-8305098.

REFERENCES

Antonucci, E., D. Marocchi, and G. E. Perona: 1978, Astrophys. J., 220,
    712.
Bell, B.: 1961, Smithsonian Contributions to Astrophys., 5, 69.
Dodson, H. W., and E. R. Hedeman: 1971, UAG-14, WDC-A, NOAA, U.S. Dept.
    of Commerce, Boulder, Colorado.
Dodson, H. W., and E. R. Hedeman: 1972, in Solar-Terrestrial Physics/
    1970, E. R. Dyer, Ed., Part I, p. 151.
Dodson, H. W., and E. R. Hedeman: 1975, UAG-52, WDC-A, NOAA, U.S. Dept.
    of Commerce, Boulder, Colorado.
Dodson, H. W., and E. R. Hedeman: 1981, UAG-80, WDC-A, EDIS, NOAA, U.S.
    Dept. of Commerce, Boulder, Colorado.
Harvey, G., and B. Bell: 1968, Smithsonian Contr. Astrophys., 10, 197.
Hashim, A., and M. Bercovitch: 1972, Planet. Space Sci., 20, 791.
Kota, J., and J. R. Jokipii: 1983, Astrophys. J., 265, 573.
Roy, J.-R.: 1977, Solar Phys., 52, 53.
Swinson, D. B.: 1970, J. Geophys. Res., 75, 7303.
Swinson, D. B., M. A. Shea, and J. E. Humble: 1985, J. Geophys. Res.,
    (in press).
Waldmeier, M.: 1971, Solar Phys., 20, 332.

# THE ANOMALOUS COMPONENT,
# ITS VARIATION WITH LATITUDE
# AND RELATED ASPECTS OF MODULATION

L. A. Fisk
Space Science Center
University of New Hampshire
Durham, NH  03824, USA

ABSTRACT:   The theory for the origin of the anomalous cosmic ray component and several possible mechanisms for its acceleration are reviewed.
The predictions of these mechanisms for the latitude variations of the
anomalous component that should be seen by Ulysses are discussed.

## 1. INTRODUCTION

Starting in 1972, a variety of groups began to observe a new
component of the energetic particle flux, the so-called anomalous component (Garcia-Munoz et al., 1973; Hovestadt et al., 1973; McDonald et
al., 1974).  The origin of this component now appears to be clear--it
is a result of interstellar neutral gas that is ionized and accelerated
in the solar wind, as suggested by Fisk, Kozlovsky, and Ramaty (1974).
However, the mechanism and location for its acceleration is uncertain.
As we discuss below, Ulysses should provide definitive tests of several
of the theories for the acceleration that have been proposed.

Shown in Figure 1 are the results of Cummings et al. (1984)
of Voyager observations of the anomalous component from September 1977
through February 1978.  The dashed lines are galactic cosmic rays; the
solid lines are the estimated solar and/or interplanetary component.
The anomalous component, then, is observed at energies where these two
components converge.  In the case of helium, the galactic and solar/
interplanetary components are subtracted to reveal the anomalous component; for the other species the anomalous component is clearly exhibited without this subtraction.

The composition of the anomalous component, and hence the
origin of its name, is restricted to only the elements shown in Figure
1.  It is seen in helium, but absent in carbon.  It contains nitrogen,
oxygen, and neon, but has not been seen in any other element.

R. G. Marsden (ed.), The Sun and the Heliosphere in Three Dimensions, 401–411.
© 1986 by D. Reidel Publishing Company.

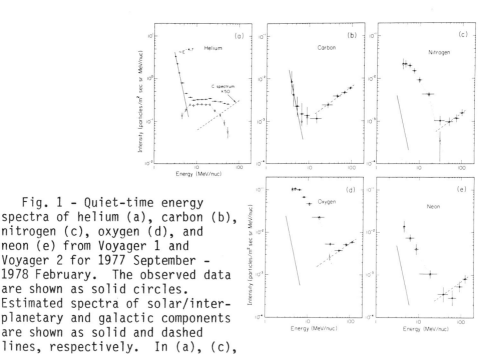

Fig. 1 - Quiet-time energy spectra of helium (a), carbon (b), nitrogen (c), oxygen (d), and neon (e) from Voyager 1 and Voyager 2 for 1977 September - 1978 February. The observed data are shown as solid circles. Estimated spectra of solar/inter-planetary and galactic components are shown as solid and dashed lines, respectively. In (a), (c), (d), and (e) the spectra of anomalous cosmic rays are indicated by o's which are joined by the dotted lines. In (b) the △'s are obtained by subtraction of solid line from data (from Cummings et al., 1984).

## 2. THE ORIGIN

The principal theory for the origin of the anomalous component is due to Fisk et al. (1974), who argue that these particles are
due to interstellar neutral gas which is ionized and accelerated in the
solar wind. The principal advantage of this theory is that it can
account for the unusual composition. Interstellar neutral gas should
consist primarily of H, He, N, O, and Ne; all other species should be
ionized. The interstellar neutral particles are then swept into the
heliosphere by the motion of the Sun through the local interstellar
medium. Once in the heliosphere, the neutral particles are ionized by
photoionization and by charge exchange. The now interstellar ions are
then picked up by the solar wind and convected back out of the heliosphere. It is argued (see section 3) that by some mechanism the ions
are accelerated to energies of the order of 10 MeV/nucleon during this
passage. Having obtained these higher energies, the interstellar ions
are then relatively mobile in the solar wind, and will tend to diffuse
back into the inner heliosphere, where they are observed as the anomalous component. Indeed, the higher-mass interstellar ions should be
quite mobile. They will be singly charged; there is time to ionize
them only once. Thus, the He, N, O, and Ne are of relatively high

rigidity, and, as with high-rigidity galactic cosmic rays, can propagate easily in the solar wind, and can diffuse back into the inner heliosphere. However, interstellar H, as with its galactic counterpart, has a low rigidity at 10 MeV, and may be excluded from the inner heliosphere. Thus, by this mechanism, a component which has only He, N, O, and Ne in the inner heliosphere is produced, as is observed.

The principal prediction, and thus the principal test of this theory for the origin of the anomalous component, is that the particles should be singly charged, which distinguishes them from fully stripped galactic particles. Unfortunately, a direct verification of the charge state of the anomalous particles has not been possible. However, over the years a fair amount of circumstantial evidence has accumulated that, indeed, the anomalous particles are singly charged. Webber et al. (1977), for example, argued that the spatial gradients of the anomalous component are compatible with other measured gradients only if the anomalous O is singly charged. Similarly, McKibben (1977) concluded that the time variation of the anomalous He is compatible with the time variations of other species only if the anomalous He is singly charged. A similar conclusion, based on time-variation studies, has also been drawn by Klecker et al. (1980).

The most recent argument on the charge state of the anomalous component was presented by Cummings et al. (1984). They note that if the spectra of the various anomalous species are similar in the outer heliosphere, where the particles are accelerated, then simple modulation theory predicts that the observed spectra should all occur at approximately the same value of the spatial diffusion coefficient. Thus, as can be seen in Figure 1, the diffusion coefficient should be the same for He at approximately 10 MeV/nucleon, for O at 6 MeV/nucleon, for N at a somewhat lower energy, and for Ne at a lower energy still, but not determined from these observations. Cummings et al. note also that the diffusion coefficient must be of the form, particle velocity times a function of rigidity. Thus, if the anomalous particles are fully stripped, or equivalently all at the same rigidity for a given energy per nucleon, it is not possible for the diffusion coefficient to be the same at each of these peaks. However, if the ions are singly charged, with different charge-to-mass ratios, then the diffusion coefficient can be the same at each peak. Indeed, as Cummings et al. note, the rigidity dependence of the diffusion coefficient that is needed to account for these observations is completely compatible with many other studies of diffusion coefficients. Once again, then, there is good circumstantial evidence that the anomalous component is singly charged, and that the origin is interstellar neutral gas that is ionized and accelerated in the solar wind.


## 3. ACCELERATION MECHANISMS

Although the origin of the anomalous component appears reasonably well-established, the mechanism and location for the acceleration of these particles is unknown. Indeed, the mechanism must provide for an extensive acceleration. When the recently ionized interstellar particles are first picked up by the solar wind, they acquire energy of approximately 1 keV/nucleon. When they are observed, they are at approximately 10 MeV/nucleon. The location of this acceleration will be the prime determinant of the latitude variation that Ulysses can be expected to see, or, conversely, observations from Ulysses should provide important constraints on likely locations for the acceleration.

Essentially three theories have been developed for the acceleration of the anomalous component:

## 3.1. Transit-time Damping

The first theory for the acceleration of the anomalous component was developed by Fisk (1976), who noted that stream-stream interaction regions in the solar wind, and other sources of turbulence might be expected to generate large-scale fluctuations in the magnitude of the magnetic field in the outer heliosphere. These fluctuations are essentially waves, and the anomalous particles can damp these waves by a transit-time damping mechanism; they should resonate with the waves at the Landau resonance. Possible damping, or equivalently acceleration rates were calculated, and in a subsequent paper Klecker (1977) showed that these acceleration rates will yield spectra for the anomalous component that are in reasonable agreement with the observations.

In defense of this mechanism, recent observations by Pioneer and Voyager in the outer heliosphere have revealed that large-scale magnitude variations in the field magnitude do occur (e.g., Burlaga, 1984). Indeed, the principal consequence of shock interactions, of merging interaction regions, and of converging streams is to produce strong variations in the field magnitude. However, whether these variations can be thought of as simple waves, and whether the interactions of the anomalous particles with these variations can be approximated by a linear theory is unclear.

## 3.2. The Termination Shock over the Solar Poles

An alternative acceleration mechanism has been proposed by Pesses, Jokipii, and Eichler (1981). It relies on acceleration at the termination shock of the solar wind, and incorporates the effects of gradient and curvature drifts, which may be important for energetic particle propagation in the solar wind (e.g., Jokipii et al., 1977; Jokipii and Kopriva, 1979). Although there are no calculations in Pesses et al. (1981), the basic idea is as follows:

The supersonic solar wind is expected to pass through a shock transition to subsonic flow somewhere in the outer heliosphere, at between 50 and 100 AU from the Sun. Particles can be accelerated at this shock through diffusive shock acceleration, just as occurs at the Earth's bow shock and at interplanetary shocks (e.g., Fisk, 1971; Fisk and Lee, 1980; Lee, 1982).  The particles are scattered back and forth across the shock front and gain energy by being compressed in the converging solar wind flow.  Indeed, the termination shock should be a particularly good accelerator since it surrounds the heliosphere. Particles cannot escape around the sides of the termination shock; rather they must pass through the shock.

        To be accelerated at the termination shock, the nearly thermal anomalous particles must first be reflected by the shock front with sufficient velocity to propagate upstream in the solar wind.  They are then reflected back to the shock front to interact again and gain energy.  Pesses et al. (1981) argue that this injection of particles into the acceleration process should be easier over the solar poles. Here, the magnetic field in the solar wind is nearly radial and thus anomalous particles following field lines can easily propagate upstream after their first reflection from the shock.  In contrast, the field at the termination shock near the equatorial plane is nearly azimuthal, and thus nearly parallel to the shock front.  Anomalous particles may require substantially higher velocities to propagate upstream in the solar wind, which may result in a threshold on the injection process.

        Once the anomalous particles are accelerated, they will be subject to the normal modulation processes.  If gradient and curvature drifts are important in modulation, then during the last solar cycle the drifts will tend to bring the particles downward from an accleration region over the solar poles onto the equatorial plane in the inner heliosphere (e.g., Jokipii and Kopriva, 1979).  Conversely, after the field reversal near 1980, the drifts should be in the opposite direction, which tends to restrict the passage of the anomalous particles from the poles to the equatorial region.  Hence, Pesses et al. predict a dramatic change in the intensity of the anomalous component from cycle to cycle.

        Jokipii (1985) has recently performed the calculations that were not in Pesses et al. (1981) and has found some interesting effects.  In order for particles to gain substantial energy in a shock mechanism, they must make repeated interactions with the shock.  When the drifts are downward from the solar poles, as in the last solar cycle, they tend to carry particles away from the shock front.  Conversely, when the drifts are towards the poles, as in the present cycle, they tend to carry particles back to the shock.  Further, during the present cycle, the drifts along the shock front, which result from the magnitude change in the magnetic field that occurs at the shock, are downward from the poles, and will spread the particles away from the poles and keep them in contact with the shock front for a longer period.  Thus, as Jokipii (1985) points out, the shock during the current portion of the solar cycle may be a more effective accelerator than it was during the previous cycle, and result in higher-energy anomalous particles.

    Recent observations by Cummings et al. (1985), which are
shown in Figure 2, provide evidence that the acceleration of the anomalous particles does vary from cycle to cycle.  The plots at the top of
the figure are Voyager observations of anomalous oxygen taken during
six different intervals:  intervals A, B, and C are prior to the field
reversal; intervals D, E, and F are afterwards.  Intervals C and E are
at roughly comparable levels of modulation, yet the spectra during
these two intervals are substantially different.  The spectrum in E is
dominated by higher-energy particles, whereas in C the spectrum is much
softer.  The bottom two graphs are theoretical calculations by Jokipii
(1985) for (left side) drifts directed downward from the poles, and
(right side) towards the poles.  Clearly, the latter case results in
higher-energy particles in agreement with the observations.

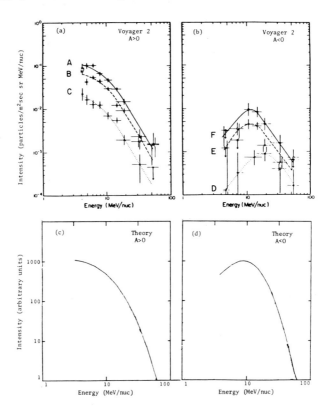

    Fig. 2 - Comparison of theoretical predictions and observations
for the spectrum of anomalous oxygen.  The data in panels (a) and
(b) are from the Voyager II spacecraft (Cummings et al., 1985) and
refer to various time periods approaching the last solar sunspot
maximum (A>0) and after the magnetic field change sign at the last
maximum (A<0).  The theoretical results (c,d) refer to spectra of
singly charged oxygen, calculated by Jokipii (1985) for 1 AU, near
the heliospheric equator, with the corresponding magnetic field
sign (figure provided by Jokipii).

However, there are also some difficulties with this mechanism for the acceleration of the anomalous component at the termination shock. It is difficult to understand how injection over the solar poles and strong drift effects can be compatible with the observed spatial gradients. The observed radial gradients for the anomalous component are reasonably strong--10 to 15%/AU (e.g., Webber et al., 1985)-- as if there is simply a source for these particles in the equatorial plane in the outer heliosphere, and the particles diffuse radially inward. With a polar source, however, and strong drifts from the poles to the equator as in the last cycle, particles flow onto the equatorial plane in the inner heliosphere, an effect which tends to make the gradients small.

Shown in Figure 3 are some recent calculations by Potgieter et al. (1985), who took a source spectrum at 50 AU that will yield approximately the observed spectrum for anomalous oxygen at Earth, used a model that included all drift effects, placed the source in 10-degree intervals around various heliographic latitudes, and calculated the radial gradient in the equatorial plane. The orientation of the drifts is from the poles downward towards the equator. As can be seen in Figure 3, when the source is near the poles (the 0° curve), the gradients in the equatorial plane are much smaller than are observed.

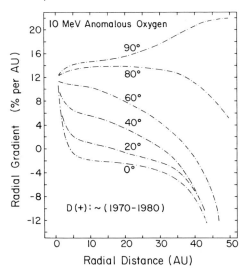

Fig. 3 - Calculated radial gradients in the equatorial plane as function of radial distance with a source at 50 AU and various polar angles. The magnetic field polarity (D(+)) is for a cycle where drifts are downward into the equatorial plane at $\Theta = 90°$ (after Potgieter et al., 1985).

Similarly, as has been reported by Webber et al. (1985), one of the curious features of the anomalous component (and also of galactic cosmic rays) is that the radial gradients in the equatorial plane do not change when the field reversal occurs; they are essentially independent of field polarity. It is difficult to imagine that the same gradients will result when, as in the previous cycle, particles drift downward from a source at the poles onto the equatorial plane, and, as in the current cycle, they either struggle upstream against outward drifts or travel along the shock front to the equatorial current sheet and enter along it.

## 3.3. The Termination Shock Near the Equatorial Plane

In the final alternative for an acceleration mechanism, we consider that gradient and curvature drifts in the solar wind, or for that matter latitude transport, may not in fact be important in the modulation process. For example, the observed insensitivity of the radial gradient of energetic particles to changes in magnetic field polarity in the solar wind (Webber et al., 1985), and the observation that the solar cycle variation of the cosmic rays appears to be the result of radial propagating disturbances in the solar wind (McDonald et al., 1981) both suggest that the transport of energetic particles is primarily in the heliocentric radial direction. However, if this is the case, the acceleration of the observed anomalous component must lie near the equatorial plane.

For this acceleration, we could of course consider the transit-time damping mechanism discussed in section 3.1. This mechanism should be most efficient near the equatorial plane since stream-stream interactions which generate the required magnitude fluctuations in the magnetic field are strongest in this region.

We can also use the termination shock near the equatorial plane. However, to do so we need to deal with two problems. First, there is the injection. As noted in section 3.2, Pesses et al. (1981) argued that the injection should be difficult near the equatorial plane since the magnetic field is nearly azimuthal and parallel to the shock front, and will impede the ability of particles to propagate upstream after their first reflection. Second, an alternative explanation to that provided by Jokipii (1985) will need to be constructed for the change in the spectra from solar cycle to solar cycle, exhibited in Figure 2.

Injection at the termination shock near the equatorial plane is probably not a problem. Although the magnetic field is on average azimuthal, it should contain relatively large directional fluctuations. Thus, the local field which intersects the shock can make an appreciable angle with the shock front. Particles, then, which are reflected by the shock can propagate upstream, to be reflected back and be accelerated again. Indeed, this latter reflection is quite simple. Particles following a field line, which fluctuates in direction but which on average is azimuthal, will automatically loop back to the shock front, or equivalently, will be reflected back to the shock front. If injection is not in fact a problem for the termination shock

near the equatorial plane, then this shock is a more promising candidate for the acceleration of the anomalous component than its counterpart over the solar poles, because the looping of the field automatically provides for many interactions of particles with the shock, and
thus for considerable energy gain.

The shift in the energy spectrum from cycle to cycle is a
more serious problem, but there are also some possible explanations
which do not involve changes in the magnetic field polarity. We need
to remember that changes in solar wind conditions, and thus the modulation conditions, propagate radially outward from the Sun. Disturbed
conditions are generated in the inner heliosphere and propagate outward. Consider, then, two cases which correspond to cases C and E in
Figure 2. In case C, when the cosmic ray intensity is declining, the
solar wind is disturbed in the inner heliosphere, but more tranquil in
the outer heliosphere; the latter region should exhibit conditions representative of those which occurred in the inner heliosphere near the
previous solar minimum. Conversely, in case E, when the cosmic ray
intensity is increasing, the outer heliosphere is disturbed, and the
inner heliosphere more tranquil. In each case the level of total modulation seen near Earth, which depends on the integral modulation from
the boundary inward, can be the same, as is observed. However, in case
C the solar wind conditions are the mirror image of those in case E.

For the acceleration of the anomalous component, the occurrence of this mirror image is important. In case C the anomalous component may be accelerated at a termination shock in tranquil solar wind
conditions; in case E, the shock should be in disturbed solar wind conditions. We should reasonably expect, then, a significant hysteresis
effect in the anomalous component, perhaps on a time scale of years.

The termination shock in disturbed solar wind conditions
(case E in Figure 2) should in fact yield higher energy particles. The
ability of a shock to accelerate particles depends on how readily the
particles are reflected back to the shock to gain energy. In disturbed
conditions, where there is considerable scattering, particles, particularly higher-energy particles, will be more readily scattered back to
the shock, and will gain more energy, thus yielding a peak at higher
energies, as is observed.

If this explanation for the shift in the spectrum from cycle
to cycle is correct, the occurrence of the change in the spectrum
around the field reversal is coincidental. The field reversal occurs
at solar maximum, when solar wind conditions begin to change from disturbed to tranquil in the inner heliosphere. It is this latter change
which is the important event and which ultimately results in the
changes in the anomalous spectra. Of course, it follows that as the
current solar cycle progresses and more tranquil conditions propagate
outward with the solar wind, then the observed anomalous spectra should
return to shapes similar to those seen at the previous solar minimum.

## 3.4. The Acceleration of Interstellar Hydrogen

One interesting point concerning all of these acceleration theories is the possibility that interstellar H could be accelerated along with the He, N, O, and Ne. As was discussed in section 2, Cummings et al. (1984) argue that the peaks of the observed anomalous spectra should all occur at approximately the same value of the spatial diffusion coefficient. If we ask where this argument implies that the observed peak of anomalous H should occur, we conclude that it is at energies above 100 MeV. At energies below this value, the spectrum of anomalous H should have a slope near unity, since here modulation will be the dominant effect in determining the spectral shape. Thus, the observed spectrum of anomalous H may be very similar to and diffi-cult to distinguish from the spectrum of modulated galactic cosmic ray H.

Indeed, a flux of anomalous H is a possible explanation for the so-called superfluxes of H seen during the last solar minimum (Garcia-Munoz et al., 1977). However, Beatty et al. (1985) have ana-lyzed this possibility and concluded that the superfluxes of H are not due to anomalous H, but rather are the result of changes in the modula-tion of galactic cosmic ray H. Nonetheless, a significant component of anomalous H could still be present, but difficult to detect relative to the galactic particles.

Acceleration of interstellar H, along with the He, N, O, and Ne could have serious consequences for possible acceleration mecha-nisms. For example, in acceleration theories that use the termination shock, the energy density in accelerated H could be appreciable and may effect both the shock location and its structure.

## 4. CONCLUDING REMARKS

The latitude variations of the anomalous component that Ulys-ses will see will depend primarily on the location of the acceleration region. If the acceleration occurs by transit-time damping of stream-stream interaction regions, which is an equatorial phenomenon, or at the termination shock near the equatorial plane, then the intensity of the anomalous component should decrease with increasing heliographic latitude. Conversely, if the acceleration occurs primarily at the ter-mination shock over the solar poles, the intensity should increase with latitude. However, it should be noted also that if gradient and curva-ture drifts are important, they will be in a direction towards the ter-mination shock during much of the Ulysses mission, which can restrict the inward passage of the anomalous particles from the shock. In any case, the latitude variation of the anomalous component that will be seen by Ulysses, and the behavior of the spectra of all the different anomalous species will provide important tests for the likely location and the mechanism for the acceleration of these particles.

ACKNOWLEDGMENT

This work was supported in part by NASA/JPL Contract 955461; the NASA Solar Terrestrial Theory Program, Grant NAGW-76; NASA Grant NSG-7411; and NSF Grant ATM 8311241.

REFERENCES

Beatty, J. J., Garcia-Munoz, M., and Simpson, J. A. 1985, Astrophys. J. (Letters), **294**, 455.
Burlaga, L. F. 1984, NASA Tech. Memo. 86137.
Cummings, A. C., Stone, E. C., and Webber, W. R. 1984, Astrophys. J. (Letters), **287**, L99.
Cummings, A. C., Stone, E. C., and Webber, W. R. 1985, preprint.
Fisk, L. A. 1971, J. Geophys. Res., **76**, 1662.
Fisk, L. A. 1976, J. Geophys. Res., **82**, 4633.
Fisk, L. A., Kozlovsky, B., and Ramaty, R. 1974, Astrophys J. (Letters), **190**, L35.
Fisk, L. A., and Lee, M. A. 1980, Astrophys. J., **237**, 620.
Garcia-Munoz, M., Mason, G. M., and Simpson, J. A. 1973, Astrophys. J. (Letters), **182**, L81.
Garcia-Munoz, M., Mason, G. M., and Simpson, J. A. 1977, Astrophys. J., **213**, 263.
Hovestadt, D., Vollmer, O., Gloeckler, G., and Fan, C. Y. 1973, Phys. Rev. Letters, **31**, 650.
Jokipii, J. R. 1985, J. Geophys. Res., in press.
Jokipii, J. R., and Kopriva, D. A. 1979, Astrophys. J., **234**, 384.
Jokipii, J. R., Levy, E. H., and Hubbard, W. R. 1977, Astrophys. J., **213**, 861.
Klecker, B. 1977, J. Geophys. Res., **82**, 5287.
Klecker, B., Hovestadt, D., Gloeckler, G., and Fan, C. Y. 1980, Geophys. Res. Letters, **7**, 1033.
Lee, M. A. 1982, J. Geophys. Res., **87**, 5063.
McDonald, F. B., Teegarden, B. J., Trainor, J. H., and Webber, W. R. 1974, Astrophys. J. (Letters), **187**, L105.
McDonald, F. B., Lal, N., Trainor, J. H., Van Hollebeke, M. A. I., and Webber, W. R. 1981, Astrophys. J. (Letters), **249**, L71.
McKibben, R. B. 1977, Astrophys. J. (Letters), **217**, L113.
Pesses, M. E., Jokipii, J. R., and Eichler, D. 1981, Astrophys. J. (Letters), **246**, L85.
Potgieter, M. S., Fisk, L. A., and Lee, M. A. 1985, Proc. 19th Inter. Cosmic Ray Conf., in press.
Webber, W. R., McDonald, F. B., and Trainor, J. H. 1977, Proc. 15th Inter. Cosmic Ray Conf., **3**, 233.
Webber, W. R., Cummings, A. C., and Stone, E. C. 1985, Proc. 19th Inter. Cosmic Ray Conf., in press.

# PICK-UP IONS IN THE SOLAR WIND AS A SOURCE OF SUPRATHERMAL PARTICLES

D. HOVESTADT[1], E. MÖBIUS[1], B. KLECKER[1], G. GLOECKLER[2],
F.M. IPAVICH[2], and    M. SCHOLER[1]
[1]Max-Planck-Institut    für    extraterrestrische    Physik,
   8046 GARCHING,    Germany
[2]Dept.   of   Physics   and   Astronomy,   University   of   Maryland,
   COLLEGE PARK , MD 20742, USA

ABSTRACT.    Singly   ionized   energetic   helium   has   been   observed   in   the   solar
wind   by   using   the   time-of-flight   spectrometer   SULEICA   on   the   AMPTE/IRM
satellite   between   September   and   December,   1984.   The   energy   flux-density
spectrum   shows   a   sharp   cut-off   which   is   strongly   correlated   with   and   nearly
equal   to   the   four-fold   solar   wind   translation   energy.   The   absolute   flux   of   the
$He^+$ ions   of   about   $10^4$   ions/cm·s   is   present   independent   of   the   IPL   magnetic
field   orientation.   The   most   likely   source   is   the   neutral   helium   of   the
interstellar   wind   which   is   ionized   by   solar   UV-radiation.   It   is   suggested   that
these   particles   represent   the   source   of   the   anomalous   cosmic   ray   component.

## 1. INTRODUCTION.

In   1972     anomalous   features   in   the   low   energy   quiet   time   cosmic   ray   energy
spectrum      have   been   detected   for   helium,   oxygen,   nitrogen,   and   neon   by
*Garcia-Munoz   et   al.*(1972),   *Hovestadt   et   al.*(1973),   and   *Mc   Donald   et   al.*(1974).
These   four   elements   are   known   to   have   a   high   first   ionization   potential
compared   to   other   elements   like   carbon,   magnesium,   silicon   and   iron   (e.g
*Allen*   (1973)).   This   fact   lead   *Fisk   et   al.*(1974)   to   suggest   that   the   source   of   the
particles   is   the   interstellar   neutral   wind   which   becomes   ionized   in   the   inner
heliosphere   by   interaction   of   the   atoms   with   solar   ultra-violet   radiation
and/or   solar   wind   ions   and   electrons.   The   newly   created   ions   then   are
picked-up   by   the   interplanetary   magnetic   field.   With   their   gyro-motion   in   the
solar   wind   frame   they   represent   a   distinguishable   population   which   is
subsequently   convected   into   the   outer   heliosphere   while   being   accelerated
within   the   turbulent   magnetic   fields   of   the   heliosphere   (e.g.   *Fisk*   (1976   a,   b))
or   at   the   terminating   shock   (*Pesses   et   al.*   (1981)).   The   resulting   energy
spectrum   is   then   reshaped   to   the   observed   spectrum   by   modulation   and
propagation   effects   in   interplanetary   space.
This   paper   presents   first   direct   observations   of   the   distribution   function   of
freshly   ionized   helium   in   the   solar   wind   which   likely   has   its   origin   in   the
neutral   interstellar   wind   and   probably   represent   the   source   of   the   anomalous
helium   component   in   cosmic   rays.

*R. G. Marsden (ed.), The Sun and the Heliosphere in Three Dimensions, 413–418.*
© *1986 by D. Reidel Publishing Company.*

## 2. THE PICK-UP PROCESS.

Freshly ionized helium atoms are locally subjected to the combined forces of the interplanetary $V_{sw} \times B$ electric field of the solar wind and the Lorentz-force $v_{ion} \times B$ of the magnetic field, B, where $V_{sw}$ is the solar wind velocity and $v_{ion}$ the velocity of the ion in the solar wind frame. In the <u>inertial-system</u> (which nearly coincides with the spacecraft-system) the particles undergo initially a cycloidal motion in a plane perpendicular to the local magnetic field, B, with a minimum velocity which is equal to the velocity of the neutral wind(assumed to be < 20 km/s and therefore being neglected in the following).The maximum velocity which these ions can obtain in the pick-up process is determined by the velocity component of the solar wind perpendicular to B. With $\alpha$ being the angle between the solar wind flow direction and the direction of the local magnetic field the maximum initial ion velocity is given by:

$$v_{ion}(max,loc) = 2 \cdot V_{sw} \cdot \sin \alpha \qquad (1) .$$

This leads to an energy of

$$E_{ion}(max,loc) = 4 \cdot M/2 \cdot V_{sw}^2 \cdot \sin^2 \alpha \qquad (2) ,$$

where M is the mass of the ion.

In the <u>solar wind</u> frame the motion initially leads to a conical pitch-angle distribution with the pitch angle $\alpha$ and velocity $|V_{sw}|$. In the three-dimensional velocity space the particles are distributed like a ring. If the interplanetary medium were homogeneous with parallel magnetic field lines and if there were no scattering imposed by intrinsic or self-generated waves, the distribution function would remain gyrotropic and ring-shaped.

In the real world, however, the motion of the particles is subjected to effects generated by temporal and spatial magnetic irregularities in the expanding interplanetary medium. If energy-changing wave-particle interactions can be neglected, pitch-angle scattering and adiabatic deceleration probably greatly influence the particle distribution. Within a scattering mean free path-length $\lambda$ the initially ring-type velocity distribution is expected to be reshaped by pitch-angle diffusion into a spherical-shell type distribution which is fully convected with the solar wind. The orbital velocity $|v| = |V_{sw}|$ remains constant in the solar wind frame as long as adiabatic deceleration does not play a significant role. In the inertial system the expected velocities range from zero to twice the solar wind velocity with corresponding

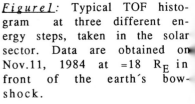

<u>Figure1</u>: Typical TOF histogram at three different energy steps, taken in the solar sector. Data are obtained on Nov.11, 1984 at =18 $R_E$ in front of the earth´s bow-shock.

observable energies between zero and four times the solar wind translation energy $(0 < E_{pick-up} < 4 \cdot M/2 \cdot V_{sw}^2)$. The energy spectrum should show a clean cut-off at that energy value.

The spectrum below cut-off should reflect the effect of adiabatic deceleration in the expanding interplanetary medium upstream of the observer.

## 3. INSTRUMENTATION and SATELLITE.

The data presented here are obtained with the suprathermal particle spectrometer SULEICA of the Max-Planck-Institute/ University of Maryland onboard the IRM spacecraft of the Active- Magnetospheric- Particle- Tracer- Explorer project (AMPTE), launched on 16. August 1984 into a highly elliptical orbit with an apogee of 18.9 earthradii. During the time period from launch until December 1984 the S/C spent a large fraction of each orbit in the solar wind upstream of the bow-shock of the earth. The SULEICA spectrometer is based on the techniques of electrostatic deflection followed by a time-of-flight and residual energy measurement (for details see $M\ddot{O}BIUS$ et al.1985).

The electrostatic deflection analyser, represented by two concentric 75 x 40 degree segments of a sphere, selects incoming ions according to their energy per charge in 24 logarithmically spaced voltage steps corresponding to an energy range from 5 to 269 keV/charge. After passing through the analyser the ions enter the time-of-flight section where the velocity of the ions is measured.

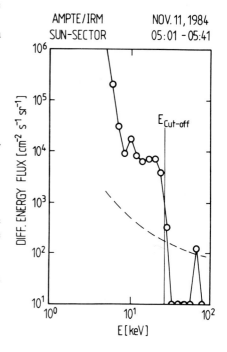

AMPTE/IRM         NOV. 11, 1984
SUN-SECTOR        05:01 - 05:41

*Figure 2:* Example of an energy-density spectrum of the M/Q= 4 ion channel accumulated over 40 minutes. Dashed curved line represents the 1 count/step level.

The ions are stopped in a silicon surface barrier detector where the residual energy is determined. The geometrical factor of the instrument is $4.3 \times 10^{-2}$ cm$^2$ sr and the energy resolution is $\Delta E/E = 0.097$.

For the investigation presented here the energy of the pick-up ions is too low to create a sufficiently high energy signal in the solid-state detectors. Therefore we identify the ion species only by combining the electrostatic deflection (E/Q ) and the time of flight signal (TOF). For a given E/Q step the TOF histogram ( c.f. *Figure 1* ) therefore represents a mass-per-charge histogram which is taken in the direction of the solar wind.

## 4. OBSERVATIONS.

A limited number (ten) of observational periods were chosen in the solar wind between launch of the S/C and 15. December, 1984. During all periods we observed a peak in the TOF-histograms at M/Q = 4 at E/Q steps which are significantly higher than we would expect for genuine solar wind particles.

Depending on the E/Q step and the solar wind velocity, we also observe in addition a broad and variable peak at TOF values which correspond to the solar wind velocity. These ions with M/Q values above 5 correspond to solar wind heavy ions.

*Figure 1* shows a typical example of TOF histograms for three different E/Q steps obtained on November 11, 1984.

It should be noted, that the $He^+$ peak is visible in high speed streams at all orientations of the interplanetary magnetic field. This is true also when bow-shock accelerated ions are present.

*Figure 2* shows the differential energy- density spectrum of the M/Q = 4 peak as measured during a sample period of 40 minute duration on November 11, 1984, when there are no bow-shock accelerated particles present. The spectra are taken from the directional sector which also contains the solar wind. Except the two neighbouring sectors all other sectors do not show any counts in the M/Q = 4 bin. There is a sharp cut-off at an energy of about 25 keV/e which corresponds to four time the translation-energy of the solar wind as measured with the plasma experiment on AMPTE-IRM at the same time. Below the cut-off energy we observe a more or less flat distribution which ends at a sharp rise of the spectrum at about 8 keV/e. We attribute the rise to solar wind particles of M/Q = 4 which will not be discussed in this paper.

## 5. DISCUSSION.

Following the arguments in section 2 we correlated the experimental cut-off energies, firstly (*Figure 3* ) with the <u>full</u> four-fold solar wind translation-energy for mass M = 4 (Helium), and secondly (*Figure 4* ) with the four-fold <u>component</u>

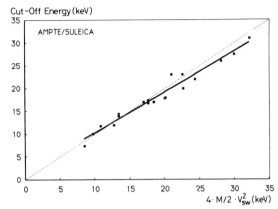

*Figure 3:* Correlation between experimental cut-off energies and the four-fold solar wind translation-energy, $4 \cdot M/2 \cdot V_{SW}^2$, (Correlation coefficient = 0.98).

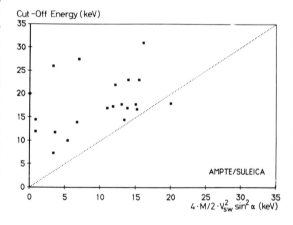

*Figure 4:* Same as *Figure 3*, but abscissa is the four-fold vertical-to-B component of the SW translation-energy, $4 \cdot M/2 \cdot V_{sw}^2 \cdot \sin^2 \alpha$ (Correlation coefficient = 0.36).

perpendicular to the local interplanetary magnetic field (eq. (2)). There is an excellent correlation with the full energy (correlation- coefficient = 0.98) while the correlation with the perpen-dicular component is rather poor (correlation coefficient = 0.36).

This result suggests that the ions originally injected perpendicular to B have lost their directional information in the solar wind frame due to pitch angle scattering, while maintaining their injection energy. The poor correlation with the perpendicular component shows that the source region of the ions extends significantly beyond one mean free scattering length upstream of the observer. The magnetic rigidity of suprathermal singly ionized helium of a few keV/nucleon is in the order of 5 to 10 MV. The mean free scattering path length $\lambda$ for that rigidity is known from cosmic ray data to be of the order of 0.05 AU during quiet interplanetary conditions (e.g.compilation in *Mason et al.* (1983)). The apparently large extent of the source region excludes a terrestrial origin of the ions.

The effect of adiabatic deceleration is expected to be high due to strong coupling of the particles to the solar wind. This sets an upper limit for the distance between the point of origin and the observer. Ions originating too far upstream (or too close to the sun) lose so much energy that they become indistinguishable from the solar wind itself.

To get a rough estimate of the source strength of the pick-up ions we use the energy flux spectrum (*Figure 2* ) and argue that the energy channels below the cutoff energy are populated via adiabatic deceleration, which leads to an energy variation E proportional to $r^{4/3}$ with the heliocentric distance. The measured quantity $E \cdot dJ/dE$ in each energy channel in *Figure 3* is then related to the relevant source quantities as:

$$E \cdot dJ/dE = E \cdot S(r) \cdot \Delta r \cdot 4 \cdot \Delta\Omega/(4\pi \cdot \Delta E \, \Delta\Omega) = S(r) \cdot \Delta r/\pi \, (E / \Delta E)$$

where $S(r) = N_{0 \, He} \cdot v_{ion}$ is the source strength of singly ionized helium. For a relative energy width of the instrument of $\Delta E/E = 0.1$ the length of the source column upstream of the S/C translates into $\Delta r = 0.2$ AU $= 3 \times 10^{12}$ cm. Using an average value of $7 \times 10^3$ ions/cm$^2$ sr s for the energy-flux density of the pick-up ions (from *Figure 2* ) we obtain a source strength $S(\, r = 1AU) = 8 \times 10^{-10}$/cm$^3$s. For an ionization rate of $v_{ion} = 5 \times 10^{-8}$/s (e.g.*Holzer* 1977) we arrive at a neutral interstellar helium density $N_0 = 1.6 \times 10^{-2}$/cm$^3$, a value which is fully compatible with results from optical EUV measurements of He I resonantly scattered EUV-lines (*Weller and Meier* 1974; *Dalaudier et al.* 1984) in the heliosphere. An attempt of *Paresce et al.* 1983 to observe He II UV lines in a rocket flight remained rather inconclusive. Our measurements therefore represent the first direct observational evidence of energetic singly ionized helium picked-up by the solar wind from the interstellar neutral gas. The shape of the distribution function remains clearly distinguishable from the solar wind and therefore may represent the source of the anomalous cosmic ray helium component in the frame of *Fisk's et al.* 1974 model. Possibly these observations are filling one more gap in the understanding of the anomalous cosmic ray component.

## 6.ACKNOWLEDGEMENT.

The authors are grateful to the many individuals at the Max- Planck-Institut and the University of Maryland who contributed to the success of the experiment and to the AMPTE    project as a whole. We acknowledge stimulating discussion with M.A. Lee. We thank G. Paschmann and    G. Lühr for making available to us the AMPTE Plasma and magnetic field data.

## 7. REFERENCES

Allen, W.A., 1973, ASTROPHYSICAL QUANTITIES,
The  Athlone  Press,  London

Dalaudier,F., I.L. Bertaux, V.G. Kurt, E.N. Mironova, 1984,
Astron.  Astrophys.,134, p.171

Fisk, L.A., B. Kozlovsky, R. Ramaty, 1974, Ap. J. Letters
190, p.L35

Fisk, L.A., 1976a, Ap. J. 206, p.333; 1976b, J. Geophys.
Res. 81, p.4633,

Garcia-Munoz, M., G.M. Mason, J.A. Simpson1973,
Ap.J.Letters 182, p.L81

Holzer, 1977, Rev. Geophys. Space Phys. 15, p.467

Hovestadt, D., O. Vollmer, G. Gloeckler, C.Y. Fan, 1973,
Phys. Rev. Letters 31, p.650

Mason, G.M., G. Gloeckler, D. Hovestadt, 1983, Ap. J. 267,
p.844

Mc Donald, F.B., B.J. Teegarden, J.H. Trainor, 1974, Ap.J.
Letters 187,p.L105

Möbius, E., G. Gloeckler, D. Hovestadt, F.M. Ipavich, B.
Klecker, M. Scholer, H. Arbinger, H. Höfner,
E.Künneth, P. Laeverenz, A. Luhn, E.O. Tums, H.
Waldleben, 1985, IEEE Transactions on Geosc. and
Remote Sensing, GE-23, p.274

Paresce, F., H. I. Fahr, G. Lay, 1983, J. Geophys. Res. 86,
p.10038

Pesses,M.E., J.R. Jokipii, D. Eichler, 1981, Ap. J. Letters,
246, p.L85

Weller, C.S.,  and R.R. Meier, 1974, Ap. J. 193, p.471

SECTION VIII:    INTERSTELLAR GAS AND INTERPLANETARY DUST

NEUTRAL INTERSTELLAR GASES IN THE HELIOSPHERE:
NEW ASPECTS OF THE PROBLEM

H.J. Fahr
Institute for Astrophysics
University of Bonn
Auf dem Hügel 71
D-5300 Bonn 1, F.R. Germany

ABSTRACT. For more than a decade, it has been attempted to derive
thermodynamical parameters of the local interstellar medium (LISM) from
observations of the interplanetary glow patterns in helium and hydrogen
resonance lines. However, reviewing the current status of these
efforts, one has to admit that, besides perhaps the vector orientation
of the interstellar wind and the unperturbed helium density, no other
parameters of the LISM could be derived without ambiguity. In the following, we shall give theoretical and observational reasons for the
present situation in this field.

## 1. REVIEW OF EARLIER THEORETICAL MODELS

In a long series of papers (Blum and Fahr, 1970; Fahr, 1971; Fahr,
1974, 1978, 1979; Axford, 1972; Holzer, 1977; Thomas, 1978; Wu and
Judge, 1978), the basics of the theoretical modelling of neutral gas
distributions of LISM species within the heliosphere have been presented. In these papers, numerical procedures for calculating interplanetary resonance glow patterns connected with the resonant scattering of solar photons by neutrals are documented. It was clear from the
very beginning (Fahr, 1968) that due to the inefficient interaction of
neutrals with the solar wind plasma and the very large mean free paths
with respect to neutral-neutral collisions, a kinetic rather than a
hydrodynamic approach to the problem is appropriate. Single neutral
test particles are followed along their dynamical trajectories from
outside the solar system to specific points in the heliosphere. The
local velocity distribution function can then be synthesized at each
place with the help of the Liouville theorem and the individual ionization histories of the test particles (for a recent review, see Fahr,
1983,).
  As an example for this kind of calculation, we give in Figure 1
the densities, i.e. the first velocity moments of the distribution function, for neutral LISM hydrogen and deuterium in the heliosphere. It
can be seen in this figure that the density contours of these two
neutral species are very different due to the differential action of

R. G. Marsden (ed.), The Sun and the Heliosphere in Three Dimensions, 421–434.
© 1986 by D. Reidel Publishing Company.

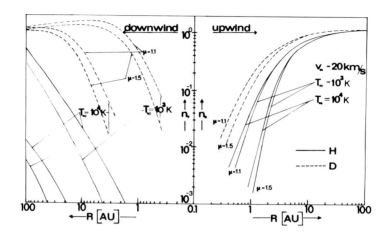

<u>Figure 1.</u> The density distribution of H and D is given as a function of the solar distance in AU along an axis defined by the LISM-wind vector. Dashed lines give deuterium, full lines hydrogen density contours.

solar gravity and solar radiation pressure. In particular, on the down-wind side a strong depletion of hydrogen as compared to deuterium can be confirmed, giving rise to a strong heliographic longitude gradient of the neutral abundance ratio D/H. In regard to this ratio, the strong influence of the LISM temperature and of the effective solar Lyman-alpha radiation pressure can be seen in Figure 1.

## 2. PROBLEMS IN THE RESONANCE GLOW INTERPRETATION

The interplanetary resonance glow patterns which have been observed by several spacecraft such as OGO V, STP-78-1, Mariner 9/10, Pioneer 10/11, Voyager 1/2, Prognoz 5/6 etc. were interpreted by multi-parameter best fit procedures. A theoretical fit to the observations is aimed for using different samples of LISM parameters. Unfortunately, the mean square deviation hypersurface in the parameter space does not show clearly pronounced "local" minima that would allow for a straight-forward best fit to the observations. The patterns depend only fairly weakly on some LISM parameters. The fit procedure becomes even more difficult because several parameters enter the theory only in the form of certain algebraic combinations. Thus only best-fitting values for these combinations can be extracted from the observations.

This is for instance the case for the speed of the LISM approach and the ionization rate. Only their quotient enters the loss integrals, i.e. higher ionization rates have the same effect as correspondingly lower velocities of approach. In interpreting the helium glow iso-photes, especially in the downwind cone region, the problem arises in that the half angle of the cone structure (FWHM angle) depends on the

ratio of the mean thermal and the LISM wind velocities (see, e.g.
Feldman et al., 1972). Therefore a best-fit value can be found only for
a specific algebraic combination of the LISM temperature and wind
velocity. This clearly supports the tendency in which different authors
derive higher helium temperatures with higher bulk flow velocities, as
is for instance well documented in a recent paper by Dalaudier et al.
(1984).

With concern to the specific problem in the H-Lyman-alpha resonance isophotes, some additional difficulties arise from the fact that
hydrogen atoms are strongly influenced by charge-exchange interactions
with the solar wind plasma. Thus all three-dimensional spatial and
temporal solar wind structures are in principle imprinted onto the
neutral gas distribution. Though longitudinal solar wind asymmetries
are fairly well smoothed out, heliographic latitude asymmetries in the
solar wind flow should show up clearly in a breakdown of cylindrical
symmetries of the density and the glow pattern with respect to the LISM
wind axis (Ajello et al., 1979; Witt, 1979). These asymmetries have in
fact been detected in the interplanetary Lyman-alpha resonance iso-
photes taken by Mariner 10 and by the Prognoz 5/6 UV photometers (Witt
et al., 1979; Lallement et al., 1985) and have been interpreted as
pointing to a clear heliographic latitude dependence of the solar wind
plasma flow. Lallement et al. (1985), for example, have derived a
30 percent increase in the solar wind plasma flow from the ecliptic
plane towards the ecliptic poles.

There appears to be a need for some process giving rise to an
enhanced depletion of hydrogen atoms at high ecliptic latitudes, as was
proven by the authors just mentioned. Nevertheless, the explanation for
such a process is not at all evident. The authors ascribe this asymmetric ionization process to a heliographic latitude variation of the
solar wind plasma flow. On the other hand, for ecliptic regions it has
been shown that the solar wind plasma flow intensity and the solar wind
momentum flow intensity behave as a perfectly invariant quantity, dis-
regarding the different signatures of the solar wind type that is
present, i.e. high-speed or low-speed streams, or intermediate stream
structures (Schwenn, 1983; Steinitz and Eyni, 1980; Steinitz, 1983;
Klemens, 1985). It can be said that the only difference between solar
wind flows in polar and ecliptic regions is that polar regions are more
strongly dominated by central coronal hole flows. Thus it can be concluded that solar wind velocities are higher at polar regions compared
to ecliptic regions. On the other hand, polar mass and momentum flows
may be identical with those in the ecliptic. Thus it is felt that this
asymmetry phenomenon needs further investigation.

## 3. IONIZATION PROCESSES

Up to now, chiefly ionizations of the LISM neutrals due to photoionization and charge-exchange ionization have been taken into account in the
models. In addition, both ionization processes have been modelled by
$1/r^2$-laws in view of the better mathematical tractability in the framework of the density distribution calculations (r is the solar distance).

As was mentioned in the previous paragraph, this has been recognized as
an unsatisfactory approximation due to heliographic latitude dependences. The impossibility to stay with the $1/r^2$-representation of the
ionization rates is even more manifest. In addition to the processes
mentioned, electron impact ionizations have to be taken into account.
The corresponding rate for these processes depends not only on the
solar wind electron density, but even more, on the electron temperatures, or in other words, on the local electron velocity distribution
function, i.e. the core/halo structure (Pilipp, 1983). At places in the
solar system where these ionization processes have to be taken into
account, it is evident that a $1/r^2$-representation of the corresponding
ionization rate is totally unsatisfactory.

Regions where electron impact ionization has to be considered are
within r=0.6 AU, where the electron temperatures are at about $10^6$ K or
larger, and possibly at large distances r>5 AU, where the solar wind
temperatures are expected to rise again to high values due to plasmas-
gas interactions (Fahr, 1973; Holzer and Leer, 1973; Fahr et al., 1978;
Petelski et al., 1980; Fahr et al., 1981). Though a physical reason for
the ion-plasma temperature increase is at hand, it is not yet clear
whether the solar wind electrons participate in this temperature
increase. Nevertheless, in Figure 2 we have shown the difference in the
density profiles of interplanetary helium and hydrogen with and without
electron impact ionizations. The bold lines give density profiles calculated with electron impact ionizations excluded, the thin lines

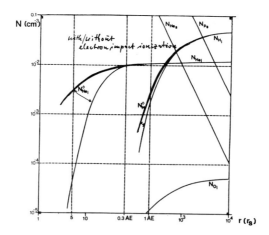

Figure 2. Upwind helium and hydrogen densities are shown versus solar
distance r in units of solar radii. Heavy lines represent calculations
without electron impact ionizations, thin lines with electron impact
ionizations. The theoretical calculations of the corresponding ionization rates are based on the solar wind models used in Fahr et al.
(1981). In addition to densities for interstellar H and He, densities
for interstellar O-atoms are shown.

corresponding densities with electron impact ionizations included that
are calculated on the basis of the solar wind model given in Fahr et
al. (1981). As one can notice, electron impact ionizations are especially important for the densities of helium and hydrogen within r<0.5 AU
or at more distant places in the helium cone region which can be
reached mainly by helium atoms with perihelia inside 0.5 AU.

Electron impact ionization processes could also come into play via
the so-called critical velocity effect proposed by Alfvén (1954),
tested in the laboratory by Danielson (1970), Danielson and Brenning
(1975) and Möbius et al. (1979), and proven by Haerendel (1982) to
operate in space. In the context of plasma-gas interactions that are
discussed in this review, the critical velocity effect is likely to
produce a high energy solar electron halo distribution by transforma-
tion of the relative kinetic energy between primary and secondary solar
wind ions via electrostatic/electromagnetic wave turbulence into electron thermal energies (Petelski et al., 1980). This complicated process
of energy conversion works only if the relative velocity between the
two solar wind ion species has a component perpendicular to the frozen-
in solar wind magnetic field. This is clearly the case in ecliptic
regions beyond 1 AU where the azimuthal component of the magnetic field
dominates. It was shown by Petelski et al. (1980) that here the electron impact ionization due to electrons energized by the critical
velocity effect competes with photoionization and charge exchange
beyond 3 AU.

What is interesting, however, is that this critical ionization
effect is impeded at higher heliographic latitudes, since here the
magnetic fields are likely to be oriented radially. Thus the relative
velocity vector has a vanishing component perpendicular to this field.
This would cause an asymmetry of the corresponding ionization rate
identical with the one needed for the explanation of H-Lyman-alpha glow
asymmetries analyzed by Witt et al. (1979) and Lallement et al. (1985).

Another important point to consider is the temporal variabilities
in the ionization rates and their influence on the neutral gas densities and the resonance glow distributions. Especially helium atoms are
subject to photoionization by solar EUV-photons. The flux of these
photons is shown to vary strongly with a long-time periodicity, following more or less the solar activity cycle. Due to this variability, the
helium atoms approaching a specific space point in the heliosphere on
their trajectories from outside the solar system are subjected to
temporarily varying photoionization rates. In the upper half of
Figure 3, the ionization frequency for helium atoms at 1 AU is given as
a function of time on the basis of 3-month averages of integrated solar
EUV-fluxes weighted with the appropriate photoionization cross section.
Solar flux observations of several authors from early 1974 until late
1979 are taken as input data.

Rucinski (1985) has developed a numerical code by which this form
of time variability in the photoionization rate can be taken into
account for the calculation of interplanetary helium densities. In the
lower half of Figure 3, we give his results. Taking a constant photoionization rate of $7.5 \, 10^{-8} \, sec^{-1}$, downwind helium densities are
obtained, as are given by the dotted curve. In comparison, the upper

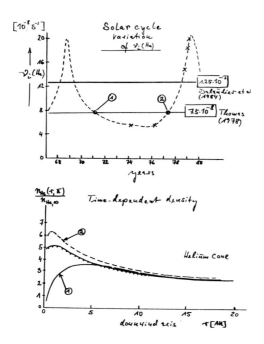

**Figure 3.** The time variation of the helium photoionization rate versus time is shown in the upper half of the figure. In the lower half, helium density curves for the upwind axis are given as a function of the solar distance. Dotted curve: constant ionization rate of $7.5 \cdot 10^{-8}$ $\sec^{-1}$. Curve 1: Time-dependent model density calculation for event 1. Curve 2: Corresponding time-dependent calculation for event 2 marked in the upper curve.

and the lower curves in this figure give corresponding solutions for the helium densities based on time-dependent model calculations carried out for two time events 1 and 2 during the solar cycle where the same actual ionization frequency of $7.5 \cdot 10^{-8}$ $\sec^{-1}$ is reached. However, event 1 corresponds to decreasing and event 2 to increasing ionization rates during the solar cycle. It is evident from a comparison of these curves that within 5 AU the time history of the photoionization rate matters substantially.


## 4. DYNAMICS OF LISM NEUTRALS IN THE HELIOSPHERE

Up to a recent publication by Fahr et al., (1985), only solar gravitational forces and solar radiation pressure forces have been considered as relevant forces determining the dynamical trajectories of LISM neutrals in the heliosphere. For helium, the radiation pressure is negligible, whereas for hydrogen it plays an important role, sometimes even leading to an overcompensation of the solar gravity acting upon H-atoms. Nevertheless, if the solar Lyman–alpha emission line profile is considered as sufficiently flat in the frequency range relevant for resonant scattering of the solar photons by the H-atoms, this still gives rise to a net conservative force field, leading to Keplerian hyperbolae as the resulting trajectories.

However, in a recent paper by Fahr et al. (1985) it is claimed that, in addition, some nonconservative forces which result from atom-ion interactions via elastic collisions have to be considered in this context. The neutrals moving through the solar wind regime are swept over by solar wind electrons and ions which, by virtue of their Coulomb fields, interact with the electrical charge distribution of the atomic system and thus give rise to a net average momentum transfer to the neutral atom. This interaction is of a monopole–monopole type for small impact parameters and high energies, of a monopole–induced dipole type for intermediate conditions, and of a monopole–constant dipole type with a $1/r^2$-interaction potential for large impact parameters (see e.g. Tschernetskij, 1969). The latter fact implies that large impact parameter collisions make non-negligible contributions to the net momentum transfer from the ion plasma to the neutral atom. By statistical

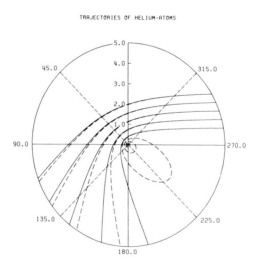

TRAJECTORIES OF HELIUM-ATOMS

Figure 4. The dynamical trajectories of helium atoms entering the solar system with impact parameters of 1.0; 1.5; 2.0; 2.5; 3.0 AU are displayed in a plane with polar coordinates r and $\Theta$ containing the LISM wind vector and the sun. In the solid curves, only the radial component of the elastic drag force has been taken into account, whereas in the dashed curves the latitudinal component of this force has been added. A change-over from a hyperbolic orbit into an elliptic orbit can be seen for an impact parameter of 1 AU. (Calculations based on a total momentum transfer cross section of $10^{-16}$ cm$^2$ were carried out by Nass and Fahr (1985).)

evaluation, this leads to a net drag force acting upon neutrals in the direction of the relative velocity between the neutral atom and the solar wind ions. This drag force has two components, one radial and one latitudinal (or orthoradial). In Figure 4, helium atom trajectories are shown which result from the combined action of this drag force component and the solar gravitational force. Since the drag force is non-conservative, the orbits are non-Keplerian, i.e. the angular momentum is not conserved, and transitions from "free" to "bound" orbits do in fact occur.

On the basis of this drag force between the solar wind ion flow and the LISM helium atoms, Fahr et al. (1985) have recalculated the interplanetary helium densities and have obtained interesting deviations from earlier density calculations. The most important feature in these deviations is a substantial broadening of the downwind helium cone structure from which the helium temperatures have usually been derived. Thus the authors feel that the large discrepancy between LISM hydrogen and helium temperatures derived up to now has its origin in an unsatisfactory modelling of the helium densities.

## 5. THE HELIOSPHERIC INTERFACE EFFECT

In recent papers by Ripken and Fahr (1983), Wallis (1984) and Fahr and Ripken (1984), the question has been studied whether the neutral gases that reach the inner region of our heliosphere permit unambiguous conclusions about the state of the unperturbed LISM. It was found that the shocked solar wind beyond the heliospheric shockfront does not substantially influence the neutral LISM atoms traversing this region on their way to the inner heliosphere. However, the much denser LISM plasma that is compressed at the heliopause strongly affects the neutral gas flow by charge exchange interactions between the perturbed LISM plasma and the neutral atoms. Especially H and O atoms are subject to these charge exchange interactions, since the relevant cross sections are fairly high ( $10^{-15}$ $cm^2$ and larger).

For H atoms, production processes caused by transcharged LISM protons with identical velocity vectors are of importance and compete with loss processes. However, for O atoms, almost no production processes exist due to the nearly complete absence of $O^+$-LISM ions with appropriate velocities (Fahr and Ripken, 1984). This leads to the fact that oxygen atoms traversing this LISM interface region are depleted substantially, possibly by a factor of $10^{-1}$ or more. In contrast, the hydrogen atom depletion is not as strong. Depending on the interface model that one may prefer (supersonic or subsonic plasma approach), depletion factors of 0.7 to 0.5 may occur here.

In Figure 5, we have shown some interesting consequences of this interface effect on the neutral LISM hydrogen distribution in the heliosphere. In conventional interpretations of the interplanetary H-Lyman-alpha glow, the hydrogen density at about 10 AU, $N_H(10 \text{ AU})$, was found to scale with the glow intensities. Therefore, if $N_H(10 \text{ AU})$ was taken as identical to the unperturbed LISM hydrogen density outside the solar system, $N_{H\infty}$, it is evident that a strict proportionality between the upwind H-Lyman-alpha intensity, e.g. seen from 1 AU, and $N_{H\infty}$ can be

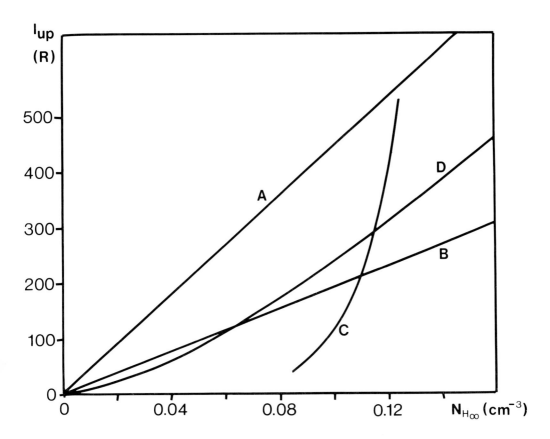

<u>Figure 5.</u> Different forms of the dependence of the interplanetary H-
Lyman-alpha intensity $I_{up}$ seen from 1 AU in an upwind direction on the
LISM hydrogen atoms density $N_{H\infty}$ are shown. Curve A: Conventional model
interpretation with $N_H(10\text{ AU})=N_{H\infty}$. Curve B: Constant depletion factor
due to an invariable LISM proton interface. Curve C: Variable interface
according to the assumption $N_{H\infty}+N_{p\infty}=N_{He\infty}\times10$. Curve D: Interface by
Baranov et al. (1979).

expected, as is shown in curve A of Figure 5. If a constant interface
with a fixed depletion factor would exist, a proportionality corresponding to curve B could be expected. More realistic relations between $N_{H\infty}$
and the upwind Lyman-alpha intensity, $I_{up}$, however, are given by
curves D or C. These nonlinear relationships result from the fact that
a variation of $N_{H\infty}$ is connected with a variation of the LISM interface
system that determines the actual depletion factor. For instance, if it
is assumed that the sum of the LISM H-atom-and proton-densities,
$N_{H\infty}+N_{p\infty}$, equals the tenfold of the LISM helium density,
$N_{He\infty}=0,0125\text{ cm}^{-3}$, then a variation of $N_{H\infty}$ would be connected with a
variation of $N_{p\infty}$, influencing the interface system and the depletion
factor. This results in a relation between $I_{up}$ and $N_{H\infty}$, as is given in

curve C of Figure 5, showing that the measured upwind Lyman-alpha intensity in fact is no sensitive indicator for the LISM hydrogen density.

   The relation between these quantities will become even more complicated and confusing due to the interface effect if in connection with a
variation of $N_{H\infty}$ and $N_{P\infty}$ the change of the outside LISM pressure on the
heliopause is considered, which influences the whole geometrical configuration of the perturbed plasma region. The results that have to be
expected under these aspects have been displayed in Figure 6 taken from
a recent work by Fahr and Ripken (1985). The change of the whole interface structure caused by a change in the proton density $N_{P\infty}$ is modelled
with the plasma-plasma-interface type described by Parker (1963). On
the basis of this model, Figure 6 shows different relations between the
LISM proton density $N_{P\infty}$ and the hydrogen density $N_{H(HP)}$ at the heliopause for different conditions in the LISM, as for example in a) constant total density $N_{H\infty}+N_{P\infty}$ or b) constant degrees of ionization X.
From this, one may notice that the derivation of the heliopause hydrogen density in some cases may not allow conclusions to be drawn with
respect to the LISM proton density.

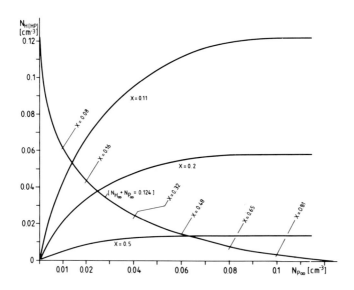

Figure 6. Based on Parker's (1963) model of pressure modulation in the
LISM plasma interface, relations between the hydrogen density at the
heliopause, $N_{H(HP)}$ and the LISM proton density $N_{P\infty}$ are given for different conditions of the LISM, as, for example, constant total density
$N_{H\infty}+N_{P\infty}=0.125$ cm$^{-3}$, or constant degrees of ionization X=0.11; 0.2; 0.5
(figure taken from Fahr and Ripken, 1985).

## 6. DIRECT DETECTION OF INTERSTELLAR GASES

The investigation of the neutral interstellar gases within the heliosphere by observing their EUV resonance glow emissions has the obvious
disadvantage that only line-of-sight integrated information can be
obtained about the state of the media with all the inherent complica-
tions of disentangling the various influences of solar and interstellar
parameters. Therefore it was planned to search for these neutral gases
in situ by locally analysing their velocity distribution functions
measured with appropriate particle detectors (Rosenbauer and Fahr,
1980). Meanwhile, this early idea was converted into hardware reality
in the form of a neutral gas experiment that soon will be flown on the
ULYSSES space probe (Rosenbauer et al., 1984). As described by
Rosenbauer et al. (1984), the detector registers neutral particles with
energies above 40 eV from a predefined angle of acceptance. The neutrals then hit a specific conversion plate made out of LiF-coated lead
glass and are converted into ions and electrons which are counted.
      In view of the threshold energies of the detector, only neutral
interstellar helium or higher atomic mass species can be registered by
the detector. Rosenbauer et al. (1984) have given the total count rate
that is to be expected from interstellar helium hitting the ULYSSES
detector as a function of the flight time of the ULYSSES probe. This
total count rate, however, could be expected only if all helium atoms
described by the local velocity distribution function were collected by
the detector. However, since different helium atoms in the distribution

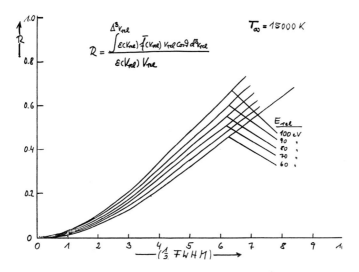

Figure 7. Ratio of real count rates over total count rates for interstellar helium atoms counted by the GAS detector on board ULYSSES for
different relative bulk energies $E_{rel}$.

function approach the detector in flight with different velocity
vectors, not all of the local helium atoms can be registered; some are
below the energy threshold of the detector, others are not within the
angle of acceptance.

     This is illustrated by Figure 7, where the ratio of the real count
rate and the total count rate is displayed as a function of the FWHM
angle of the instrument for different relative energies of the bulk
helium flow. These calculations are carried out taking into account the
particle conversion efficiency of the detector and the realistic helium
velocity distribution function in space for LISM helium with a temperature of 15 000 K.

     It is evident that the GAS detector will receive the maximum count
rate if it is pointed exactly into the anti-direction of the relative
velocity of the local interstellar helium atoms with respect to the
spacecraft. The structure of the local velocity distribution function,
however, can be derived only from an analysis of the count rate variation with the offset angle from this direction. This variation of the
count rate as expected from recent theoretical models for the interstellar helium flow is shown in Figures 8 and 9.

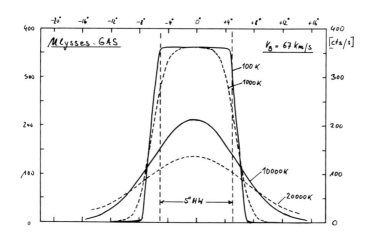

     In Figure 8, we have displayed the variation of the count rates
with this offset angle for a detector with a FWHM angle of acceptance
of 5° at a position of the ULYSSES spacecraft where the relative bulk
velocity of the helium atoms with respect to the detector is 67 km/sec.
The temperature of the LISM helium has been varied between 100 K and
20 000 K.

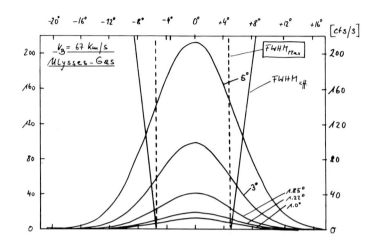

In Figure 9, we have displayed the variation of the count rate with the offset angle from the relative bulk velocity vector that detectors with different FWHM angles of acceptance between $1^{\circ}$ and $5^{\circ}$ would register if the interstellar helium would have a LISM temperature of 15 000 K.

We hope very much that the forthcoming measurements of the ULYSSES GAS experiment, scheduled to begin in May 1986, will give us important information needed for a better understanding of the neutral LISM gases in the heliosphere.

References

Alfvén, H.: 1954, On the Origin of the Solar System  Oxford, Clarendon
    Press
Axford, W.I.: 1972, Solar Wind II, NASA SPR 308
Baranov, V.B., Lebedev, M.G., Ruderman, M.S.: 1979, Astrophys. Space
    Sci. 66, 441
Blum, P.W. and Fahr, H.J.: 1970, Astron. Astrophys. 4, 280
Dalaudier, F., Bertaux, J.L., Kurt, V.G., Mironva, E.N.: 1984, Astron.
    Astrophys.
Danielson, L.R., 1970, Phys. Fluids 13, 2288
Danielson, L.R. and Brenning, N.: 1975, Phys. Fluids 18, 661
Fahr, H.J.: 1968, Astrophys. Space Sci. 2, 474
Fahr, H.J.: 1971, Astron. Astrophys. 14, 263
Fahr, H.J.: 1973, Solar Phys. 30, 193
Fahr, H.J.: 1974, Space Sci. Rev. 15, 483
Fahr, H.J.: 1978, Astron. Astrophys. 66, 103
Fahr, H.J.: 1979, Astron. Astrophys. 77, 101
Fahr, H.J.: 1983, Solar Wind V, NASA SPR 2280, 541
Fahr, H.J., Petelski, E.F. and  Ripken, H.W.: Solar Wind IV,
    NASA SPR W-100-81-31, 543

Fahr, H.J., Ripken, H.W. and Lay, G.: 1981, Astron. Astrophys. 102, 359
Fahr, H.J. and Ripken, H.W.: 1984, Astron. Astrophys. 139, 551
Fahr, H.J. and Ripken, H.W.: 1985, preprint Univ. Bonn
Fahr, H.J., Nass, H.U. and Rucinski, D.: 1985, Astron. Astrophys. 142,
    476
Feldman, W.C., Lange, J.J. and Scherb, F.: 1972, Solar Wind II,
    NASA SPR 308, 684
Haerendel, G., 1982, Z. Naturforsch. A37, 728
Holzer, T.E.: 1977, Rev. Geophys. Space Phys. 15, 467
Holzer, T.E. and Leer, E.: 1973, Astrophys. Space Sci. 24, 335
Klemens, Y.: 1985, Astron. Astrophys. 148, L5
Lallement, R., Bertaux, J.L. and Kurt, V.G.: 1985, J. Geophys. Res. 90,
    1413
Möbius, E., Piel, A. and Himmel, G.: 1979, Z. Naturforsch. A34, 405
Nass, H.U. and Fahr, H.J.: 1985, preprint Univ. Bonn
Parker, E.N.: 1963, Interplanetary Dynamical Processes, Interscience,
    New York
Petelski, E.F., Fahr, H.J., Ripken, H.W., Brenning, N. and Axnas, I.:
    1980, Astron. Astrophys. 87, 20
Pilipp, W.G.: 1983, Solar Wind V, NASA SPR 2280, 413
Ripken, H.J. and Fahr, H.J.: 1983, Astron. Astrophys. 122, 181
Rosenbauer, H. and Fahr, H.J.: 1978, Technical Proposal ISPM
Rosenbauer, H., Fahr, H.J., Keppler, E., Witte, M., Hemmerich, P.,
    Lauche, H., Leidl, A. and Zwick, R.: 1984, ESA SPR 1050, 125
Rucinski, D.: 1985, PhD Thesis, Warsaw
Steinitz, R.: 1983, Solar Phys. 83, 379
Steinitz, R. and Eyni, M.: 1980, Astrophys. J. 241, 417
Schwenn, R.: 1983, Solar Wind V, NASA SPR 2280, 489
Thomas, G.E.: 1978, Ann. Rev. Earth Planet. Sci. 6, 173
Tschernetskij: 1969, Vredenije v fiziku plasmy, Atomizdat, Moscow
Wallis, M.K.: 1984, Astron. Astrophys. 130, 200
Witt, N., Ajello, J.M. and Blum, P.W.: 1979, Astron. Astrophys. 73, 272
Wu, F.M. and Judge, D.L.: 1978, Astrophys. J. 225. 1045

INTERSTELLAR GAS PARAMETERS AND SOLAR WIND ANISOTROPIES DEDUCED FROM H

AND HE OBSERVATIONS IN THE SOLAR SYSTEM

J.L. BERTAUX, R. LALLEMENT, E. CHASSEFIERE
Service d'Aéronomie du CNRS,
91371 - Verrières-le-Buisson, France

ABSTRACT - The results of four space experiments looking at resonance scattering of H and He atoms in the solar system are reported. The characteristics of the Local Interstellar Medium are determined to be $T = 8000 \pm 500$ K, and the wind velocity is $20 \pm 1$ km.s$^{-1}$ . Larger values are found for Helium as a result of an artefact of the modeling. The degree of ionization is about 30 % in the LISM. Such conditions were predicted to prevail by Mc Kee and Ostriker (1977), at the interface between a dense and cold cloud and a hot and tenuous ISM. In addition, the solar wind mass flux was measured at all latitudes, showing a 50% decrease when going from the ecliptic to the pole in the 1976-1977 period, linked to the presence of a polar coronal hole. Predictions are made concerning some results which will be collected with ULYSSES.

1. INTRODUCTION

The flow of interstellar gas through the solar system can be detected by resonance scattering of solar photons on H atoms (Lyman $\alpha$ , 121.6 nm) and He atoms (He I 58.4 nm). Two identical photometers were placed on PROGNOZ 5 and 6 spacecrafts in 1976 and 1977, and two EUV spectrophotometers were flown aboard VENERA 11 and 12, collecting measurements after the encounter with Venus in December 1978. The analysis of these 4 experiments has been recently completed, and the purpose of this paper is to present a summary of the results, as well as to draw some conclusions and make predictions about what will be observed during the ULYSSES mission.

Outside the heliosphere (beyond the heliopause) the Local Interstellar Medium (LISM) is characterized by the densities of H, He, $H^+$, $He^+$ , the temperature T and the velocity $- \vec{V}_W$ in respect to the solar system. Ions are prevented from penetrating the heliopause, whereas neutrals can invade the heliosphere. There, they suffer a multiform interaction : ionization by charge exchange with solar wind protons and EUV photoionization (characterized for H by Td, the lifetime of one H atom at 1 AU), gravitation Fg, counteracted by solar

435

*R. G. Marsden (ed.), The Sun and the Heliosphere in Three Dimensions, 435–440.*

Lα radiation pressure Fr = μFg for H atoms. This interaction will modify the space density distribution and the velocity distribution. An ionization cavity of elongated shape is formed for H, whereas a spectacular focusing cone is observed along the axis $-\vec{V_w}$ , opposite to the motion of the sun with respect to the LISM (Figure 1). A complete model of this interaction has been computerized (Lallement et al., 1985a).

The study of the density and velocity distributions of H and He convey therefore information on three different topics, interesting three different scientific communities :
  - characteristics of the LISM just outside the heliosphere
  - solar parameters : radiation pressure μ and ionization power of the sun
  - mechanism of interaction between the heliosphere and the LISM.

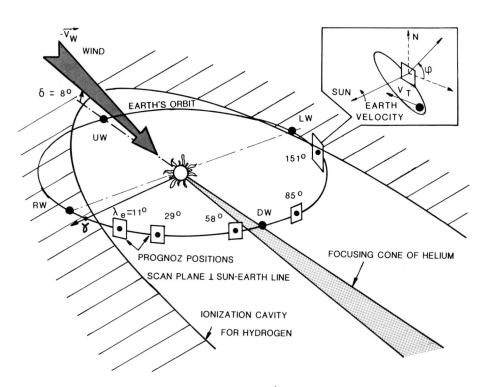

Fig. 1 The flow of interstellar gas - $\vec{Vw}$ results in a focusing cone of helium and in a ionization cavity for hydrogen, elongated in the down wind region. Five positions of PROGNOZ (in orbit around the Earth) are indicated in the solar system.

## 2. INTERSTELLAR RESULTS

Characteristics of the LISM in the vicinity of the sun derived from PROGNOZ and VENERA observations are summarized in Table I. Temperature and velocity of the LISM were derived independently from Helium and Hydrogen observations. The photometric distribution of He I 58.4 nm, recorded from PROGNOZ photometers, and in particular the position and the angular size of the focusing come of Helium were analysed by Dalaudier et al (1984). The L $\alpha$ photometers on PROGNOZ 5 and 6 were equipped with a hydrogen absorption cell, working as a "negative" spectrometer, since it absorbs all photons within $2 \times 10^{-2}$ Å of the L$\alpha$ resonance wavelength. The spectral scanning is provided by a spatial scanning, which modulates the Doppler shift according to the observation geometry, in a way described as the Doppler Angular Spectral Scanning (DASS) method devised by Bertaux and Lallement (1984). The **reduction factor** R (the ratio by which the L$\alpha$ intensity is reduced when the H cell is activated) measured in five planes perpendicular to the Sun-Earth line was compared to a model (Lallement et al., 1985 a) and characteristics of interstellar Hydrogen were thus derived (Bertaux et al., 1985) as summarized in Table I. On Figure 2 is an example of fitting the measured R with models at 6000, 8000 and 10000 K.

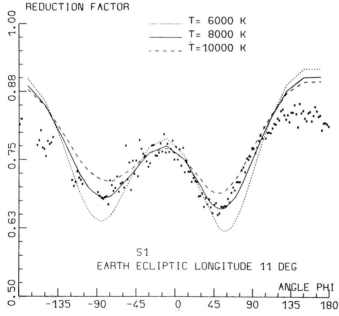

Fig. 2 The reduction factor (black dots) measured in a plane is compared to a model at three temperatures. Together with observations at other places, the best fit temperature is T = 8000 + 500 k. The fit is poorer around the down wind hemisphere (90 < $\phi$ < 270°), probably because of elastic collisions with solar wind protons, a mechanism not included in the model.

TABLE I : LISM CHARACTERISTICS

| Parameter | | Hydrogen | Helium |
|---|---|---|---|
| Temperature T(k) | | 8000± 500 | 16000 ± 5000 |
| Velocity Vw(km/s) | | 20± 1 | 27 ± 3 |
| Ecliptic longitude | $\lambda$w(°) | 251± 2 | 254.5± 3 |
| Ecliptic latitude | $\beta$w(°)$_3$ | 7.5± 3 | 6± 3 |
| Density n(H), n(He)( cm$^{-3}$ ) | | 0.065± 0.01 | 0.01± 0.005 |

In order to derive the densities from photometric observations, knowledge is required of the calibration factor of the instrument and the exact value of the solar flux, in addition to the knowledge of the other parameters T, Vw, Td. Thanks to observations of the upper atmosphere of Venus with VENERA 11 and 12 EUV spectrophotometers, which enabled the subsequent photometric observations of the LISM with the same instrument to be calibrated, absolute densities could be derived.

The **direction** of Vw, from which the interstellar wind blows, is identical if determined either from H or from He observations, whereas there is a strong discrepancy for the modulus of Vw and the temperature T. It has been pointed out (Lallement, 1983 ; Bertaux, 1984) that in order to have the same direction of $\vec{V}_w$ but different velocity modulus would require an extraordinary coincidence, that the intrinsic difference of flow vectors in the Local Standard of Rest Vlsr (H) - Vlsr (He) would be parallel to $\vec{V}_w$ , which depends on the particular velocity of the sun with respect to the LSR (Figure 3).

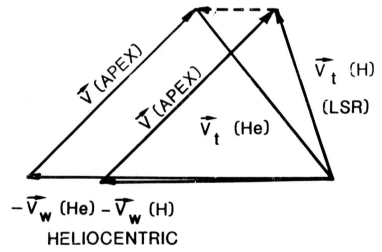

Fig. 3. In the solar system, the vectors Vw(H) and Vw(He) have the same direction, but seem to have different modulus. Therefore, the difference $\vec{V}$w(H)-$\vec{V}$w(He) , which is independant of the coordinate system, is parallel to Vw(H), which includes the particular motion of the sun $\vec{V}$ (APEX). This suggests an artefact in the determination of the modulus of Vw(He).

Therefore, the difference in modulus is probably spurious. Spectroscopic measurements for H are more accurate than photometric measurements for He, and we believe the H results to represent reality. Since the shape of the focusing Helium cone depends roughly on √T/Vw , reducing Vw will also tend to reduce T for helium as determined from the angular size of the cone, also alleviating the temperature discrepancy (Bertaux et al., 1985). Elastic collisions with solar wind protons, not accounted for in the model, could be the source of the enlarged helium cone (Fahr, 1985).

The fact that the density ratio n(H)/n(He) is smaller than the cosmic ratio 10 indicates that the LISM is partially ionized ( ≃ 30%). All these characteristics point toward a phase of the LISM at the interface between a cold and neutral cloud, and a hot and tenuous inter-cloud medium (Mc Kee and Ostriker, 1977).

## 3. SOLAR WIND ANISOTROPIES

The distribution of Lα intensity as recorded by PROGNOZ showed the clear signature that the ionization rate of H atoms was smaller at high ecliptic latitudes than in the ecliptic plane (Figure 4), demonstrating indisputably for the first time that the solar wind mass flux (which is responsible for most of the ionization by charge exchange) is about a factor of 2 smaller at the solar pole, for the solar minimum conditions of 1976 (Lallement et al., 1985b). When coupled with observations of density in a polar coronal hole, taking a relevant reduced solar flux of $\simeq 1.5 \times 10^8$ protons $cm^{-2} s^{-1}$ changes very much the coronal temperature profile (Lallement et al., 1985c).

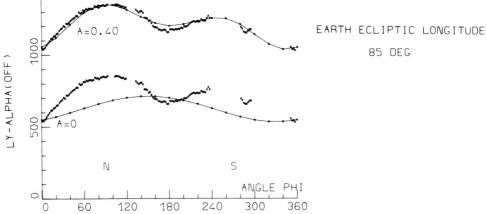

Fig. 4 : The measured Lα intensity (black dots) in a plane is compared to two models of the ionization rate (solid lines). The letter A = 0 designates an isotropic model, and A = 0.40 a model in which the ionization rate decreases by 40% at the ecliptic pole. Data are reproduced twice, being vertically displaced by 500 counts. There is more Lα intensity, therefore less ionization, in the data than in the model with A = 0, in the directions of north and south ecliptic ples (after Lallement et al, 1985b).

Therefore, a new method to monitor the 3-D structure of the solar wind
mass flux through Lα mapping can be used in future space
experiments, without leaving the ecliptic plane.

## 4. CONCLUSIONS AND PREDICTIONS FOR ULYSSES

The GAS instrument placed on board ULYSSES (Rosenbauer et
al., 1983) devoted to the direct detection of neutral Helium atoms
should be able to measure both the velocity and the temperature of
interstellar He atoms. We predict values identical to our H
determinations of $20 \pm 1$ km s$^{-1}$ and $8000 \pm 500$ k. If the instrument
is sensitive to Lα stray light, despite all precautions taken in the
design, it could still be used to map the Lα intensity and detect
the solar wind latitude anisotropies, early in the mission (1986).
When ULYSSES goes out of the ecliptic plane and come back over the
poles, in situ measurements of the solar wind flux will be made at all
heliographic latitudes. If the feature that we found in 1976 is a
permanent one, it will be confirmed by the **in situ** measurements. If it
is only a solar minimum feature, the solar wind mass flux measured by
ULYSSES will remain constant at its ecliptic value throughout the
mission, since the spacecraft climb out of ecliptic during the
increase toward solar maximum.

## REFERENCES

Bertaux J.L. : 1984, Proceedings of IAU colloquium n° 81, "The Local
Interstellar Medium", p. 3-23, NASA CP 2345
Bertaux J.L., Lallement R. : 1984, Astron. Astrophys. 140, 230
Bertaux J.L., Lallement R., Kurt V.G., Mironova E.N. : 1985, Astron.
Astrophys., in press
Chassefière E., Bertaux J.L., Lallement R. : 1985, Astron. Astrophys.,
submitted
Dalaudier F., Bertaux J.L., Kurt V.G., Mironova E.N. : 1984, Astron.
Astrophys., 134, 171
Fahr H.J. : 1985, Astron. Astrophys., 142, 476
Lallement R. : 1983, Thèse de 3è cycle, Université P et M. Curie,
Paris
Lallement R., Bertaux J.L., Kurt V.G., Mironova E.N. : 1984, Astron.
Astrophys., 140, 243
Lallement R., Bertaux J.L., Dalaudier F. : 1985a, Astron. Astrophys.,
in press
Lallement R., Bertaux J.L., Kurt V.G. : 1985b, Journal Geophys. Res.,
90, 1413
Lallement R., Holzer T.E., Munro R. : 1985c, Journal Geophys. Res.,
submitted
Mc Kee C.F., Ostriker J.P. : 1977, Astrophys. J., 218, 148
Rosenbauer H., Fahr H.J., Keppler E., Witte M., Hemmerich P.,
Lauche H., Loidl A., Zwick R. : 1983, in "The International Solar
Polar Mission - Its Scientific Investigations", Wenzel K-P, Marsden
R.G., Battrich B. (Eds.), ESA SP-1050, 123.

# THE 3-DIMENSIONAL STRUCTURE OF THE INTERPLANETARY DUST CLOUD

R.H. Giese and G.Kinateder
Ruhr-Universität Bochum
Bereich Extraterrestrische Physik
P.O. Box 10 21 48
4630 Bochum 1, F.R.G.

ABSTRACT. The three dimensional structure of the interplanetary dust cloud can be investigated by analysis of particle impacts and of zodiacal light. Optical models are presented and modifications by spatial variations of albedo are discussed. Relations between the spatial density and orbital inclinations of micrometeoroids provide the connection to the ULYSSES dust experiment. In future missions optical probing at low elongations and at altitudes up to about 0.5 AU above the ecliptic plane are most promising for investigations of 3D-problems.

## 1. INTRODUCTION

The 3-dimensional distribution (3D-distribution) of particle number densities, particle sizes, and of the physical properties of interplanetary dust grains can be obtained in principle by two methods:
The first method implies direct measurements and analysis of particle impacts on spaceborne detectors and evaluation of meteoritic impacts on the earth's atmosphere. Orbital elements can be derived from impact velocities for different particle classes. An advantage of this method is direct access to the dynamics of the interplanetary dust cloud. The disadvantages are selection effects and the limited number of impact events which put serious limits on the statistical conclusions.

The second method is based on sunlight scattered by interplanetary dust grains (zodiacal light). In this case (see Fig.1) a photometer (at 0) collects the integrated contributions of all dust particles contained in the viewing cone along its line of sight (LOS). Therefore a huge number of particles is involved, which allows to derive the 3D-distribution without statistical problems. However, due to the integration along the line of sight and to the involvement of optical scattering properties, local number densities can only be derived indirectly i.e. by model computations using some simplifying assumptions, except in a few special cases. While neither of the two methods is satisfying by itself alone, combination of both is powerful

441

*R. G. Marsden (ed.), The Sun and the Heliosphere in Three Dimensions, 441–454.*
© *1986 by D. Reidel Publishing Company.*

due to the involvement of aspects from two very different fields
(dynamics, optics). In the following sections results  obtained by the
optical method will be presented and discussed. Then it will be
sketched how both methods can be combined especially in the case of
ULYSSES. Furthermore it will be shown what type of observations should
be performed on future missions investigating the interplanetary dust
cloud in three dimensions.

## 2.   OPTICAL MODELS

### 2.1  Principle and Geometry

The principle of the optical method is shown in Fig.1 for an observer
(O) in the ecliptic plane looking along his line of sight (LOS) in a
direction defined by the ecliptic longitude with respect to the sun
$(\lambda-\lambda_O)$ and by the ecliptic latitude ß. Alternatively the viewing
direction may be presented by the angle of elongation ($\varepsilon$) between the
LOS and the direction from the observer to the sun and by the angle of
inclination (j) between the scattering plane (SPO) and the ecliptic
plane.

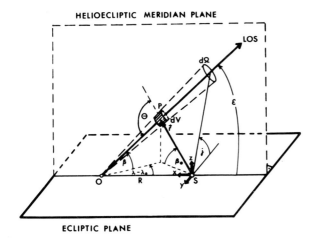

Figure 1. Basic Geometry
0     Observer
S     Sun
'LOS' Line of Sight
SPO Scattering Plane
P(x,y,z) Location of dV
$\vec{r}$     Position Vector (SP)

No distinction is made
hereafter between the
Ecliptic and the Symmetry
Plane.

A volume element dV located on the LOS at a distance $\ell$ from the
observer and subtending the spatial angle $d\Omega$ of his viewing cone
contains $n(\vec{r}) \cdot \ell^2 d\Omega\ d\ell$ particles, where n is the number density at
the location P. These particles are illuminated by the spectral flux
density $F_O \cdot (AU/r)^2$, where $F_O$ is the solar spectral irradiance (Watt
$m^{-2}nm^{-1}$) at 1 astronomical unit (AU). If $\sigma(\Theta)$ is an average scattering
function(differential scattering cross section,  $m^2/sr$) per one
particle with respect to the scattering angle ($\Theta$), the irradiance from
LOS at the observers position is $n\ell^2\sigma\ F_O(AU/r)^2 \cdot (1/\ell^2)\ d\Omega\ d\ell$, and
therefore the surface brightness (radiance $Wm^{-2}sr^{-1}nm^{-1}$) observable at
the observers position in the direction of $\varepsilon$, j is given by the
"brightness integral".

$$(1) \qquad I(R,\varepsilon j) \;=\; F_o \cdot (AU)^2 \int_{\ell=0}^{\infty} \frac{n(\vec{r}) \; \sigma(\Theta, \vec{r})}{r^2} \, d\ell$$

It is evident, that the contribution of each volume element is strongly biased by the distance r from the sun due to the illumination factor $(AU/r)^2$ and that inversion of the brightness integral can provide only the 'volume scattering function' VSF $= n(\vec{r}) \cdot \sigma(\vec{r},\Theta)$ and not without further assumptions n and $\sigma$ separately. This means, that any spatial surfaces of equal VSF derived from measurements of the zodiacal light can be considered as surfaces of equal number densities n only in the case, if $\sigma(\vec{r},\Theta) = \sigma(\Theta)$, i.e. if the average physical properties of the grains are independent of location in the solar system. Nevertheless this special assumption will be adopted in the following sections as a first approach to the possible types of 3D-distributions for $n(\vec{r})$. Later it will be discussed, how such simple interpretations have eventually to be modified.

2.2  Scattering Function $\sigma(\Theta)$

Although in principle $\sigma(\Theta)$ can only be derived from (1) if it is independent of location and if $n(\vec{r})$ is known, there exists already some knowledge about the scattering function of interplanetary dust particles. This information stems partially from laboratory measurements of irregularly shaped particles (cf. Weiss-Wrana 1983) and partially from earlier analysis of the zodiacal light (Dumont and Sanchez 1975, Leinert et al. 1976, Mujica et al. 1980).
    If the particle number density follows a power law $n \sim r^{-\nu}$ it can be shown (Giese and Dziembowski 1969, Leinert 1975) that the surface brightness observable at a fixed elongation from a spacecraft at solar distance R in the ecliptic plane changes with R according to $I(\varepsilon,R) \sim R^{-(\nu+1)}$ .

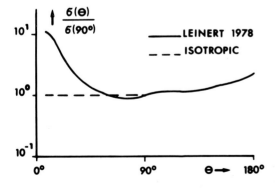

Figure 2.
Scattering function
of interplanetary
dust (relative units)

——— Leinert 1978
--- Isotropic.

Zodiacal light observations by Helios are in good agreement with a power law with $\nu = 1.3$ for the region of r = 0.08 to 1.5 AU (Leinert et al. 1981). This allowed Leinert et al. 1980 to derive an improved average scattering function, which is shown in Fig.2. This curve is

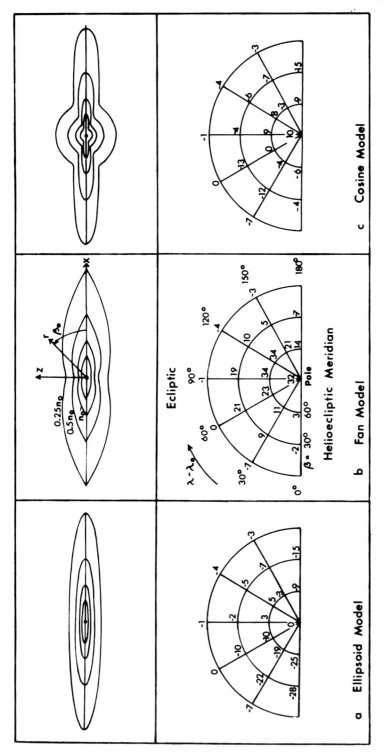

Figure 3. Upper part: Lines of equal particle number density for $n/n_0$ = 2.0, 1.0, 0.5, 0.25 in the
helioecliptic meridian plane; $n_0$ density at 1 AU in the ecliptic plane; $\beta_0$ helioecliptic
latitude; r solar distance.
Lower part: Relative deviations (in percent) of the predicted intensities (Rittich 1985)
from the observational tables of Levasseur-Regourd and Dumont 1980; $\beta$ ecliptic latitude;
$\lambda-\lambda_0$ ecliptic longitude with respect to the sun as seen from the observer.

also sufficiently in agreement with the laboratory measurements and with the independent results obtained by Dumont and Levasseur-Regourd 1985a,b using their new method of "nodes of lesser uncertainty". Although a more isotropic scattering function can not be excluded for scattering angles $\Theta$ < 90°, we adopt the function of Fig. 2 for further discussion.

## 2.3   3D-Models

Very often the 3D-dependence of particle number densities on r is approximated by a product $n = n_0 \cdot (r/AU)^{-\nu} f(r,z)$, where $n_0$ is the number density at the earths orbit and z the altitude 'above' the ecliptic plane (Fig.1). Models of this type are shown in the upper part of Fig.3. In each case the isodensity lines of n are drawn in the helioecliptic meridian plane (x-z plane) of the coordinate system (Fig. 1) which is centered at the sun with z pointing towards the northern pole of the ecliptic. In all cases rotational symmetry is assumed about the z axis and no distinction will be made between the ecliptic and the plane of symmetry of the zodiacal light, which slightly deviates   from the ecliptic plane (see e.g. Morfill and Grün 1979, Leinert et al. 1980). Due to the rotational symmetry it is convenient, to use polar coordinates (Fig.1), e.g. the solar distance r and the heliocentric ecliptic latitude $\beta_\odot = \arcsin(z/r)$.

### 2.3.1   Ellipsoid Models (Fig. 3a)   This type of models was used by Giese and Dziembowski 1969 in their discussions of optical probing the 3D-structure of interplanetary dust cloud and by Dumont 1976 to explain his observations in terms of space densities. In such models, the surfaces of equal number density are spheroids (i.e. ellipses in the x-z plane)

$$(2) \qquad n = n_0\, r^{-\nu}\, (\, 1 + (\, \gamma\, \sin\beta_\odot\,)^2\,)^{-\nu/2}.$$

The first parameter $\nu$ determining the decrease of number density in the ecliptic plane with solar distance is between 1.0 and 1.5 (Dumont and Sanchez 1975, Hanner et al. 1976, Leinert et al. 1981). It will be adopted here as $\nu = 1.3$.
        The second parameter $\gamma$ determining the concentration of dust towards the symmetry plane was derived by Dumont 1976 from the pole to ecliptic ratio of intensity of the zodiacal light which   led   to a ratio 7:1 of the major to the minor axes of the ellipsoids, equivalent to $\gamma = 6.9$. By using a least square fit along the circle $\varepsilon = 90°$ perpendicular to the helioecliptic meridian (Fig.1) one arrives with $\nu$ = 1.3 at $\gamma = 6.5$. Along the helioecliptic meridian a better fit was achieved for $\gamma = 4.5$. Since, however, $\gamma = 6.5$ is exclusively based on the more reliable portions $\Theta \geq \varepsilon = 90°$ of the scattering function, we adopt this value for further discussion.

### 2.3.2   Fan Models (Fig. 3b)   The analytical form of this type is

$$(3) \qquad n = n_0\, r^{-\nu}\, \exp\,(-\gamma|\,\sin\beta_\odot|).$$

Lines of equal n are fan-like in the x-z projection indeed for the
special case $\nu = 0$. More recent work is usually based on $\nu = 1.3$ and
$\gamma = 2.1$ as adopted by Leinert et al. 1980, 1981 with respect to Helios
results. This approach was also used for theoretical investigations
(Fahr et al. 1981) and in comparison with infrared results (Hauser et
al. 1984). If predictions with $\nu = 1.3$ are compared with observations
along the helioecliptic meridian (Levasseur-Regourd and Dumont 1980)
the best least square fit is achieved quite in agreement with Leinert
et al. by values of $\gamma$ close to 2.1. Along a circle $\varepsilon = 90°$ the
dependence of the deviations on $\gamma$ is weak and the best fit value of
$\gamma \approx 3.0$ would dramatically detoriate the approximation in the
helioecliptic meridian (Kinateder 1984). We adopt hereafter $\gamma = 2.1$.

### 2.3.3 Sombrero Models and Cosine Models (Fig. 3c) These models,
resembling in the x-z projection a sombrero hat, were proposed 1976
during the IAU General Assembly at Grenoble by Dumont to improve the
fit in the helioecliptic meridian compared to the ellipsoid models.
They are based on the superposition of a spherically symmetric
distribution of dust and a component producing a concentration of n
towards the symmetry plane. Dumonts four parametrical form $n = n_0 r^{-\nu}$
$(k + (1-k) \exp(-\gamma_1 |\sin\beta_\odot|^{\gamma_2})$ leads with $k = 0.15$, $\gamma_1 = 13$, and $\gamma_2 = 1.9$
to practically identical surfaces of equal density as a three
parametrical subgroup of sombrero models named hereafter cosine models

$$(4) \qquad n = n_0 \ r^{-\nu} \ (k + (1-k) \ (\cos\beta_\odot)^\gamma), \qquad \gamma \text{ even}$$

with the special choice of $k = 0.15$ and $\gamma = 28$. This was shown by
Rittich (1985) who, however, achieved her best fit to the
observational table of Levasseur-Regourd and Dumont 1980 by using $k = 0.2$ and $\gamma = 44$. These values will be adopted in the following
discussions.

### 2.3.4 Layer Models Spatial distributions of the type $n = n_0$
$f_1(r) . f_2(z)$ ("layer models") produce layers of constant number
densities which are approximately parallel to the x,y plane as long as
$f_1$ is nearly constant. Examples are $n = n_0 r^{-\nu} \exp(-\gamma|z|)$ or the "Gauß
models" $n = n_0 r^{-\nu} \exp(-(\gamma z)^2)$. These models predict for scans of the
zodiacal light at low ($\varepsilon \leq 15°$) elongation around the Sun nearly
circular isophotes (Giese 1971, 1975) since the line of sight remains
in the well illuminated regions mainly at small z values where the
exponential factor is close to one. Rocket observations of the inner
zodiacal light are in contradiction to this behavior (Leinert et al.
1976). Therefore this type of models can be excluded from further
consideration.

### 2.3.5 Lobe Models All models referred to in the previous sections,
have above any point (x,y = const.) of the symmetry plane a monotonous
decrease of particle number densities with increasing $|z|$.Contrary to
this Buitrago et al. 1983 proposed a model, which shows an additional
part of the particle population concentrated about larger angles

($\beta_\odot \approx 75°$) of heliocentric latitude. Taking into account the rotational symmetry around the z-axis this second "lobe" in the $n(r,\beta_\odot)$ diagram would present in three dimensions a cone of high number densities producing a bimodal function $f(\beta_\odot)$. As Rittich (1985) has demonstrated, lobe models can be generated analytically by the expression

$$(5) \qquad n = n_o\ r^{-\nu}\ (\cos\beta_\odot)^2\ (\gamma_1\ \sin^2\beta_\odot - 1)^{\gamma_2}\ ,$$

This approach was used by Giese et al. 1984 for further investigations. It could be shown, that lobe models are in obvious contradiction to some observational data: If one scans around the sun at low elongation (e.g. $\varepsilon = 15°$), the line of sight misses at medium ($j = 45$) angles between the scattering plane and the plane of symmetry to penetrate the characteristic cone of high concentration. This results in a pronounced minimum of intensity, which was definitely not observed in the photometry of the inner zodiacal light (Leinert et al. 1976).

## 2.4   Interpretation

The lower part of Fig. 3 presents for the corresponding models the deviations (percent) of the intensities $I(\lambda-\lambda_0,\beta)$ predicted by Rittich 1985 from the oservational data (Levasseur-Regourd and Dumont 1980). They show the same features as found earlier: The ellipsoid model approximates in a nearly perfect way the intensities along a great circle at $\varepsilon = 90°$ of elongation but produces much too low intensity at small elongations in the helioecliptic meridian (Dumont 1976). On the other hand the fan model (Leinert et al. 1980) and its modification (Leinert et al. 1976) fit very well at small ($\varepsilon \lesssim 30°$) elongations in the helioecliptic meridian. It produces however untolerable high intensities at the ecliptic pole and also along the circle with $\varepsilon = 90°$. The reason is obvious from Fig. 4, which shows the isodensity lines $n(r,\beta_\odot)$ for $n/n_0 = 0.8$ and $0.2$ in the helioecliptic plane for the ellipsoid-, fan-, and cosine model, respectively. The lines of sight for some viewing directions are also shown. It is evident, that for $\varepsilon = 30°$ and $60°$ the ellipsoid model must produce a much lower intensity due to its rapid decrease of density in the well illuminated inner parts of the zodiacal cloud in comparison to the fan model. This effect is partially compensated in the case of the cosine model by the additional bulge of the isodensity lines. At the ecliptic pole ($\varepsilon = 90°$), this bulge is not as much efficient, since it contributes only by its thin and faintly illuminated outer regions. Therefore the cosine model like similar sombrero models produces only a tolerable (10%) excess of intensity compared to the ellipsoid model (0%), quite contrary to the fan model (32%). At larger elongation ($\varepsilon = 150°$) the isodensity curves of the ellipsoid- and of the cosine model become very similar as do the predicted intensities, which are lower than the observed ones. There are also lower intensities for the fan model, but due to the higher density the deficiency is not as pronounced. Obviously all models underestimate the density somewhat in this

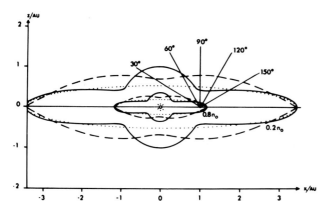

Figur 4.
Explanation of
differences predicted
in the helioecliptic
meridian by the
ellipsoid (•••),
fan (---), and
cosine (——) model.

region. The good overall fit of the sombrero type models is of course
not limited to the special analytical form of the cosine model
preferred here for convenience (see 4.1.1). Analytically quite
different forms such as Dumonts sombrero model or the model of Lumme
and Bowell 1985 do a similar job, however in all cases some
enhancement of density above the sun (bulge) seems to be necessary.

Although the models of Fig. 3 are very different in detail, they
all predict a decrease of density to $n_0/2$ "above" the earths orbit
within $z \lesssim 0.3$ AU. In all cases the decrease is monotonous.

## 3.  SPATIAL VARIATION OF PARTICLE PROPERTIES

In the previous sections it was adopted, that the physical properties
of interplanetary dust (size distribution, particle materials, shape)
are not changing with location. This simplification has been strongly
questioned (Schuerman 1980, Dumont and Levasseur-Regourd 1985a,b),
especially for the region beyond $r = 1$ AU. Inside 1 AU Leinert et al.
1981 observed a decrease of polarization with decreasing r but no
other evidence of spatial variations. Since $\sigma(\theta)$ depends on the
particle size distribution and on the physical properties of each
particle, spatial changes can be expected from dynamical selection
effects or from physical processes changing albedo and structure
(Fechtig 1984, Mukai and Fechtig 1983). A decrease of the average
albedo with increasing solar distance was proposed by Cook 1977,
Fechtig 1984, Lumme and Bowell 1985. Interpretation of infrared (IRAS)
results lead Levasseur-Regourd and Dumont 1985 to the conclusion that
the absorption factor of interplanetary grains increases (i.e. albedo
decreases) between 1 and 1.5 AU and that the decrease of number
density $n \sim r^{-\nu}$ should be described rather more by $\nu \approx 1$ instead of $\nu =$
1.3. Fig. 5 shows the effect of changes in albedo on the surfaces of
equal number density for the fan model. In this case we assumed, that
the generalized geometric albedo A $= \pi\sigma(\theta)/G$ as defined by Hanner et
al. 1981 is proportional to $r^{+0.5}$. To reproduce the zodiacal light
such a change requires lower dust densities in the inner regions of
the solar system (r < 1 AU) and higher densities in the outer parts.

Quantitatively: If G is the average geometrical cross section of one particle, the VSF is proportional to $n(r,\beta) \cdot G \cdot A(r)$. Therefore one arrives for VSF $\sim r^{-1.3}$ with $A \sim r^{-0.5}$ at $n \sim r^{-0.8} f(\beta_\odot)$ (Fig. 5, dashed curves) and not at $n \sim r^{-1.3} f(\beta_\odot)$ as with A being independent of r (Fig.5, solid curves).

Here it was assumed, that the normalized scattering function (i.e. $\sigma(\Theta)/\sigma(90°)$ does not change with albedo as a rough approximation for dark grains. In the general case changes in all physical properties and in the size distribution should be considered. Forthcoming analyis of infrared data and improved dynamical analysis should provide the necessary information to justify such advanced modelling.

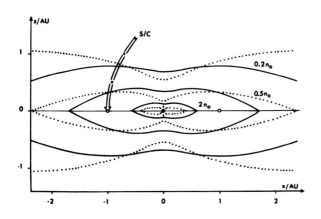

Figure 5.
Isodensity lines of the fan model with
—— constant albedo A
... $A \sim r^{-0.5}$
S/C Spacecraft trajectory.

## 4.  3D – DIAGNOSTICS BY SPACECRAFT

Since interplanetary spacecraft (S/C) change their location (R solar distance, Z altitude above the symmetry plane) they can inspect the interplanetary dust cloud from various positions. Helios 1 and 2, for example, even following trajectories very close to the symmetry plane, provided important information concerning the orientation of this plane and confirmed, that the separation $r^{-\nu} \cdot f(\sin\beta_\odot)$ with $\nu = 1.3$ is applicable inside about r = 1.5 AU (Leinert et al. 1980, 1981).

Much more may be expected from out-of-ecliptic missions. Fig. 5 shows schematically a trajectory entering the zodiacal cloud near 1 AU. In this case (cf. Giese 1979) the S/C starts to penetrate into the inner regions (z < 0.5) about 3 weeks before the perihelion passage. Measurements during this period will be therefore especially important for investigations of the 3D-structure.

### 4.1  Probing by impact measurements

Impact measurements hitherto available e.g. from Helios and Pioneer are not sufficient to derive by themselves the 3D-structure since they provide only poor statistics of orbital elements due to the limited number of impact events, to the low directivity of the sensors, and to selection effects imposed by the location and velocity of the S/C (Schmidt 1980). The same might be true for ULYSSES as far as global 3D-problems are concerned. Therefore analysis of the existing optical data base remains to be necessary also for correct interpretation of impact measurements out of the ecliptic.

4.1.1 <u>Relations to Dynamics</u>  Since it turned out, that at least for a first approach the spatial number density can be expressed by a product $n \sim r^{-\nu} \cdot f(\beta_\odot)$, it is possible to take advantage of a formalism derived by Haug 1958 which relates n with the distribution of the orbital elements a (semimajor axis), e (excentricity), and i (inclination) of interplanetary particles in the case of a random distribution of the nodes and the arguments of perihelion. Provided a separation of the distribution function F(a,e,i) da de di in the form $F_1$ (a,e) $\cdot$ $F_2$ (i) is possible, one obtains $n(r,\beta_\odot) = f_1(r) \cdot f(\beta_\odot)$, where (cf. Leinert et al. 1983)

$$(6) \qquad f(\beta_\odot) \;=\; const \int_{i=\beta_\odot}^{\pi/2} \left[ F_2(i)\ di\ /\ (sin^2 i - sin^2 \beta_\odot)^{1/2} \right].$$

For a completely random orientation of orbital planes a spherically symmetric distribution $n(r,\beta_\odot) = n(r)$ can be expected. The normal vectors on the orbital planes are equaly distributed over the sphere which leads to a distribution function of i as $F_2 \sim sin i\ di$. It also follows (Haug 1958) that a distribution of $F_2 \sim (sin i)(cos i)^\gamma$ where $\gamma$ is an integer leads to a distribution of heliocentric latitudes $f(\beta_\odot)$ which is just proportional to an integer power of $cos\beta_\odot$. Therefore the cosine models offer a convenient direct way to infer from the spatial number density on the distribution of orbital inclinations, although numerical inversion of Haugs integral is possible in more general cases.

4.1.2 <u>Analysis of out-of-ecliptic Impacts</u>  The approach being followed is to derive via Haugs integral from models for $n(\vec{r})$ which are compatible with optical (and infrared) observations the corresponding distributions F(a,e,i) of orbital elements. These can be checked by out-of-ecliptic impact measurements of ULYSSES. For this it is necessary to predict the probability of impacts on the sensors and the expected relative impact velocity vectors taking into account both, the orbital velocity of the S/C and the dynamics of the dust particles as predicted by the models in a heliocentric frame. The impact rates and direction observed by ULYSSES should be an important clue to corroborate or to refute 3D-models.

4.2  Optical Probing

Due to the cancellation of the NASA-S/C there is no opportunity for optical probing on the ULYSSES (ISPM) mission. Preparing a zodiacal light photopolarimeter experiment (ZLE) for the NASA-S/C the ZLE-team obtained results which will be summarized here. They are helpful for

any forthcoming out-of-ecliptic or solar mission (photometer, wide
angle coronograph).

Typical values of brightness of the zodiacal cloud as seen from a
spacecraft during its journey from Jupiters orbit up to z = 2.2 AU and
back to the symmetry plane near R = 1 AU were predicted by Giese 1979.
It turned out, that the brightness of the zodiacal light will be a few
S10V (stars of 10th magnitude/square degree) in most regions of the
sky "above" the S/C as long as it remains at z > 1 AU. On the other
hand, from about two months before perihelion passage, there is an
increase of brightness, which will be appropriate for optical probing.

4.2.1 <u>Inversion Methods:</u>   If the LOS is tangential to the trajectory
of the S/C, the intensity of zodiacal light observable in this
direction changes within a distance ds according to dI ~ n· σ· ds,
which allows to obtain nσ directly without any additional assumption,
however only for one angle of θ and π-θ. This method of "inversion"
(cf. Dumont and Levasseur-Regourd 1980) can be somewhat generalized by
adopting the assumption of rotational symmetry about z and applied for
an out-of-ecliptic mission to viewing directions in the "inversion
plane", which is defined by the velocity vector of the S/C and a
vector parallel to the ecliptic and perpendicular to the line Sun -
S/C (Dumont et al. 1979, Schuerman 1979 and 1980). Furthermore, if
observations are available from two locations $P_1$ and $P_2$ on a straight
line    in the viewing direction (e.g. a secant with the earths
orbit) there exist regions ("nodes") of lesser incertainty, where
parts of the VSF can be determined rather accurately with a few
plausible mathematical assumptions (Dumont and Levasseur-Regourd
1985a, b).

4.2.2 <u>Discrimination of Models</u>  The possibilities to distinguish by
photometric observation between 3D-distributions such as the ellipsoid
model and the fan model where investigated by Kinateder 1984 for a S/C
trajectory perpendicular to the ecliptic plane and penetrating this
plane at Jupiters orbit and at r = 1.2 AU. Observational criteria such
as surface brightness I(Z), intensity ratios or changes of intensity
(dI/dZ) as a function of Z were predicted and considered as useful if
they allowed distinction even in the presence of a superimposed random
"noise" of $|\Delta I| \leq$ 10 S 10. Therefore obvious observing directions like
the ecliptic pole or viewing angles towards zones of maximum
differences in number densitiy between the models turned out to be
unsuitable due to their low absolute intensity, which does not allow
distinction within the observational accuracy.

On the other hand observations in a plane parallel to the
ecliptic plane are promising. If the position of the S/C at an
"altitude" Z ≲ 0.6 AU is in this plane and when the viewing directions
are defined as $\lambda'-\lambda'_o$ , β' in an analog way as in the ecliptic plane,
observations with β' = 0 and $|\lambda' - \lambda'_o| \leq$  60° provide remarkable
differences between the ellipsoid and the fan model. At $|\lambda' -\lambda'_o|$ =
30° for example a difference of 160 S 10 can be expected for Z = 0.3
AU. There are also diagnostic features in the change  ΔI/ΔZ  of
intensity while the S/C is approaching the symmetry plane (Fig. 6).

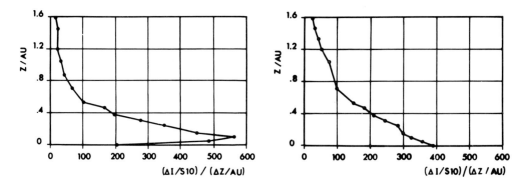

Figure 6.   Differential rate of change of brightness per change of
altitude Z during approach of the S/C to the symmetry plane (left:
ellipsoid model, right: fan model). Viewing direction $\lambda' - \lambda'_o$ = 60°
parallel to the symmetry plane.

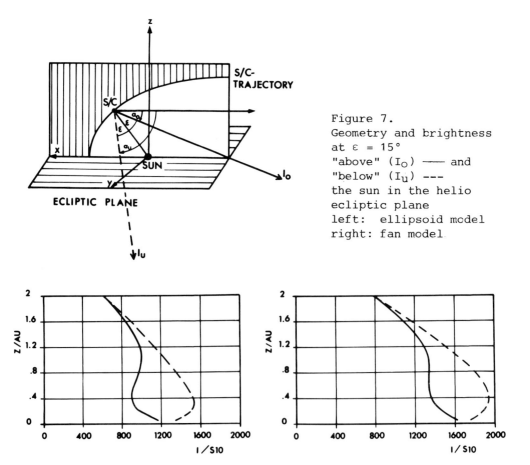

Figure 7.
Geometry and brightness
at $\varepsilon$ = 15°
"above" ($I_o$) —— and
"below" ($I_u$) ---
the sun in the helio
ecliptic plane
left:  ellipsoid model
right: fan model

In the helioecliptic meridian plane observations of the intensities $I_o$ ("above") and $I_u$ ("below") the sun as seen from the S/C (Fig. 7) at the same elongation $\varepsilon$ are rather significant, since the two LOS have the same conditions of illumination and any differences of I are due to the spatial distribution $n(\vec{r})$ provided there are no spatial changes in $\sigma(\Theta)$. Fig. 7 shows for $\varepsilon$ = 15° the run of $I_o$ and $I_u$ for the ellipsoid model and for the fan model, respectively. Not just the absolute intensities, but also the Z value for the location of the maximum of $I_u$ and the relative minimum of $I_o$ are characteristic.

Generally it can be seen, that the most diagnostic features are observed at rather low positions Z < 0.5 AU above the symmetry plane and that low elongations ($\varepsilon$ < 30°) are most favorable due to higher intensity and to more significant differences between the observable intensity curves of the two models. Furthermore most relative features such as ratios of intensity turned out to be rather independent of the scattering function. These results can provide useful suggestions for future mission planning.

Acknowledgements  We thank B. Kneissel for very helpful discussions and Mrs U. Rittich for the permission to use some results of her forthcoming thesis. This work was suported by the Bundesminister für Forschung und Technologie (FKZ 010 NO 142).

5.    REFERENCES

Buitrago, J., Gomez, R., Sanchez, F.: 1983, Planet. Space Sci. 31, 373.

Cook, A.f.: 1977, Icarus 33, 349.

Dumont, R., Sanchez, F.: 1975, Astron. & Astrophys. 38, 405.

Dumont, R.: 1976, Lecture Notes in Physics 48 (H .Elsässer and H. Fechtig, eds.), Springer Verlag Heidelberg, 85.

Dumont, R., Rapaport, M., Schuerman, D., LevasseurRegourd, A.C.: 1979, Space Research XIX, 452.

Dumont, R. and Levasseur-Regourd, A.C.: 1980, IAU Symp. No.90: Solid Particles in the Solar System (I. Halliday and B.A. McIntosh eds.), D.Reidel Publishing Comp. Dordrecht, 67.

Dumont, R. and Levasseur-Regourd, A.C.: 1985a, Planet. Space Sci. 33, 1.

Dumont, R. and Levasseur-Regourd, A.C.: 1985b, IAU Coll. No. 85: Properties and Interactions of Interplanetary Dust (R.H. Giese and P. Lamy, eds.), D. Reidel Publishing Comp., Dordrecht, 207.

Fahr, H.J., Ripken, H.W., and Lay, W.: 1981, Astron. & Astrophys. 102, 359.

Fechtig, H.: 1984, Adv. Space Res. 4, No.9, 5.

Giese, R.H. and v.Dziembowski, C.: 1969, Planet. Space Sci. 17, 949.

Giese, R.H.: 1971, Space Research XI, 255.

Giese, R.H.: 1975, Eldo-Cecles/Esro-Vers. Scient. and Techn. Rev. 7, 43.

Giese, R.H.: 1979, Astron. & Astrophys. 77, 223.

Giese, R.H., Kinateder, G., Kneissel, B., Rittich, U.: 1985, IAU Coll. No.85, 255.

Hanner, M.S., Sparrow, J.G., Weinberg, J.L., Beeson, D.E.: 1976,
    Lecture Notes in Physics 48, 29.
Hanner, M.S., Giese, R.H., Weiss, K., Zerull, R.: 1981, Astron. &
    Astrophys. 104, 42.
Haug, U.: 1958, Zeitschr. f. Astrophysik 44, 71.
Hauser. M.G., Gillet, F.C., Low, F.J., Gautier, T.N., Beichman, C.A.,
    Neugebauer, G., Aumann, H.H., Baud, B., Boggess, N., Emerson,
    J.P., Houck, J.R., Soifer, B.T., and Walker, R.G.: 1984, The
    Astrophysical J. 278, L15.
Kinateder, G.: 1984, Unterschiede von Zodiakallichtmodellen im
    Hinblick auf photometrische Messungen einer künftigen
    "Out-Of-Ecliptic" Raumsonde, Diplomarbeit Ruhr-Univ. Bochum.
Leinert, C.: 1975, Space Sci. Rev. 18, 281.
Leinert, C., Link, H., Pitz, E. Giese, R.H.: 1976, Astron. &
    Astrophys. 47, 221.
Leinert, C.: 1978, personal communication.
Leinert, C., Hanner, M., Richter, I., Pitz, E.: 1980, Astron. &
    Astrophys. 82, 328.
Leinert, C., Richter, I., Pitz, E., and Planck, B.: 1981, Astron. &
    Astrophys. 103, 177.
Leinert, C., Röser, S.,and Buitrago,J.:1983, Astron. & Astrophys. 118,
    345.
Levasseur-Regourd, A.C., Dumont, R.: 1980, Astron. & Astrophys. 84,
    277.
Levasseur-Regourd, A.C., Dumont, R.: 1985, C.R.Acad. Sc. Paris   t.
    300, Serie II, 109.
Lumme, K. and Bowell, E.: 1985, Icarus 62, 54.
Morfill, G.E., Grün, E.: 1979, Planet. Space Sci. 27, 1269.
Mujica, A., Lopez, G., and Sanchez, F.: 1980, Planet. Space Sci. 28,
    657.
Mukai, T., Fechtig, H.: 1983, Planet. Space Sci. 31, 655.
Rittich, U.: 1985, personal communication (Diplomarbeit Bochum).
Schmidt, K.D.: 1980, Bahnelemente von Mikrometeoriten: Analyse von
    Messungen der Sonnensonde Helios 1, BMFT-FB-W 80-036.
Schuerman, D.W.: 1979, Planet. Space. Sci. 27, 551.
Schuerman, D.W.: 1980, IAU Symp. No.90, 71.
Weiss-Wrana, K.: 1983, Astron. & Astrophys. 126, 240.

# THE INTERACTION OF SOLID PARTICLES WITH THE INTERPLANETARY MEDIUM

G.E. Morfill*, E. Grün+ and C. Leinert[o]
*Max-Planck-Institut für extraterrestrische Physik, D-8046 Garching,
+Max-Planck-Institut für Kernphysik, D-6900 Heidelberg,
[o]Max-Planck-Institut für Astronomie, D-6900 Heidelberg

## 1. INTRODUCTION

Craters on the lunar surface proved that a continuous spectrum of interplanetary particulates exists from km-sized boulders (small asteroids and comets) to submicron sized dust grains. The dominant force for all particles, except for the smallest dust grains is the solar gravitational attraction

$$F_{grav} (r) = \gamma \frac{M_\odot m}{r^2} = 2.5 \, s^3 \, \rho \left(\frac{r}{r_o}\right)^{-2} \qquad (1)$$

where $\gamma = 6.67 \times 10^{-8}$ $g^{-1}$ $cm^{-3}$ $s^{-2}$ is the gravitational constant and $M_\odot = 1.99 \times 10^{33}$ g the solar mass. With particle radius s in (cm), its density $\rho$ in (g $cm^{-3}$) and $r_o = 1$ AU the gravitational force is given in dynes. Scattering and absorption of solar radiation by an interplanetary particle leads to a radiation pressure force $F_{rad}$ directed almost radially outward (cf. Burns et al., 1979). The ratio of radiation pressure to gravitational attraction (both forces have the same dependence with solar distance r) has for spherical particles the value (Dohnanyi, 1978)

$$\beta = \frac{F_{rad}}{F_{grav}} = 5.7 \times 10^{-5} \frac{Q_{pr}}{\rho s} \qquad (2)$$

With $Q_{pr}$ being an efficiency factor for the momentum transfer. $Q_{pr} = 1$ for a perfectly absorbing sphere, for real particles, however, $Q_{pr}$ decreases for decreasing s below $10^{-5}$ to $10^{-4}$ cm (i.e. order of the effective wavelength of the solar light). Therefore $\beta$ has its maximum value for absorbing particles (like carbon or magnetite) at $\beta = 2$ to 5 and for dielectric particles (like silicates) at $\beta = 0.5$ to 1. Radiation pressure may dominate in the size regime from $10^{-5}$ to $10^{-4}$ cm, below that it becomes less important again.

*R. G. Marsden (ed.), The Sun and the Heliosphere in Three Dimensions, 455–474.*
© *1986 by D. Reidel Publishing Company.*

The photoelectric effect from the solar UV radiation is expected to develop a net positive charge Q (e.s.u) on interplanetary particles corresponding to a surface potential $\phi$ of about + 5 V

$$Q = \frac{s\phi}{300} = 4.8 \times 10^{-10} \, Ne \qquad (3)$$

where N is the number of elementary charges e on the grain. Once charged, the interplanetary magnetic field (IMF) will exert a force on these particles and hence influence their orbit. The Lorentz force $F_L$ is given by

$$\underline{F_L} = \frac{Q}{c} (\underline{v} \times \underline{B}) \qquad (4)$$

where $\underline{v} = \underline{v_{sw}} + \underline{u}$. Here $\underline{v_{sw}}$ is the solar wind speed (400 km s$^{-1}$ in the radial direction) and $\underline{u}$ is the orbital velocity of the particle (typically 30 km/s) and $\underline{B}$ is the magnetic field strength (about $5 \times 10^{-5}$ g at an angle of 45° with respect to the radial direction at 1 AU). With these values at 1 AU equ. (4) becomes

$$F_L = 1.4 \times 10^{-10} \, s \, \phi \qquad (5)$$

Comparison of (5) with (1) shows ($\phi$ = 5 V and $\rho$ = 2.5 g cm$^{-3}$) that for $s \leq 10^{-5}$ cm the Lorentz force becomes dominant. Besides the direct effects which radiation pressure and electromagnetic force have on submicron sized particles, we will see below that secular effects of these forces influence the distribution of even larger particles.

Dynamics of small interplanetary particles including gravity and radiation pressure have been discussed by Robertson (1937); Wyatt and Whipple (1950); Briggs (1962); Whipple (1967); Singer and Bandermann (1967); Dohnanyi (1978). Electromagnetic effects on particles orbiting the sun have been included by Parker (1964); Morfill and Grün ((1979a); Consolmagno (1979); Barge et al. (1982); Hassan and Wallis (1983). The dynamic effects on interstellar particles entering the solar system were treated by Levy and Jokipii (1976) and Morfill and Grün (1979b). In the following two sections the structure of the interplanetary medium both electromagnetic (II) and particulate (III) are reviewed in order to set the stage for discussion of their interactions in the following sections. In section IV the effects on the small particle population are discussed. Properties of the zodiacal dust cloud are described in section V which may be explained by electromagnetic interactions. Stochastic perturbation theory is derived in section VI and conclusions on the large scale distribution of interplanetary particles are drawn in section VII.

## 2. CHARACTERISTICS OF THE INTERPLANETARY MAGNETIC FIELD AND THE SOLAR WIND

The hot plasma of the solar corona expands into interplanetary space, where the plasma flow, referred to as the solar wind, becomes supersonic a few solar radii above the solar surface. The plasma is fully ionized and consists of electrons, protons, some heavier elements mostly $\alpha$-particles. The coronal magnetic flux is carried together with the coronal plasma into interplanetary space giving rise to the IMF. The field lines are stretched radially outwards and twisted into a spiral by the combined action of the solar wind motion and solar rotation. The expansion of the solar wind occurs in streams of either high or low speeds where each high speed stream is usually embedded within a magnetic polarity domain (magnetic field direction points either towards the sun along the spiral direction or away from it). The characteristic plasma parameters of both types of streams are described in Table I (from Feldmann et al., 1977). Large scale models of the heliospheric magnetic field topology suggest that an equatorial current sheet (which bounds the solar field regions of opposite polarity, see e.g. Pneumann and Kopp, 1971; Schulz, 1973; Svalgaard and Wilcox, 1976; Smith et al., 1978) is wraped, up to solar magnetic latitudes of $\theta_0 \sim 20°$, giving rise to the well known magnetic sector structure (Wilcox and Ness, 1965).

Table I

Typical plasma parameters of the solar wind at 1 AU

| Parameter | low speed | high speed |
|---|---|---|
| proton number density $n_i$ $(cm^{-3})$ | 12 | 4 |
| bulk speed $v_{sw}$ $(km\ s^{-1})$ | 330 | 700 |
| $\alpha$-particle to proton number density $n_\alpha$ $(cm^{-3})$ | 0.04 | 0.05 |
| mean thermal proton energy $E_p$ (eV) | 3 | 20 |
| mean thermal electron energy $E_e$ (eV) | 11 | 9 |

Over a large part of the 11-year solar cycle the polar regions of the sun retain a unidirectional field directed towards the sun in on hemisphere and away in the other. At sunspot maximum this polarity changes (Babcock, 1961; Wilcox and Scherrer, 1972) and the polarity change is preceded and followed by about one year of weak disordered and sectored field regions even at high solar latitudes. In the previous solar cycle 21 from 1969 to 1980 the magnetic field polarity at the solar North pole was positive (field directed away from the sun), during the next cycle (from 1980 to 1991) it is negative and so on.

The heliosphere in which the solar wind plasma and the IMF are dominant extends to $\sim 50$ AU (Morfill et al., 1976; O'Gallagher, 1975). Outside an interaction zone around the heliosphere the interstellar medium governs the plasma and magnetic field conditions. The average IMF at latitude $\theta$, longitude $\phi$ and distance r is described by (Parker, 1985).

$$B_r = \pm B_0 \left(\frac{r_0}{r}\right)^2$$

$$B_\phi = \pm B_0 \frac{r_0}{r} \cos \theta \qquad\qquad\qquad (6)$$

$$B_\theta = 0$$

The coordinate system is solar magnetic, which is inclined with respect to the ecliptic system by an angle $7.3^0$. The constants are $r_0 = 1$ AU and $B_0 \sim 3\times10^{-5}$ G.

The charging of a dust grain is mainly governed by the photoelectric effect. At 1 AU flux of photoelectrons from a metal surface is $2.5\times10^{10}$ electrons $cm^{-2}$ $s^{-1}$ (Wyatt, 1969). For other materials like silica and graphite Feuerbacher and Fitton (1972) found a yield which was up to an order of magnitude lower. This flux has to be compared with the flux of electrons and positive ions from the solar wind plasma onto the same particle. For the solar wind parameters given in Table I one finds a thermal electron flux of $2.4\times10^9$ and $7\times10^8$ $cm^{-2}$ $s^{-1}$ for the low and high speed stream conditions, respectively. The flux of the ions is determined by the bulk motion of the solar wind and amounts to $4\times10^8$ and $3\times10^8$ $cm^{-2}$ $s^{-1}$, respectively. Of the charged particle fluxes the eletron flux dominates but it does not reach the photoelectron flux except for low photoelectron yield materials under extreme high density low speed stream conditions. Therefore the charge of an interplanetary dust particle is generally positive at an electrostatic potential of about $+ 5$ V (Rhee, 1967; Wyatt, 1969; Mukai, 1981). This potential is independent of position in the heliosphere because solar UV radiation, which causes photoelectron emission, decreases as $r^{-2}$ and so does the electron density which governs the recombination.

Besides affecting the charge state of interplanetary dust grains, solar wind ions exert a drag onto the dust grains. For supersonic conditions (i.e. the relative speed between the ions and the grain is much larger than the thermal speed of the plasma ions, which is generally true for the solar wind) the drag force is given by

$$F_D \simeq n_i m_i v_{sw}^2 \pi s^2 \qquad\qquad\qquad (7)$$

where $m_i$ is the proton mass. For a more detailed discussion of this force see Morfill and Grün (1979a), Mukai and Yamamoto (1982). Generally this force can be neglected compared to gravity and radiation pressure.


## 3. PARTICULATES IN INTERPLANETARY SPACE

The sources of solid particles in the solar system are decaying comets, asteroidal debris and interstellar grains penetrating the solar system. Inside the Earth's orbit comets are believed to be the main source for interplanetary dust. However, they don't feed directly into the zodiacal dust cloud but rather release mm-sized and larger grains into bound orbit which eventually get ground-up by mutual collisions into smaller dust grains (Whipple, 1967; Leinert et al., 1983; Grün et al., 1985). Micron-sized dust particles, visible in cometary dust tails, will mostly be expelled from the solar system directly by radiation pressure (Röser, 1976; Delsemme, 1976).

The mass density of the interplanetary meteoritic cloud (micron- to cm-sized objects) at 1 AU is $10^{-22}$ g cm$^{-3}$ which is about 10 times the solar wind mass density there.

The life time of meteor sized particles (s $\geq$ 0.01 cm) is governed by collisions, see Fig. 1 (Grün et al., 1985). In contrast to that the life times of smaller particles are dominated by the Poynting-Robertson drag (Robertson, 1937; Wyatt and Whipple, 1950; Burns et al., 1979) which causes the particles to lose orbital angular momentum by the interaction with solar

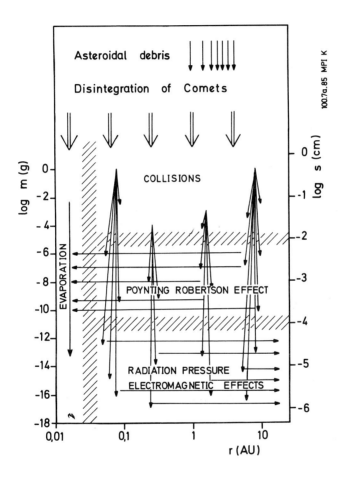

Figure 1. Schematic diagram of dynamical effects which change mass and heliocentric distance of interplanetary meteoroids. Sources for meteoroids are comets and asteroids, loss processes are ejection from the solar system (β-meteoroids which eventually become interstellar grains) and evaporation near the sun (this material is ionized and carried away with the solar wind).

photons and solar wind ions. The life time due to radiation pressure drag of a particle to spiral from an initial orbit (with semi-major axis $a^o$ and eccentricity $e^o$) into the sun is

$$T_{PR} = 7 \times 10^6 \; s\rho Q_{pr} \; a^{o2} \eta \; (e^o) \; \text{years} \tag{8}$$

where $\eta \; (e^o)$ is a factor $(\eta(0) = 1)$ which decreases with increasing $e^o$ (Wyatt and Whipple, 1950). Solar wind ion drag reduces these life times additionally by about 30%.

The inward mass flux of micron to mm-sized particles due to the Poynting-Robertson effect in the ecliptic plane at 1 AU is ~ $2 \times 10^{-22}$ g cm$^{-2}$ s$^{-1}$ (Grün et al., 1985). Because of the increase of the spatial density towards the sun (Leinert et al., 1981) the inward flux at the inner edge of the zodiacal cloud will increase by a fator 2 to 3, depending on where the inner edge is assumed to be (0.1 or 0.03 AU). Sublimation of this meteoritic material occurs in the distance range 0.02 to 0.05 AU from the sun (cf. Fig. 1). The evaporated atoms (mostly C, O, Mg, Si, S, and Fe) will be ionized by photoionization and charge exchange reactions and will subsequently be convected along with the solar wind and the IMF. The initially large gyromotion of the freshly created ions around the magnetic field lines will be adiabatically decelerated during the transit from the source region to 1 AU. But here they will still be detectable at a flux level of about 10 ions cm$^{-2}$ s$^{-1}$ because of their low ionization state in the atomic mass range corresponding to the meteoritic composition (high mass coronal ions have a much higher ionization state than the meteoritic ions, see also Geiss, 1985). So far they have not directly been observed.

Occasionally a comet gets close to the sun and sublimes as a whole. Because of their related orbits sungrazing comets form the Kreutz group of comets. A recent example of this type comet (Comet Howard - Komen - Michels 1979 XI) has been reported by Michels et al. (1982). These comets introduce large amounts of material (~ $10^{16}$ g) about half of which contains high mass meteoritic elements. The short term enhancement of the heavy solar wind ions may be tremendous.

Outside 1 AU asteroidal debris will strongly contribute to the interplanetary particulate cloud. Observations of the thermal radiation with the infrared satellite IRAS (Hauser et al., 1984) showed high intensities at wavelength of 12 to 100 $\mu$m in the ecliptic plane and in bands on both sides about $10°$ away from it. These observations have been interpreted by Dermott et al. (1985) as being due to impact ejecta particles from the known asteroids.

Collisional destruction is the major loss process for mm and large sized meteoroids. Inside 1 AU about 9 t/s are lost from the large particle population due to collisions and have to be replenished by comets (Grün et al., 1985). These 9 t/s of material are redistributed into smaller fragments. About 1.5 to 3 t/s of micron and submicron fragment particles are produced that way. These particles are affected by radiation pressure. A small fragment particle which is generated by a collision between a larger parent meteoroid and another meteoroid will move on an unbound trajectory if its reduced potential energy (gravitation minus radiation pressure) exceeds its kinetic energy. This is especially effective at the perihelion of an eccentric parent particle's orbit, where the kinetic energy and the collision rate are highest. Since the eccentricities of the parent particles are significant even fragment particles of several microns in size can get on hyperbolic trajectories and become β-meteoroids (Zook and Berg, 1975). β-meteoroids have been observed by the dust experiments on the Pioneer 8 and 9 spacecrafts (Berg and Grün, 1973) as well as by the micrometeoroid detectors on board the Helios spaceprobes (Grün et al., 1980). However, the latter was at a much reduced frequency.

The same experiments which observed the flux of β-meteoroids from the general direction of the sun detected also particles on hyperbolic orbits (determined by their excessive speed) but approaching the sun (Wolfe et al., 1976; Grün, 1981). These hyperbolic particles could be interstellar particles, emitted by comets during their approach trajectory to the sun or products of collisions occurring outside 1 AU and involving particles on high eccentric inbound orbits.

## 4. ELECTROMAGNETIC EFFECTS ON HYPERBOLIC PARTICLE ORBITS

There are two types of particles which move on hyperbolic orbits inside the solar system: (1) interstellar dust grains which traverse the inner parts of our system and (2) small particles which are generated from larger bodies either by collisions (β-meteoroids) or by emission from comets when they get close to the sun and sublimation of the icy bonding material takes place (generally inside 5 AU from the sun). The latter particles are brought from bound orbits of their parent bodies onto hyperbolic orbits by the additional significant action of radiation pressure.

The fluxes of these different particle populations on hyperbolic orbits show quite different characteristics. Interstellar dust grains in the vicinity of the solar system amount to $10^{-27}$ g cm$^{-3}$ or $10^{-26}$ g cm$^{-3}$. The higher value is derived from the average interstellar extinction of 1 mag per kpc (Greenberg, 1973) while the lower value assumes that the solar system is currently surrounded by low density warm interstellar material of density $\sim 10^{-25}$ g cm$^{-3}$, one percent of which is in dust (Wood et al., 1985). We will use this lower value in the following discussion. The sun moves with respect to this material at a speed of 20 km/s which amounts to a mass flux of $2 \times 10^{-17}$ g m$^{-2}$ s$^{-1}$. If we assume that most mass is in dust particles of s = 0.1 µm (m ~ $10^{-14}$ g) and s = 0.01 µm (m ~ $10^{-17}$ g) then a flux of $2 \times 10^{-3}$ and 2 particles m$^{-2}$ s$^{-1}$, respectively, is expected to arrive from the solar apex direction. This high flux has not been observed. There are several reasons to explain this deficit in the flux of interstellar particles. These small particles may have β-values in excess of 1 especially if they are made out of absorbing material. For example interstellar particles with β = 2 (s = 0.01 µm graphite grains) will reach just as close as the distance of Jupiter before radiation pressure deflects them out again. Volatile icy constituents will sublimate inside 3 to 5 AU leaving only refractory elements to penetrate the inner solar system. Electromagnetic effects may also be efficient in preventing interstellar particles from reaching the inner part of the solar system, as we will discuss below.

Long period or "new" comets emit on the average 20 t s$^{-1}$ dust by evaporation in the inner solar system (Delsemme, 1976). This amounts to a mass outflux of about $7 \times 10^{-17}$ g m$^{-2}$ s$^{-1}$ at 1 AU or $7 \times 10^{-3}$ particles m$^{-2}$ s$^{-1}$ if most mass is in s = 0.1 μm particles. This flux should be very variable in time and space. One comet per several years contributes significantly to this flux. Only once the in situ detection of such an extended comet tail has been reported (Hoffman et al., 1976).

A steady stream of particles on hyperbolic orbits originates from collisions of larger meteoroids in the inner solar system. Grün et al. (1985) estimate that this flux of β-meteoroids from the solar direction is $10^{-1}$ and $3 \times 10^{-4}$ particles m$^{-2}$ s$^{-1}$ for particles of s ~ 0.01 μm and s ~ 0.1 μm, respectively, and it corresponds to a mass flux of ~ $3 \times 10^{-27}$ g m$^{-2}$ s$^{-1}$ at 1 AU in the ecliptic plane.

Levy and Jokipii (1976) studied the effects of the interplanetary magnetic field on charged interstellar grains penetrating the solar system. They suggested that submicron sized interstellar grains will be largely excluded from the solar system by the sweeping action of the solar wind magnetic field. A more detailed study by Morfill and Grün (1979b) showed that the unipolar field regimes at high latitudes lead either to a "focusing" or "defocusing" of interstellar dust particles with respect to the solar magnetic equator. The stochastic magnetic fluctuations in the equatorial region caused by the warping of the current sheet which separates the polar fields, leads to a diffusive transport of particles of sizes s $<$ $10^{-5}$ cm. They concluded dust that particles with radii s $>$ $10^{-5}$ cm can penetrate deeply into the heliosphere if their incidence direction at the heliopause is almost radially inward and close to the solar magnetic equatorial plane (i.e. approx. within 20° ecliptic latitude). Particle trajetories coming from high solar magnetic latitudes are focused towards the solar magnetic equatorial plane during solar cycles of negative field polarity in the Northern hemisphere. During other solar cycles the inner heliosphere is shielded from interstellar grains approaching from high latitudes.

Numerical trajectory calculations have been performed using the code described by Morfill and Grün (1979a) in order to study the effects of the interplanetary plasma and magnetic field on β-meteorids. The basis was a realistic model of the sector structure of the IMS which extended to 15° ecliptic latitude. As initial conditions it was assumed that particles were created on orbits with inclinations i° = 0° and 30° and with the local Keplerian velocity for circular orbits. Because of their radiation pressure values β $\geq$ 0.5 these particles would leave the solar system on hyperbolic orbits just because of the radiation pressure alone. Fig. 2 shows trajectories of charged (surface potential of 6 V) olivine particles (s = $10^{-5}$ cm) for both possible configurations of the IMF. Depending on the initial phase with respect to the secture structure, particles on initially low inclination orbits are defocused to high ecliptic latitudes during positive Northern magnetic polarity, whereas β-meteoroids produced in the ecliptic and at i° = 30° are focused towards the solar equatorial plane during negative Northern polarities. However, the electromagnetic effect depends strongly on the charge-to-mass ratio of the β-meteoroids. The heaviest particles considered were magnetite particles of s = $4 \times 10^{-5}$ cm which had a charge-to-mass ratio of 1/24 of the olivine particles shown. In this case the deflection by electromagnetic forces from the orbit determined solely by gravitation and radiation pressure was less than 1° in latitude.

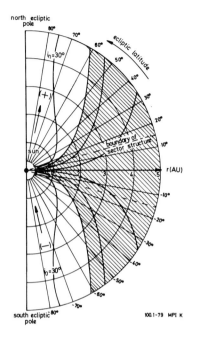 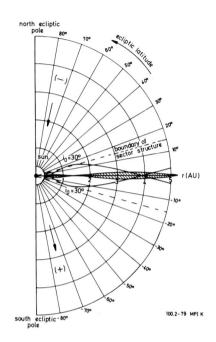

Figure 2.  Trajectories of β-meteoroids (olivine particles, s = $10^{-5}$ cm, β = 0.51, Q/m = 4.7 C/kg) under the influence of the interplanetary magnetic field (IMF). Two initial orbit inclinations have been considered $i^0$ = 0°, +30°. The hatched area corresponds to trajectories with $i^0$ = 0°, but different phases to the sector structure of the IMF.
a) magnetic field polarity at the solar North pole is positive (field directed away from the sun).
b) magnetic field polarity at the solar North pole is negative (field direted towards the sun).

## 5. SURVEY OF RELEVANT ZODIACAL LIGHT OBSERVATIONS

Four types of zodiacal light observations are of particular interest with respect to dust particle dynamics:

- the variation of observed brightness with heliocentric distance of the observer, from which the radial gradient of the dust distribution can be inferred
- the decrease of zodiacal light brightness outside the ecliptic, which is an image of the inclination distribution of dust particle orbits
- the plane of symmetry of the dust cloud, usually inferred from the position of the brightness maxima in the zodiacal light

- unusual orbital velocities of interplanetary dust found in particular by Fried (1978).

Of these, the first one is determined by the interplay of dust production, Poynting-Robertson effect and dust destruction in collisions, as was discussed at length by Leinert et al. (1983). The last one did not reproduce in the latest measurements (East and Reay, 1984). However, these authors find a line shift in the Gegenschein of 0.04 Å, corresponding to a recession of the material with an average velocity of 2.5 cm/sec. This probably constitutes the first optical detection of Beta meteoroids. When decomposed into the main contribution from material on bound orbits and a small addition due to a fast-moving receding component, it will constrain number and size distribution of escaping dust particles.

The plane of symmetry of interplanetary dust has been found to deviate slightly from the ecliptic plane. Inside 1 AU its ascending node and inclination were determined by Helios to be $\Omega = 87 \pm 4^\circ$, $i = 3.0 \pm 0.3^\circ$ (Leinert et al. 1980), a result with which the earthbound observations of Misconi and Weinberg (1978) fully agree. There is evidence that outside 1 AU the symmetry is closer to the oribt of Mars or the invariable plane of the solar system (Misconi 1980, Dumont and Levasseur-Reyourd, 1978), however, it is not yet clear whether this will be supported by the final analysis of IRAS infrared measurements (Hauser et al., 1985).

In any case, if the interplanetary dust cloud is perturbed by forces in addition to solar gravitation, it tends to pick up the symmetry of the additional force. This means that for gravitational perturbation by the larger planets, symmetry to the invariable plane ($\Omega = 108^\circ$, $i = 1.6^\circ$) should occur while for a predominance of perturbations by interplanetary plasma the symmetry should be with the solar equator ($\Omega = 75^\circ$, $i = 7.3^\circ$). The observed values for the inner solar system are intermediate, indicating the possibility of electromagnetic effects. However, if the sources of dust, e.g. short period comets, were already to possess the observed symmetry, no significant action of perturbing forces would be required.

The average inclination of interplanetary dust particle orbits was derived from zodiacal light brightness profiles perpendicular to the ecliptic to be about $i = 30^\circ$ (Bandermann, 1968, Leinert et al., 1976). This is far more than the corresponding values for possible parent bodies like short-period comets ($i = 17^\circ$, Southworth and Sekanina, 1973). Such numbers suggest an evolutionary sequence from larger to smaller particles, where the orbital inclinations are increasing due to perturbing effects which still have to be identified, but among which collisional processes (Trulsen and Wikan, 1980) and Lorentz scattering (Morfill and Grün, 1979a; Cosolmagno, 1979; Barge et al., 1982) are the most promising.

## 6. STOCHASTIC PERTURBATIONS

The work on Lorentz scattering of interplanetary dust is presented in various publications (e.g. Parker 1964, Morfill and Grün 1979a, b; Consolmagno 1979, Barge et al. 1982, Hassan and Wallis 1983, Pellat et al. 1984, Hassan 1985). We shall give a brief general description of the problem based on the method of Pellat et al. (1984), followed by a simplified description of the effect of the 11-year solar cycle variations in the large scale solar magnetic field polarity.

We assume particles in Keplerian orbits, subject to a stochastic force $\underline{F}_1(\underline{r},t)$. The stochastic processes are assumed to be strictly Markovian, so that the system can be described by the Fokker-Planck approximation. We define three constants of the unperturbed motion ($K_1$, $K_2$, $K_3$) as dynamical action variables of the Fokker-Planck equation, describing the distribution function $f(\underline{K},t)$

$$\frac{\partial f}{\partial t} = (-a_i f + b_{ij}\frac{\partial f}{\partial K_j}) \tag{9}$$

where $a_i$ is a friction coefficient, $b_{ij}$ a diffusion coefficient. In the quasilinear approximation (small perturbations around the mean orbit) it can be shown that equation (9) reduces to a pure diffusion-convection type equation, where the convection (or friction) term is the sum of Poynting-Robertson and gas (Coulomb) drag and the diffusion term is due to stochastic electromagnetic forces. In the Kepler problem, the constants of the unperturbed motion are total energy, total angular momentum, and the z-component of angular momentum. (In the case of circular orbits of radius r these are for a particle of mass m $K_1 = -1/2\ mr^2\Omega^2$, $K_2 = mr^2\Omega$, $K_3 = K_2 \cos i$ - here $\Omega^2 = GM/r^3$ is the Keplerian angular velocity and i the orbit inclination, M the mass of the sun and G the gravitational constant). The Poynting-Robertson drag term for spherical particles of radius $s \geq 0.3\ \mu m$ can be written as

$$a_{PR} = -2\ (\frac{0.19\ \Omega^4 r^3}{s})\ \frac{}{c} \tag{10}$$

where s is given in $\mu m$. The diffusion term in this case is given by the trace of the tensor $b_{ij}$ and can be evaluated.

To do this, a Fourier analysis of the stochastic forces $\underline{F}_1$ is performed. For circular orbits of zero inclination we require $F_{1\phi}$ and then

$$b_{ii} = \pi\ r^2 \int d\omega \sum_m |F_{1\phi}\ (m,\omega)|^2\ \delta(\omega-n\Omega) \tag{11}$$

The force is frequency ($\omega$) dependent, and the interaction is "resonant", i.e. most efficient orbit perturbations occur for $\omega = n\Omega$ ( = 1, 2, 3 ...), i.e. multiples of the orbit frequency. Provided we have a physical model for the power spectrum ($\sim |F_{1\phi}|^2$) we can solve the transport problem. It is at this point, where different assumptions are made in the literature, and it is at this stage where values may differ by an order of magnitude or more (see e.g. Morfill and Grün 1979, Barge et al. 1982, Wallis and Hassan 1985). We shall not concern ourselves with the details here - they involve questions whether the power decreases below $\sim 10^{-6}$ Hz, or whether it remains constant to much lower frequencies. Observations are not very reliable because of the long data acquisition times needed. They show a flattening of the power curve below $\sim 10^{-5}$ Hz. It is our belief that long term fluctuations in the magnetic field - on a decreasing frequency scale they are the sector structure, solar wind streams, interplanetary transients (shocks from flares), 11-year solar cycle variations, ensure a significant fluctuating power level in the range $10^{-9}$-$10^{-5}$ Hz. The diffusion coefficient is then calculated approximately (from Morfill and Grün, 1979a) by assuming field polarity variations to dominate the power spectrum. The equation of motion, considering only electromagnetic forces and assuming the solar wind speed $v_{sw}$ to be much larger than the particle's Keplerian velocity $r\Omega$, becomes

$$m \, dv_\theta/dt = \pm \frac{1}{c} |v_{sw} Q B_\phi| \qquad (12)$$

where $v_\theta$ is the particle velocity (perpendicular to the orbit plane). The sign depends on the direction of the field polarity. The change in the orbit inclination, i, is given from

$$\frac{di}{dt} = \frac{1}{v_\phi} \cdot \frac{dv_\theta}{dt} \cos\theta \, \sin\phi \qquad (13)$$

where $\phi$ is measured from the line of nodes between the particle orbit (inclination i) and the magnetic equator, $\theta$ is the instantaneous latitude of the particle position.

The stochastic element responsible for a net diffusion in particle inclination is the random variation in the length of the magnetic sectors. The mean duration of a sector is $\tau_0$, and the deviation from this mean $\delta\tau$ is assumed to be given by a Gaussian probability distribution with a standard deviation $\delta\tau_0$.

The diffusion coefficient ($radian^2$/sec) is then calculated (see Morfill and Grün 1979a) by first evaluating the mean square deviation in i, and then averaging over longitude. The result is

$$D(i) \sim \delta\tau_0^2/4\tau_0 (\frac{2}{\pi})^2 |v Q B_0 r_0/mc \sqrt{\mu r}|^2 \{1 - \frac{2}{3} \sin^2(i)\}^2 \qquad (14)$$

where $\mu$ is the radiation pressure modified gravity $= (1-\beta) \, GM_\odot$. When considering the sector structure, we have to restrict ourselves to low inclination orbits (i $\leq$ 15°), and the values for $\tau_0$ and $\delta\tau_0$ are measured to be 8 days and 1.5 days, respectively.

In the case of solar cycle variations, higher inclinations (i $\geq$ 15°) have to be considered, where the more rapid sector structure caused by the warping of the equatorial current sheet is unimportant. For solar cycle variations $\tau_0$ and $\delta\tau_0$ are 11 years and ~ 2 years, respectively. Solar cycle changes in the field polarity are illustrated in Fig. 3.

An important quantity for estimating the strength of the Lorentz scattering is the RMS inclination change during the Poynting-Robertson life time

$$\langle \Delta i^2 \rangle_{PR}^{1/2} = |D(i) \, \tau_{PR}|^{1/2} \qquad (15)$$

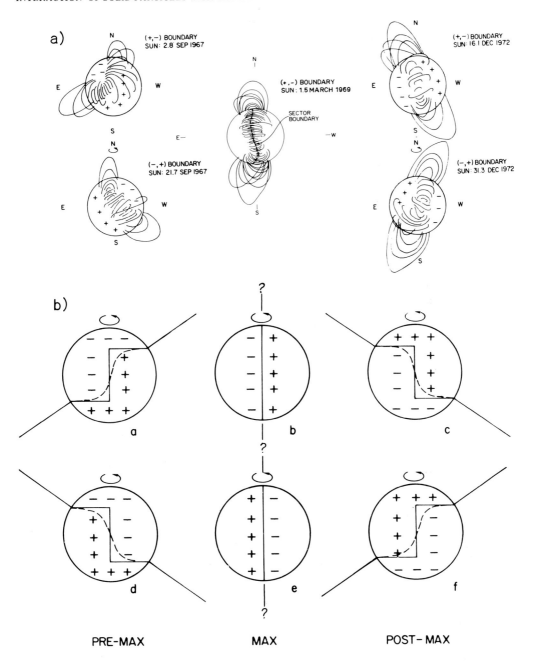

Figure 3. Illustration of solar cycle field polarity changes.

TABLE II

| s | $i$ (deg) | $\langle \Delta i^2 \rangle^{1/2}_{PR}$ (deg) (SECTOR) | $\langle \Delta i^2 \rangle^{1/2}_{PR}$ (deg) (SOL. CYCLE) |
|---|---|---|---|
| | 0 | $0.25 \times \sqrt{\dfrac{r}{r_0}}$ | |
| | 30 | | 22.5 |
| 1 $\mu$m | 45 | | $14.3 \times \sqrt{\dfrac{r}{r_0}}$ |
| $(\tau_{PR} \sim 3 \times 10^3 \, \mathrm{y})$ | 60 | | 11.4 |
| | 0 | $8.0 \times 10^{-3} \times \sqrt{\dfrac{r}{r_0}}$ | |
| | 30 | | 0.71 |
| 10 $\mu$m | 45 | | $0.45 \times \sqrt{\dfrac{r}{r_0}}$ |
| $(\tau_{PR} \sim 3 \times 10^4 \, \mathrm{y})$ | 60 | | 0.36 |
| | 0 | $2.5 \times 10^{-4} \times \sqrt{\dfrac{r}{r_0}}$ | |
| | 30 | | $2.25 \times 10^{-2}$ |
| 100 $\mu$m | 45 | | $1.43 \times 10^{-2} \times \sqrt{\dfrac{r}{r_0}}$ |
| $(\tau_{PR} \sim 3 \times 10^5 \, \mathrm{y})$ | 60 | | $1.14 \times 10^{-2}$ |
| | 0 | $8.0 \times 10^{-6} \times \sqrt{\dfrac{r}{r_0}}$ | |
| | 30 | | $7.1 \times 10^{-4}$ |
| 1 mm | 45 | | $4.5 \times 10^{-4} \times \sqrt{\dfrac{r}{r_0}}$ |
| $(\tau_{PR} \sim 3 \times 10^6 \, \mathrm{y})$ | 60 | | $3.6 \times 10^{-4}$ |

In table II, this quantity is calculated for scattering due to field polarity changes both in sectors and solar cycles. Numerical values are given for $r_o = 1$ AU, and the scaling to other distances is indicated. As can be seen, for zodiacal light particles (10 µm to 1 mm) magnetic scattering is very weak. At the lowest mass range (~ 10 µm) solar cycle effects account for a stochastic spread in inclinations of a few degrees at large distances from the sun. The sector field appears to be largely irrelevant.

Note, however, the rapid increase in scattering for particles below ~ 10 µm. There is no reason why these small particles should not be produced by the same source which injects the bigger particles (e.g. comets). According to the calculations of table II, stochastic scattering by Lorentz forces, mainly caused by solar cycle effects, should disperse these small particles throughout the heliosphere. This does not take into consideration any systematic orbit tilts during single solar cycles, which for the correct polarity may inject particles into the low latitude sector dominated region, where they then remain due to the much weaker scattering.

In order to illustrate the systematic orbit modification introduced by the high latitude unidirectional field, we have performed trajectory integrations. The dust particles were taken to be olivine, the magnetic sector structure was assumed to extend to $\pm$ 15° magnetic latitude. Poynting-Robertson drag, plasma drag (including the effects due to the slight corotation of the solar wind) both by direct and distant Coulomb collisions, radiation pressure and gravity were taken into account.

The results are shown in Figs. 4, 5, and 6. Particles were injected into the solar system at 50 AU, with the velocities ($v_o$ ~ 10-30) km/sec) indicated. Trajectories drawn with solid lines correspond to a solar cycle where the field in the northern hemisphere points towards the sun, broken curves correspond to the opposite polarity.

We see that for high incliantion orbits and very small particles of ~ 0.1 µm, Lorentz forces are dominant, whereas for 1 µm particles the orbit deflection is much smaller. (Note the changed scale in Fig. 5 - particles were injected, as those in Fig. 4, along a radius vector at 45° inclination.) 0.1 µm particles injected in the magnetic equatorial plane remained there during the "toward" cycle, whereas during an "away" cycle (all this refers to the N. pole) they managed to reach high latitudes.

# 7. CONCLUSIONS ON THE LARGE SCALE DISTRIBUTION OF INTERPLANETARY PARTICLES

Based on the orbit perturbation analysis (results of table I) and direct particle trajectory integrations (a selection of which were shown in Figs. 4-6) we have arrived at the following conclusions:

1. Submicron particles are dispersed to high latitudes during "away" cycles, and focussed into the magnetic equatorial plane during "toward" cycles of the solar magnetic field. Interstellar dust can therefore become highly modulated by the solar cycle, and cannot be expected to follow the neutral gas trajectories.

2. There is a systematic decrease in the inclinations of large particles during "toward" cycles. This will bring these particles into the low latitude sector regime, where much more rapid field polarity fluctuations reduce the effectiveness of the Lorentz forces. This leads to a kind of "trapping" of these particles at inclinations below ~ 15°, an effect which should become more important as we get closer to the sun.

3. Stochastic spreading in particle inclinations in the low latitude "sectored" field regime is unimportant for particles of size greater than ~ 10 μm.

4. Stochastic orbit perturbations for high inclination particles (i ≥ 15°), caused by solar cycle variations, is more important, affecting particles up to ~ 10 μm.

5. There should be a general tendency for electromagnetic forces to bring the zodiacal light cloud into symmetry around the magnetic equator. This could become noticeable for the most evolved inner region of the dust cloud, inside the Earth's orbit.

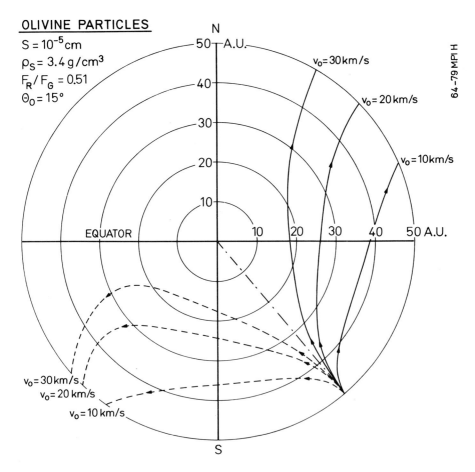

Figure 4.  Olivine particle trajectories in the heliosphere.

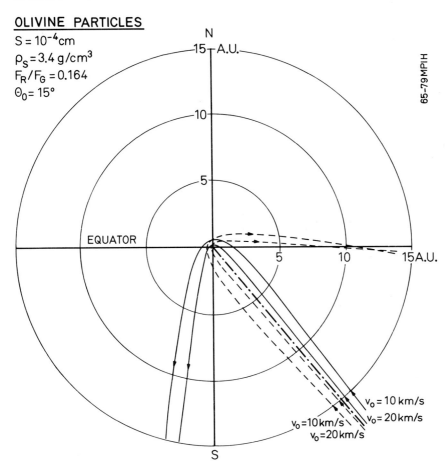

Figure 5.   Olivine particle trajectories in the heliosphere.

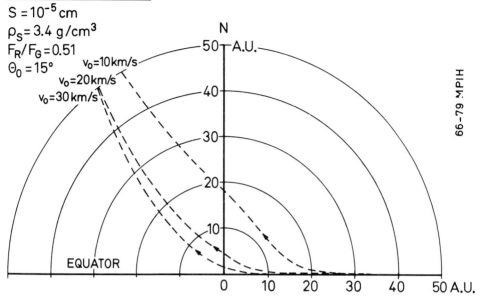

Figure 6.   Olivine particle trajectories in the heliosphere.

## REFERENCES

Babbock, H.W.: 1961, Astrophys. J. 133, 572.
Bandermann, L.W.: 1965, Thesis, University of Maryland, College Park
Barge, P., Pellat, R., Millet, J.: 1982, Astron. Astrophys. 115, 8.
Berg, O.E. and Grün, E.: 1973, in Space Research XIII, Akademie Verlag,
    Berlin, 1047.
Briggs, R.E.: 1962, Astron. J. 67, 710.
Burns, J.A:, Lamy, P.L., and Soter, S.: 1979, Icarus 40, 1.
Consolmagno, G.J.: 1979, Icarus 38, 398.
Delsemme, A.H.: 1976, Lecture Notes in Phys. 48, 314.
Dermott, S.F., Nicholson, P.D., Burns, J.A., and Houck, J.R.: 1985, in
    Properties and Interactions of Interplanetary Dust (Eds. R.H. Giese and
    P.L. Lamy), Reidel, Dordrecht, 395.
Dohnanyi, J.S.: 1978, in Cosmic Dust (Ed. J.A.M. McDonnell), Wiley,
    Chichester, 527.
Dumont, R. and Levasseur-Reyourd, A.Ch.: 1978: Astron. Astrophys. 64, 9.
East, I.R. and Reay, N.K.: 1984, Astron. Astrophys. 139, 512.
Feldman, W.C., Asbridge, J.R., Bame, S.J., Gosling, J.T.: 1977, in The Solar
    Output and its Variation (ed. O.R. White), Colorado Associated
    University Press, Boulder, 351.
Feuerbacher, B. and Fitton, B.: 1972, J. Appl. Phys. 43, 1563.
Fried, J.W.: 1978, Astron. Astrophys. 68, 295.
Geiss, J.: 1986, this volume.

Greenberg, J.M.: 1973, in Evolutionary and Physical Properties of Meteoroids (eds. C.L. Hemenway, P.M. Millman and A.F. Cook), NASA-SP 319, 375.

Grün, E.: 1981, Bundesministerium für Forschung und Technologie, Forschungsbericht W81-034.

Grün, E., Pailer, N., Fechtig, H., and Kissel, J.: 1980, Planet. Space Sci. 28, 333.

Grün, E., Zook, H.A., Fechtig, H., and Giese, R.H.: 1985, Icarus 62, 244.

Hassan, M., Wallis, M.K.: 1983, Planet. Space Sci. 31, 1.

Hauser, M.G., Gillet, F.C., Low, F.J., Gautier, T.N., Beichman, C.A., Neugebauer, G., Aumann, H.H., Baud, B., Boggess, N., Emerson, J.P., Houck, J.R., Soifer, B.T., and Walker, R.G.: 1984, Astrophys. J. Lett. 278, L15.

Hauser, M.G., Gautier, T.N., Good, J., and Low, F.J.: 1985, in Properties and Interactions of Interplanetary Dust (eds. L. Lamy and R.H. Giese, Reidel, Dordrecht, 43.

Hoffmann, H.J., Fechtig, H., Grün, E., and Kissel, J.: 1976, in The Study of Comets (eds. B. Donn, M. Mumma, W. Jackson, M. A'Hearn, and R. Harrington), NASA SP-393, 949.

Leinert, Ch., Röser, S., and Buitrago, J.: 1983, Astron. Astrophys. 118, 345.

Leinert, Ch., Hauner, M., Richter, I., and Pilz. E.: 1980, Astron. Astrophys. 82, 328.

Leinert, Ch., Link, H., Pilz, E. and Giese, R.H.: ev. 1976?, Astron. Astrophys. 47, 221.

Leinert, C., Richter, I., Pitz, E., and Planck, B.: 1981, Astron. Astrophys. 103, 177.

Leinert, C., Röser, S., and Buitrago, J.: 1983, Astron. Astrophys. 118, 345.

Levy, E.H. and Jokipii, J.R.: 1976, Nature 264, 423.

Michels, D.J., Sheeley, N.R., Jr., Howard, R.A., and Koomen, M.J.: 1982, Science 215, 1097.

Misconi, N.Y. and Weinberg, J.L.: 1978, Science 200, 1484.

Misconi, N.Y.: 1980, in Solid Particles in the Solar System (eds. I. Halliday and B.A. McIntosh) Reidel, Dordrecht, 49.

Morfill, G.E. and Grün, E.: 1979a, Planet. Space Sci. 27, 1269.

Morfill, G.E. and Grün, E.: 1979b, Planet. Space Sci. 27, 1283.

Morfill, G.E., Völk, H.J., and Lee, M.A.: 1976, J. Geophys. Res. 81, 5841.

Mukai, T.: 1981, Astron. Astrophys. 99, 1.

Mukai, T., and Yamamoto, T.: 1982, Astron. Astrophys. 107, 97.

O'Gallagher, J.J.: 1975, Astrophys. J. 197, 495.

Parker, E.N.: 1985, Astrophys. J. 128, 664.

Parker, E.N.: 1964, Astrophys. J. 139, 951.

Pellat, R., Barge, P., Hornung, P., and Millet, J.: 1984, Adv. Space Res. 4, 95

Pneumann, G.W. and Kopp, R.A.: 1971, Solar Phys. 18, 258.

Rhee, J.W.: 1967, in The Zodiacal Light and the Interplanetary Medium (Ed. J.C. Weinberg), NASA SP-150, 291.

Robertson, H.P.: 1937, Mon. Not. Roy. Astron. Soc. 97, 423.
Röser, S.: 1976, Lecture Notes in Physics 48, 319.
Schulz, M.: 1975, Space Sci. Rev. 17, 481.
Singer, S.F: and Bandermann, L.W.: 1967, in The Zodiacal Light and the
    Interplanetary Medium (Ed. J.L. Weinberg), NASA SP-150, 379.
Smith, E.J., Tsurutani, B.T., and Rosenberg, R.L.: 1978, J. Geophys. Res. 83,
    717.
Southworth, R.B. and Lekanina, Z.: 1973, NASA CR-2316, Washington D.C.
Svalgaard, L. and Wilcox, J.M.: 1976, Nature 262, 766.
Tralsen, J. and Wikan, A.: 1980, Astron. Astrophys. 91, 155.
Wallis, M.K., and Hassan, M.: 1985 Astron. Astrophys. (in press)
Wallis, M.K.: 1985, Nature, submitted
Whipple, F.L.: 1967, in The Zodiacal Light and the Interplanetary Medium
    (Ed. J.L. Weinberg), NASA-SP 150, 409.
Wilcox, J.M. and Ness, N., 1965, J. Geophys. Res. 70, 5793.
Wilcox, J.M. and Scherrer, P.H.: 1972, J. Geophys. Res. 77, 5385.
Wolfe, H., Rhee, J.W., and Berg, O.E.: 1976, Lecture Notes in Physics 48,
    165.
Wood, J.A. and Working-group: 1985, in Interrelationships Among
    Circumstellar, Interstellar and Interplanetary Dust (Eds. J.A. Nuth III,
    and R.E. Stencel), NASA-SP, in press.
Wyatt, S.P.: 1969, Planet. Space Sci. 17, 155.
Wyatt, S.P. and Whipple, F.L.: 1950, Astrophys. J. 111, 134.
Zook, H.A. and Berg, O.E.: 1975, Planet. Space Sci. 23, 183.

SECTION IX:  ULYSSES

# THE ULYSSES MISSION

R. G. Marsden, K-P. Wenzel
Space Science Dept. of ESA
ESTEC
2201 AG Noordwijk
The Netherlands

E. J. Smith
Jet Propulsion Lab.
California Institute of
Technology
Pasadena, CA 91103, USA

**ABSTRACT:** The Ulysses mission is unique in the history of the exploration of our solar system by spacecraft. The path followed by Ulysses will enable us, for the first time, to explore the heliosphere within a few astronomical units of the Sun over the full range of heliographic latitudes, thereby providing the first characterisation of the uncharted third heliospheric dimension. Highly-sophisticated scientific instrumentation carried on board the spacecraft is designed to measure the properties of the solar wind, the Sun/wind interface, the heliospheric magnetic field, solar radio bursts and plasma waves, solar X-rays, solar and galactic cosmic rays, and interplanetary/interstellar neutral gas and dust. Ulysses will also be used to detect cosmic gamma-ray bursts and search for gravitational waves. The mission is a collaboration between ESA and NASA, to be launched in May 1986 and utilising a Jupiter gravity-assist to achieve a high-solar-latitude trajectory.

## 1. INTRODUCTION

Our current understanding of the physics of the heliosphere, that vast region of space dominated by the radial outflow of the solar wind plasma from the Sun, is largely founded on observations made by spacecraft confined to the ecliptic. In terms of heliographic latitude, these spacecraft are only able to sample a narrow belt extending 7.25 degrees above and below the heliographic equator. To date, only two spacecraft have escaped the confines of the ecliptic, Pioneer 11 and Voyager 1, climbing to 16 and 30 degrees latitude, respectively. On the other hand, the global, three-dimensional structure of the heliosphere clearly does not possess a simple symmetry, indicating extreme caution is needed when making extrapolations of physical conditions encountered in the essentially two-dimensional world of the ecliptic.

The need for comprehensive, in-situ measurements of the heliosphere covering the full range of heliographic latitudes has long been

477

*R. G. Marsden (ed.), The Sun and the Heliosphere in Three Dimensions, 477–490.*
© *1986 by D. Reidel Publishing Company.*

recognised (e.g., Simpson et al., 1959). The Ulysses mission
(Wenzel, 1980; Marsden and Wenzel, 1981; Wenzel et al., 1983) will
provide the first exploratory step in achieving this objective.

Originally conceived and approved as a two-spacecraft mission, the
joint ESA-NASA collaborative project has suffered a number of
set-backs during its lifetime. The most far-reaching of these was
the cancellation in 1981 by NASA of the spacecraft it was to have
provided, owing to budgetary constraints. For the same reason the
launch, originally planned for 1983, has been delayed by 3 years, and
is now scheduled for May 1986. Within the cooperative programme, ESA
provides the spacecraft and its operation, and NASA the spacecraft
power supply, the launch, the Mission Ooperations Centre at JPL and
tracking via the Deep Space Network (DSN).

The Ulysses spacecraft (Wenzel, 1983) will be launched by NASA's
Space Transportation System, comprising the Shuttle orbiter
Challenger and a Centaur upper-stage motor. After passage by
Jupiter, the spacecraft will be travelling in an elliptical orbit
with aphelion near 5 AU, perihelion near 1 AU and an inclination to
the ecliptic of approximately 90°. The spacecraft will reach high
solar latitudes in the southern hemisphere in late 1989, then return
to the ecliptic and travel to high solar latitudes again in the
opposite solar hemisphere. In this mission overview, a summary of
the aims of the Ulysses scientific investigations will be given,
followed by a description of the mission time-line and spacecraft
trajectory. The final section will deal with technical aspects of
the spacecraft itself and the mission operations.

## 2. SCIENTIFIC OBJECTIVES

The primary objective of the Ulysses mission is to investigate, for
the first time, as a function of solar latitude, the properties of
the solar wind, the structure of the Sun/wind interface, the
heliospheric magnetic field, solar radio bursts and plasma waves,
solar X-rays, solar and galactic cosmic rays and the
interstellar/interplanetary neutral gas and dust. Secondary
objectives of the mission include interplanetary-physics
investigations during the in-ecliptic Earth-Jupiter phase,
measurements of the Jovian magnetosphere during the Jupiter flyby
phase, the detection of cosmic gamma-ray bursts and a search for
gravitational waves.

The detailed objectives of the various Ulysses scientific
investigations have been reported elsewhere (Wenzel et al. (eds),
1983). Here we give a summary of the major aims of these
investigations which will contribute to the achievement of the
overall Ulysses objective :
- to provide an accurate assessment of the global
  three-dimensional properties of the interplanetary magnetic
  field and the solar wind

- to improve our knowledge of the composition of the solar atmosphere and the origin and acceleration of the solar wind by systematically studying the composition of the solar-wind plasma and solar energetic particles at different heliographic latitudes
- to provide new insight into the acceleration of energetic particles in solar flares and into storage and transport of these particles in the corona by observing the X-ray and particle emission from solar active regions and from other magnetic configurations which are more accessible for study from out of the ecliptic
- to further our knowledge of the internal dynamics of the solar wind, of the waves, shocks and other discontinuities, and of the heliospheric propagation and acceleraton of energetic particles, by sampling plasma conditions that are expected to be different from those available for study near the eclintic
- to improve our understanding of the spectra and composition of galactic cosmic rays in interstellar space by measuring the solar modulation of these particles as a function of heliographic latitude and by sampling these particles over the solar poles, where low-energy cosmic rays may have an easier access to the inner solar system than near the ecliptic plane
- to advance our knowledge of the neutral component of interstellar gas by measuring as a function of heliographic latitude the properties and distribution of neutral gas that enters the heliosphere
- to improve our understanding of interplanetary dust by measuring its properties and distribution as a function of heliographic latitude
- to search for gamma-ray-burst sources and, in conjunction with observations from other spacecraft, to identify them with known celestial objects or phenomena
- to search for low-frequency gravitational waves by recording very precise two-way Doppler tracking data at the ground stations.

Experience with previous missions, particularly in the realm of the Earth's magnetosphere, has shown that an interdisciplinary approach is essential if the complex interactions between plasmas, magnetic fields and charged particles are to be studied on a global scale. In the case of an exploratory mission like Ulysses, this aspect is clearly of crucial importance. To this end, a Common Data Record (Wenzel and Smith, 1983) comprising key scientific parameters from different investigations, will be produced after acquisition of the data, and distributed to all investigators. Theoretical scientists are already included as co-investigators in most teams. Collaborative studies amongst the different investigations will also be encouraged by organising post-launch workshops, potentially also involving guest investigators.

## 3. SCIENTIFIC PAYLOAD

A summary of the nine flight experiments which make up the spacecraft

Table 1: SUMMARY OF SCIENTIFIC INVESTIGATIONS AND ASSOCIATED INSTRUMENTATION FOR THE ULYSSES SPACECRAFT

| Investigation | Principal Investigator | Measurement | Instrumentation |
|---|---|---|---|
| Magnetic Field (HED) | A. Balogh Imperial College London | Spatial and temporal variations of the heliospheric and Jovian magnetic field in the range $\pm$ 0.01 nT to $\pm$ 44000 nT | Triaxial vector helium and fluxgate magnetometers |
| Solar-Wind (BAM) | S.J. Bame Los Alamos Nat. Lab. | Solar-wind ions between 257 eV/Q and 35 keV/Q; Solar-wind electrons between 1 eV and 903 eV | Two electrostatic analysers with channel electron multipliers (CEMs) |
| Solar-Wind Ion Composition (GLG) | G. Gloeckler Univ. Maryland; J. Geiss Univ. Bern | Elemental and ionic-charge composition, temperature and mean velocity of solar-wind ions for speeds from 145 km/s($H^+$) to 1352 km/s($Fe^{+8}$) | Electrostatic analyser with time-of-flight and energy measurement |
| Low-Energy Ions and Electrons (LAN) | L. Lanzerotti Bell Laboratories | Energetic ions from 50 keV to 5 MeV. Electrons from 30 keV to 300 keV | Two sensor heads with five solid-state detector telescopes |
| Energetic-Particle Composition an Interstellar Gas (KEP) | E. Keppler MPI Lindau | Composition of energetic ions from 80 keV to 15 MeV/nuc; Interstellar neutral helium | Four solid-state detector telescopes; LiF-coated conversion plates with channel electron multipliers |

**Table 1:** SUMMARY OF SCIENTIFIC INVESTIGATIONS AND ASSOCIATED INSTRUMENTATION FOR THE ULYSSES SPACECRAFT (continued)

| Investigation | Principal Investigator | Measurement | Instrumentation |
|---|---|---|---|
| Cosmic Rays/ Solar Particles (SIM) | J.A. Simpson Univ. Chicago | Cosmic rays and energetic solar particles in the range 0.3–600 MeV/nuc Electrons in the range 4–2000 MeV | Five solid-state detector telescopes one double Cerenkov and semi-conductor telescope for electrons |
| Unified Radio and Plasma Waves (STO) | R.G. Stone NASA/GSFC | Plasma waves; remote sensing of travelling solar radio bursts, and electron density Electric field: Plasma waves:0–60 kHz Radio receiver:1–940 kHz Magnetic field: 10–500 Hz | 72 m radial dipole antenna 7.5 m axial monopole antenna Two-axis search coil |
| Solar X-rays Cosmic Gamma-Ray Bursts (HUS) | K. Hurley CESR Toulouse M. Sommer MPI Garching | Solar-flare X-rays and cosmic gamma-ray bursts in the energy range 5–150 keV | Two Si solid-state detectors two CsI scintillation crystals |
| Cosmic Dust (GRU) | E. Grün MPI Heidelberg | Direct measurement of particulate matter in mass range $10^{-16}$–$10^{-7}$g | Multicoincidence impact detector with channeltron |

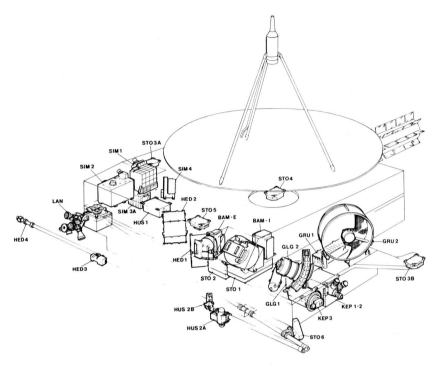

Figure 1a.  Layout of the scientific instruments on the Ulysses spacecraft, shown here in launch configuration.  The experiment codes are listed in Table 1.

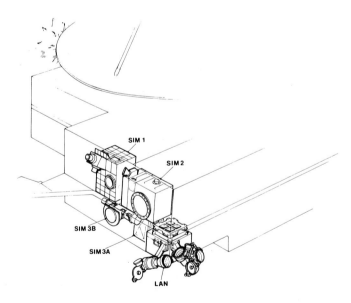

Figure 1b.  Detail of the Ulysses spacecraft, showing the layout of the SIM and LAN experiment packages.

payload is presented in Table 1.  These hardware investigations will provide the comprehensive measurements needed to fulfil the objectives outlined above, utilising instrumentation which represents the state of the art.  They have been designed to cover the full spectrum of changing conditions on their long-duration mission, including some margins to allow for unexpected phenomena.  Figure 1 shows the layout of the scientific instruments on the spacecraft. The body-mounted experiments are located around the periphery on one side of the spacecraft to provide the required fields of view, good EMC shielding and minimum radiation background from the Radio-isotope Thermoelectric Generator (RTG).

Besides the hardware experiments, radio-science investigations will be conducted on Ulysses which, at specific times during the mission, will make use of the spacecraft- and ground-communications systems to perform scientific measurements.  In addition, there are interdisciplinary investigations that will use data from more than one Ulysses experiment to address specific scientific problems.  Both these categories of investigations are summarised in Table 2.

**Table 2: ULYSSES RADIO-SCIENCE AND INTERDISCIPLINARY INVESTIGATIONS**

| Investigation/<br>Principal Investigator | Objectives |
| --- | --- |
| | **Radio Science** |
| **Coronal Sounding**<br>H. Volland<br>Univ. Bonn | Electron density, turbulence spectrum and velocity of coronal plasma during superior conjunction using dual-frequency ranging and Doppler data |
| **Gravitational Waves**<br>B. Bertotti<br>Univ. Pavia, Italy | Search for low-frequency, wideband gravitational waves using precise Doppler link |
| | **Interdisciplinary** |
| **Directional Discontinuities**<br>J. Lemaire<br>Inst. d'Aeronomie Spatiale<br>de Belgique, Brussels | Modelling of discontinuities in solar wind and comparision with Ulysses data |
| **Mass Loss and Ion Composition**<br>G. Noci<br>Osservatorio Astrofisico<br>di Arcetri, Florence, Italy | Dependence of mass loss and ion composition of the solar wind on heliographic latitude |

## 4.  THE ULYSSES TRAJECTORY AND MISSION TIMELINE

The Ulysses spacecraft is to be launched in May 1986 by means of the
Space Shuttle orbiter Challenger, using a Centaur upper-stage motor
to inject the spacecraft into an ecliptic transfer orbit.   The
trajectory will be such that the spacecraft, arriving at Jupiter 14
months after launch, will pass the planet slightly above its
equatorial plane.  At this point, the Jovian gravitational field will
deflect the spacecraft into a high-inclination orbit, taking it south
of the ecliptic plane.   The mission profile is shown in Figure 2.
The elliptical out-of-ecliptic trajectory, which has aphelion near 5
AU and perihelion near 1 AU, is such that, 2.5 years after Jupiter
encounter, the spacecraft will pass over the southern solar pole at a
distance of ca. 2 AU above the ecliptic.   Following this polar
passage, it then crosses the plane of the ecliptic, headed for a
second (opposite) polar passage some eight months after the first.
The baseline mission is completed when the heliographic latitude of
the spacecraft, on its way to aphelion near Jupiter's orbit, falls
below 70°.

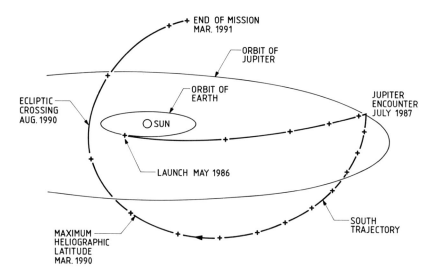

Figure 2.   Nominal Ulysses spacecraft trajectory, viewed from
15 degrees above the ecliptic plane.   Tick marks are shown at
100-day intervals.

There is no clear scientific rationale for choosing to explore one
solar hemisphere before the other; however, a key mission design
requirement is that the spacecraft spends as long as possible (at
least 2 solar rotation periods) at high latitudes during each of the
two polar passages.  This requirement is strengthened by the fact
that the period of observations over the solar poles is likely to
occur near the maximum of the solar-activity cycle, so the need to

Table 3.  Typical Ulysses mission time-line.

| Event | Calendar date | Average time elapsed (months) |
|---|---|---|
| Launch | 15 May 1986 | 0 |
| Jupiter encounter | July 1987 | 14 |
| First polar pass | | |
|   start | December 1989 | 43 |
|   max.latitude | February 1990 | 45 |
|   end | April 1990 | 47 |
| Ecliptic crossing | July 1990 | 50 |
| Second polar pass | | |
|   start | November 1990 | 54 |
|   max.latitude | January 1991 | 56 |
|   end | March 1991 | 58 |
| End of mission | 31 March 1991 | |

maximise the observing time at the highest latitudes above 70° is apparent if the best possible data base is to be obtained in order to take account of the rapidly changing physical conditions.  Given the restrictions on available launch energy, and the relative inclination of the solar equatorial plane with respect to the ecliptic, the south-going trajectory provides approximately 30% more time above 70° than the north-bound mission.

An added constraint arises from the fact that the length of time spent at high latitudes depends rather critically on which day within the ca. 20 day launch window the mission is launched.  Ulysses will have to share this launch window with Galileo, NASA's Jupiter orbiter and probe, and the restrictions imposed by the complex joint operations at the launch site effectively remove the north-going mission.  Table 3 presents a summary of the different phases of the mission and their nominal duration.  In this context, a "polar pass" is defined to be the segment of the solar polar orbit above 70° heliographic latitude.

## 5.  THE SPACECRAFT

The Ulysses spacecraft (Wenzel, 1983) is shown in its operational configuration in Figure 3.  Dictated by the long distances from the Earth and the Sun at which the spacecraft will be operating, the configuration of the spinning spacecraft (5 rpm) is dominated by the large-diameter (1.65 m), parabolic, High-Gain Antenna (HGA) providing the communication link to Earth and by the RTG supplying the spacecraft's electrical power.  Experiment requirements for electromagnetic cleanliness (EMC) and for minimisation of the RTG radiation environment resulted in a 5.6 m-long radial boom,  It carries several experiment sensors and is mounted on the opposite side of the spacecraft to the RTG.  A 72 m tip-to-tip dipole wire

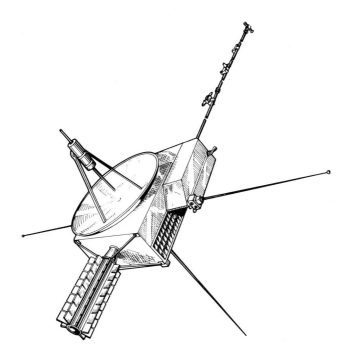

Figure 3.    Artist's impression of the Ulysses spacecraft in operational configuration. The electrical wire-boom antennae are deployed to a total length of 72m tip-to-tip; the axial boom which is located on the rear of the spacecraft along the spin axis extends 7.5m.    The radial boom, carrying five experiment packages, has a total length of 5.6m.

boom and a 7.5 m axial boom serve as electrical antennas for the Unified Radio and Plasma-Wave Experiment (Stone et al., 1983).  Most of the scientific instruments are mounted on the main body, as far as possible removed from the RTG, and in compliance with the field-of-view requirements of the experiment sensors.  The HGA meets the radio-link requirement to Earth with 20 W X-band and 5 W S-band transmitters.  The uplink S-band carries commands and ranging code. The downlinks in X- and S-band carry telemetry and turnaround ranging codes, respectively.  This simultaneous ranging and telemetry is a basic feature of the spacecraft communication system.

At launch, Ulysses will weigh 366 Kg.  Spacecraft mass properties and balance have been a driver in the spacecraft design to meet the requirements both for the launch configuration (stowed booms) and for the deployed boom configuration (HGA pointing towards Earth).  The spin axis of the launch configuration will be the geometric centre line.  The electrical axis of the HGA is aligned with the theoretical spin axis in deployed configuration.

Near-continuous data throughout the mission is a prime scientific requirement. Since continuous coverage by ground stations is impossible for such a long-duration mission, data will be stored on-board and replayed, interleaved with real-time data, during periods of coverage. The nominal tracking coverage is 8 h in every 24 h, using the 34 m DSN network. In special situations, 4 h of coverage every 48 h by the 64 m network may be substituted. Based on this nominal coverage and the calculated telemetry-link performance, a variety of downlink bit rates between 128 and 8192 bit/s is selectable, which can provide real-time data rates between 128 and 1024 bit/s and stored data rates between 128 and 512 bit/s. The prime data rates are 1024 bit/s for real-time data ('tracking mode') and 512 bit/s for stored data ('storage mode').

The spacecraft also provides automonous system capabilities for failure-mode detection and for safe spacecraft reconfiguration. This is required during unexpected and/or predicted periods of nontracking and because of the long signal travel-time between ground and spacecraft. The preprogrammable functions include search-mode initiation to reacquire the Earth if no command is received after a preselectable time of up to 30 days, switch-over to redundant units, and preprogrammed attitude manoeuvres at superior conjunction.

The Jupiter gravity assist necessitates the spacecraft's passage through the Jovian radiation belts. All subsystems and experiments have therefore been designed to survive this environment and radiation-hardened parts (design dose rate 60 krad) have been used throughout the spacecraft.

As a result of the launch delays, ESA adopted a "build-and-store" approach for the spacecraft in 1980. The flight spacecraft integration and test programme was completed at the end of 1983, following which the spacecraft was put into storage. In early 1985, the re-certification programme was started, including a cold-case thermal vacuum test and a vacuum spin-balance, ending in a successful Shipment Review held in September. Transport of the spacecraft to the Eastern Test Range at Cape Canaveral is scheduled to take place in the first week of 1986, leading into the launch preparation activities.

## 6. MISSION OPERATIONS

Within the project, the responsibilities for Ulysses misson operations have been defined as follows :
- NASA will launch the spacecraft and provide telemetry acquisition, tracking, and command transmission through its DSN ground stations
- ESA will provide data processing for spacecraft control, spacecraft manoeuvring, and scientific quick-look data examination, and provide spacecraft control personnel to control the mission from the JPL Mission Operations Centre

-    NASA will provide the infrastructure for ESA's facilities and
     staff at JPL.   It will process tracking data to provide
     trajectory determination, and will supply the Principal
     Investigators (PIs) with the scientific data generated by their
     experiments on board the spacecraft and with the radio-science
     data.
The implementation of these responsibilities will be performed by an
integrated NASA-ESA Mission Operations Team, lead by the Mission
Operations Manager and located at JPL.

Key events during the first 30 days following separation from the
Centaur upper stage include spacecraft subsystem and experiment
check-out, deployment of radial and axial booms and the execution of
two trajectory correction manoeuvres (TCM).   The mission then enters
its routine operation phase, with nominal 8 hr/day tracking
coverage.   This phase lasts until ca. 10 days prior to Jupiter
closest approach, at which time a third TCM is performed in
preparation for the encounter.   After exposure to the severe Jovian
radiation environment, the health of the spacecraft and experiments
will once again be thoroughly evaluated, as Ulysses starts its
journey away from the ecliptic.

## 7.   UYLSSES AND THE SOLAR CYCLE

Observations of the interplanetary medium from spacecraft in or very
near to the ecliptic, as well as extensive interplanetary
scintillation measurements, seem to indicate generally higher
solar-wind velocities and nearly mono-polar magnetic regions at
higher helio-latitudes.   This trend is particularly evident at solar
minimum, when the polar regions are dominated by extensive coronal
holes.   The 39 months delay in the launch of Ulysses, from February
1983 to May 1986, has however shifted the high-latitude passes from a
period around solar minimum (1986/87) to an advanced stage of the
ascending phase of the next solar cycle (1989/90).

At that time, solar activity is expected to be at a high level and to
extend to high latitudes, in excess of 70° heliographic latitude.
Polar coronal holes will be undergoing rapid changes leading to their
near-disappearance at solar maximum.   The rate of decrease of the
coronal-hole area may result in a significant sporadic sweeping of
irregular hole boundaries over the spacecraft.   It may thus be quite
possible that, should they reflect the isotropy of the corona at
solar maximum, the polar regions of the heliosphere would not look
too different from the everyday interplanetary medium in the
ecliptic.

The change in the mission period with respect to the solar cycle, and
the loss of the simultaneity and equality of observations in the two
solar hemispheres due to the cancellation of the second spacecraft,
will certainly complicate the separation of spatial and temporal
processes in the heliosphere.   It will be more difficult to

understand whether measured changes are indeed due to the change in latitude or to solar disturbances. For example, the reversal of the Sun's polar field that is expected to occur around or following solar sunspot maximum will be harder to detect than would have been the case with simultaneous measurements from two spacecraft, one in each hemisphere. The availability of in-ecliptic baseline observations at the time of the Ulysses high-latitude measurements would greatly improve the chances of distinguishing between spatial and temporal features.

## 8. CONCLUSION

The exploratory mission described in this paper is unique in the history of the exploration of our solar system by spacecraft. The name "Ulysses", which was proposed by Professor Bruno Bertotti, a radio-science investigator on the mission, was inspired by a passage from Dante's "Inferno". Dante's story says that Ulysses, after returning home to Ithaca, soon became restless, finally setting off on a new journey to explore that part of the world which lay beyond Gibraltar (at that time completely unknown and unexplored). In a speech to his companions, Ulysses exhorts them "To venture the uncharted distances ... of the uninhabited world behind the Sun ... to follow after knowledge and excellence." These lines capture in a striking way the spirit of the first space mission to fly over the polar regions of the Sun. Ulysses will provide the means to help solve outstanding prcblems in solar and heliospheric physics, while undoubtedly bringing new and unanticipated phenomena to light.

## REFERENCES

Marsden, R.G. and Wenzel K.-P.: 1981, ESA SP-164, 51 and ESA SP-161, 167.

Simpson, J.A., Rossi, B., Hibbs, A.R., Jastrow, R., Whipple F.L., Gold, T., Parker, E., Christofilos, N., and Van Allen, J.A.: 1956, J. Geophys. Res., 64, 1691.

Stone, R.G., Caldwell, J., de Conchy, Y., Deschanciaux, C., Ebett, R., Epstein, G., Groetz, K., Harvey, C.C., Hoang, S., Howard, R., Hulin, R., Huntzinger, G., Kellogg, P., Klein, B., Knoll, R., Lokerson, D., Manning, R., Mengué, J.P., Meyer, A., Monge, N., Monson, S., Nicol, G., Phan, V., Steinberg, J.L., Tilloles, P., Torres, E., and Wouters, F.: 1983, in The International Solar Polar Mission - its scientific investigations, ESA SP-1050, 185.

Wenzel, K.-P., : 1980, Phil. Trans. R. Soc. Lond., A297, 565.

Wenzel, K.-P., : 1983, in The International Solar Polar Mission - its scientific investigations, ESA SP-1050, 277.

Wenzel, K.-P., Marsden, R.G. and Battrick, B. (eds.): 1983, The International Solar Polar Mission - its scientific investigations, ESA SP-1050.

Wenzel, K.-P., Marsden, R.G. and Smith, E.J.: 1983, in the International Solar Polar Mission - its scientific investigations, ESA SP-1050, 9.

Wenzel, K.-P. and Smith, E.J. : 1983, in The International Solar
    Polar Mission - its scientific investigations, ESA SP-1050, 305.

# SECTION X:  SUMMARY

SUMMARY REMARKS

L. A. Fisk
Space Science Center
University of New Hampshire
Durham, NH  03824, USA

ABSTRACT:  This paper is intended to provide a summary of some of the
highlights of the 19th ESLAB Symposium on the Sun and the Heliosphere
in Three Dimensions.  Particular emphasis is placed on the conditions
and processes that Ulysses is expected to observe.

## 1. INTRODUCTION

The exploration of the heliosphere in three dimensions has
been a quest shared by many for many years.  Possible missions were
proposed as early as the 1960's.  Ulysses, the first mission to explore
the heliosphere in three dimensions, had its origin in the joint ESA/
NASA study of 1974.

For all of us who participated in these early studies and in
the planning for Ulysses, the dominant motivation has been the realiza-
tion that the heliosphere is intrinsically three-dimensional; we cannot
hope to understand it by observing only near the equatorial plane of
the Sun.  Analogies abound here.  We could not hope to understand the
Earth by exploring only near the Earth's equator.  We could not attempt
to understand the magnetosphere by flying only equatorial satellites.

The papers presented in this symposium document the many different scientific questions that can be addressed only with detailed
knowledge of the conditions and processes occurring on the Sun and in
the heliosphere in three dimensions, and the contribution that Ulysses
is expected to make to addressing these questions.  I have attempted
below to summarize the results of these papers.  This summary is not
intended to be comprehensive, but rather emphasizes only some of the
highlights.

In preparing this summary presentation, I examined the proceedings of the first symposium on the study of the Sun and the heliosphere in three dimensions, which was held at the Goddard Space Flight
Center almost exactly one decade ago, in May 1975.  Some of the basic

*R. G. Marsden (ed.), The Sun and the Heliosphere in Three Dimensions, 493–501.*
© *1986 by D. Reidel Publishing Company.*

understandings of what the three-dimensional structure of the heliosphere is likely to be were not appreciably different then.  The magnetic field in the solar wind was known to be primarily azimuthal near
the equatorial plane, and was expected to be nearly radial over the
poles.  The solar wind, at least the high-speed streams, was known to
originate in coronal holes, and coronal holes were recognized as being
important for the polar flow of the solar wind.  However, some effects
were not considered.  The 1975 symposium was prior to the recognition
that at least during solar-minimum conditions, the heliospheric magnetic field is unipolar in each hemisphere and separated by a wavy current sheet, whose excursions from the equator depend on the evolution
of coronal holes.  It was prior to the recognition, but not the general
acceptance, that unipolar fields in each hemisphere could make gradient
and curvature drifts important for the modulation of galactic cosmic
rays, and could profoundly affect predictions of energetic particle
fluxes at high latitudes.  It was prior also to the recognition that
stream-stream interactions and other transient effects can have an
important role in particle acceleration in the solar wind and in the
modulation of cosmic rays.

It is comforting to know, by comparing the 1975 symposium
with the present talks, that progress is being made in our field.  However, if we were to characterize this work over the last decade, it has
been work that has provided an increasing awareness of the processes
that are possible in the heliosphere, increasing speculation on how
things might work.  However, closure has not yet been possible: we
still do not know how the solar wind is heated and accelerated; we
still do not know what causes the solar modulation of galactic cosmic
rays.

The problem remains that we are still hampered by the lack of
definitive knowledge of conditions and processes occurring in the
heliosphere in three dimensions.  There is never any substitute for in
situ observations, which is what Ulysses is to provide.


## 2. THE SUN

The basic division on the Sun is between open and closed magnetic field lines.  Open field lines give rise to the solar wind; the
most dramatic example is coronal holes which give rise to high-speed
solar wind.  Closed field lines give rise to active regions, flares,
ejecta, etc.

As MacQueen (1986), Hoeksema (1986), and others discussed,
the magnetic field configuration on the Sun evolves throughout the
solar cycle.  During solar-minimum conditions, the polar coronal holes
extend downward from the poles to form large, well-defined open magnetic field regions.  In turn, as Withbroe (1986) discussed, well-
defined high-speed solar wind streams should develop in the heliosphere, particularly at high latitudes, and unipolar regions should

form north and south of an equatorial current sheet. At the time of least solar activity, the polar coronal holes have their maximum extension, and the equatorial current sheet is nearly flat, with few warps.

High-speed solar wind flows at high latitudes have been confirmed by radio scintillation observations, as were described by, e.g., Coles (1986). Stream structure in the solar wind and the behavior of the equatorial current sheet can be confirmed by comparing K-coronograph observations with data taken from near Earth. Although the Earth only has an excursion of seven degrees in heliographic latitude, the excursion in heliomagnetic latitude is much larger, and thus observation from near Earth can see the systematic behavior of solar wind streams and the current sheet near solar minimum (e.g., Newkirk and Fisk, 1986).

As solar-maximum conditions approach, the pattern of solar wind streams and the heliospheric magnetic field becomes less ordered. Well-defined holes, at least long-lasting holes are less prevalent, and, as discussed by Hoeksema (1986), a similar, less well-ordered structure should develop in the heliosphere. As mentioned by Smith, et al. (1986) and Hoeksema (1986), there is a possibility of multiple current sheets. Indeed, as the polarity of the Sun changes, current sheets, regions of mixed polarity, and/or closed field lines separating the polarity of the previous cycle from the polarity of the next cycle must propagate out through the heliosphere.

Ulysses will fly in these less well-ordered times. The transit to Jupiter and to the first polar pass will occur in approaching solar-maximum conditions, as the ordered solar wind and magnetic field structures are beginning to break up. The pole-to-pole passage will occur in full solar-maximum conditions. Although perhaps a confusing time, the breakup of the ordered fields and the emergence of new fields is one of the most interesting and dynamic parts of the solar cycle, and includes processes that are essential to comprehend if we are to understand the evolution of the solar magnetic field, the modulation of galactic cosmic rays, etc.

## 3. SOLAR WIND

One of the continuing problems in solar and heliospheric physics is our lack of complete understanding of the mechanism for the heating and the acceleration of the solar wind. As Withbroe (1986) pointed out, the role of the fine-scale structure on the Sun, e.g., spicules, macrospicules, and bullets, in the heating of the corona is unclear. These phenomena contain an interesting amount of energy, momentum, and mass, relative to the solar wind, but whether they are important for the acceleration of the wind is unknown. As Parker (1986) noted, Alfven waves may be a natural way to heat and accelerate the wind; yet whether there is observational support for this mechanism is uncertain. Similarly, the full range of possible sources of the solar wind is unknown. Coronal holes are clearly a source of high-speed streams; however, possible sources of low-speed wind, such as streamers, are less clear.

One of the exciting prospects for Ulysses is that not only will it explore a full range of solar wind sources and conditions, but also it will fly with and in an era of new instrumentation. As Withbroe (1986) noted, it should be possible, coincident with the flight of Ulysses, to have spectroscopic measurements of the composition and Doppler shifts, and thus of the systematic and random velocities, in the corona which is the source of the solar wind seen by Ulysses. A combined study which uses these data from the acceleration region, and in situ data of the resulting flow from Ulysses, should thus be possible. Further, as Geiss (1986) noted, we are entering with Ulysses into a new era of solar wind studies, an era of detailed compositional measurements. As has been known for decades in energetic particle studies, there is a wealth of information available in measurements of the composition, information which will be fully exploited for solar wind studies only with Ulysses. Indeed, for the solar wind, there is not only the chemical composition but also the ionic charge composition. The latter is frozen-in in the corona and thus provides information on conditions in the acceleration region and on likely source locations for the wind. Clearly, correlated studies of spectroscopic observations of possible source regions of the solar wind seen by Ulysses, and of the ionic charge composition measured by Ulysses will be of particular value.

However, the complexity of the conditions on the Sun and in the heliosphere during the flight of Ulysses may be a problem for solar wind studies. The coronal sources will not be steady; there may be a considerable number of transients. In the original plans for Ulysses, with the launch earlier than 1986, it had been hoped that Ulysses would observe the solar wind from a well-defined polar coronal hole. The flow then would have been particularly simple and easy to understand. Whether such simplified conditions will prevail at any time during the Ulysses flight is unclear.

## 4. SMALL-SCALE VARIATIONS IN THE SOLAR WIND

The closed magnetic field structures on the Sun give rise to many of the transient disturbances that affect the heliosphere. Flares and coronal mass ejecta, which come in different forms, cause disturbances which affect the steady flow of the solar wind. High-speed streams collide with slower flows. Waves propagate outward from the Sun. All of which produce a potpourri of plasma interactions in the heliosphere--shocks, turbulence, particle acceleration, etc.

One of the motivations of Ulysses is to extend the range of our observations of plasma interactions. We use the solar wind as a giant plasma physics laboratory. What we see and understand in the solar wind has applications to other astrophysical plasmas; e.g., theories for collisionless shocks and for particle acceleration at shocks had their origins in detailed solar wind studies.

In this context, solar-minimum conditions would not be a good time to observe the solar wind at high latitudes. During solar-minimum conditions, the high-latitude wind should be relatively tranquil. Also, as Schwenn (1986) noted, coronal mass ejecta tend to be concentrated near the equatorial plane, and may produce few or weaker effects at high latitudes. Similarly, there should be few stream-stream inter- actions at high latitudes, with a steady coronal hole at the solar poles and few rotation effects.

In contrast, solar-maximum conditions, when Ulysses is to fly, may be an ideal time to observe the small-scale structure of the solar wind at high latitudes. At this time, coronal mass ejecta are more broadly distributed in latitude (Schwenn, 1986), the coronal holes on the Sun are more transient, and the wind at high latitudes should be disturbed. Further, the plasma conditions over the poles, in which these disturbances occur, may be somewhat different than those near the equatorial plane. The underlying field is nearly radial over the poles, and perhaps somewhat weaker than near the equatorial plane; the fluid approximation for the solar wind may be suspect over the poles; and, as Lee (1986) noted, parallel shocks should be more prevalent than perpendicular ones. Ulysses should thus accomplish the desired goal. It should extend the range of plasma phenomena we can study.

## 5. ENERGETIC PARTICLES

In typical conditions near the equatorial plane of the Sun, solar and interplanetary-accelerated particles are observed at energies less than 10 MeV/nucleon, galactic cosmic rays are observed at energies greater than 10 MeV/nucleon, and the so-called anomalous component is seen at energies where the solar/interplanetary and the galactic compo- nents merge. However, this division with energy is likely to be a peculiarity of the equatorial plane.

As discussed by Wibberenz (1986), it is unclear how coronal propagation occurs at high latitudes. It could be that solar flare particles originate and propagate primarily in the complex magnetic field regions at low solar latitudes. However, as noted by Roelof (1986) the complex magnetic field structures could conceivably extend to high latitudes at the time when Ulysses flies. Gold et al. (1986) noted that, although interplanetary acceleration processes appear to occur at the latitudes obtained by Voyager, approximately 20 degrees, nonetheless, the processes seem weaker. Thus, one possibility is that, as Ulysses obtains high latitudes, the solar/interplanetary component could be reduced, exposing more of the anomalous and galactic compo- nents.

Similarly, we should expect the relative strength of the anomalous component, which is believed to result from interstellar neutral gas that is ionized and accelerated in the solar wind, to change with latitude. As Fisk (1986) noted, if the acceleration of the anomalous component occurs near the poles, e.g., at the termination shock of the solar wind, then the strength of the anomalous component should increase with latitude. Conversely, if the acceleration occurs in stream-stream interactions or at the termination shock near the equatorial plane, the anomalous component should decrease with latitude.

The behavior of the galactic cosmic ray flux with latitude is particularly interesting. If gradient and curvature drifts are not important in the modulation process, then we can invoke the old argument that the magnetic fields should be nearly radial over the solar poles, and should provide easier access to the galactic particles, with high intensities resulting (e.g., Fisk, 1976). Conversely, if gradient and curvature drifts are important, then, until the next field reversal in 1991-92, the drifts are outward over the solar poles, should restrict the access of the galactic particles, and tend to reduce the intensities (e.g., Jokipii and Kopriva, 1979). Of course, since Ulysses will fly near the field reversal, the magnetic field structure may be confused, with mixed polarity and a reduction in systematic drift effects.

The importance of drift effects in the modulation process is still uncertain. As Jokipii (1986) argued, we should expect on simple theoretical grounds that drifts will be very important. Yet the observations are unclear as to their importance.

On the negative side for the importance of drift effects, as noted by McKibben (1986), solar cycle variations in the cosmic ray flux appear to be the result of the outward radial propagation of large disturbances in the solar wind. Such observations suggest that the cosmic ray particles propagate primarily in the radial direction, and that drifts in latitude, or indeed any latitude transport is not particularly significant. Similarly, McKibben (1986) reported that the observed radial gradients of galactic cosmic rays continue to be small, independent of the sign of the field polarity. One of the basic predictions of models in which drifts are important (e.g., Jokipii and Kopriva, 1979) is that the gradients are small in the previous solar cycle, when drifts are downward from the poles, but should increase dramatically when the field reversal occurs in 1980-81, when particles enter the heliosphere along the equatorial current sheet, and then drift upward towards the solar poles.

On the positive side for drifts, Garcia-Munoz et al. (1986) reported that, at least after 1981, there appears to be a charge-dependent effect in modulation in which electrons behave differently from ions. The simplest explanation for this observation is some form of drift effect, which is the only known modulation effect to treat electrons differently from ions.

Clearly, simply observing the spectra and composition of energetic particles from Ulysses will settle many different issues. We should be able to tell the extent of coronal propagation of solar particles, the latitude extent of interplanetary acceleration, the likely source region for the acceleration of the anomalous component, and the importance of gradient and curvature drifts in the modulation of galactic cosmic rays. Indeed, observing galactic cosmic rays at high latitudes near solar-maximum conditions should be particularly interesting. Although perhaps confusing, these observations may provide us with an understanding of what for many of us has been a long quest--the true cause of the solar cycle variation in galactic cosmic rays.

## 6. INTERSTELLAR GAS AND DUST

As was described by Fahr (1986), Ulysses presents some exciting prospects for direct observations of neutral interstellar gas. These observations, combined with global measurements of the Ly $\alpha$ backscatter of the neutral gas, should provide information on the local interstellar medium, and on its interaction with the heliosphere.

Indeed one of the opportunities provided by Ulysses is for a multipronged attack on the behavior of interstellar neutral and ionized gas in the heliosphere. Experiments on Ulysses should be able to observe the neutral gas directly, to observe it with the solar wind composition detectors just after it is ionized by photoionization and charge exchange, and, then, upon assuming that the anomalous component has its origin as ionized interstellar gas, to observe the interstellar gas as energetic particles.

As described by Geise (1986) and Morfill (1986), Ulysses will provide important information on the interplanetary dust cloud, surveying it for the first time in latitude.

## 7. CONCLUDING REMARKS

With the launch of Ulysses approaching, and the long wait for measurements of the heliosphere in three dimensions now about to end, we should ask ourselves what final efforts should we be pursuing prior to launch. Several efforts seem to be of particular importance:

First, the onus is on the theoreticians and modelers to put the theories for solar and heliospheric phenomena in order so that they can readily be compared with the observations from Ulysses. Unexpected phenomena are bound to be observed by Ulysses. The question will arise as to whether these unexpected results are a fundamental departure from the way we have understood the heliosphere, or whether they can be explained by some tinkering with the parameters of the standard theory. Theoreticians and modelers should be challenged to speculate on what are the definitive tests of current theories; what are the most crucial observations for Ulysses to pursue.

Second, we should recognize that prior-to-launch is an ideal time in which to plan for collaborative efforts. The problems that Ulysses will be studying are inherently multidisciplinary. They will require data from a number of different instruments, and inputs from theoreticians and modelers that are not now part of the mission. We need to be concerned with how best we can foster the collaborations that will be needed for an effective pursuit of these problems.

We need also to be concerned with the missions that will be needed to support Ulysses. As others have noted, Ulysses might better be called Cyclops, for with the loss of the US spacecraft there is only one eye left. And now it is not clear that there will be an adequate near-Earth monitor of heliospheric conditions, from which to determine, together with Ulysses, latitude gradients in the solar wind and energetic particles. Indeed, one of the motivations of Ulysses is to determine how phenomena at high latitudes affect conditions near Earth; to explore this aspect of the solar-terrestrial connection. Clearly, this will not be possible without simultaneous near-Earth measurements. It is to be hoped that NASA will make every effort to track existing spacecraft that can provide measurements of solar wind conditions near Earth, and that the WIND spacecraft of the International Solar Terrestrial Program can be flown in time to support Ulysses.

Finally, we should recognize that Ulysses is only the beginning of the exploration of the heliosphere in three dimensions. We would be naive to think that Ulysses will provide all the answers, any more than the first Mariners and Pioneers provided all the answers on conditions and processes in the solar wind near the equatorial plane. We need to think at some point--perhaps it is a little early now--about an appropriate follow-on to Ulysses; perhaps a Solar Polar Orbiter, to make repeated measurements at all latitudes.

## REFERENCES

Coles, W. A. 1986, presentation to the 19th ESLAB Symposium.
Fahr, H. J. 1986, presentation to the 19th ESLAB Symposium.
Fisk, L. A. 1976, J. Geophys. Res., 81, 4646.
Fisk, L. A. 1986, presentation to the 19th ESLAB Symposium.
Garcia-Munoz, M., Meyer, P., Pyle, K. R., and Simpson, J. A. 1986, presentation to the 19th ESLAB Symposium.
Geiss, J. 1986, presentation to the 19th ESLAB Symposium.
Giese, R. H. 1986, presentation to the 19th ESLAB Symposium.
Gold, R. E., Lanzerotti, L. J., Maclennan, C. G., and Krimigis, S. M. 1986, presentation to the 19th ESLAB Symposium.
Hoeksema, J. T. 1986, presentation to the 19th ESLAB Symposium.

Jokipii, J. R. 1986, presentation to the 19th ESLAB Symposium.
Jokipii, J. R., and Kopriva, D. A. 1979, Astrophys. J., **234**, 384.
Lee, M. A. 1986, presentation to the 19th ESLAB Symposium.
MacQueen, R. M. 1986, presentation to the 19th ESLAB Symposium.
McKibben, R. B. 1986, presentation to the 19th ESLAB Symposium.
Morfill, G. 1986, presentation to the 19th ESLAB Symposium.
Newkirk, G., and Fisk, L. A. 1985, J. Geophys. Res., **90**, 3391.
Parker, E. N. 1986, presentation to the 19th ESLAB Symposium.
Roelof, E. C. 1986, presentation to the 19th ESLAB Symposium.
Schwenn, R. 1986, presentation to the 19th ESLAB Symposium.
Smith, E. J., Slavin, J. A., and Thomas, B. T. 1986, presentation to
       the 19th ESLAB Symposium.
Wibberenz, G. 1986, presentation to the 19th ESLAB Symposium.
Withbroe, G. L. 1986, presentation to the 19th ESLAB Symposium.

# LIST OF PARTICIPANTS

ANDERSON, K.            Space Sciences Laboratory
                       Univ. of California, Berkeley CA 94720
                       USA

ANGLIN, J. D.          Herzberg Inst. of Astrophysics
                       100 Sussex Drive, Ottawa, Ontario K1A OR6
                       Canada

BALOGH, A.             The Blackett Laboratory
                       Imperial College, Prince Consort Road
                       London SW7 2BZ, England

BAME, S.               ESS-8, MS D438
                       Los Alamos National Lab., Los Alamos NM 87545
                       USA

BARNES, A.             245-3, NASA-AMES Research Center
                       Moffett Field, CA 94035
                       USA

BERCOVITCH, M.         Herzberg Inst. of Astrophysics
                       100 Sussex Drive, Ottawa, Ontario K1A OR6
                       Canada

BERTAUX, J-L.          Institution Service d'Aéronomie
                       du CNRS, BP 3 91370 Verrières-le-Buisson
                       France

BHONSLE, R.            Room No. 758
                       Physical Research Laboratory, Ahmedabad-380009
                       India

BIRD, M.               Radioastronomisches Institut
                       University of Bonn, Auf dem Hügel 71
                       5300 Bonn, West Germany

BOHLIN, J. D.          Code EZ - NASA Headquarters
                       Washington DC 20546,
                       USA

BOUGERET, J-L.         Observatoire de Paris
                       92195 Meudon Principal Cedex,
                       France

BURROWS, J. R.         Herzberg Inst. of Astrophysics
                       100 Sussex Drive, Ottawa, Ontario K1A OR6
                       Canada

BURLAGA, L.            Code 692
                       NASA/GSFC, Greenbelt, MD 20771
                       USA

BÜRGI, A.              Physikalisches Inst.
                       University of Bern, Sidlerstrasse 5
                       CH-3012 Bern, Switzerland

CHASSEFIÈRE, E.        Institution Service d'Aeronomie
                       du CNRS, BP 3 91370 Verrieres-le-Buisson
                       France

CINI-CASTAGNOLI, G.    Istituto di Cosmo-Geofisica
                       Corso Fiume 4, 10133 Torino
                       Italy
CLIVER, E.             Air Force Geophysics Laboratory (PHP)
                       Hanscom AFB, MA 01731
                       USA
COLES, W.              EE/CS Dept. Mail Code C-014
                       University of California,
                       La Jolla CA 92093, USA
CORNILLEAU, N.         CRPE/CNET
                       3 Ave. de la République, 92131 Issy les Moulineaux
                       France
DENZEL, R.             University Graz (Astronomy)
                       Kasmanhuberstrasse 13, A-9500 Villach
                       Austria
DOMINGO, V.            Space Science Dept. of ESA
                       ESTEC, Postbus 299, 2200 AG  Noordwijk
                       The Netherlands
DRYER, M.              NOAA
                       Space Environment Laboratory, 325 Broadway
                       Boulder, CO 80303, USA
DULK, G.               Dept. of Astrophysical, Planetary
                       and Atmospheric Sciences, University of Colorado
                       Boulder CO 80309, USA
ENGELMANN, J-J.        DPHG/SAP
                       91191 Gif-sur-Yvette Cedex,
                       France
ETCHETO, J..           CRPE/CNET
                       3 Ave. de la République, 92131 Issy les Moulineaux
                       France
FAHR, H.               Institut für Astrophysik
                       Universität Bonn, Auf dem Hügel 71
                       D53 Bonn, West Germany
FISK, L.               Univ. of New Hampshire
                       Room 210 Thompson Hall, Durham, NH 03824
                       USA
FLÜCKIGER, E.          Physikalisches Institut
                       Universität Bern, Sidlerstrasse 5
                       CH-3012 Bern, Switzerland
FRY, C.                Geophysical Institute
                       University of Alaska, Fairbanks
                       Alaska 99701, USA
GARCIA-MUNOZ, G.       Univ. of Chicago, LASR
                       933 E. 56th Street,
                       Chicago, IL 60637, USA
GEISS, J.              Physikalisches Institut
                       University of Bern, Sidlerstrasse 5
                       CH-3012 Bern, Switzerland
GIANI, F.              Aer Italia Space Systems Group
                       125 Via Servaise, 10146 Torino
                       Italy

GIESE, R.              Bereich Extraterrestrische Physik
                       Ruhr-Universität Bochum, P. O. Box 102148
                       D-4630 Bochum, West Germany
GOLLIEZ, F.            University of Bern
                       Physikalisches Institut, Sidlerstrasse 5
                       CH 3012 Bern, Switzerland
GOLD, R.               APL/JHU
                       Johns Hopkins Road, Laurel, MD 20707
                       USA
GRÜN, E.               Max-Planck-Institut für Kernphysik
                       Saupfercherckweg,
                       69 Heidelberg 1, West Germany
HARVEY, C.             DESPA
                       Observatoire de Paris-Meudon,
                       Meudon 92195, Principal Cedex, France
HATTON, C.             Department of Combined Studies
                       University of Leeds, Leeds LS2 9JT
                       England
HICK, P.               SRON Space Research Utrecht
                       Beneluxlaan 21, 3527 HS Utrecht
                       The Netherlands
HOANG, S.              DESPA
                       Observatoire de Paris-Meudon,
                       Meudon 92195, Principal Cedex, France
HOEKSEMA, J. T.        Center for Space Science and
                       Astrophysics, ERL 328 Via Crespi
                       Stanford, CA 943065, USA
HOVESTADT, D.          Max-Planck-Institut
                       für Physik & Astrophysik,
                       D-8046 Garching, West Germany
HOWARD, R.             Code 4173 H
                       Naval Research Lab, Washington DC 20375
                       USA
HURLEY, K.             CESR BP 4346
                       31029 Toulouse Cedex,
                       France
HYNDS, R.              The Blackett Laboratory
                       Imperial College, Prince Consort Road
                       London SW7 2BZ, England
JACKSON, B.            Department of EE/CS C-014 UCSD
                       Univ. of California,
                       La Jolla CA 92093, USA
JOKIPII, J.            University of Arizona
                       Dept. of Planetary Sciences,
                       Tucson, AZ 85721,USA
KAHLER, S.             Emmanuel College
                       AFGL/PHP, Hanscom AFB
                       MA 01731, USA
KANE, S.               Space Sciences Laboratory
                       University of California, Berkeley
                       CA 94720, USA

KELLOGG, P.           Section d'Astrophysique
                      Observatoire de Paris-Meudon, Meudon
                      France
KEPPLER, E.           Max-Planck-Institut für Aeronomie
                      Postfach 20, D-3411 Katlenburg-Lindau 3
                      West Germany
KLEIN, K-L.           Observatoire de Meudon
                      F-92195 Meudon Principal Cedex,
                      France
KOCH-MIRAMOND, L.     C.E.N. Saclay
                      SAP, 91191 Gif-sur-Yvette Cedex
                      France
KOJIMA, M.            The Research Inst. of Atmospherics
                      Nagoya University, 3-13 Honohara
                      Toyokawa, 442, Japan
KÖMLE, N.             Space Research Institute
                      Lustbühel Observatory , A-8042 Graz
                      Austria
KUNOW, H.             Universität Kiel, Institut für
                      Reine & Angewandte Kernphysik, Olshausenstrasse
                      D-2300 Kiel 1, West Germany
LACOMBE, C.           DESPA
                      Observatoire de Paris-Meudon, Meudon 92195
                      Principal Cedex, France
LALLEMENT, R.         Inst. Service d'Aeronomie de CNRS
                      B.P. 3,
                      91370 Verrieres-le-Buisson, France
LANZEROTTI, L.        MH 1E-439
                      Bell Laboratories, 600 Mountain Avenue
                      Murray Hill, NJ 07974, USA
LECACHEUX, A.         Observatoire de Paris
                      92195 Meudon Principal Cedex,
                      France
LEE, M.               Physics Dept.
                      Univ. of New Hampshire, Durham, NH 03824
                      USA
MACQUEEN, R.          High Altitude Observatory, National
                      Center for Atmospheric Research, P.O. Box 3000
                      CO 80307, USA
MARSCH, E.            Max-Planck-Institut für Aeronomie
                      Postfach 20, D-3411 Katlenburg-Lindau 3
                      W. Germany
MARSDEN, R.           Space Science Dept. of ESA
                      ESTEC, Postbus 299, 2200 AG  Noordwijk
                      The Netherlands
MCDONALD, F.          NASA Headquarters - Code P
                      Washington, DC 20546,
                      USA
MCGUIRE, R.           Code 661
                      NASA/GSFC, Greenbelt, MD 20771
                      USA

MCKIBBEN, B.        Univ. of Chicago, LASR
                    933 E. 56th Street,
                    Chicago, IL 60637, USA
MORFILL, G.         Max-Planck-Institut
                    für Physik & Astrophysik,
                    D-8046 Garching, West Germany
MÜLLER-MELLIN, R.   Universität Kiel, Institut für
                    Reine & Angewandte Kernphysik, Olshausenstrasse
                    D-2300 Kiel 1, West Germany
NASS, U.            Universität Bonn
                    Institut für Astrophysik, Auf dem Hügel 71
                    D53 Bonn, West Germany
NOCI, G.            Arcetri Astrophysical Observatory
                    Largo E. Fermi 5, Florence
                    Italy
PAGE, D. E.         Space Science Dept. of ESA
                    ESTEC, Postbus 299, 2200 AG   Noordwijk
                    The Netherlands
PAIZIS, C.          Dipartimento di Fisica
                    Universita di Milano, Via Celoria 16
                    20133 Milano, Italy
PARKER, E.          Univ. of Chicago, LASR
                    933 E. 56th Street,
                    Chicago, IL 60637, USA
PICK, M.            DESPA
                    Observatoire de Paris-Meudon,
                    92195 - Meudon Principal Cedex, France
POLETTO, G.         Arcetri Astrophysical Observatory
                    Largo E. Fermi 5, Florence
                    Italy
PYLE, K. R.         Univ. of Chicago
                    LASR, 933 E. 56th Street
                    Chicago, IL 60637, USA
REINHARD, R.        Space Science Dept. of ESA
                    ESTEC, Postbus 299, 2200 AG   Noordwijk
                    The Netherlands
RICHTER, A.         Max-Planck-Institut für Aeronomie
                    Postfach 20, D-3411 Katlenburg-Lindau 3
                    W. Germany
ROELOF, E.          APL/JHU
                    Johns Hopkins Road, Laurel, Maryland 20707
                    USA
ROSENBAUER, H.      Max-Planck-Institut für Aeronomie
                    Postfach 20, D-3411 Katlenburg-Lindau 3
                    W. Germany
ROTH, M.            Institute for Space Aeronomy
                    Avenue Circulaire 3, B-1180 Brussels
                    Belgium.
SANDERSON, T.       Space Science Dept. of ESA
                    ESTEC, Postbus 299, 2200 AG   Noordwijk
                    The Netherlands

SCHWENN, R.              Max-Planck-Institut für Aeronomie
                        Postfach 20, D-3411 Katlenburg-Lindau 3
                        West Germany
SCHERER, M.             Inst. for Space Aeronomy
                        Avenue Circulaire 3, B-1180 Brussels
                        Belgium
SHEA, M.                Air Force Geophysics Laboratory (PHP)
                        Hanscom AFB,
                        Bedford, MA 01731, USA
SIME. D.                High Altitude Observatory
                        P.O. Box 3000, Boulder
                        CO. 80307, USA
SIMPSON, J.             Univ. of Chicago, LASR
                        933 E. 56th Street,
                        Chicago, IL 60637, USA
SMART, D.               Air Force Geophysics Laboratory (PHP)
                        Hanscom AFB,
                        Bedford, MA 01731, USA
SMITH, E.               MS 169-506
                        Jet Propulsion Laboratory, 4800 Oak Grove Drive
                        Pasadena, CA 91109, USA
SOMMER, M.              Max-Planck-Institut für
                        Physik & Astrophysik,
                        D-8046 Garching, West Germany
SOUTHWOOD, D.           The Blackett Laboratory
                        Imperial College, Prince Consort Road
                        London SW7 2BZ, England
STEINBERG, J-L.         Section d'Astrophysique
                        Observatoire de Paris - Meudon,
                        92195 Meudon Principal Cedex, France
STEVENS, G.             SRON Space Research Utrecht
                        Beneluxlaan 21, 3527 HS Utrecht
                        The Netherlands
SUESS, S.               Space Science Lab/ES52
                        NASA-Marshall Space Flight Center, Huntsville
                        Alabama 35812, USA
TRANQUILLE, C.          Space Science Dept. of ESA
                        ESTEC, Postbus 299, 2200 AG   Noordwijk
                        The Netherlands
TREGUER, L.             C.E.N. Saclay
                        SAP, 91191 Gif-sur-Yvette Cedex
                        France
TROTTET, G.             Section d'Astrophysique
                        Observatoire de Paris, DASOP 5 Place Jansen
                        France 92195
TSURUTANI, B.           MS 169-506
                        Jet Propulsion Laboratory, 4800 Oak Grove Drive
                        Pasadena, CA 91109, USA
van ROOIJEN, J.         SRON Space Research Utrecht
                        Beneluxlaan 21, 3527 HS Utrecht
                        The Netherlands

VENKATESAN, D.        Dept. of Physics
                     Univ. of Calgary, Calgary, Alberta
                     Canada
VILMER, N.           Section d'Astrophysique
                     Observatoire de Paris - Meudon,
                     92195 Meudon Principal Cedex, France
VILLEDARY de, C.     CRPE/CNET
                     3 Ave. de la République, 92131 Issy les Moulineaux
                     France
WATANABE, T.         Research Institue of Atmospherics
                     Nagoya University, Honohara, Toyokawa 442
                     Japan
WENZEL, K-P.         Space Science Dept. of ESA
                     ESTEC, Postbus 299, 2200 AG  Noordwijk
                     The Netherlands
WIBBERENZ, G.        Universität Kiel, Institut für
                     Reine & Angewandte Kernphysik, Olshausenstrasse
                     D-2300 Kiel 1, West Germany
WITTE, M.            Max-Planck-Institut für Aeronomie
                     Postfach 20, D-3411 Katlenburg-Lindau 3
                     West Germany
WITHBROE, G.         Center for Astrophysics
                     60 Garden Street, Cambridge
                     MA 02138, USA
WRIGHT, C.           IPS, P.O. Box 702
                     Darlinghurst, NSW 2010
                     Australia

# SUBJECT INDEX

Active region, coronal                33-37, 74, 77, 82, 214, 215,
                                      217, 219, 220, 229, 233, 257,
                                      341, 344, 350, 479, 494
Albedo                                441, 448, 449
Alfvén speed                          34-36, 308, 315
Alfvén wave                           33-37, 59, 62, 135, 160, 308,
                                      495
Alfvenic fluctuation                  160 ff.
Anomalous component                   306, 361, 363, 368, 369, 371,
                                      372, 384-387, 413, 417, 498
  acceleration of                     368, 369, 372, 384, 401,
                                      403 ff., 499
  origin of                           402 ff.
  radial gradient of                  407, 408
  variation with latitude             401 ff.
Archimedes' spiral                    49, 231, 259, 261, 364, 376,
                                      378, 379
Asymmetry, north-south, in solar
        activity                      395 ff.
Atom-ion separation                   173, 174, 179-181


Ballistic propagation model           205 ff.
Beta-meteoroid                        459, 461-464
Blast wave                            120
Bow shock                             305-308, 312, 325
Bremsstrahlung                        73-75, 78


Centaur                               478, 484
Chromosphere                          173, 174, 179, 181
Comet                                 182, 187-189, 458-460, 462,
                                      464, 469
  ion tail, orientation of            187-189
Compact radio source, interplanetary
        scintillation observations of 153
Comprehensive flare index            395-397
Corona                                5, 8-11, 16, 39, 40, 43, 45-50,
                                      53, 101, 104, 105, 107, 119,
                                      173-176, 179, 184, 213, 214,
                                      217, 220, 221, 255-258, 341,
                                      346, 456
  density of                          45, 49, 50, 257
  heating of                          33-37, 93, 173, 495
  potential field of                  5-7, 11, 14, 241-243, 247-249,
                                      258, 260, 261

propagation of energetic particle
  in                                        304, 497
rotation of                               45 ff., 252-254
source region of solar wind in            57, 174, 184
source surface model of                   260, 275-277, 280, 287, 288,
                                          290-292
3-dimensional structure of                213 ff.
white light structures in                 5, 6
X-ray emission of                         19-22, 25, 93 ff., 479
Coronagraph
  Solwind                                 107, 108, 115, 116, 124, 125
  white-light                             10, 12, 28, 30, 39, 40, 102,
                                          107, 113, 116, 117, 119, 129,
                                          136, 148, 275
Coronal active region                     33-37, 74, 77, 82, 214, 215,
                                          217, 219, 220, 229, 233, 257,
                                          341, 344, 350, 479, 494
Coronal arch (see Coronal loop)
Coronal current sheet                     11, 34, 148, 150
Coronal hole                              5, 9, 19-30, 33-37, 48, 57,
                                          103, 148, 176, 177, 181, 182,
                                          192, 193, 195, 214-218, 241,
                                          252, 253, 258, 260, 264, 272,
                                          282, 283, 319, 321, 322, 423,
                                          496, 497
  polar                                   9-11, 19-22, 26, 29-31, 34, 36,
                                          37, 40, 50, 148, 158, 193, 195,
                                          196, 247, 259, 315, 435, 439,
                                          488, 494-497
  solar cyclic evolution of               12,13
Coronal loop                              80, 93, 95, 98, 103, 119, 121,
                                          214, 217-219, 257, 258
Coronal magnetic field                    5 ff., 22, 24, 34, 47, 65-70,
                                          93, 119, 219, 220, 241-243,
                                          252-256, 261, 265, 274, 346,
                                          457
  model of                                65-70, 188, 276
  solar cyclic variation of               5, 12, 13, 148-150, 176, 177
Coronal mass ejection                     5, 13-16, 19, 22-24, 26, 30,
                                          101 ff., 119-121, 123-128, 199,
                                          496, 497
  edge-on                                 124, 125, 128
  halo                                    125, 126, 128
  Helios images of                        113 ff.
  rate of occurrence                      108, 109
  solar cycle dependence of               107 ff.
  3-dimensional shape of                  110, 113, 114, 119
Coronal streamer (see Streamer)
Coronal transient                         65, 119-121, 176, 218, 268
Co-rotating interaction region            49, 305, 319 ff., 325-329
Co-rotating interplanetary stream         191 ff.

Co-rotating shock wave                     319, 323-325, 327, 349
Cosine model                               444, 446-448, 450
Cosmic ray
    galactic                               173, 185, 201, 349, 350, 353,
                                           354, 361 ff., 389-394, 395,
                                           397-400, 401-403, 407, 410,
                                           477-479, 481, 494, 495, 497-499
        north-south anisotropy of          392, 395
    solar (see Solar flare particle)
Cosmic ray diurnal variation               395, 398-400
Cosmic ray gradient
    azimuthal                              191
    latitudinal                           363, 369-372, 389 ff., 500
    north-south asymmetrical              398-400
    north-south symmetrical               399, 400
    radial                                191, 363, 367-369, 372, 378,
                                           386, 389 ff., 498
Cosmic ray modulation                      201, 350, 354, 361 ff., 389,
                                           391, 392, 394, 403, 405, 406,
                                           408-410, 479, 494, 495, 498,
                                           499
    effect of 3-dimensional helios-
        pheric structure on                375 ff.
Cosmic ray transport, theory of            307, 355 ff., 375 ff., 389, 392
Cross-helicity                             165
Current sheet
    coronal                               11, 34, 148, 150
    heliospheric                          48, 167 ff., 238, 239, 241,
                                           242, 245-247, 251-256, 259-261,
                                           264, 275 ff., 320, 322, 324,
                                           345, 363, 367, 370, 376-384,
                                           386, 457, 494
        changes in, with solar cycle      267 ff.

Differential rotation                      47, 252-254
Diffusion approximation                    355, 358, 362, 364, 372, 377,
                                           378
Dipole, solar                             257, 272, 273
    tilted                                273
Disappearing filament                      125 ff.
Discontinuity, tangential                  167 ff., 265
Disturbance, travelling interplanetary     123-128, 153, 215, 223, 224,
                                           268, 328, 353, 366, 367, 381,
                                           384
    3-dimensional configuration of         123 ff.
Diurnal variation, cosmic ray              395, 398-400
Doppler Angular Spectral Scanning          437
Doppler-dimming                            28, 29, 43, 53

Drift, charged particle                          313, 361-364, 367-369, 371,
                                                 372, 375 ff., 404-408, 410,
                                                 494, 498, 499

Drift velocity, charged particle                 378, 384
Driver gas                                       120, 174, 175
Dust (see Interplanetary dust)


Ecliptic baseline measurement                    373, 489, 500
Electron
   acceleration of                               73 ff.
   energetic                                     73, 74, 87 ff., 229, 235, 302,
                                                 303, 312, 349-352
Electron density variations                      39, 40, 42, 43, 115, 129, 143,
                                                 144, 157, 214, 230, 235
Ellipsoid model                                  444-448, 451-453
Energetic particle                               120, 305 ff., 325-328, 345,
                                                 349 ff., 497-500

   acceleration of                               298, 305 ff., 325, 326, 328,
                                                 479, 497-499

   coronal propagation of                        304, 497
   enhancements, latitude dependence of          320, 325 ff.
Energetic storm particle                         306, 310, 313, 315, 316, 354
Energy cascade                                   163, 164
Eruptive prominences                             14, 19, 22, 23, 30, 120


Fan model                                        444-449, 451-453
Fermi acceleration                               307
First ionization potential                       174, 179-182


Galactic cosmic ray                              173, 185, 201, 349, 350, 353,
                                                 354, 361 ff., 389-394, 395,
                                                 397-400, 401-403, 407, 410,
                                                 477-479, 481, 494, 495, 497-499

   north-south anisotropy of                     392, 395
Gamma-ray burst                                  477-479, 481
Geomagnetic activity                             129, 130, 132-134
Giant region                                     281, 283-285
Gradient, cosmic ray
   azimuthal                                     191
   latitudinal                                   361, 369-372, 389 ff., 500
   north-south asymmetrical                      398-400
   north-south symmetrical                       399, 400
   radial                                        191, 363, 367-369, 372, 378,
                                                 386, 389 ff., 498
Gradient, solar flare particle                   297, 300, 301, 331 ff.
Gravitational wave                               477-479, 483

Heliopause                                  428, 430, 435, 436
Heliosphere                                 191, 205, 375, 376, 386, 387,
                                            389, 391, 402, 405, 421, 425,
                                            426, 431, 433, 435, 436, 493,
                                            499, 500
    inner                                   162, 403, 409, 413
    magnetic structure of                   363, 371, 380
    origin of                               33-35, 37
    outer                                   162, 305, 306, 313, 315, 325,
                                            328, 361, 368, 377, 392, 394,
                                            403, 404, 407, 409, 413
    physical processes in                   16, 184, 361, 373, 488
    solar cycle evolution of                281, 282
    3-dimensional structure of              45, 48, 153, 205, 213 ff., 235,
                                            238, 239, 254, 267, 281 ff.,
                                            372, 477, 493-496
Heliospheric current sheet                  48, 167 ff., 238, 239, 241,
                                            242, 245-247, 251-256, 259-261,
                                            264, 275 ff., 320, 322, 324,
                                            345, 363, 367, 370, 376-384,
                                            386, 457, 494
    changes in, with solar cycle            267 ff.
Heliospheric energy source                  33 ff.
Heliospheric magnetic field (see Inter-
        planetary magnetic field)
Heliospheric termination shock              306, 368, 375, 377, 384-386,
                                            413, 428
    equatorial                              408-410, 498
    polar                                   369, 404 ff., 498
Helium, focusing cone of                    422, 425, 428, 436, 437, 439
High speed jet                              19, 27, 31
HXIS                                        66, 70, 93-97
Hydrodynamic propagation model              205 ff.
Hydromagnetic wave                          62, 159 ff., 312
Hysteresis                                  409

Interaction region                          192-195, 197, 201
    co-rotating                             49, 305, 319 ff.
    merged                                  194, 197, 366
Interface                                   193-195
Interplanetary disturbance, travelling      123-128, 153, 215, 223, 224,
                                            268, 328, 353, 366, 367, 381,.
                                            384
    3-dimensional configuration of          123 ff.
Interplanetary dust cloud                   456, 458, 460, 463, 464, 470
    optical model of                        441-448
    3-dimensional structure of              441 ff., 499
interplanetary dust particle                455 ff., 477-479, 481
    density, 3-dimensional distribution
        of                                  441-450, 453, 479

  dynamics of                                 182, 450, 455 ff.
  hyperbolic orbit of                         461-463
  impact measurements of                      441, 449, 450
  spatial variation of properties of          448, 449, 453
Interplanetary grain (see Inter-
    planetary dust particle)
Interplanetary magnetic field                 11, 49, 50, 130, 133, 135-138,
                                              142, 220, 235, 238, 241 ff.,
                                              277, 281 ff., 287-289, 297,
                                              300, 319-323, 341-343, 345,
                                              349, 362, 413, 416, 417,
                                              456-458, 462, 477, 478, 480,
                                              495
  multipole components of                     249, 251, 273
  polarity of                                 183, 243, 244, 246, 248, 249,
                                              267-274, 281, 283-285, 363,
                                              364, 371, 383, 385, 386,
                                              398-400, 457, 469
  power spectrum of                           162-164
  3-dimensional structure of                  242-247, 252, 255 ff., 375-379
Interplanetary magnetic field lines,
    latitude distribution of                  229 ff.
Interplanetary propagation                    297, 301, 302, 413, 479
Interplanetary resonance glow                 422, 428, 431
Interplanetary scintillation                  20, 48-50, 59, 60, 123 ff.,
                                              136, 143 ff., 164, 187-189,
                                              213, 215, 222-225, 241, 247,
                                              324, 488, 495
  observation of, from compact radio
    source                                    153
Interplanetary shock front                    120, 123
Interplanetary shock wave                     22, 105, 109, 119-121, 192,
                                              194, 196, 198 ff., 205, 206,
                                              208, 288, 302-304, 305 ff.,
                                              325, 349, 353, 354, 358, 366,
                                              372, 466, 496, 497
  co-rotating                                 319, 323-325, 327, 349
  directional diagram of speed of            124-127
  forward                                     194, 196, 305, 319-321
  reverse                                     194, 196, 305, 319-321
  speed of                                    124-127
  3-dimensional MHD model of                  135 ff.
Interplanetary stream
  corotating                                  191 ff.
  transient                                   191 ff., 366, 466
Interstellar gas                              173, 184, 401-403, 413, 417,
                                              421 ff., 435-440, 469, 477-480,
                                              498, 499
  dynamics of                                 426-428
  flow velocity of                            435-440
  ionisation of, by charge exchange           413, 423-425, 428, 435, 439

parameters deduced from H and He
   observations in the solar system      435 ff.
  temperature of                       423, 428, 435-440
Interstellar grain                 458, 459, 461, 462, 469
Interstellar medium               33, 328, 402, 421-423, 426-430,
                                    432, 433, 435-440, 457
Interstellar wind                  421-423, 438
Ion, pick-up                       413 ff.
Ion, solar wind                    173 ff.  309, 310, 413, 416,
                                    425, 428, 458, 460, 480
  charge-state of                 173-175, 177, 184


Jayne's principle (see Maximum entropy
     principle)
Jupiter                           477, 478, 484, 485, 487


Kinematic method                 287-289


Landau damping                   36
Laplace equation                 6, 66, 242
Layer model                     446
Legendre function                6, 67, 68
Lobe model                      446, 447
Lyman- $\alpha$                      29, 40-43, 187, 423, 425,
                                    427-430, 435-440, 499
Lyman-$\alpha$ photometer          423, 435, 437


Magnetic cloud                  105, 120, 192, 199
Magnetic cycle, solar          256, 257, 267, 379, 380, 383,
                                    384, 386, 465, 467, 469, 470
Magnetic field                  59, 61, 63, 88-91, 160, 171,
                                  191, 197, 199, 201, 213, 229,
                                  306, 308, 310-312
  corona.                      5 ff., 22, 24, 34, 47, 65-70,
                                  93, 119, 219, 220, 241-243,
                                  252-256, 261, 265, 274, 346,
                                  457
    model of                  65-70, 188, 276
    solar cyclic variation of   5, 12, 13, 148-150, 176, 177
  heliospheric (see Interplanetary
     magnetic field)
    interplanetary            11, 49, 50, 130, 133, 135-138,
                                140, 220, 235, 238, 241 ff.,
                                277, 281 ff., 287-289, 297,
                                300, 319-323, 341-343, 345,
                                349, 362, 413, 416, 417,

|  |  |
|---|---|
|  | 456-458, 462, 477, 478, 480, 495 |
| multipole component of | 249, 251, 272 |
| polarity of | 183, 243, 244, 246, 248, 249, 267-274, 281, 283-285, 363, 364, 371, 383, 385, 386, 398-400, 457, 469 |
| power spectrum of | 162-164 |
| 3-dimensional structure of | 242-247, 252, 255 ff., |
| photospheric | 5, 6, 8, 9, 24, 47, 65, 66, 70, 241 ff., 255, 256, 274, 281, 283, 284 |
| solar polar | 257, 265, 315, 405, 457, 463 |
| polarity reversal of | 246, 247, 250-252, 257, 267, 268, 274, 365, 367-369, 371, 372, 399, 400, 457, 489 |
| solar surface (see Photospsheric magnetic field) |  |
| Magnetic helicity | 165 |
| Magnetic reconnection, 3-dimensional, after a prominence eruption | 65 ff., 119 |
| Magnetic sector boundary | 187, 188, 238, 276, 280, 285, 320, 322 |
| Magnetic sector structure | 16, 187, 188, 220, 241, 244, 246-249, 253, 256, 259, 260, 264, 267-269, 274, 283-285, 289, 320, 457, 462, 463, 466, 469 |
| Magnetograph, solar | 148, 273 |
| Magneto-hydrodynamics (MHD) | 163, 164 |
| Maximum entropy approximation | 355-358 |
| Maximum entropy principle | 355 ff. |
| Mean free path | 307, 310, 343, 377, 421 |
| Merged interaction region | 194, 197, 366 |
| Meteoroid, beta- | 459, 461-464 |
| MHD simulation | 135, 136, 140 |
| Modulation (see Cosmic ray modulation) |  |
| Multispacecraft observations |  |
| of solar flare particle | 297 ff. |
| of solar hard X-rays | 78, 80, 81, 83, 84 |
| of type III radio burst | 235 ff. |
| Neutral sheet (see Current sheet) |  |
| O VI resonance doublet | 53 ff. |
| Optical probing | 450-452 |

Parker model                                          192, 255, 256, 259, 261-263,
                                                      265, 379

Particle number density, dust,
     3-dimensional distribution of                    441-450, 453
Photoionisation                                       423, 425, 426, 460
Photometer
    Lyman- $\alpha$                                   423, 435, 437
    zodiacal light                                    113-117, 129, 441
Photosphere                                           35, 37, 47, 49, 50, 257, 258
Photospheric magnetic field                           5, 6, 8, 9, 24, 47, 65, 66, 70,
                                                      241 ff., 255, 256, 274, 281,
                                                      283, 284
Pick-up ion                                           414 ff.
Pinhole/Occulter Facility                             30
Plasma turbulence (see Turbulence,
     solar wind)
Plasma wave                                           59, 62, 63, 159 ff., 265, 308-
                                                      312, 328, 329, 477-479, 481
Polar coronal hole                                    9-11, 19-22, 26, 29-31, 34, 36,
                                                      37, 40, 50, 148, 158, 193, 195,
                                                      196, 247, 259, 315, 435, 439,
                                                      488, 494-497
Polar magnetic field, solar                           257, 265, 315, 405, 457, 463
    polarity reversal of                              246, 247, 250-252, 257, 267,
                                                      268, 274, 365, 367-369, 371,
                                                      372, 399, 400, 457, 489
Polar plume                                           26, 27, 31
Polarity reversal, polar magnetic field               246, 247, 250-252, 257, 267,
                                                      268, 274, 365, 367-369, 371,
                                                      372, 399, 400, 457, 489
Polarity, solar magnetic field                        255, 257, 259, 465, 467, 469
Potential field, coronal                              5-7, 11, 14, 241-243, 247-249,
                                                      258, 260, 261
Power spectrum, interplanetary magnetic               162-164
Prominence eruption                                   65, 103
Propagation channel                                   342, 344, 346, 347
Propagation model
    ballistic                                         205 ff.
    hydrodynamic                                      205 ff.

Quadrupole                                            257, 273, 281, 282
Quasilinear theory                                    310, 311

Radiation pressure                                    436, 455, 456, 458, 460, 461,
                                                      463, 466, 469
Radio burst                                           213 ff., 235 ff., 477, 478, 481
    type I                                            217, 219, 220
    type III                                          25, 214, 220-226
        multispacecraft observations of               235 ff.

type U                                          214, 218, 219, 229
Radio emission
  non-thermal                                   213 ff., 224-226
  thermal                                       213, 217, 218
Radio heliograph                                136, 214-221, 226
Radio-isotope Thermoelectric Generator          483, 485
Radio propagation observations                  143
Remote sensing techniques                       28-31
Resonance scattering                            435


Scattering function                             442-445, 448, 449, 453
  volume                                        443, 449, 451
Scintillation (see Interplanetary scint-
    illation)
Sector boundary, magnetic                       187, 188, 238, 276, 280, 285,
                                                320, 322
Sector structure, magnetic                      16, 187, 188, 220, 241, 244,
                                                246-249, 253, 256, 259, 260,
                                                264, 267-269, 274, 283-285,
                                                289, 320, 457, 462, 463, 466,
                                                469
Shock acceleration                              185, 303, 305 ff., 325, 353,
                                                404-410
Shock front, interplanetary                     120, 123
Shock speed, interplanetary                     124-127
  directional diagram of                        124-127
Shock wave
  interplanetary                                22, 105, 109, 119-121, 192,
                                                194-196, 198 ff., 205, 206,
                                                208, 288, 302-304, 305 ff.,
                                                325, 349, 353, 358, 366, 372,
                                                466, 496, 497
    co-rotating                                 319, 323-325, 327, 349
    forward                                     194, 196, 305, 319-321
    reverse                                     194, 196, 305, 319-321
    3-dimensional MHD model of                  135 ff.
Solar activity                                  121, 255, 261, 262, 302, 361,
                                                368, 399, 400, 488
  north-south asymmetry in                      395 ff.
Solar and Heliospheric Observatory              30, 136
Solar corona (see Corona)
Solar cosmic ray (see Solar flare
    particle)
Solar cycle (see Sunspot cycle)
Solar dipole                                    257, 272, 273
  tilted                                        273
Solar flare                                     14, 15, 19, 22, 30, 73, 76-81,
                                                84, 87, 92 ff., 103, 119, 120,
                                                129, 135, 136, 140, 199, 268,
                                                298, 301, 302, 304, 313, 315,

| | |
|---|---|
| | 349, 350, 352-354, 395-397, 400, 479, 494 |
| multiple-ribbon | 66 |
| two-ribbon | 65, 66 |
| Solar flare particle | 174, 179-182, 306, 307, 310, 312, 315, 316, 325-328, 349, 352, 372, 477-479, 481 |
| gradient of | 297, 300, 301, 331 ff. |
| long-lived streaming of | 341 ff. |
| multispacecraft observations of | 297 ff. |
| Solar hard X-rays | 87 ff., 94-98, 477-479, 481 |
| centre-to-limb measurements of | 76, 78, 79, 82, 84 |
| energy spectrum of | 75, 76, 91, 92 |
| fine time structure of | 78, 79 |
| multispacecraft observations of | 78, 80, 81, 83, 84 |
| non-thermal model of | 73-76, 78 |
| occultation measurements of | 76-78, 80-84, 92 |
| spatial distribution of | 77, 88 |
| stereoscopic measurements of | 73 ff., 87, 91, 92 |
| thermal model of | 73-76 |
| time history of | 75, 78, 94, 95, 97 |
| Solar magnetic cycle | 256, 257, 267, 379, 380, 383, 384, 386, 465, 467, 469, 470 |
| Solar magnetic field polarity | 255, 257, 259, 465, 467, 469 |
| Solar magnetograph | 148, 273 |
| Solar Optical Telescope | 28, 35, 37 |
| Solar surface magnetic field (see Photospheric magnetic field) | |
| Solar wind | 33-37, 40, 42, 45, 48, 49, 53 ff., 130, 135-137, 154, 159 ff., 205 ff., 241, 242, 244, 247, 252, 254-256, 261, 264, 268, 275, 309, 316, 319, 328, 350, 361, 362, 366, 375, 377, 378, 386, 387, 401-405, 408, 409, 413-417, 421, 423-425, 428, 435, 439, 456-460, 477, 478, 494-497, 500 |
| acceleration of | 39, 53, 159, 161, 173-175, 479, 495 |
| acceleration region of | 59, 206 |
| bulk flow speed of (see Solar wind velocity) | |
| composition of | 173 ff. 479, 480, 483, 496 . |
| coronal origin of | 19 ff. 479 |
| coronal source region of | 57, 174, 184 |
| density of | 130, 133, 207-210, 457 |
| freezing-in temperature | 173, 175 |
| high speed stream | 20, 22, 30, 34, 36, 48, 50, 145, 148, 150, 158, 159, 161, 174-177, 192, 195. 206. 207, |

|                                         | 258,259, 282, 288, 319, 320, |
|                                         | 322, 423, 457, 494-500 |
| mass flux, anisotropy of                | 435, 439, 440 |
| observations of, at high latitude       | 143 ff. |
| observations of, near the sun           | 59 ff. |
| random velocity                         | 39, 42, 59, 62, 63, 496 |
| rotation of                             | 45 ff. |
| slow                                    | 19, 24, 25, 30, 31, 34, 53, |
|                                         | 174, 175, 177, 206, 207, 258, |
|                                         | 288, 423, 457, 496 |
|                                         | |
| 3-dimensional structure of              | 187-189 |
| turbulence in                           | 59-63, 143, 144, 153, 157, |
|                                         | 159 ff., 198, 201, 308, 377, |
|                                         | 496 |
| velocity gradient of                    | 275 ff. |
| velocity of                             | 39, 40, 42, 43, 53, 55, 59-63, |
|                                         | 130, 135, 137, 139, 143-150, |
|                                         | 153, 154, 160, 183, 184, 187, |
|                                         | 188, 191, 193, 206-210, 225, |
|                                         | 259, 275, 280, 287-289, 297, |
|                                         | 300, 304, 344, 366, 370, 378, |
|                                         | 379, 390-393, 423, 456, 457, |
|                                         | 466, 488 |
| Solar wind ion                          | 173 ff., 309, 310, 413, 416, |
|                                         | 425, 428, 458, 460, 480 |
| charge-state of                         | 173-175, 177, 184 |
| Solwind coronagraph                     | 107, 108, 115, 116, 124, 125 |
| Sombrero model                          | 446-448 |
| Source surface model, coronal           | 260, 275-277, 280, 287, 288, |
|                                         | 290-292 |
| Spacecraft                              | |
| AMPTE                                   | 179, 413-417 |
| Explorer 34                             | 297-300, 302, 303 |
| Helios 1-2                              | 22, 113-117, 124-126, 161, 193, |
|                                         | 194, 199, 222, 225, 259, 350- |
|                                         | 354, 443, 446, 449, 461, 464 |
| Hinotori                                | 77, 80, 81, 87 |
| IMP 6                                   | 222 |
| IMP 7                                   | 192, 268, 269, 365 |
| IMP 8                                   | 192, 268, 269, 351-354, 365, |
|                                         | 373, 389-393 |
| ISEE-3                                  | 77, 81, 126, 127, 183, 198, |
|                                         | 215, 222, 223, 225, 226, 229, |
|                                         | 230, 235-239, 267-272, 305, 306 |
| Mariner 9-10                            | 422, 423 |
| OGO 5                                   | 79-81, 422 |
| OSO 7                                   | 20, 79-81, 119 |
| Phobos                                  | 80, 83, 84 |
| Pioneer 6-9                             | 297-300, 302, 303 |

Pioneer 10                                  161, 260, 261, 305, 313, 314,
                                            319-324, 364, 365, 367-369,
                                            376, 390, 404, 422
Pioneer 11                                  161, 259, 262-263, 267-269,
                                            271, 305, 319-324, 364, 365,
                                            369, 370, 390, 404, 422, 477
Prognoz 5-6                                 422, 423, 435-437
Prognoz 7                                   77, 79, 81
PVO                                         77, 81
P78-1                                       22, 107, 119, 422
Sakigake                                    281-285
Skylab                                      5, 12, 14, 15, 20, 23, 26, 101-
                                            104, 107, 109, 119, 192, 217,
                                            219, 258
SMM                                         93, 101-104, 107, 119
Venera 11-12                                77, 79, 81, 435, 437, 438
Voyager 1                                   191, 192, 197, 205-210, 226,
                                            235-239, 261, 275-280, 306,
                                            325-329, 364-366, 369, 370,
                                            386, 389-394, 401, 402, 406,
                                            422, 477, 497
Voyager 2                                   197, 198, 205-210, 226, 235-
                                            239, 261, 275-280, 306, 325-
                                            329, 364-366, 386, 389-394,
                                            401, 402, 406, 422, 497
Space Shuttle                               478, 484
Spartan                                     30, 39 ff.
Spectral broadening                         143
Spectrophotometer, EUV                      435, 438
Spherical harmonics                         65-67
Spicule                                     19, 27, 31, 34, 493
Stream-stream interaction                   183, 184, 206, 288, 289, 404,
                                            408, 410, 494, 497, 498
Streamer                                    9, 19, 24, 25, 30, 31, 104,
                                            110, 111, 177, 214, 217, 220,
                                            226, 258, 496
Sun (see Chromosphere,Corona,Coronal..,
      Photosphere,Solar..)
Sunspot cycle                               101, 107-109, 111, 121,
                                            243 ff., 257, 263, 264, 319,
                                            361, 363, 365, 367, 376, 379,
                                            383, 386, 408, 426, 457, 465,
                                            469, 470, 484, 488
      21                                    12, 13, 248, 249, 260, 350,
                                            389, 394, 457
Super-event                                 349 ff.
Synchrotron radiation                       213
Synoptic map                                10-12, 145-148, 260

Tangential discontinuity — 167 ff., 265

Termination shock, heliospheric — 306, 368, 375, 377, 384-386, 413, 428

  equatorial — 408-410, 498

  polar — 369, 404 ff., 498

Thomson scattering — 113, 115

Transit-time damping — 404, 408, 410

Travelling interplanetary disturbance — 123-128, 153, 215, 223, 224, 268, 328, 353, 366, 367, 381, 384

  3-dimensional configuration of — 123 ff.

Turbulence, solar wind — 59-63, 143, 144, 153, 157, 159 ff., 198, 201, 308, 377, 496


Ulysses — 19, 30, 31, 33, 36, 37, 39, 40, 45, 49, 73-75, 80, 82-84, 92, 101, 105, 121, 128, 136, 140, 143, 144, 150, 159, 167, 170, 171, 173, 176-178, 181-185, 196, 199, 201, 227, 241, 242, 244, 246, 247, 252, 254-256, 263-265, 287, 304, 306, 316, 319, 320, 324, 328, 329, 362, 372, 373, 387, 401, 404, 410, 431-433, 435, 440, 450, 477 ff., 493-500

  hardware investigation

    cosmic dust — 441, 450, 481

    cosmic ray and solar particle — 481

    energetic particle composition and interstellar gas — 431-433, 480

    low energy ion and electron — 480

    magnetic field — 480

    solar hard X-ray and gamma-burst — 80, 481

    solar wind ion composition — 173, 177-179, 184, 480

    solar wind plasma — 173, 177, 184, 480

    unified radio and plasma wave — 481, 486

  interdisciplinary investigation

    directional discontinuity — 170, 483

    mass loss and ion composition — 483

  mission operations — 487, 488

  radio-science investigation

    coronal sounding — 483

    gravitational wave — 477-479, 483

  spacecraft — 478, 483-488

    trajectory — 478, 484, 485

UV coronal emission — 39-43

UV coronal spectrometer — 28, 30, 39-43

Very Large Array                          59, 60, 63, 221, 222


Wave kinetic equation                     309
White light coronagraph                   10, 12, 28, 30, 39, 40, 102,
                                          107, 113, 116, 117, 119, 129,
                                          136, 148, 275
Wilcox Solar Observatory                  241, 242, 277
WKB theory                                158-160


X-rays (see Solar hard X-rays)


Zodiacal light                            182, 441, 443, 445-451, 463,
                                          464
    photometer                            113-117, 129, 441

# ASTROPHYSICS AND SPACE SCIENCE LIBRARY

Edited by

J. E. Blamont, R. L. F. Boyd, L. Goldberg, C. de Jager, Z. Kopal, G. H. Ludwig, R. Lüst,
B. M. McCormac, H. E. Newell, L. I. Sedov, Z. Švestka

1. C. de Jager (ed.), *The Solar Spectrum, Proceedings of the Symposium held at the University of Utrecht, 26–31 August, 1963.* 1965, XIV + 417 pp.
2. J. Orthner and H. Maseland (eds.), *Introduction to Solar Terrestrial Relations, Proceedings of the Summer School in Space Physics held in Alpbach, Austria, July 15–August 10, 1963 and Organized by the European Preparatory Commission for Space Research.* 1965, IX + 506 pp.
3. C. C. Chang and S. S. Huang (eds.), *Proceedings of the Plasma Space Science Symposium, held at the Catholic University of America, Washington, D.C., June 11–14, 1963.* 1965, IX + 377 pp.
4. Zdeněk Kopal, *An Introduction to the Study of the Moon.* 1966, XII + 464 pp.
5. B. M. McCormac (ed.), *Radiation Trapped in the Earth's Magnetic Field. Proceedings of the Advanced Study Institute, held at the Chr. Michelsen Institute, Bergen, Norway, August 16–September 3, 1965.* 1966, XII + 901 pp.
6. A. B. Underhill, *The Early Type Stars.* 1966, XII + 282 pp.
7. Jean Kovalevsky, *Introduction to Celestial Mechanics.* 1967, VIII + 427 pp.
8. Zdeněk Kopal and Constantine L. Goudas (eds.), *Measure of the Moon. Proceedings of the 2nd International Conference on Selenodesy and Lunar Topography, held in the University of Manchester, England, May 30–June 4, 1966.* 1967, XVIII + 479 pp.
9. J. G. Emming (ed.), *Electromagnetic Radiation in Space. Proceedings of the 3rd ESRO Summer School in Space Physics, held in Alpbach, Austria, from 19 July to 13 August, 1965.* 1968, VIII + 307 pp.
10. R. L. Carovillano, John F. McClay, and Henry R. Radoski (eds.), *Physics of the Magnetosphere, Based upon the Proceedings of the Conference held at Boston College, June 19–28, 1967.* 1968, X + 686 pp.
11. Syun-Ichi Akasofu, *Polar and Magnetospheric Substorms.* 1968, XVIII + 280 pp.
12. Peter M. Millman (ed.), *Meteorite Research. Proceedings of a Symposium on Meteorite Research, held in Vienna, Austria, 7–13 August, 1968.* 1969, XV + 941 pp.
13. Margherita Hack (ed.), *Mass Loss from Stars. Proceedings of the 2nd Trieste Colloquium on Astrophysics, 12–17 September, 1968.* 1969, XII + 345 pp.
14. N. D'Angelo (ed.), *Low-Frequency Waves and Irregularities in the Ionosphere. Proceedings of the 2nd ESRIN-ESLAB Symposium, held in Frascati, Italy, 23–27 September, 1968.* 1969, VII + 218 pp.
15. G. A. Partel (ed.), *Space Engineering. Proceedings of the 2nd International Conference on Space Engineering, held at the Fondazione Giorgio Cini, Isola di San Giorgio, Venice, Italy, May 7–10, 1969.* 1970, XI + 728 pp.
16. S. Fred Singer (ed.), *Manned Laboratories in Space. Second International Orbital Laboratory Symposium.* 1969, XIII + 133 pp.
17. B. M. McCormac (ed.), *Particles and Fields in the Magnetosphere. Symposium Organized by the Summer Advanced Study Institute, held at the University of California, Santa Barbara, Calif., August 4–15, 1969.* 1970, XI + 450 pp.
18. Jean-Claude Pecker, *Experimental Astronomy.* 1970, X + 105 pp.
19. V. Manno and D. E. Page (eds.), *Intercorrelated Satellite Observations related to Solar Events. Proceedings of the 3rd ESLAB/ESRIN Symposium held in Noordwijk, The Netherlands, September 16–19, 1969.* 1970, XVI + 627 pp.
20. L. Mansinha, D. E. Smylie, and A. E. Beck, *Earthquake Displacement Fields and the Rotation of the Earth, A NATO Advanced Study Institute Conference Organized by the Department of Geophysics, University of Western Ontario, London, Canada, June 22–28, 1969.* 1970, XI + 308 pp.
21. Jean-Claude Pecker, *Space Observatories.* 1970, XI + 120 pp.
22. L. N. Mavridis (ed.), *Structure and Evolution of the Galaxy. Proceedings of the NATO Advanced Study Institute, held in Athens, September 8–19, 1969.* 1971, VII + 312 pp.

23. A. Muller (ed.), *The Magellanic Clouds. A European Southern Observatory Presentation: Principal Prospects, Current Observational and Theoretical Approaches, and Prospects for Future Research, Based on the Symposium on the Magellanic Clouds, held in Santiago de Chile, March 1969, on the Occasion of the Dedication of the European Southern Observatory.* 1971, XII + 189 pp.

24. B. M. McCormac (ed.), *The Radiating Atmosphere. Proceedings of a Symposium Organized by the Summer Advanced Study Institute, held at Queen's University, Kingston, Ontario, August 3–14, 1970.* 1971, XI + 455 pp.

25. G. Fiocco (ed.), *Mesospheric Models and Related Experiments. Proceedings of the 4th ESRIN-ESLAB Symposium, held at Frascati, Italy, July 6–10, 1970.* 1971, VIII + 298 pp.

26. I. Atanasijević, *Selected Exercises in Galactic Astronomy.* 1971, XII + 144 pp.

27. C. J. Macris (ed.), *Physics of the Solar Corona. Proceedings of the NATO Advanced Study Institute on Physics of the Solar Corona, held at Cavouri-Vouliagmeni, Athens, Greece, 6–17 September 1970.* 1971, XII + 345 pp.

28. F. Delobeau, *The Environment of the Earth.* 1971, IX + 113 pp.

29. E. R. Dyer (general ed.), *Solar-Terrestrial Physics/1970. Proceedings of the International Symposium on Solar-Terrestrial Physics, held in Leningrad, U.S.S.R., 12–19 May 1970.* 1972, VIII + 938 pp.

30. V. Manno and J. Ring (eds.), *Infrared Detection Techniques for Space Research. Proceedings of the 5th ESLAB-ESRIN Symposium, held in Noordwijk, The Netherlands, June 8–11, 1971.* 1972, XII + 344 pp.

31. M. Lecar (ed.), *Gravitational N-Body Problem. Proceedings of IAU Colloquium No. 10, held in Cambridge, England, August 12–15, 1970.* 1972, XI + 441 pp.

32. B. M. McCormac (ed.), *Earth's Magnetospheric Processes. Proceedings of a Symposium Organized by the Summer Advanced Study Institute and Ninth ESRO Summer School, held in Cortina, Italy, August 30–September 10, 1971.* 1972, VIII + 417 pp.

33. Antonin Rükl, *Maps of Lunar Hemispheres.* 1972, V + 24 pp.

34. V. Kourganoff, *Introduction to the Physics of Stellar Interiors.* 1973, XI + 115 pp.

35. B. M. McCormac (ed.), *Physics and Chemistry of Upper Atmospheres. Proceedings of a Symposium Organized by the Summer Advanced Study Institute, held at the University of Orléans, France, July 31–August 11, 1972.* 1973, VIII + 389 pp.

36. J. D. Fernie (ed.), *Variable Stars in Globular Clusters and in Related Systems. Proceedings of the IAU Colloquium No. 21, held at the University of Toronto, Toronto, Canada, August 29–31, 1972.* 1973, IX + 234 pp.

37. R. J. L. Grard (ed.), *Photon and Particle Interaction with Surfaces in Space. Proceedings of the 6th ESLAB Symposium, held at Noordwijk, The Netherlands, 26–29 September, 1972.* 1973, XV + 577 pp.

38. Werner Israel (ed.), *Relativity, Astrophysics and Cosmology. Proceedings of the Summer School, held 14–26 August 1972, at the BANFF Centre, BANFF, Alberta, Canada.* 1973, IX + 323 pp.

39. B. D. Tapley and V. Szebehely (eds.), *Recent Advances in Dynamical Astronomy. Proceedings of the NATO Advanced Study Institute in Dynamical Astronomy, held in Cortina d'Ampezzo, Italy, August 9–12, 1972.* 1973, XIII + 468 pp.

40. A. G. W. Cameron (ed.), *Cosmochemistry. Proceedings of the Symposium on Cosmochemistry, held at the Smithsonian Astrophysical Observatory, Cambridge, Mass., August 14–16, 1972.* 1973, X + 173 pp.

41. M. Golay, *Introduction to Astronomical Photometry.* 1974, IX + 364 pp.

42. D. E. Page (ed.), *Correlated Interplanetary and Magnetospheric Observations. Proceedings of the 7th ESLAB Symposium, held at Saulgau, W. Germany, 22–25 May, 1973.* 1974, XIV + 662 pp.

43. Riccardo Giacconi and Herbert Gursky (eds.), *X-Ray Astronomy.* 1974, X + 450 pp.

44. B. M. McCormac (ed.), *Magnetospheric Physics. Proceedings of the Advanced Summer Institute, held in Sheffield, U.K., August 1973.* 1974, VII + 399 pp.

45. C. B. Cosmovici (ed.), *Supernovae and Supernova Remnants. Proceedings of the International Conference on Supernovae, held in Lecce, Italy, May 7–11, 1973.* 1974, XVII + 387 pp.

46. A. P. Mitra, *Ionospheric Effects of Solar Flares.* 1974, XI + 294 pp.

47. S.-I. Akasofu, *Physics of Magnetospheric Substorms.* 1977, XVIII + 599 pp.

48. H. Gursky and R. Ruffini (eds.), *Neutron Stars, Black Holes and Binary X-Ray Sources*. 1975, XII + 441 pp.

49. Z. Švestka and P. Simon (eds.), *Catalog of Solar Particle Events 1955–1969. Prepared under the Auspices of Working Group 2 of the Inter-Union Commission on Solar-Terrestrial Physics*. 1975, IX + 428 pp.

50. Zdeněk Kopal and Robert W. Carder, *Mapping of the Moon*. 1974, VIII + 237 pp.

51. B. M. McCormac (ed.), *Atmospheres of Earth and the Planets. Proceedings of the Summer Advanced Study Institute, held at the University of Liège, Belgium, July 29–August 8, 1974*. 1975, VII + 454 pp.

52. V. Formisano (ed.), *The Magnetospheres of the Earth and Jupiter. Proceedings of the Neil Brice Memorial Symposium, held in Frascati, May 28–June 1, 1974*. 1975, XI + 485 pp.

53. R. Grant Athay, *The Solar Chromosphere and Corona: Quiet Sun*. 1976, XI + 504 pp.

54. C. de Jager and H. Nieuwenhuijzen (eds.), *Image Processing Techniques in Astronomy. Proceedings of a Conference, held in Utrecht on March 25–27, 1975*. 1976, XI + 418 pp.

55. N. C. Wickramasinghe and D. J. Morgan (eds.), *Solid State Astrophysics. Proceedings of a Symposium, held at the University College, Cardiff, Wales, 9–12 July, 1974*. 1976, XII + 314 pp.

56. John Meaburn, *Detection and Spectrometry of Faint Light*. 1976, IX + 270 pp.

57. K. Knott and B. Battrick (eds.), *The Scientific Satellite Programme during the International Magnetospheric Study. Proceedings of the 10th ESLAB Symposium, held at Vienna, Austria, 10–13 June 1975*. 1976, XV + 464 pp.

58. B. M. McCormac (ed.), *Magnetospheric Particles and Fields. Proceedings of the Summer Advanced Study School, held in Graz, Austria, August 4–15, 1975*. 1976, VII + 331 pp.

59. B. S. P. Shen and M. Merker (eds.), *Spallation Nuclear Reactions and Their Applications*. 1976, VIII + 235 pp.

60. Walter S. Fitch (ed.), *Multiple Periodic Variable Stars. Proceedings of the International Astronomical Union Colloquium No. 29, held at Budapest, Hungary, 1–5 September 1976*. 1976, XIV + 348 pp.

61. J. J. Burger, A. Pedersen, and B. Battrick (eds.), *Atmospheric Physics from Spacelab. Proceedings of the 11th ESLAB Symposium, Organized by the Space Science Department of the European Space Agency, held at Frascati, Italy, 11–14 May 1976*. 1976, XX + 409 pp.

62. J. Derral Mulholland (ed.), *Scientific Applications of Lunar Laser Ranging. Proceedings of a Symposium held in Austin, Tex., U.S.A., 8–10 June, 1976*. 1977, XVII + 302 pp.

63. Giovanni G. Fazio (ed.), *Infrared and Submillimeter Astronomy. Proceedings of a Symposium held in Philadelphia, Penn., U.S.A., 8–10 June, 1976*. 1977, X + 226 pp.

64. C. Jaschek and G. A. Wilkins (eds.), *Compilation, Critical Evaluation and Distribution of Stellar Data. Proceedings of the International Astronomical Union Colloquium No. 35, held at Strasbourg, France, 19–21 August, 1976*. 1977, XIV + 316 pp.

65. M. Friedjung (ed.), *Novae and Related Stars. Proceedings of an International Conference held by the Institut d'Astrophysique, Paris, France, 7–9 September, 1976*. 1977, XIV + 228 pp.

66. David N. Schramm (ed.), *Supernovae. Proceedings of a Special IAU-Session on Supernovae held in Grenoble, France, 1 September, 1976*. 1977, X + 192 pp.

67. Jean Audouze (ed.), *CNO Isotopes in Astrophysics. Proceedings of a Special IAU Session held in Grenoble, France, 30 August, 1976*. 1977, XIII + 195 pp.

68. Z. Kopal, *Dynamics of Close Binary Systems*. XIII + 510 pp.

69. A. Bruzek and C. J. Durrant (eds.), *Illustrated Glossary for Solar and Solar-Terrestrial Physics*. 1977, XVIII + 204 pp.

70. H. van Woerden (ed.), *Topics in Interstellar Matter*. 1977, VIII + 295 pp.

71. M. A. Shea, D. F. Smart, and T. S. Wu (eds.), *Study of Travelling Interplanetary Phenomena*. 1977, XII + 439 pp.

72. V. Szebehely (ed.), *Dynamics of Planets and Satellites and Theories of Their Motion. Proceedings of IAU Colloquium No. 41, held in Cambridge, England, 17–19 August 1976*. 1978, XII + 375 pp.

73. James R. Wertz (ed.), *Spacecraft Attitude Determination and Control*. 1978, XVI + 858 pp.

74. Peter J. Palmadesso and K. Papadopoulos (eds.), *Wave Instabilities in Space Plasmas. Proceedings of a Symposium Organized Within the XIX URSI General Assembly held in Helsinki, Finland, July 31–August 8, 1978.* 1979, VII + 309 pp.

75. Bengt E. Westerlund (ed.), *Stars and Star Systems. Proceedings of the Fourth European Regional Meeting in Astronomy held in Uppsala, Sweden, 7–12 August, 1978.* 1979, XVIII + 264 pp.

76. Cornelis van Schooneveld (ed.), *Image Formation from Coherence Functions in Astronomy. Proceedings of IAU Colloquium No. 49 on the Formation of Images from Spatial Coherence Functions in Astronomy, held at Groningen, The Netherlands, 10–12 August 1978.* 1979, XII + 338 pp.

77. Zdeněk Kopal, *Language of the Stars. A Discourse on the Theory of the Light Changes of Eclipsing Variables.* 1979, VIII + 280 pp.

78. S.-I. Akasofu (ed.), *Dynamics of the Magnetosphere. Proceedings of the A.G.U. Chapman Conference 'Magnetospheric Substorms and Related Plasma Processes' held at Los Alamos Scientific Laboratory, N.M., U.S.A., October 9–13, 1978.* 1980, XII + 658 pp.

79. Paul S. Wesson, *Gravity, Particles, and Astrophysics. A Review of Modern Theories of Gravity and G-variability, and their Relation to Elementary Particle Physics and Astrophysics.* 1980, VIII + 188 pp.

80. Peter A. Shaver (ed.), *Radio Recombination Lines. Proceedings of a Workshop held in Ottawa, Ontario, Canada, August 24–25, 1979.* 1980, X + 284 pp.

81. Pier Luigi Bernacca and Remo Ruffini (eds.), *Astrophysics from Spacelab.* 1980, XI + 664 pp.

82. Hannes Alfvén, *Cosmic Plasma,* 1981, X + 160 pp.

83. Michael D. Papagiannis (ed.), *Strategies for the Search for Life in the Universe,* 1980, XVI + 254 pp.

84. H. Kikuchi (ed.), *Relation between Laboratory and Space Plasmas,* 1981, XII + 386 pp.

85. Peter van der Kamp, *Stellar Paths,* 1981, XXII + 155 pp.

86. E. M. Gaposchkin and B. Kołaczek (eds.), *Reference Coordinate Systems for Earth Dynamics,* 1981, XIV + 396 pp.

87. R. Giacconi (ed.), *X-Ray Astronomy with the Einstein Satellite. Proceedings of the High Energy Astrophysics Division of the American Astronomical Society Meeting on X-Ray Astronomy held at the Harvard-Smithsonian Center for Astrophysics, Cambridge, Mass., U.S.A., January 28–30, 1980.* 1981, VII + 330 pp.

88. Icko Iben Jr. and Alvio Renzini (eds.), *Physical Processes in Red Giants. Proceedings of the Second Workshop, helt at the Ettore Majorana Centre for Scientific Culture, Advanced School of Agronomy, in Erice, Sicily, Italy, September 3–13, 1980.* 1981, XV + 488 pp.

89. C. Chiosi and R. Stalio (eds.), *Effect of Mass Loss on Stellar Evolution. IAU Colloquium No. 59 held in Miramare, Trieste, Italy, September 15–19, 1980.* 1981, XXII + 532 pp.

90. C. Goudis, *The Orion Complex: A Case Study of Interstellar Matter.* 1982, XIV + 306 pp.

91. F. D. Kahn (ed.), *Investigating the Universe. Papers Presented to Zdenek Kopal on the Occasion of his retirement, September 1981.* 1981, X + 458 pp.

92. C. M. Humphries (ed.), *Instrumentation for Astronomy with Large Optical Telescopes, Proceedings of IAU Colloquium No. 67.* 1981, XVII + 321 pp.

93. R. S. Roger and P. E. Dewdney (eds.), *Regions of Recent Star Formation, Proceedings of the Symposium on "Neutral Clouds Near HII Regions – Dynamics and Photochemistry", held in Penticton, B.C., June 24–26, 1981.* 1982, XVI + 496 pp.

94. O. Calame (ed.), *High-Precision Earth Rotation and Earth-Moon Dynamics. Lunar Distances and Related Observations.* 1982, XX + 354 pp.

95. M. Friedjung and R. Viotti (eds.), *The Nature of Symbiotic Stars,* 1982, XX + 310 pp.

96. W. Fricke and G. Teleki (eds.), *Sun and Planetary System,* 1982, XIV + 538 pp.

97. C. Jaschek and W. Heintz (eds.), *Automated Data Retrieval in Astronomy,* 1982, XX + 324 pp.

98. Z. Kopal and J. Rahe (eds.), *Binary and Multiple Stars as Tracers of Stellar Evolution,* 1982, XXX + 503 pp.

99. A. W. Wolfendale (ed.), *Progress in Cosmology,* 1982, VIII + 360 pp.

100. W. L. H. Shuter (ed.), *Kinematics, Dynamics and Structure of the Milky Way,* 1983, XII + 392 pp.

101.  M. Livio and G. Shaviv (eds.), *Cataclysmic Variables and Related Objects,* 1983, XII + 351 pp.
102.  P. B. Byrne and M. Rodonò (eds.), *Activity in Red-Dwarf Stars,* 1983, XXVI + 670 pp.
103.  A. Ferrari and A. G. Pacholczyk (eds.), *Astrophysical Jets,* 1983, XVI + 328 pp.
104.  R. L. Carovillano and J. M. Forbes (eds.), *Solar-Terrestrial Physics,* 1983, XVIII + 860 pp.
105.  W. B. Burton and F. P. Israel (eds.), *Surveys of the Southern Galaxy,* 1983, XIV + 310 pp.
106.  V. V. Markellos and Y. Kozai (eds.), *Dynamical Trapping and Evolution on the Solar System,* 1983, XVI + 424 pp.
107.  S. R. Pottasch, *Planetary Nebulae,* 1984, X + 322 pp.
108.  M. F. Kessler and J. P. Phillips (eds.), *Galactic and Extragalactic Infrared Spectroscopy,* 1984, XII + 472 pp.
109.  C. Chiosi and A. Renzini (eds.), *Stellar Nucleosynthesis,* 1984, XIV + 398 pp.
110.  M. Capaccioli (ed.), *Astronomy with Schmidt-type Telescopes,* 1984, XXII + 620 pp.
111.  F. Mardirossian, G. Giuricin, and M. Mezzetti (eds.), *Clusters and Groups of Galaxies,* 1984, XXII + 659 pp.
112.  L. H. Aller, *Physics of Thermal Gaseous Nebulae,* 1984, X + 350 pp.
113.  D. Q. Lamb and J. Patterson (eds.), *Cataclysmic Variables and Low-Mass X-Ray Binaries,* 1985, XII + 452 pp.
114.  M. Jaschek and P. C. Keenan (eds.), *Cool Stars with Excesses of Heavy Elements,* 1985, XVI + 398 pp.
115.  A. Carusi and G. B. Valsecchi (eds.), *Dynamics of Comets: Their Origin and Evolution,* 1985, XII + 442 pp.
116.  R. M. Hjellming and D. M. Gibson (eds.), *Radio Stars,* 1985, XI + 411 pp.
117.  M. Morris and B. Zuckermann (eds.), *Mass Loss from Red Giants,* 1985, xvi +320 pp.
118.  Y. Sekido and H. Elliot (eds.), *Early History of Cosmic Ray Studies,* 1985, xvi +444 pp.
119.  R. H. Giese and P. Lamy (eds.), *Properties and Interactions of Interplanetary Dust,* 1985, xxvi + 444 pp.
120.  W. Boland and H. van Woerden (eds.), *Birth and Evolution of Massive Stars and Stellar Groups,* 1985, xiv + 377 pp.
121.  G. Giuricin, F. Mardirossian, M. Mezzetti, and M. Ramella (eds.), *Structure and Evolution of Active Galactic Nuclei,* 1986, xxvi +772 pp.
122.  C. Chiosi and A. Renzini (eds.), *Spectral Evolution of Galaxies,* 1986, xii + 490 pp.